Introduction to
General Relativity

WSPC-PKUSP PHYSICS SERIES FOR POSTGRADUATES

Published

Vol. 1 *Introduction to General Relativity*
by Bin Chen

WSPC-PKUSP Physics Series
for Postgraduates – Vol. 1

Introduction to General Relativity

Bin Chen
Ningbo University, China

北京大学出版社
PEKING UNIVERSITY PRESS

World Scientific

Published by

World Scientific Publishing Co. Pte. Ltd.
5 Toh Tuck Link, Singapore 596224
USA office: 27 Warren Street, Suite 401-402, Hackensack, NJ 07601
UK office: 57 Shelton Street, Covent Garden, London WC2H 9HE

Library of Congress Control Number: 2026932967

British Library Cataloguing-in-Publication Data
A catalogue record for this book is available from the British Library.

广义相对论导论
Originally published in Chinese by Peking University Press

WSPC-PKUSP Physics Series for Postgraduates — Vol. 1
INTRODUCTION TO GENERAL RELATIVITY

ISBN 978-981-98-2729-9 (hardcover)
ISBN 978-981-98-2730-5 (ebook for institutions)
ISBN 978-981-98-2731-2 (ebook for individuals)

For any available supplementary material, please visit
https://www.worldscientific.com/worldscibooks/10.1142/14703#t=suppl

Preface

Relativity and quantum physics are the two pillars of modern physics. In 1905, Albert Einstein's groundbreaking special theory of relativity—Special Relativity—laid the foundation for contemporary physical theories, while his subsequent formulation of the relativistic theory of gravity—General Relativity in late 1915—has been widely regarded as the most beautiful theory in classical physics. The development of General Relativity represents a remarkable synthesis of Einstein's profound physical intuition and exceptional mathematical prowess. Historically, this theory appeared as an intellectual summit inaccessible to many physicists, with its seemingly abstract notions considerably removed from observable physical phenomena. After more than a century of development, particularly through significant discoveries in cosmology and astrophysical observations in recent decades, General Relativity has experienced a renaissance of interest, and its influence has extended to diverse fields including cosmology, astrophysics, and high-energy physics.

General Relativity can be succinctly summarized in the following key points. First, General Relativity is Einstein's theory about space, time, and gravity. Second, in General Relativity, gravity is not treated as a distinct force but rather as an intrinsic property of spacetime itself. This relationship can be encapsulated in a simple yet profound relation:

$$\text{Gravity} = \text{Curvature of Spacetime}.$$

In General Relativity, spacetime is dynamic and curved, with the magnitude of this curvature directly determining the strength of gravitational effects. Finally, unlike many other physical theories that were primarily through experimental evidence, General Relativity emerged largely from Einstein's profound theoretical insights and physical intuition, providing a novel methodology for studying fundamental physics.

As a relativistic theory of gravity, General Relativity is particularly important in situations where relativistic effects are significant. In Special Relativity, these effects are quantified by the ratio of a particle's velocity to the speed of light, v/c. When the particle's velocity approaches the speed of light, relativistic effects become apparent. To illustrate this in the context of gravity, consider a sphere with mass M and radius R. The escape velocity at its surface is given by $\sqrt{2GM/R}$. If this velocity is comparable to the speed of light, relativistic effects become substantial. Consequently, in gravitational systems, the dimensionless parameter $GM/(Rc^2)$ serves as a measure of the significance of relativistic effects. When $GM/(Rc^2) = O(1)$, Newtonian gravity ceases to be applicable, and General Relativity must be employed to accurately describe the system. Conversely, when $GM/(Rc^2) \ll 1$, General Relativity reduces to Newtonian gravity, with relativistic corrections being negligible. For instance, on Earth, $GM/(Rc^2) \sim 10^{-9}$. Although this value is extremely small, it must be taken into account in the Global Positioning System (GPS), which requires high precision. For the Sun, $GM/(Rc^2) \sim 10^{-6}$, and the physical effects caused by the relativistic corrections are observable in experiments. Historically, three critical experimental tests of General Relativity—perihelion precession of Mercury, light deflection, and time delay—were all conducted within the solar system.

General Relativity serves as a cornerstone in the realm of relativistic astrophysics, providing an essential framework for understanding various cosmic phenomena. This theory proves particularly indispensable in investigating black holes, pulsars, quasars, and stellar evolution processes. The significance of General Relativity becomes especially evident when examining compact objects like neutron stars and black holes. Consider a neutron star with one solar mass: its radius measures a mere 10 km, resulting in a substantial dimensionless parameter $GM/(Rc^2) \sim 0.1$. Therefore, systems related to neutron stars, such as binary systems, are important objects of study in relativistic celestial mechanics. Black hole systems present even more extreme conditions, with $GM/(Rc^2) \sim 0.5$, where relativistic effects dominate the system's behavior. Theoretically, black hole formation through stellar collapse was firmly established in the 1960s. Observationally, after arduous efforts, evidence for the existence of supermassive black holes at galactic centers was obtained.[1] Recent years have

[1]The 2020 Nobel Prize in Physics recognized Roger Penrose, Reinhard Genzel, and Andrea Ghez for their pivotal work on black hole theory and observational confirmation.

witnessed groundbreaking advancements in the field, marked by the successful detection of gravitational waves and the direct imaging of black holes.[2] These achievements have opened unprecedented opportunities to test gravitational theories in intense relativistic regimes and investigate astrophysical processes under extreme conditions. This progress has ushered in a new era of gravitational physics research. Looking ahead, the study of relativistic celestial objects is poised to remain at the forefront of physics and astronomy for the coming decades, promising exciting discoveries and deeper insights into the fundamental nature of our universe.

General Relativity also forms the theoretical cornerstone of modern cosmology. In the context of the universe, the presence of any form of matter leads to relativistic effects that scale with size—since mass increases with the cube of the radius. This scaling makes General Relativity the fundamental framework for understanding the evolution and development of the universe, as well as the formation of its large-scale structures. Over the past three decades, cosmological observations have yielded a series of groundbreaking discoveries. These include the precise measurement of the cosmic microwave background (CMB) radiation spectrum and the revelation of the universe's current accelerated expansion.[3] These observational advances have propelled cosmological research into the era of precision cosmology, where theoretical predictions can be tested against highly accurate observational data. Among the most profound questions raised by these advances are the nature of dark energy and dark matter, which have emerged as central challenges in contemporary physics. These enigmatic components, which together constitute the vast majority of the universe's energy and matter content, continue to drive both theoretical and observational efforts to uncover their true nature.

While numerous excellent monographs on General Relativity exist in

---

[2]The 2017 Nobel Prize in Physics honored Rainer Weiss, Barry Barish, and Kip Thorne for their decisive contributions to the Laser Interferometer Gravitational-Wave Observatory (LIGO) project and the detection of gravitational waves. In 2020, the Event Horizon Telescope (EHT) team received the Breakthrough Prize in Fundamental Physics for their groundbreaking black hole imaging achievement.

[3]The 2006 Nobel Prize in Physics was awarded to John C. Mather and George F. Smoot, for their discovery of the blackbody form and anisotropy of the cosmic microwave background radiation, achieved through the Cosmic Background Explorer (COBE) satellite mission. The 2011 Nobel Prize in Physics was awarded to Saul Perlmutter, Brian P. Schmidt, and Adam G. Riess, for their discovery of the accelerating expansion of the universe via observations of distant supernovae. Half of the 2019 Nobel Prize in Physics was awarded to James Peebles, for his foundational contributions to physical cosmology.

both Chinese and English, beginners often find it challenging to engage directly with English works. In terms of Chinese literature, there is a noticeable gap in the availability of books that strike a balance in difficulty—those that not only provide foundational knowledge but also effectively guide readers to the forefront of current research. Given that research papers in gravitational studies predominantly employ the language of differential geometry, it is essential to equip readers with this technical framework as early as possible. To address this need, from 2006–2024, I have structured the "General Relativity" courses for both graduate and undergraduate students at Peking University to introduce the fundamental concepts of General Relativity using the language of differential geometry within a single semester. While this approach has proven feasible, the limited timeframe of 64 class hours has necessitated some compromises in mathematical rigor. Nevertheless, the core objectives of enabling students to grasp essential techniques and concepts have been achieved. This book encapsulates the content of this course, covering a range of topics essential to understanding General Relativity. These include the geometric formulation of Special Relativity, the mathematical foundations of General Relativity, the derivation of Einstein's equations, the study of spherically symmetric spacetimes, classical experimental tests of General Relativity, black hole physics, and gravitational waves. By integrating these topics, the book aims to bridge the gap between foundational knowledge and advanced research, providing readers with a comprehensive and accessible pathway into the field of gravitational physics.

The structure of this book is organized as follows. The first part comprises two foundational chapters that systematically introduce the core concepts of Special Relativity. To help readers become familiar with geometric language as soon as possible, in the first chapter, we use Minkowski spacetime to explain the basic content of Special Relativity, including fundamental principles, physical effects, kinematics and dynamics in Special Relativity, and introduce how to establish the reference frame and the meaning of measurement for an observer. In the second chapter, we preliminarily introduce the definitions of coordinate-independent vectors, dual vectors, and general tensors, the basic techniques of tensor operations, the definition of the energy-momentum tensor, and review the covariant form of electrodynamics using geometric language.

The second part of this book presents the rigorous mathematical foundation of General Relativity. First, in Chapter 3, we briefly revisit Newtonian gravity, highlighting its theoretical limitations and incompatibility with the principle of relativity. Then, we introduce the equivalence principle, encompassing both the weak equivalence principle and Einstein's equivalence principle. We further examines the phenomenon of gravitational redshift, its implications for the GPS system, and its profound consequence that spacetime geometry is intrinsically non-flat. In the subsequent chapters, we systematically introduce the mathematical foundation of General Relativity in a progressive manner: In Chapter 4, we establish the fundamental concepts of manifolds, laying the groundwork for understanding spacetime as a differentiable manifold. In this chapter we introduce tensor fields and develops the essential tools of tensor analysis on manifolds. In Chapter 5, we delve into the critical concepts of connections, covariant derivatives, and parallel transport, and further explores geodesics, elucidating their mathematical formulation and physical significance. We also introduce advanced topics including Lie derivatives, Fermi–Walker transport, and the precise mathematical description of observer reference frames in the chapter. In Chapter 6, we complete the mathematical foundation by introducing curvature tensors and torsion tensors, essential for characterizing the geometry of spacetime. We examine Killing vectors and their profound physical significance in General Relativity.

The third part of this book provides a comprehensive exploration of Einstein's equations and their physically significant spherically symmetric solutions. In Chapter 7, we systematically introduce Einstein's equations as the cornerstone of General Relativity. This includes a detailed discussion of the weak field approximation, which bridges the gap between Einstein's theory and Newtonian gravity, and the elegant formulation of the theory through the action principle. The chapter also provides an overview of modified gravity theories, offering insights into potential extensions and alternatives to Einstein's framework. Additionally, we examine crucial related topics such as energy conditions and causality, which are fundamental to understanding the physical implications and consistency of solutions to Einstein's equations.

In Chapter 8, we focus on the most renowned spherically symmetric solution of Einstein's equations: the Schwarzschild spacetime. We thoroughly discuss the three classical experimental tests of General Relativity that confirmed the theory's validity: gravitational redshift, perihelion precession of Mercury, and light deflection. Beyond these historic tests, we delve into modern experimental verifications, including gravitational clock experiments and geodetic precession experiments. Furthermore, we introduce some key concepts from relativistic astrophysics, demonstrating how the Schwarzschild solution provides crucial insights into the physics of compact objects and the behavior of matter and radiation in strong gravitational fields.

The fourth part of this book is dedicated to the physics of black holes. In Chapter 9, we systematically examine the physics of Schwarzschild black holes, including the observational phenomena experienced by different observers, the maximum extension of spacetime, and the astrophysical processes leading to black hole formation. Furthermore, we extend our analysis to two additional spherically symmetric spacetimes: the internal geometry of stellar objects and the structure of charged spherically symmetric black holes (Reissner–Nordström solutions). In Chapter 10, we present a detailed examination of rotating black hole systems. We commence our discussion with an analysis of rotational effects on spacetime within the weak-field approximation, paying particular attention to the physical manifestations of the gravitomagnetic field. Subsequently, we introduce the Kerr solution, focusing on its distinctive physical characteristics. Our discussion highlights several remarkable features, including the Penrose process of energy extraction from rotating black holes. We conclude the chapter with an introductory exploration of black hole thermodynamics and the quantum mechanical phenomenon of Hawking radiation, which represents a profound connection between quantum physics, thermodynamics, and general relativity.

In the concluding chapter of this book, we present a comprehensive introduction to gravitational waves, one of the most remarkable predictions of Einstein's theory of general relativity. Within the framework of linearized gravity, we examine three fundamental aspects: (1) the propagation and characteristics of gravitational waves in vacuum, including their observable effects on spacetime; (2) the astrophysical mechanisms responsible for generating gravitational waves; and (3) the energy loss associated with gravitational radiation. We also briefly introduce the detection of gravitational waves.

This book is a condensed version of my work "General Relativity" (Peking University Press, 2018), specifically designed to provide a more accessible entry point for beginners. In addition, this book corrects some errors in "General Relativity" and improves certain discussions. The scope of this book only covers the relevant content of the course, and some other important topics, including the initial value problem, $3+1$ decomposition and Hamiltonian form, large-scale structure of spacetime, general discussion of black holes, black hole mechanics, perturbations of black holes, etc., have to be reluctantly omitted. Furthermore, the semiclassical aspects of gravity—such as quantum field theory in curved spacetime, Euclidean gravity, and black hole thermodynamics—are not covered here. Cosmology, while introduced briefly in Chapter 5, is treated in a relatively simplified manner due to space constraints. These advanced and frontier topics will be explored in greater depth in another book. Upon completing this book, readers are encouraged to delve into these areas based on their research interests. For those seeking to stay abreast of the latest developments, a wealth of cutting-edge research can be found at the following online resource: `http://www.springer.com/gp/livingreviews/relativity/lrr-articles`.

Although General Relativity is a relatively independent course, it also requires a sufficient understanding of classical physics. Due to space constraints, this book does not encompass all prerequisite knowledge. Prior to engaging with this material, readers are expected to possess a solid foundation in Special Relativity, along with completed coursework in electrodynamics and analytical mechanics. A thorough grasp of electromagnetic field theory is essential, coupled with familiarity with Lagrangian mechanics, Euler–Lagrange equations, and the principle of least action. From a mathematical perspective, readers should have a good understanding of multivariable calculus, linear algebra, and differential equations from advanced mathematics courses. While this course was originally designed for graduate students and occasionally some basic knowledge of quantum field theory and group theory is used, these elements are not fundamental to the core material. Years of teaching has demonstrated that senior undergraduate students can successfully navigate this course without significant difficulty. Certain chapters, particularly those marked with an asterisk (*) at their beginning, present more advanced material that might be challenging for novice learners. Readers are advised to approach these sections in accordance with their individual capabilities and background knowledge.

General Relativity is an inherently technical and mathematically rigorous discipline. To achieve a thorough comprehension of its intricate details, it is indispensable to engage with practical exercises. Recognizing this fundamental requirement, this book incorporates carefully crafted problems at the end of each chapter.

The completion of this manuscript has been made possible through the invaluable support and contributions of numerous colleagues, students, and collaborators, to whom I express my deepest gratitude. Foremost, I extend my sincere appreciation to all the students who have participated in the "General Relativity" course over the past decade. Their intellectual curiosity, insightful questions, and active engagement have served as the primary impetus and inspiration for refining and completing this work. I am profoundly grateful to my research collaborators, whose stimulating discussions and innovative perspectives have significantly deepened my understanding of gravitational physics and related phenomena. My special thanks go to my esteemed senior colleagues who have generously shared their wisdom and expertise in General Relativity. In particular, I wish to acknowledge the profound influence of Professor Wu Ke, Professor Huang Chaoguang, and the deceased Professors Guo Hanying, Yan Mullin, and Zhang Yuanzhong, whose insights and dedication to physics have left an indelible mark on my academic journey. I would be remiss not to highlight the significant impact of the "Dark Energy and Its Fundamental Theory" advanced seminar series. The thought-provoking presentations and stimulating discussions by participants, including but not limited to Professors Cai Ronggen, Gao Yihong, Li Miao, Lu Jianxin, Lv Hong, Wang Bin, Zhao Liu, Jing Jiliang, Yu Hongwei, Gong Yunjiu, Guo Zongkuan, Huang Qingguo, Piao Yunsong, Liu Yuxiao, Li Mingzhe, Wu Puxun, Cao Liming, Zhang Xin, and Song Wei, have greatly enriched the content and depth of this work. Finally, I acknowledge with gratitude the financial support provided by the National Natural Science Foundation of China (Grant Nos. 11735001, 12275004) during the preparation of this manuscript.

Owing to the limitations of my scholarly capabilities, I acknowledge that this work may contain inadvertent errors and omissions. I sincerely welcome the critical feedback and constructive corrections from experts and discerning readers alike. Should you identify any inaccuracies or have suggestions for improvement, I would be most grateful if you could contact me via email at chenbin1@nbu.edu.cn.

Bin Chen
August 2025, at Ningbo

Contents

Chapter 1

Special Relativity

In this chapter, we will review and discuss special relativity using geometric language. We begin by analyzing the concept of spacetime in special relativity, introducing the Lorentz–FitzGerald transformation, and discussing light cones. We will also delve into the motion of both massive and massless particles, along with several related physical phenomena. These include length contraction, time dilation, Einstein's mass-energy relation, the Doppler effect, and the aberration of light, among others. Furthermore, we will provide a systematic discussion of kinematics and dynamics in special relativity, and clearly define the concepts of observers and observations.

1.1 Geometric Interpretation of Special Relativity

Special relativity encompasses numerous aspects that are counterintuitive to our everyday experience. Relying solely on our daily observations makes it challenging to comprehend the physical phenomena predicted by special relativity, such as length contraction and time dilation, often leading to various paradoxes. This intuitive disconnect stems from our conventional perceptions of time and space. While It is easy for us to accept that space has three dimensions, and it is also relatively easy to accept Galileo's principle of relativity—which posits that motion is relative to different inertial frames—these notions are underpinned by a critical assumption: the absoluteness of time. This assumption implies that time possesses an absolute significance, progressing uniformly across all inertial frames. Consequently, different temporal slices of spacetime can be delineated based on varying temporal coordinates, as illustrated in Figure 1.1. This inherent disparity between the treatment of time and space is a cornerstone of Newtonian mechanics, predating the advent of special relativity.

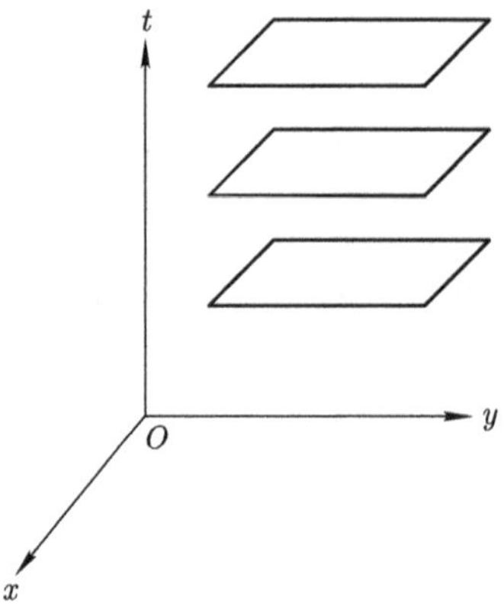

Fig. 1.1 In Newtonian mechanics, time has an absolute significance, and spacetime slices can be obtained using different times.

The Galileo Galilei's principle of relativity in Newtonian mechanics asserts that the fundamental laws of motion remain independent under the choice of inertial reference frame. An inertial reference frame, or simply an inertial frame, is defined as a coordinate system in which an object free from external forces move with a constant velocity along a straight line, thereby satisfying Newton's first law of motion. In both Newtonian mechanics and special relativity, it is assumed that such inertial frames exist. However, from a rigorous perspective, the omnipresence of gravitational forces implies that true inertial frames cannot exist in an absolute sense. Nevertheless, in many practical scenarios, it is operationally valid to approximate certain reference frames as inertial. For instance, the Earth's reference frame is often treated as an inertial frame for local experiments, despite the presence of gravitational effects. If a fundamental inertial frame is established, any other reference frame moving at a constant velocity relative to this frame also qualifies as an inertial frame. The principle of Galilean relativity can thus be formally articulated as follows:

Principle of Galilean Relativity All laws of motion have the same form in different inertial frames.

If an observer moves with a constant velocity relative to the fundamental inertial frame, then this observer defines another inertial frame and is referred to as an inertial observer. However, if the observer's motion is more complex, such that their velocity relative to the fundamental inertial frame is not constant, it becomes impossible to define a global inertial reference frame for the observer. Nevertheless, at any given instant, the observer's velocity relative to the fundamental inertial frame can still be treated as constant. This allows for the definition of an instantaneous inertial reference frame at each moment in time. The concept of an instantaneous inertial frame is particularly useful in analyzing non-inertial motion, as it permits the application of inertial frame principles—such as those of Newtonian mechanics or special relativity—on a momentary basis.

In the framework of Galilean–Newtonian mechanics, time is endowed with an absolute significance, implying that it flows uniformly and independently of the state of motion of observers or reference frames. Consequently, from a geometric point of view, different inertial frames are connected through Galilean transformations:

$$\begin{pmatrix} x \\ y \\ z \end{pmatrix} = H \begin{pmatrix} x' \\ y' \\ z' \end{pmatrix} + \begin{pmatrix} a + ut \\ b + vt \\ c + wt \end{pmatrix}, \tag{1.1}$$

where H is a constant $SO(3)$ matrix, representing the arbitrariness in the choice of spatial orientation of the coordinate system. The constant (a, b, c) correspond to the translational freedom in selecting the origin of the coordinate system, while the constants (u, v, w) represent the components of the relative velocity between the two frames, indicating that different inertial frames can differ by a constant velocity. From this transformation, it is evident that both the spatial interval and the time interval

$$\Delta r = \sqrt{\Delta x^2 + \Delta y^2 + \Delta z^2}, \qquad \Delta t \tag{1.2}$$

remain invariant under Galilean transformations. The invariance of the temporal interval is inherently guaranteed by the absolute nature of time. Regarding spatial intervals, it should be noted that they remain invariant under both translational transformations of the coordinate origin and three-dimensional spatial rotations, as illustrated in Figure 1.2.

Galilean relativity posits that physical laws maintain their forms across different inertial frames of reference that differ solely by a constant relative velocity. Within the framework of Galilean–Newtonian mechanics, while kinematic quantities such as velocity exhibit frame-dependence, the fundamental physical laws demonstrate frame invariance. Newton's three laws of mechanics retain their identical form in all inertial reference frames. For example, Newton's second law of mechanics $\boldsymbol{F} = m\boldsymbol{a}$ is the same for all observers. Although the measured velocity of a particle varies across different inertial frames, its acceleration remains invariant, thereby preserving

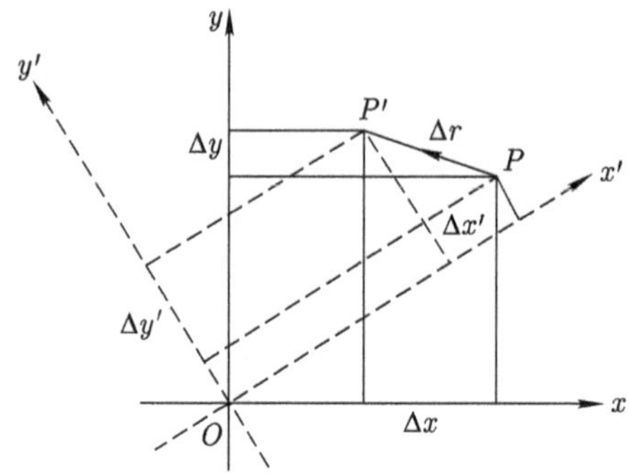

Fig. 1.2 A rotation between two reference frames (coordinate systems) does not change the distance (spatial interval) between two points in space. The figure represents a two-dimensional rotation.

the form of this fundamental law. More fundamentally, the principle of Galilean relativity extends beyond mechanical laws, postulating that all physical laws should exhibit invariance under Galilean transformations.

By the end of the 19th century, the formulation of Maxwell's electromagnetic theory revealed a fundamental incompatibility with Galilean transformations. Specifically, it became evident that Maxwell's equations were not invariant under Galilean relativity, thereby challenging the classical Newtonian framework. This inconsistency implied that the principle of Galilean relativity could not be reconciled with Maxwell's theory as it stood. In response to this theoretical conflict, physicists explored various resolutions, including the historically prominent luminiferous ether hypothesis, which posited an absolute reference frame for electromagnetic waves. However, these efforts ultimately proved unsatisfactory. The issue was decisively resolved in 1905 by Einstein's groundbreaking special theory of relativity. Einstein's framework not only retained the principle of relativity but also introduced the principle of the constancy of the speed of light, thereby providing an elegant and self-consistent foundation for electrodynamics and mechanics.

Principle of the Constancy of the Speed of Light The speed of light in a vacuum is the same in different inertial frames.

The principle of the constancy of the speed of light received empirical confirmation through the Michelson–Morley experiment, which failed to detect any variation in light's velocity due to Earth's motion through the hypothetical ether. This experimental result strongly supports the postulate that the speed of light is invariant across all inertial reference frames. From a theoretical perspective, if we demand that Maxwell's equations retain their form in every inertial frame—a requirement essential for the consistency of electrodynamics—then the invariance of the speed of light emerges as a necessary consequence. This foundational assumption leads directly to the derivation of the Lorentz transformation, which preserves the structure of Maxwell's equations while reconciling them with the principles of relativity. Thus, the framework of special relativity arises naturally from these two postulates: the universality of physical laws across inertial frames and the constancy of the speed of light.

In essence, Einstein's special relativity preserves the principle of relativity, mandating that the fundamental laws of physics maintain identical forms across all inertial reference frames. However, this framework fundamentally redefines classical notions of time and space. The Newtonian conception of time as an absolute, universal parameter is relinquished, as is the privileged status of simultaneity. A key consequence is that two events judged simultaneous in one inertial frame may no longer appear simultaneous when observed from another inertial frame in relative motion. This relativity of simultaneity underscores the profound conceptual shift introduced by special relativity. Most fundamentally, special relativity unifies time and space into a single four-dimensional continuum—spacetime, in which temporal and spatial dimensions are treated on an equal footing. This represents a radical departure from Newtonian mechanics, where time and space were considered independent and absolute. The theory thus subverts the classical dichotomy between space and time, establishing instead an interwoven geometric structure that governs all physical phenomena.

In 1908, Hermann Minkowski introduced a geometric framework for special relativity, demonstrating that spacetime phenomena are most naturally described using four-dimensional geometry. This approach not only elegantly unifies time and space as coordinates of equal standing in a pseudo-Euclidean manifold but also provides a powerful mathematical language for expressing relativistic physics. The geometric formulation proved particularly significant as it laid the conceptual foundation for Einstein's subsequent development of general relativity. Indeed, the tensor calculus and differential geometry employed in Minkowski spacetime became essential tools for describing curved spacetime in the general theory. In the following discussion, we shall adopt this geometric perspective to systematically reconstruct the foundations of special relativity. For comprehensive treatments of special relativity, we refer the reader to [3, 7, 25].

1.1.1 *Lorentz transformation*

Special relativity unifies temporal and spatial dimensions into a single continuum called spacetime. This unification requires a four-dimensional geometric description where time and space coordinates hold equal status as components of a spacetime event. Formally, we define:

(1) **Spacetime events:** An event represents a single point in spacetime, specified by coordinates (t, x, y, z) in a given inertial reference frame. The complete collection of all possible events forms the spacetime manifold.
(2) **Worldlines:** The trajectory of a point particle through spacetime is described by its worldline, which represents the complete history of the particle's position in both space and time.

A crucial aspect of this formulation is the coordinate-independent nature of physical reality:

- While event coordinates are frame-dependent, the events themselves represent absolute physical occurrences.
- The spacetime interval between two events is an invariant quantity, independent of any particular coordinate system.

This geometric perspective reveals the intrinsic structure of spacetime, where physical laws must be formulated in terms of invariants that transcend any specific reference frame.

The foundation of describing geometry is to define the distance between two points through differential and integral operations. For two points that are infinitely close, it can be assumed that they are connected by a straight line, and the line element is defined as

$$\mathrm{d}s^2 = \sum_{i,j} g_{ij}\mathrm{d}x^i\mathrm{d}x^j, \tag{1.3}$$

where i, j depend on the dimensions of spacetime, g_{ij} is called the metric, and its specific form depends on the choice of coordinates. This metric serves as the conceptual equivalent of a ruler for measuring spatial distances or a clock for quantifying temporal intervals between two points.

Example 1.1. Two-dimensional Euclidean space $\mathbf{R}^2$.

(1) In Cartesian coordinates, $\mathrm{d}s^2 = \mathrm{d}x^2 + \mathrm{d}y^2$.
(2) In polar coordinates, let $x = r\cos\phi$, $y = r\sin\phi$, then $\mathrm{d}s^2 = \mathrm{d}r^2 + r^2\mathrm{d}\phi^2$.

No matter what coordinate system is used, the line element $\mathrm{d}s^2$ is invariant.

In special relativity, spacetime is flat, called Minkowski spacetime, or simply Minkowski space, denoted as $\mathbf{R}^{1,3}$. If we choose the Cartesian coordinates

$$x^0 = ct, \quad x^1 = x, \quad x^2 = y, \quad x^3 = z, \tag{1.4}$$

then the line element of Minkowski space is

$$\mathrm{d}s^2 = \sum_{\mu,\nu} \eta_{\mu\nu} \mathrm{d}x^\mu \mathrm{d}x^\nu, \quad \mu, \nu = 0, 1, 2, 3, \tag{1.5}$$

where the metric $\eta_{\mu\nu}$ takes the following diagonal form:[1]

$$\eta_{\mu\nu} = \begin{pmatrix} -1 & 0 & 0 & 0 \\ 0 & 1 & 0 & 0 \\ 0 & 0 & 1 & 0 \\ 0 & 0 & 0 & 1 \end{pmatrix}. \tag{1.6}$$

In the subsequent discussions in this book, we shall adopt natural units ($\hbar = c = 1$) without explicit declaration. This conventional choice simplifies the mathematical formalism. When required for physical interpretation or experimental comparison, the fundamental constants $\hbar, c$ may be systematically restored in relevant expressions through dimensional analysis.

Throughout this book, we employ Einstein's summation convention: repeated indices appearing once as a superscript and once as a subscript on the same side of an equation are implicitly summed over. In the context of flat Euclidean space with Cartesian coordinates, where the metric tensor reduces to the Kronecker delta, it is conventional, though formally an abuse of notation, to relax this rule and allow summation over repeated indices even when both appear in the same (upper or lower) position. However, in more general settings, e.g., curved spaces or non-Cartesian coordinate systems, it is essential to strictly adhere to the contravariant-covariant index pairing. By this convention, the line element (1.5) can be simplified to

$$\mathrm{d}s^2 = \eta_{\mu\nu} \mathrm{d}x^\mu \mathrm{d}x^\nu. \tag{1.7}$$

Furthermore, we adhere to the following notational conventions for tensor indices:

- Greek indices (e.g., μ, ν) denote spacetime components, with $\mu = 0$ corresponding to the temporal dimension and $\mu = 1, 2, 3$ representing spatial coordinates in a (3+1)-dimensional spacetime.

[1] In this book, our convention for the metric signature is the one most commonly used in gravitational research, also known as the east coast convention, which is $(-+++)$. In the field of high-energy physics, the metric signature frequently used is $(+---)$, also known as the west coast convention.

- Latin indices (e.g., i, j) are reserved for purely spatial components, spanning $i = 1, 2, 3$ in three-dimensional space.

In special relativity, two inertial frames are connected through Lorentz transformations rather than Galilean transformations. As mentioned above, the interval between two adjacent events, if connected by a straight line, is

$$\Delta s^2 = -\Delta t^2 + \Delta x^2 + \Delta y^2 + \Delta z^2, \tag{1.8}$$

which is independent of the choice of inertial frame, and therefore is invariant under Lorentz transformations. In the geometric formulation of relativity, the selection of an inertial frame corresponds to the choice of a global Cartesian coordinate system in Minkowski spacetime. Transformations between different inertial frames are then represented by transitions between such rectangular coordinate systems.

Given that the spacetime interval must remain invariant under coordinate transformations, we now derive the permissible class of such transformations. Using the above flat metric, the interval between events can be written as

$$\Delta s^2 = \eta_{\mu\nu}(\Delta x^\mu)(\Delta x^\nu). \tag{1.9}$$

If written in matrix multiplication, the interval should be

$$\Delta s^2 = (\Delta x)^{\mathrm{T}} \cdot \eta \cdot \Delta x, \tag{1.10}$$

that is, Δx is a column vector, η is a 4×4 matrix, and $(\Delta x)^{\mathrm{T}}$ is a row vector, where the upper index T represents the transpose of the matrix. First, this interval remains invariant under global spacetime translation of the form: $x^\mu \to x^{\mu'} = x^\mu + a^\mu$, where a^μ denotes a constant 4-vector. This transformation symmetry corresponds physically to the freedom in choosing the origin of our coordinate system in spacetime. Secondly, consider a rotation

$$x^{\mu'} = \Lambda^{\mu'}{}_{\nu} x^\nu, \tag{1.11}$$

in which $\Lambda^{\mu'}{}_{\nu}$ is a constant matrix, the infinitesimal displacement vector is transformed into $\Delta x^{\mu'} = \Lambda^{\mu'}{}_{\nu} \Delta x^\nu$. In order to keep the interval between spacetime invariant, it is necessary to require

$$\eta_{\rho\sigma} = \Lambda^{\mu'}{}_{\rho} \Lambda^{\nu'}{}_{\sigma} \eta_{\mu'\nu'}. \tag{1.12}$$

In matrix language, the above equation can be compactly written as

$$\Lambda^{\mathrm{T}} \eta \Lambda = \eta. \tag{1.13}$$

This is very similar to the relation satisfied by the orthogonal group $O(3)$ in three-dimensional Euclidean space $\mathbf{R}^3$

$$I = R^{\mathrm{T}} I R, \tag{1.14}$$

where I is a unit matrix.

In mathematical terms, a group $\mathbf{G}$ is a set equipped with a binary operation (called group multiplication) that satisfies the following axioms:

(1) Closure: For all $g_1, g_2 \in G$, $g_1 g_2 \in \mathbf{G}$.
(2) Associativity: $(g_1 g_2) g_3 = g_1 (g_2 g_3)$ for all $g_1, g_2, g_3 \in \mathbf{G}$.
(3) Identity: There exists $e \in \mathbf{G}$ such that $eg = ge = g$ for all $g \in \mathbf{G}$.
(4) Inverse: For each $g \in \mathbf{G}$, there exists $g^{-1} \in \mathbf{G}$ such that $gg^{-1} = g^{-1} g = e$.

The set of all Lorentz rotations (or homogeneous Lorentz transformations) preserving the spacetime interval forms the Lorentz group, denoted $O(1,3)$. As seen from transformation (1.12), these linear transformations satisfy $\mathrm{Det}(\Lambda) = \pm 1$.

In addition to the above continuous transformations, the interval between events is also invariant under two discrete symmetry operations:

- Time reversal ($\mathcal{T}$): $t \to -t$;
- Space inversion ($\mathcal{P}$): $x^i \to -x^i$.

These discrete symmetries, combined with translations and Lorentz rotations, complete the full symmetry group of Minkowski spacetime. To remove this redundancy and select a connected component of the Lorentz group, we restrict our consideration to the proper, orthochronous Lorentz transformations satisfying:

$$\Lambda^0{}_0 \geq 1, \qquad \mathrm{Det}(\Lambda) = 1. \tag{1.15}$$

It can be proved that the collection of these transformations constitutes a subgroup, denoted as $SO(1,3)$, called the proper Lorentz group, commonly referred to as the Lorentz group. In fact, from the definition of Lorentz rotation mentioned above, it can be seen that

$$(\Lambda^0{}_0)^2 = 1 + \sum_i (\Lambda^i{}_0)^2, \tag{1.16}$$

therefore, $|\Lambda^0{}_0| \geq 1$, that is, $\Lambda^0{}_0 \geq 1$ or $\Lambda^0{}_0 \leq -1$. Since $\Lambda^0{}_0 = 1$ represents the identity transformation, the continuity requires $\Lambda^0{}_0 \geq 1$. The other branch $\Lambda^0{}_0 \leq -1$ corresponds to the transformation related to time reversal. Similarly, $\mathrm{Det}(\Lambda) = 1$ indicates that it is connected with the identity transformation, and the continuity naturally requires $\mathrm{Det}(\Lambda) = 1$. As it stands, the proper Lorentz group is the identity component of the full Lorentz group.

The two conditions defining proper Lorentz transformations in (1.15) have important physical significance:

(1) Temporal orientation preservation ($\Lambda^0{}_0 \geq 1$): the condition ensures the transformation maintains the causal structure by preserving the direction of time flow. Mathematically, it excludes time reversal transformations that would map future-directed timelike vectors to past-directed ones, thus maintaining the fundamental distinction between past and future light cones.
(2) Spacetime orientation preservation ($\mathrm{Det}(\Lambda) = 1$): this condition preserves the right-handed orientation of spacetime coordinates and therefore can preserves the light cone structure.

The proper Lorentz group consists of two classes of transformation. The first class contains the usual Lorentz transformations, that is, the so-called Lorentz boost. This kind of transformation mixes temporal and spatial coordinates through hyperbolic rotations and represents the relativistic transformation between different inertial frames. For example, we consider the transformation between the t direction and the x direction, then the rotation matrix is

$$\Lambda = \begin{pmatrix} \cosh\phi & -\sinh\phi & 0 & 0 \\ -\sinh\phi & \cosh\phi & 0 & 0 \\ 0 & 0 & 1 & 0 \\ 0 & 0 & 0 & 1 \end{pmatrix}. \tag{1.17}$$

From this, we get

$$\begin{aligned} t' &= t\cosh\phi - x\sinh\phi, \\ x' &= -t\sinh\phi + x\cosh\phi, \\ y' &= y, \\ z' &= z. \end{aligned}$$

In more familiar language, this transformation is actually the usual Lorentz transformation

$$\begin{aligned} t' &= \gamma(t - vx), \\ x' &= \gamma(x - vt), \\ y' &= y, \\ z' &= z, \end{aligned}$$

where

$$v = \tanh\phi, \quad \gamma = \frac{1}{\sqrt{1-v^2}}. \tag{1.18}$$

ϕ is called the rapidity parameter, and γ is called the Lorentz factor. If we let $\phi \to -\phi$, we get the inverse transformation of the above Lorentz transformation, that is, let the velocity be reversed. As shown in Figure 1.3, for an observer in reference frame S, another reference frame S' moves at a constant velocity v along the positive direction of the x-axis, then the above relationship tells us how the coordinates are connected. Conversely, for an observer in reference frame S', reference frame S moves at a constant velocity v along the negative direction of the x-axis, then the relationship between coordinates is still given by the Lorentz transformation, only the sign of the velocity is opposite, that is

$$\begin{aligned} t &= \gamma(t' + vx'), \\ x &= \gamma(x' - vt'), \\ y &= y', \\ z &= z'. \end{aligned}$$

In this case, the Lorentz transformation matrix is denoted as $\Lambda^{\nu}{}_{\mu'}$, which is the inverse of the original Lorentz transformation:

$$\Lambda^{\mu'}{}_{\nu}\Lambda^{\sigma}{}_{\mu'} = \delta^{\sigma}_{\nu}, \tag{1.19}$$

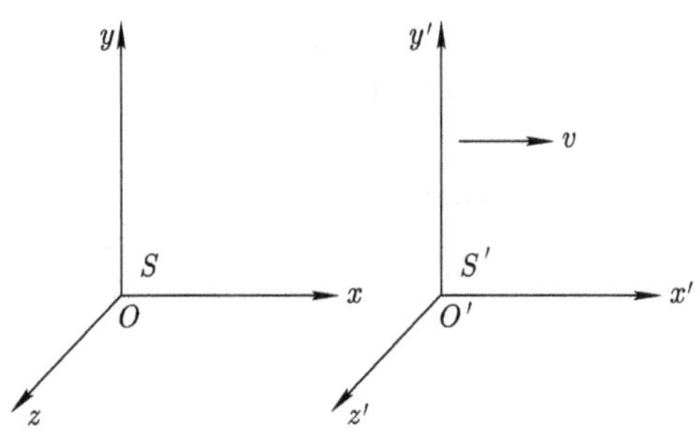

Fig. 1.3 There is a uniform linear motion along the x direction between the two reference frames S and S'.

δ^σ_ν is the Kronecker δ function, defined as

$$\delta^\sigma_\nu = \begin{cases} 1, & \text{if } \sigma = \nu, \\ 0, & \text{if } \sigma \neq \nu. \end{cases}$$

The other class of transformations in the proper Lorentz group contains spatial rotations, which form a subgroup $SO(3)$. Therefore, the proper Lorentz group has a total of 6 generators, that is, three Lorentz transformations, from the transformations between (t, x), (t, y), (t, z), and three spatial rotations, from the transformations between (x, y), (x, z), (y, z), corresponding to three Euler angles.[2] In fact, if the special nature of the time direction is ignored, the number of generators of the $SO(1, 3)$ group is the same as that of the $SO(4)$ group. Generally speaking, for an $SO(n)$ group, the number of generators is $n(n-1)/2$.

The proper Lorentz group is non-commutative, similar to the rotation group in three-dimensional Euclidean space. In mathematical language, this means that the Lorentz group is non-Abelian. This non-commutative structure means the order of successive transformations affects the physical outcome, manifesting in phenomena like Thomas precession. In addition, the Lorentz group when combined with spacetime translations extends to a new group, called the Poincaré group. For more discussions on the Lorentz group, see the Appendix at the end of this chapter.

For a Lorentz transformation caused by motion along any spatial direction, the transformation matrix is

$$\begin{aligned} \Lambda^{0'}{}_{0} &= \gamma, \\ \Lambda^{0'}{}_{i} &= \Lambda^{i'}{}_{0} - \gamma v^i, \\ \Lambda^{i'}{}_{j} &= \Lambda^{j'}{}_{i} = (\gamma - 1)\frac{v^i v^j}{|\boldsymbol{v}|^2} + \delta^{ij}, \end{aligned} \tag{1.20}$$

where the Lorentz factor $\gamma = 1/\sqrt{1 - |\boldsymbol{v}|^2}$. The inverse matrix of this transformation can be obtained by making $\boldsymbol{v} \to -\boldsymbol{v}$.

1.1.2 *Light cones and causality*

Using geometric language makes it easier to understand various physical effects in special relativity. First, consider the loss of simultaneity in special relativity. In Figure 1.4, two events, A and B, appear simultaneous in the

[2] In this book, we typically refer to the transformations related to the time direction as Lorentz transformations, which correspond to Lorentz boosts, while the transformations in the spatial direction are called spatial rotations.

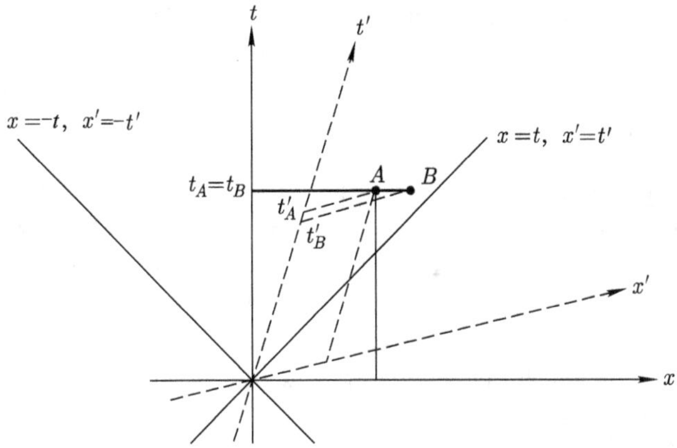

Fig. 1.4 The two reference frames (t, x) and (t', x') are related by a Lorentz transformation. From the figure, it is evident that events A and B are simultaneous in (t, x) but not in (t', x'). Additionally, the worldlines of photons—given by $x = \pm t$ in (t, x) and $x' = \pm t'$ in (t', x')—coincide, confirming that the speed of light remains invariant in both frames.

reference frame (t, x), as they lie on a line perpendicular to the t-axis. However, in the frame (t', x'), these events are no longer simultaneous. For simplicity, we restrict our analysis to the (t, x) plane and assume the Lorentz transformation acts purely along these directions—i.e., $(t, x) \to (t', x')$. Figure 1.4 also illustrates how distances and time intervals differ between inertial frames. A closer examination reveals time dilation and length contraction effects. Nevertheless, the speed of light remains invariant in all inertial frames

$$x = \pm t \to x' = \pm t'. \tag{1.21}$$

Since light-like curves remain invariant across inertial frames, they naturally define light cones. In simple terms, the light cone at a given event is the surface formed by all light rays passing through that event. By exploiting symmetry and using spherical coordinates, we can reduce the problem to the (t, r) plane. Consider two events connected by a straight-line signal transmission, i.e., $\Delta r = v\Delta t$, where v is the propagation speed. From the line element $\Delta s^2 = -\Delta t^2 + \Delta r^2$, the events are classified as follows:

(1) Spacelike separated ($\Delta s^2 > 0$): This implies $v > 1$, meaning the signal would need to travel faster than light. Since this is impossible, the events cannot be causally connected.
(2) Timelike separated ($\Delta s^2 < 0$). This means $v < 1$.

(3) Lightlike separated (null interval, ($\Delta s^2 = 0$)): This corresponds to (v = 1), meaning the signal propagates exactly at the speed of light. Such events lie on the light cone, representing the fastest possible causal connection.

Therefore, the light cone defines the causal relationship between events. If two spacetime points are spacelike connected, then there is no causal relationship between them. And if two spacetime points are timelike connected, then there is a causal relationship. As shown in Figure 1.5, for an observer, there are future light cones and past light cones, and the events inside them are causally related to him, while the events outside the light cones are not causally related to him.

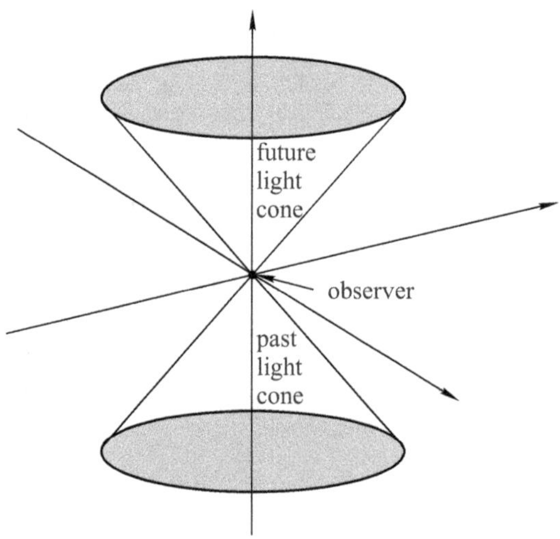

Fig. 1.5 The light cone defines causality.

The temporal ordering of events in relativity is fundamentally tied to their causal relationship. For spacelike-separated events ($\Delta s^2 > 0$), temporal ordering is frame-dependent—different inertial observers may disagree on which event occurred first. This is a direct consequence of the relativity of simultaneity. For timelike-separated events ($\Delta s^2 < 0$), the causal ordering is absolute and preserved across all inertial frames. This invariance stems from the fact that such events lie within each other's light cones. The light cone structure, being a geometric property of spacetime itself, provides an invariant distinction. This geometric perspective shows that while time ordering is relative for spacelike events, causality remains absolute for timelike-connected events.

The path of a particle through spacetime is known as its worldline, where the tangent vector at any point along this curve corresponds to the particle's instantaneous velocity. For a massive particle to preserve causality, its worldline must remain entirely confined within its light cone. This means that at every point along the trajectory, the tangent vector must lie within the local light cone. Such a curve is referred to as a timelike curve. Formally, a timelike curve is defined as a curve whose tangent vector is timelike at every point. If we use $x^\mu(\lambda)$ to describe the curve, its tangent vector at a certain point λ_0 is

$$V^\mu(\lambda = \lambda_0) = \left.\frac{\mathrm{d}x^\mu}{\mathrm{d}\lambda}\right|_{\lambda=\lambda_0}, \tag{1.22}$$

then the timelike curve means that at any point on the curve, there is

$$\eta_{\mu\nu}V^\mu V^\nu < 0. \tag{1.23}$$

For massless particles, the spacetime interval between any two events on their worldline is zero, meaning they always travel at the speed of light. Such a worldline is referred to as a null curve (or lightlike curve), and its tangent vector satisfies the condition that it lies on the light cone at every point

$$\eta_{\mu\nu}V^\mu V^\nu = 0. \tag{1.24}$$

In addition, there are spacelike curves in spacetime, which are defined as curves whose tangent vectors are always spacelike vectors, that is

$$\eta_{\mu\nu}V^\mu V^\nu > 0. \tag{1.25}$$

As shown in Figure 1.6, curve 1 represents the worldline of a massive particle, while curve 2 is the worldline of a massless particle.

The previous discussion of causality appears to assume that two events are connected by a straight worldline, but this simplification was merely for illustrative purposes. In reality, within Minkowski spacetime, if two events are causally related, any continuous curve connecting them must be timelike—a consequence of the continuity requirement. While such a curve may differ in length from the geodesic (straight-line path), its interval under the Minkowski metric remains necessarily negative. Conversely, two causally disconnected events can only be linked by a spacelike curve, as no timelike or null path exists between them.

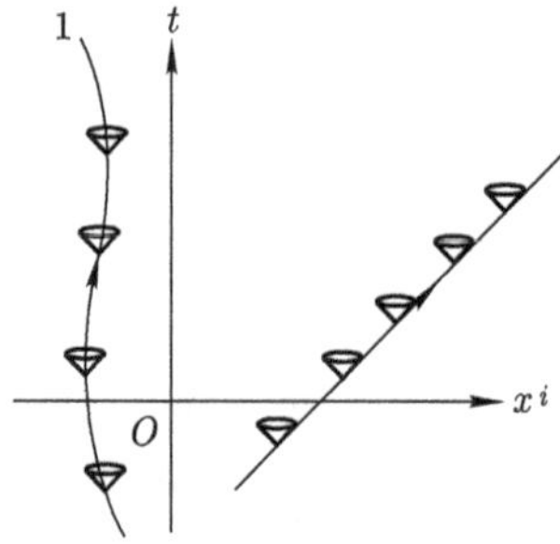

Fig. 1.6 The worldline of a massive particle requires that the tangent vector at each point on it lies within the light cone at that point, as shown by curve 1. In contrast, the tangent vector of a massless particle's worldline is contained on the surface of the light cone at that point, as shown by curve 2.

1.1.3 *Proper time*

Before studying the kinematics of special relativity, let us revisit the standard definition of a vector. In three-dimensional Euclidean space, a vector $\boldsymbol{a}$ is geometrically defined as a directed line segment connecting an initial point to a terminal point, possessing both magnitude and direction. Analogously, in Minkowski spacetime $\mathbf{R}^{1,3}$, a 4-vector $\widehat{a}$ represents a directed segment between two events. Crucially, vectors in Euclidean space and 4-vectors in spacetime are both geometric objects, independent of the choice of coordinates. However, for practical calculations, it is often convenient to introduce a coordinate system and define a set of basis vectors aligned with the coordinate axes,

$$\widehat{e}_t, \quad \widehat{e}_x, \quad \widehat{e}_y, \quad \widehat{e}_z. \tag{1.26}$$

In this coordinate system, the 4-vector $\widehat{a}$ can be expressed in terms of its projections along the coordinate axes. Specifically, it decomposes as: $\widehat{a} = a^\mu \widehat{e}_\mu$, $\mu = 0, 1, 2, 3$, where a^μ are the components of the 4-vector. It is essential to emphasize that while the basis vectors $\widehat{e}_\mu$ depend on the choice of the coordinate system (or inertial reference frame), the 4-vector $\widehat{a}$ itself is a frame-independent geometric object.

The usual operations of three-dimensional vectors, except for the cross product, can also be defined on 4-vectors. The scalar product we commonly use satisfies

(1) $\widehat{a} \cdot \widehat{b} = \widehat{b} \cdot \widehat{a}$,
(2) $\widehat{a} \cdot (\widehat{b} + \widehat{c}) = \widehat{a} \cdot \widehat{b} + \widehat{a} \cdot \widehat{c}$,
(3) $(c\widehat{a}) \cdot \widehat{b} = c(\widehat{a} \cdot \widehat{b})$, where c is a constant.

In the form of components in a certain coordinate system, there is

$$\widehat{a} \cdot \widehat{b} = (a^\mu \widehat{e}_\mu) \cdot (b^\nu \widehat{e}_\nu) = a^\mu b^\nu (\widehat{e}_\mu \cdot \widehat{e}_\nu). \tag{1.27}$$

Define the metric in this coordinate system as

$$\eta_{\mu\nu} \equiv \widehat{e}_\mu \cdot \widehat{e}_\nu, \tag{1.28}$$

then

$$\widehat{a} \cdot \widehat{b} = \eta_{\mu\nu} a^\mu b^\nu. \tag{1.29}$$

The scalar product (inner product) of two vectors must be defined through the metric tensor. In three-dimensional Euclidean space, the scalar product of two vectors depends solely on the magnitudes and relative orientation of the vectors, and is independent of the choice of coordinate system. Analogously, in Minkowski spacetime, the scalar product of two 4-vectors is independent of the choice of coordinate system as well. Sometimes, we also use the notation $\langle \widehat{a}, \widehat{b} \rangle$ to represent the scalar product of vectors $\widehat{a}$ and $\widehat{b}$.

For example, for the spacetime interval defined above, we can regard it as the scalar product of two infinitesimal displacement 4-vectors $\Delta\widehat{x}$: $(\Delta s)^2 = \Delta\widehat{x} \cdot \Delta\widehat{x}$, that is, $\mathrm{d}s^2 = \eta_{\mu\nu}\mathrm{d}x^\mu \mathrm{d}x^\nu$.

Similar to the three-dimensional Euclidean case, once the scalar product of 4-vectors is defined, we can introduce several fundamental concepts:

(1) The magnitude of the vector. The magnitude is determined by the squared norm of the vector, defined as: $||\widehat{a}||^2 \equiv \widehat{a} \cdot \widehat{a}$.
(2) In four-dimensional Minkowski spacetime, the magnitude of the vector can characterize the nature of the 4-vector: if the squared norm is positive, the 4-vector is spacelike; if the squared norm is negative, the 4-vector is timelike; if the squared norm is zero, the 4-vector is lightlike, or null.
(3) Two vectors $\widehat{a}$ and $\widehat{b}$ are orthogonal if their scalar product vanishes: $\widehat{a} \cdot \widehat{b} = 0$. Notably, a lightlike vector is orthogonal to itself.

In measurement, a ruler is a tool for measuring spacelike distances, and a clock is a tool for measuring timelike "distances". An observer carrying a clock moves along a worldline, and the "distance" measured by the clock is the observer's proper time. Usually, when discussing the motion of a massive particle, it is assumed that there is a clock accompanying the particle, so that the proper time of this particle can be measured. Suppose two events are connected by a certain timelike worldline, the proper time between two events is the time an observer spends traveling from one event to another along the worldline.

To gain a deeper understanding of proper time, it is essential to introduce the concept of an instantaneous rest frame (IRF). For a particle in motion, its trajectory is not necessarily uniform or linear but can instead follow any timelike curve in spacetime. Typically, since the particle's velocity varies over time, it is impossible to define a single, global rest frame for the entire motion. However, at any given instant, the particle's velocity can be treated as momentarily constant, allowing the definition of an instantaneous rest frame—also referred to as a comoving frame. Within this frame, the particle is instantaneously at rest, and the progression of time is governed by the proper time.

Consider a curve in spacetime, characterized by a parameter λ. In the coordinate system $\{x^\mu\}$, this curve or path is described by the function $x^\mu(\lambda)$. For an infinitesimal displacement vector,

$$\mathrm{d}x^\mu = \frac{\mathrm{d}x^\mu}{\mathrm{d}\lambda}\mathrm{d}\lambda. \tag{1.30}$$

Therefore, the distance between two points A and B connected by a spacelike curve is

$$\Delta s = \int_{\lambda_A}^{\lambda_B} \sqrt{\eta_{\mu\nu}\frac{\mathrm{d}x^\mu}{\mathrm{d}\lambda}\frac{\mathrm{d}x^\nu}{\mathrm{d}\lambda}}\mathrm{d}\lambda, \tag{1.31}$$

and the proper time between two points connected by a timelike curve is

$$\Delta\tau = \int_{\lambda_A}^{\lambda_B} \sqrt{-\eta_{\mu\nu}\frac{\mathrm{d}x^\mu}{\mathrm{d}\lambda}\frac{\mathrm{d}x^\nu}{\mathrm{d}\lambda}}\mathrm{d}\lambda. \tag{1.32}$$

Here λ_A, λ_B are the parameters corresponding to events A, B on the worldline. Obviously, the interval between two events connected by a worldline, whether spacelike or timelike, is independent of both the choice of coordinate system and the parametrization λ used to describe the worldline. Specifically, under a reparametrization $\lambda \to f(\lambda)$, where $f(\lambda)$ is an arbitrary function of λ, the interval remains unchanged. This invariance is known as the reparametrization invariance of the worldline. The origin of this property lies in the fact that both events and worldlines are fundamental geometric objects in spacetime, existing independently of any coordinate description.

We can generally discuss the effects of time dilation and length contraction. Here we discuss the effect of time dilation. This requires us to compare the relationship between proper time and coordinate time. No matter what coordinate system is taken, there is always

$$\begin{aligned}\tau_{AB} = \int_A^B \mathrm{d}\tau &= \int_A^B [\mathrm{d}t^2 - \mathrm{d}r^2/c^2]^{1/2} \\ &= \int_{t_A}^{t_B} \mathrm{d}t'[1 - v^2(t')/c^2]^{1/2},\end{aligned} \tag{1.33}$$

where v is the velocity of the particle in the coordinate system (t, r) (here for intuition, the speed of light c is explicitly written out), obviously,

$$\tau_{AB} < t_B - t_A. \tag{1.34}$$

Consequently, in any reference frame not comoving with the particle, the coordinate time is always greater than the proper time. This fundamental result has profound implications in particle physics experiments. Specifically, the observed lifetime of particles in the laboratory frame is invariably longer than their proper lifetime—a direct manifestation of relativistic time dilation. This phenomenon is routinely confirmed in cosmic ray studies and accelerator experiments, where precise measurements of high-speed unstable particles consistently demonstrate this effect.

For a certain type of unstable particle, its proper lifetime τ_p is its intrinsic property, which can be obtained through correct theoretical calculation. Consider a large collection of such particles, after time t, some of them have decayed, and the proportion of decayed particles to the total number of particles is $1 - \exp(-t/\tau_\mathrm{p})$. If this particle moves relative to the laboratory inertial frame at a velocity v, then the lifetime measured in the laboratory is

$$\tau_\mathrm{p}(\gamma) = \gamma\tau_\mathrm{p}, \tag{1.35}$$

where $\gamma = (1 - v^2/c^2)^{-1/2}$ is the Lorentz factor. Below we take the muon as an example to illustrate this phenomenon.

Example 1.2. Muon lifetime experiment.

The main decay channel of the muon is the decay to electron and neutrino:

$$\mu^- \to \nu_\mu + \mathrm{e}^- + \bar{\nu}_\mathrm{e}, \tag{1.36}$$

its proper lifetime is $\tau_\mu = 2.2$ μs. In the late 1970s, a large number of high-speed muons were produced in the colliders at CERN, their velocity is close to the speed of light, $v/c \approx 0.9994$, and the corresponding Lorentz factor is $\gamma \approx 29.3$. The lifetime of muons measured in the detector is

$$\tau_\mu^+ = (64.419 \pm 0.058)\ \mu\mathrm{s}, \qquad \tau_\mu^- = (64.368 \pm 0.029)\ \mu\mathrm{s}. \tag{1.37}$$

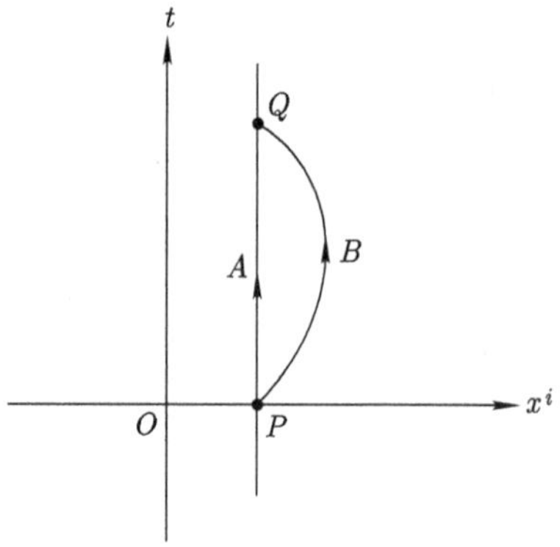

Fig. 1.7 Twins A and B part at event P and meet at event Q.

For the antimuon (μ^+),

$$[\tau_\mu^+(1) - \tau_\mu^+(\gamma)/\gamma]/\tau_\mu^+(1) = (2 \pm 9) \times 10^{-4}. \tag{1.38}$$

This is highly consistent with the prediction of special relativity.

To conclude this section, we examine the well-known twin paradox. Consider twins A and B: while A remains at rest on Earth, B embarks on a journey through space (as illustrated in Figure 1.7). Both maintain identical lifestyles and biological rhythms, ensuring that any observed differences arise purely from relativistic effects. When they reunite after several years, which twin will be younger? The answer lies in comparing the proper time elapsed on their respective wristwatches. For timelike worldlines connecting two events, the proper time (the "length" of the curve in spacetime) is maximized for inertial (unaccelerated) motion. This leads to a counterintuitive result: In Euclidean geometry, the shortest path between two points is a straight line segment. In spacetime, the longest proper time between two events corresponds to a straight (inertial) worldline. Since twin A's worldline is a straight line along the time axis (representing inertial motion), more proper time elapses for A than for B, whose worldline is curved due to acceleration. Thus, the traveling twin B returns younger than the stay-at-home twin A, demonstrating that "motion preserves youth" relativistically.

1.2 Kinematics in Special Relativity

In this section, we examine the kinematics of both massive and massless particles within the framework of special relativity, along with their associated physical phenomena. For massive particles, the motion need not be uniform or linear—the only requirement is that the particle's worldline must be a timelike curve in spacetime. Our focus will be on analyzing the diverse possible motions of particles in flat spacetime.

1.2.1 *Massive particles*

The worldline (spacetime trajectory) of a massive particle can be represented as a parametric curve $x^\mu(\lambda)$, where $\{x^\mu\}$ denotes coordinates in an inertial reference frame and λ is an arbitrary parameter characterizing the particle's evolution along its worldline. For massive particles, the proper time τ emerges as the most natural parameter choice. In other words, we can use the reparametrization invariance of the particle's motion to choose the particle's proper time as the parameter, which provides convenience for our study of the particle's motion. Under this choice, the particle's 4-velocity is

$$u^\mu \equiv \frac{\mathrm{d}x^\mu}{\mathrm{d}\tau}, \tag{1.39}$$

it satisfies the normalization condition

$$\hat{u} \cdot \hat{u} = -1, \tag{1.40}$$

showing that the 4-velocity is a timelike vector with a square of the modulus of -1. In the comoving reference frame of the particle, the particle is always at rest, and the components of the 4-velocity are

$$u^\mu = (1, 0, 0, 0), \tag{1.41}$$

while in another reference frame, the particle's motion velocity is $\boldsymbol{v}$, and the components of its 4-velocity are

$$u^\mu = \gamma_v(1, \boldsymbol{v}), \tag{1.42}$$

where γ_v is the Lorentz factor, $\gamma_v = 1/\sqrt{1 - \boldsymbol{v}^2}$.

A particle with mass m, its 4-momentum is defined as

$$\hat{p} \equiv m\hat{u}, \tag{1.43}$$

its components are

$$p^\mu = mu^\mu = (E, \boldsymbol{P}), \tag{1.44}$$

where

$$E = m\gamma = m + \frac{1}{2}mv^2 + \cdots , \tag{1.45}$$

$$\boldsymbol{P} = m\gamma \boldsymbol{v} = m\boldsymbol{v} + \cdots , \tag{1.46}$$

$\boldsymbol{v}$ is the particle's velocity. From (1.40), the above components satisfy (the following two equations clearly write out the speed of light c)

$$E^2 - \boldsymbol{P}^2 c^2 = m^2 c^4. \tag{1.47}$$

In the reference frame that is at rest relative to the particle, the so-called comoving reference frame,

$$E = mc^2. \tag{1.48}$$

This is the famous Einstein mass-energy relation. In fact, the mass-energy relation comes from the normalization relation of the 4-velocity of a massive particle. Further, we can define the particle's 4-acceleration as

$$a^\mu \equiv \frac{\mathrm{d}u^\mu}{\mathrm{d}\tau} = \frac{\mathrm{d}^2 x^\mu}{\mathrm{d}\tau^2}. \tag{1.49}$$

In special relativity, the addition of velocity is different from that in Newtonian mechanics. If the worldline of a particle in a reference frame S is described by $x^i(t)$, $i = 1, 2, 3$, then its coordinate velocity in S is

$$u^i = \frac{\mathrm{d}x^i}{\mathrm{d}t}. \tag{1.50}$$

If the reference frame S' moves at a velocity v relative to S along the x direction, as shown in Figure 1.3, then in S', the particle's velocity should be given by $\mathrm{d}x'^i/\mathrm{d}t'$. From

$$\mathrm{d}t' = \gamma(\mathrm{d}t - v\mathrm{d}x), \quad \mathrm{d}x' = \gamma(\mathrm{d}x - v\mathrm{d}t), \quad \mathrm{d}y' = \mathrm{d}y, \quad \mathrm{d}z' = \mathrm{d}z, \tag{1.51}$$

we can know

$$\begin{aligned} u'_x &= \frac{\mathrm{d}x'}{\mathrm{d}t'} = \frac{u_x - v}{1 - u_x v}, \\ u'_y &= \frac{\mathrm{d}y'}{\mathrm{d}t'} = \frac{u_y}{\gamma(1 - u_x v)}, \\ u'_z &= \frac{\mathrm{d}z'}{\mathrm{d}t'} = \frac{u_z}{\gamma(1 - u_x v)}. \end{aligned} \tag{1.52}$$

Obviously, this is different from the velocity addition in Newtonian mechanics. Of course, when the velocity is small, $v \ll 1$, the above velocity addition relationship returns to the case of Newtonian mechanics.

If we have three reference frames, in which S' moves at a velocity v along the positive direction of the x-axis relative to S, and S'' moves at a velocity v' along the positive direction of the x'-axis relative to S', then using the Lorentz transformation, we have

$$\begin{aligned} t'' &= t\cosh(\phi_v + \phi_{v'}) - x\sinh(\phi_v + \phi_{v'}), \\ x'' &= -t\sinh(\phi_v + \phi_{v'}) + x\cosh(\phi_v + \phi_{v'}), \\ y'' &= y, \\ z'' &= z, \end{aligned} \tag{1.53}$$

where ϕ_v, $\phi_{v'}$ are the rapidity parameters corresponding to the velocities v, v', $\tanh\phi_v = v$, $\tanh\phi_{v'} = v'$. The above transformation relationship indicates that the reference frame S'' moves at a velocity $u = \tanh(\phi_v + \phi_{v'})$ along the positive direction of the x-axis relative to S. That is to say, using the rapidity parameter, we can simply add: $\phi_u = \phi_v + \phi_{v'}$. Therefore

$$u = \tanh(\phi_v + \phi_{v'}) = \frac{\tanh\phi_v + \tanh\phi_{v'}}{1 + \tanh\phi_v \tanh\phi_{v'}} = \frac{v' + v}{1 + v'v}. \tag{1.54}$$

This is the formula for the relativistic addition of two velocities along the same direction.

A common misconception in special relativity is that only uniform motion can be analyzed, and it is not known how to discuss accelerated motion, particularly because the inertial frame appears to change continuously during acceleration. In reality, special relativity is fully capable of describing various particle motions in flat spacetime, including accelerated trajectories. Below, we begin with the simplest case: uniform acceleration in a straight line. In the above discussion, we have inertial frames S and S', which are connected through Lorentz transformations. If a particle moves in S, its coordinate velocity is u^i, then its coordinate acceleration is

$$a^i = \frac{\mathrm{d}u^i}{\mathrm{d}t}. \tag{1.55}$$

And for the observer in the coordinate system S', the velocity it sees for the particle is u'^i, and the acceleration is

$$a'^i = \frac{\mathrm{d}u'^i}{\mathrm{d}t'}. \tag{1.56}$$

From

$$\mathrm{d}t' = \gamma(1 - u_x v)\mathrm{d}t, \tag{1.57}$$

we can get

$$\begin{aligned} a'_x &= \frac{1}{\gamma^3(1-u_x v)^3} a_x, \\ a'_y &= \frac{1}{\gamma^2(1-u_x v)^2} a_y + \frac{u_y v}{\gamma^3(1-u_x v)^3} a_x, \\ a'_z &= \frac{1}{\gamma^2(1-u_x v)^2} a_z + \frac{u_z v}{\gamma^3(1-u_x v)^3} a_x. \end{aligned} \tag{1.58}$$

Unlike in Newtonian mechanics, where acceleration remains invariant under Galilean transformations, acceleration in special relativity is frame-dependent. Nevertheless, it retains an absolute physical meaning: if a particle has zero proper acceleration (i.e., no acceleration in its instantaneous rest frame), it will be measured as acceleration-free by all inertial observers, even though the accelerations seen by different observers are different.

For a massive particle following a timelike worldline, we can define an instantaneous inertial frame at any event along its trajectory. This frame moves with a velocity $\boldsymbol{v} = \boldsymbol{u}$ relative to the original coordinate system S, where $\boldsymbol{u}$ is the particle's velocity at that instant. In this instantaneous rest frame (IRF) S', the particle is momentarily at rest $u' = 0$, allowing us to analyze its dynamics in a locally comoving inertial system. Secondly, from the above discussion of proper time, we know that the proper time interval is the same as the coordinate time interval in S', $\Delta\tau = \Delta t'$. We consider a one-dimensional motion along the x direction, at this time there is

$$a = (1-u^2)^{3/2} a', \tag{1.59}$$

where a' is the particle's rest acceleration in the IRF, that is, the acceleration relative to the proper time. If we let $a' = a_0$ be a constant, then we have

$$\frac{\mathrm{d}u}{\mathrm{d}t} = (1-u^2)^{3/2} a_0. \tag{1.60}$$

If the initial velocity of the particle is $u(t=0) = 0$, we can integrate the above equation to get

$$\left(x - x_0 + \frac{1}{a_0}\right)^2 - t^2 = \frac{1}{a_0^2}. \tag{1.61}$$

This trajectory describes a hyperbola, indicating that a particle undergoing constant proper acceleration follows a hyperbolic worldline. Crucially, the "constant acceleration" here refers to the proper acceleration—the magnitude of the 4-acceleration vector remains invariant in the particle's instantaneous rest frame. We can take the motion origin of the particle as

$x_0 = 1/a_0$ (see Figure 1.8), then the above equation can be further simplified. For simplicity, we discuss the worldline of this particle in $1+1$ dimensions

$$t(\sigma) = a_0^{-1}\sinh\sigma, \quad x(\sigma) = a_0^{-1}\cosh\sigma, \quad -\infty < \sigma < \infty,$$

σ is a parameter that describes the worldline. Since this is a timelike curve, its metric is $\mathrm{d}\tau^2 = \mathrm{d}t^2 - \mathrm{d}x^2 = (a_0^{-1}\mathrm{d}\sigma)^2$. Since τ is related to the particle's proper time, by choosing $\sigma = 0$ when $\tau = 0$, we have

$$\tau = a_0^{-1}\sigma. \tag{1.62}$$

Using the proper time as a parameter, the coordinate functions describing the worldline become

$$t(\tau) = a_0^{-1}\sinh(a_0\tau), \quad x(\tau) = a_0^{-1}\cosh(a_0\tau). \tag{1.63}$$

The particle's 4-velocity $u^\mu = \dfrac{\mathrm{d}x_\mu}{\mathrm{d}\tau}$ is always tangent to the worldline,

$$u^t = \cosh(a_0\tau), \quad u^x = \sinh(a_0\tau), \tag{1.64}$$

satisfying $\widehat{u}\cdot\widehat{u} = -1$. And in the original rest coordinate system, the particle's coordinate velocity is

$$v^x = \frac{\mathrm{d}x}{\mathrm{d}t} = \tanh(a_0\tau), \tag{1.65}$$

which tends to the speed of light when $\tau = \pm\infty$. The particle's 4-acceleration is

$$a^t = a_0\sinh(a_0\tau), \quad a^x = a_0\cosh(a_0\tau), \tag{1.66}$$

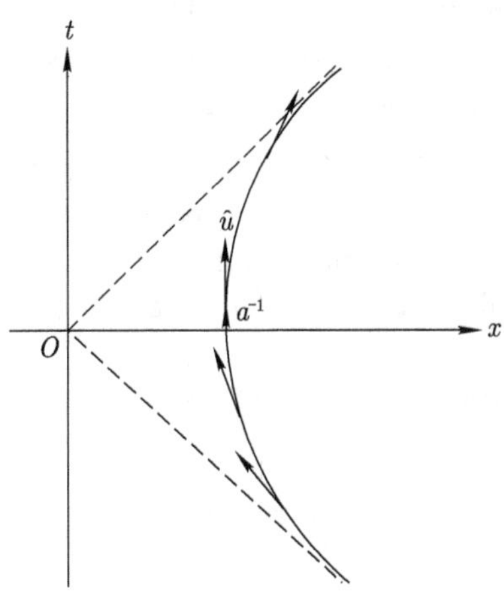

Fig. 1.8 The worldline of a particle undergoing uniform acceleration in the x direction.

with its magnitude being constant

$$a_0 = (\widehat{a} \cdot \widehat{a})^{\frac{1}{2}}. \tag{1.67}$$

Therefore, the motion of this particle is indeed a constant acceleration motion, relative to the particle's own instantaneous rest frame.

1.2.2 *Massless particles*

For massless particles ($m = 0$), which necessarily move at the speed of light, proper time vanishes identically along their worldlines. Consequently, proper time cannot serve as an affine parameter for their trajectories. This absence of a preferred temporal parameter introduces significant freedom in the choice of parametrization for null geodesics. For example, for a particle moving at the speed of light along the x direction, its worldline satisfies

$$x = t, \tag{1.68}$$

a parameter choice is

$$x^\mu = b^\mu \lambda, \quad b^\mu = (1, 1, 0, 0), \tag{1.69}$$

where λ is a real number. Of course, there are many other parameterization options, as long as $x = t$ is satisfied, such as

$$x^\mu = b^\mu \lambda^3. \tag{1.70}$$

No matter what kind of parameterization, the particle's 4-velocity always satisfies $\widehat{u} \cdot \widehat{u} = 0$, where $u^\mu = \dfrac{\mathrm{d}x^\mu}{\mathrm{d}\lambda}$. Despite the inherent freedom in parameter choice, we naturally seek the most convenient parametrization possible. A parameter λ is called an affine parameter if it satisfies the condition $\dfrac{\mathrm{d}\widehat{u}}{\mathrm{d}\lambda} = 0$, where $\widehat{u}$ represents the tangent vector to the worldline. Affine parametrization is particularly advantageous as it preserves the geodesic equation's simple form. Regarding the two parametrization options previously discussed: the first one constitutes an affine parametrization, while the second does not.

For massless particles, such as photons, their energy is proportional to the frequency of the wave,

$$E = \hbar\omega, \tag{1.71}$$

and the particle's momentum 4-vector is

$$p^\mu = (E, \boldsymbol{P}) = (\hbar\omega, \hbar\boldsymbol{k}) = \hbar k^\mu, \tag{1.72}$$

where k^μ is called the wave 4-vector. Obviously

$$\widehat{p}\cdot\widehat{p}=\widehat{k}\cdot\widehat{k}=0. \tag{1.73}$$

The momentum 4-vector and the wave 4-vector are both tangent to the photon's worldline:

$$\widehat{p},\widehat{k}\propto\widehat{u}. \tag{1.74}$$

In affine parameterization, $\dfrac{\mathrm{d}\widehat{p}}{\mathrm{d}\lambda}=0$.

Below we discuss several relativistic effects related to light, and more about relativistic optics will be introduced in the next section.

Example 1.3. Drag effect of fluid on light.

Prior to the advent of special relativity, ether-based theories faced significant challenges in explaining light propagation through moving media. Consider a transparent liquid flowing in a straight pipe, where the light wave experiences partial dragging by the moving fluid. This phenomenon was conclusively demonstrated by Fizeau's 1851 experiment. According to ether theory, just as air serves as the propagation medium for sound waves, the luminiferous ether was postulated as the medium for light waves. The observed dragging effect would therefore imply that the ether itself must be dragged by the moving fluid. However, experimental evidence revealed only partial ether drag, a result that proved deeply problematic for classical ether theories. This partial drag effect, quantitatively expressed by Fresnel's dragging coefficient, could not be fully reconciled with the notion of a stationary, universal ether. If the propagation speed of light in the stationary liquid is u_0, and the velocity of the liquid relative to the pipe or laboratory is v, then the velocity of light relative to the laboratory is

$$u=u_0+(1-1/n^2)v, \tag{1.75}$$

where $n=c/u_0$ is the refractive index of the liquid. This fact can easily be explained by the velocity addition principle in special relativity: the velocity of light relative to the liquid is u_0, and the velocity of the liquid relative to the laboratory reference frame is v, then the velocity of light relative to the laboratory reference frame is

$$u=\frac{u_0+v}{1+u_0v/c^2}\approx(u_0+v)\left(1-\frac{u_0v}{c^2}\right)\approx u_0+v\left(1-\frac{u_0^2}{c^2}\right)=u_0+(1-1/n^2)v, \tag{1.76}$$

where we ignore the term v^2/c^2. Einstein gave the velocity addition formula in special relativity in 1905, and two years later Laue used this formula to elegantly explain the drag effect of light in a fluid.

Example 1.4. Doppler shift.

We now analyze the relativistic Doppler effect. As shown in Figure 1.9, consider a light source at rest in its reference frame, emitting photons isotropically with frequency ω. Suppose a receiver moves relative to this source along the x-axis with velocity v, where $v > 0$ indicates approach and $v < 0$ indicates recession. For a photon emitted at angle α relative to the x-axis in the source frame, the receiver will detect this photon at a modified frequency ω' and angle α' in its own reference frame. The relationship between these quantities can be derived through Lorentz transformation of the photon's 4-momentum between frames. Assume that in the observer's stationary reference frame S', the observer observes the photon to have a wave 4-vector

$$k'^{\mu} = \frac{2\pi}{\lambda'}(1, \cos\alpha', \sin\alpha', 0). \tag{1.77}$$

In Minkowski spacetime, photons follow null geodesics (straight lines in inertial coordinates), with their 4-momentum $\widehat{k}$ being a constant along the worldline. This implies that the 4-momentum is identical at both the emission event and reception event when expressed in global inertial coordinates. In the coordinate system where the light source is located, the emitted photon has a 4-momentum $k^{\mu} = \widehat{k} \cdot \widehat{e}^{\mu}$, where $\widehat{e}^{\mu}$ is the local laboratory of the light source. Therefore,

$$k^{\mu} = \frac{2\pi}{\lambda}(1, \cos\alpha, \sin\alpha, 0) = \Lambda^{\mu}_{\nu} k'^{\nu}, \tag{1.78}$$

where λ is the photon's proper wavelength, and λ' is the observed wavelength, and

$$\Lambda^{\mu}{}_{\nu} = \begin{pmatrix} \gamma & -v\gamma & 0 & 0 \\ -v\gamma & \gamma & 0 & 0 \\ 0 & 0 & 1 & 0 \\ 0 & 0 & 0 & 1 \end{pmatrix}. \tag{1.79}$$

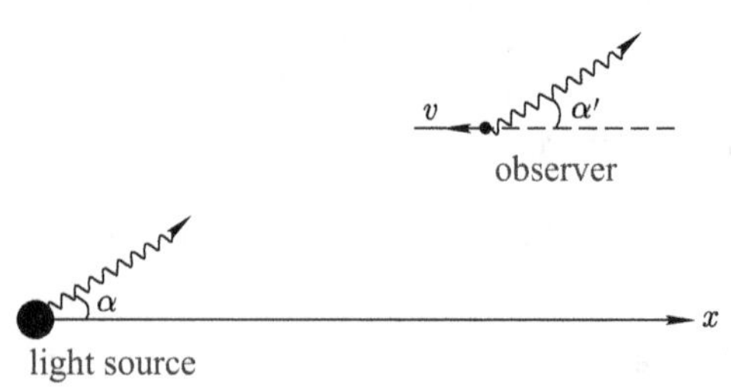

Fig. 1.9 Schematic diagram of the Doppler effect.

From this, we obtain

$$\frac{\lambda'}{\lambda} = \gamma(1 - v\cos\alpha'), \tag{1.80}$$

$$\tan\alpha = \frac{\tan\alpha'}{\gamma(1 - v\sec\alpha')}. \tag{1.81}$$

Equation (1.80) is actually the formula for the Doppler effect, which is clearer when expressed in terms of frequency:

$$\nu' = \frac{\nu}{\gamma(1 - v\cos\alpha')} = \nu\frac{\sqrt{1-v^2}}{1 - v\cos\alpha'}. \tag{1.82}$$

If $v \ll 1$, then we have

$$\nu' \approx \nu(1 + v\cos\alpha'). \tag{1.83}$$

Since $|\alpha'| \le \pi/2$, therefore

(1) If $v > 0$, when the source and receiver are moving towards each other, the frequency of the received light increases, a blue shift occurs, $\Delta v = v\nu\cos\alpha'$.
(2) If $v < 0$, when moving away from each other, the frequency of the received light decreases, a red shift occurs, $\Delta\nu = -|v|\nu\cos\alpha'$.
(3) If $\alpha' = \frac{\pi}{2}$, the frequency of the received light decreases, still resulting in a red shift, $\Delta\nu \propto v^2$. It is known as the transverse Doppler shift, entirely caused by the time dilation effect.

The relativistic Doppler effect shares some qualitative similarities with its classical counterpart for sound waves: when the source and receiver approach each other $v > 0$, the observed frequency increases (blueshift), while recession $v < 0$ leads to frequency decrease (redshift). However, unlike the Doppler effect of sound waves, due to the time dilation effect, even transverse motion can still cause a change in frequency. It is not difficult to find that the Doppler shift effect is a first-order relativistic effect, proportional to v, while the time dilation effect is a second-order relativistic effect, proportional to v^2.

In particular, when $\alpha' = 0$, we have

$$\nu' = D\nu, \tag{1.84}$$

where D is the so-called Doppler factor,

$$D = \left(\frac{1+v}{1-v}\right)^{1/2}. \tag{1.85}$$

For the general case, we have

$$\nu' = \frac{\nu}{1+z}, \tag{1.86}$$

where z is the redshift factor, defined as

$$z = \frac{\lambda_r - \lambda_e}{\lambda_e}, \tag{1.87}$$

with λ_r and λ_e being the photon wavelengths at the time of reception and emission, respectively.

For the transverse Doppler effect, $\alpha' = \pi/2$, the change in frequency is $\nu' = \nu/\gamma$. It happens when in the observer's reference frame, the direction of the photon's motion is perpendicular to the observer's velocity. If the angle in the emitter's reference frame is used, then the formula (1.82) becomes

$$\nu' = \nu \frac{1 + v\cos\alpha}{\sqrt{1 - v^2}}. \tag{1.88}$$

Consider a disk of radius r_0 rotating with constant angular velocity ω in an inertial frame. A light source at the center emits photons of proper frequency ν_0, while an observer at the rim measures their frequency. At the instant when the observer's position vector aligns with the x-axis, their velocity $\vec{v} = \omega r_0 \hat{x}$ is perpendicular to the photon's momentum $\vec{k} \parallel \hat{y}$, corresponding to $\alpha = \pi/2$. In this configuration, the classical Doppler shift vanishes (first-order effect $\propto \cos\alpha = 0$), and the observed frequency is purely determined by time dilation:

$$\nu = \nu_0 \gamma = \frac{\nu_0}{\sqrt{1 - (\omega r_0)^2/c^2}} \tag{1.89}$$

where $\gamma = (1 - v^2/c^2)^{-1/2}$ is the Lorentz factor with $v = \omega r_0$. This transverse Doppler blueshift provides a direct verification of relativistic time dilation. The 1960 experiment by H. J. Hay *et al.*[3] confirmed this prediction with $< 5\%$ deviation using Mössbauer spectroscopy with a rotating ^{57}Fe source.

More generally, we can assume that the position of the light source P_0 on the disk is at a radius of r_0, and the detector P_1 is at a position of r_1, and their connecting line is aligned along the radial direction. Since photons travel in straight lines on the disk, all signals propagating from P_0 to P_1 require the same transit time. Consequently, in the laboratory reference frame, the time interval Δt between consecutive signals remains

[3] H. J. Hay *et al.*, *"Measurement of the Transverse Doppler Effect"*, Phys. Rev. Lett. 4, 165 (1960).

identical at both the emitter and the receiver. Since the motion of the photon is strictly along the radial direction, perpendicular to the direction of the linear velocity of the emitter and detector, we only need to consider the time dilation effect, therefore, the time interval seen at P_0 is $\Delta t\gamma_0$, and the time interval seen at P_1 is $\Delta t\gamma_1$, where

$$\gamma_i = \frac{1}{\sqrt{1-(\omega r_i)^2}}, \quad i = 0, 1. \tag{1.90}$$

So, we find that the ratio of the photon frequencies observed by the emitter and receiver is

$$\frac{\nu_0}{\nu_1} = \frac{\gamma_0}{\gamma_1}. \tag{1.91}$$

The relative motion between the light source and the observer provides a means to test the time dilation effect. However, since the ordinary Doppler effect constitutes a first-order contribution, it often dominates and obscures the time dilation signal. Thus, isolating the time dilation effect requires eliminating the Doppler contribution. In 1938, H. Ives and G. R. Stilwell achieved this by employing ions in oscillatory motion, thereby canceling out the Doppler effect and enabling the first direct experimental verification of time dilation. In 1985, M. Kaivolo *et al.*[4] used the same idea, through the technology of moving nickel atom beams and lasers, to test the time dilation effect, and the deviation between theory and experiment was 4×10^{-5}. Further, in 1994, R. Grieser *et al.*[5] improved the precision to 7×10^{-7}.

Example 1.5. Relativistic aberration and beaming effect.

Below we consider the beaming phenomenon caused by relativistic effects. Still using the previous Figure 1.9, assume that the photon is emitted at an angle α with the x-axis, therefore

$$\cos\alpha = k^x/\omega, \tag{1.92}$$

where k^x is the x-component of the wave vector. And for the observer, there is

$$\cos\alpha' = k'^x/\omega'. \tag{1.93}$$

The Lorentz transformation between (ω, k^x) and (ω', k'^x) can lead to Eq. (1.81). It demonstrates a discrepancy between the observed photon

[4]Kaivolo M. *et al.*, Phys. Rev. Lett., 1985, 54: 255.
[5]Grieser R. *et al.*, Appl. Phys. B, 1994, 59: 127.

direction and its original emission direction, a phenomenon described by the relativistic aberration formula. It can actually be written as

$$\cos\alpha' = \frac{\cos\alpha + v}{1 + v\cos\alpha}, \tag{1.94}$$

or

$$\tan\frac{\alpha'}{2} = \left(\frac{1-v}{1+v}\right)^{1/2} \tan\frac{\alpha}{2}. \tag{1.95}$$

If the source and observer are moving relative to each other, and the light source has a certain size distribution, such as a star, photons emitted from different parts of it may be observed by us. If $v > 0$, then $\alpha' < \alpha$, and if $v < 0$, then $\alpha' > \alpha$. Assuming that the light source is moving towards the observer, the photons emitted at the edge are forward photons ($|\alpha| < \pi/2$), these photons have a smaller angle with the x-axis when seen by the observer, and the angle that the entire light source makes relative to the observer also becomes smaller, $|\alpha'| < \arccos v$. When the relative motion speed approaches the speed of light $v \approx 1$, this angle will be very small. Consequently, the relative motion between the source and the observer redirects photons emitted in all directions into a preferred orientation. This phenomenon, known as the headlight effect, describes how a source that emits isotropically in its comoving frame will appear to concentrate nearly all its radiation into a narrow forward cone when moving at relativistic speeds. This effect persists even for anisotropic emission patterns. It is particularly pronounced in synchrotron radiation, where highly accelerated charged particles produce a sharply collimated beam due to their extreme velocities.

Empirical observations reveal that moving vehicles and pedestrians perceive vertically falling rain or snow as approaching them obliquely rather than descending straight downward. Similarly, Eq. (1.95) demonstrates that observers in relative motion measure distinct angles of incoming light, a phenomenon known as light aberration. Historically, the English astronomer James Bradley discovered the aberration of starlight as early as 1728. By observing annual shifts in the apparent positions of stars, Bradley demonstrated that the direction of incoming starlight depends on the relative velocity between the observer and the light source, a phenomenon now known as stellar aberration. This finding offered strong support for the Copernican heliocentric model, as it confirmed that the Earth is not stationary but orbits the Sun.

Furthermore, the relative motion induces a Doppler shift: photons emitted in the direction of motion experience a blueshift, while those emitted in the opposite direction undergo a redshift. Consequently, the radiation intensity becomes highly anisotropic, peaking sharply along the direction of motion. This phenomenon, known as relativistic beaming, causes a uniformly radiating object to appear significantly brighter when approaching the observer and dimmer when receding. The effect is particularly pronounced at relativistic velocities, where the forward-directed emission dominates the observed radiation pattern. More accurately, from the above discussion, we know that if an infinitesimal angle is considered in the source's rest frame, it will be found that the angle becomes

$$\delta\alpha \approx D\delta\alpha', \tag{1.96}$$

in the laboratory observation. Since the angle of the forward photons in the laboratory becomes smaller, and the number of photons remains unchanged, the density of photons increases by a factor of D^2. In addition, considering that the photon energy is increased by a factor of D due to the Doppler motion, and the number of photons arriving per unit time is increased by a factor of D, we finally find that the brightness of the source (given by the energy flow) increases by a factor of D^4:

$$\mathcal{E} = \left(\frac{1+v}{1-v}\right)^2 \mathcal{E}_0. \tag{1.97}$$

Therefore, when the light source moves towards us, the observed light appears particularly dazzling. More generally, if the motion of the observer is considered (such as the rotation of the Earth), then the change in brightness is

$$\mathcal{E} = \left(\frac{1}{1+z}\right)^4 \mathcal{E}_0. \tag{1.98}$$

where z is the redshift factor of monochromatic light.

Example 1.6. Compton scattering.

Finally, we examine the Compton scattering of photons by electrons, a phenomenon of fundamental importance in the development of quantum mechanics. Let us consider the photon-electron scattering process in detail. We discuss the problem in the comoving reference frame S of the electron. Consider a photon is incident along the x^1-direction. Its 4-momentum and the electron's are respectively (in this example, the speed of light c is clearly written out)

$$\begin{aligned} p^\mu|_S &= (h\nu/c, h\nu/c, 0, 0), \\ q^\mu|_S &= (m_e c, 0, 0, 0). \end{aligned} \tag{1.99}$$

Due to the axial symmetry, we can assume that the scattering occurs in the x^1-x^2 plane. After the scattering, the 4-momentum of the photon and electron are

$$\begin{aligned}\bar{p}^\mu|_S &= (h\bar{\nu}/c, (h\bar{\nu}/c)\cos\theta, (h\bar{\nu}/c)\sin\theta, 0),\\ \bar{q}^\mu|_S &= (\gamma_u m_e c, \gamma_u m_e u\cos\phi, -\gamma_u m_e\sin\phi, 0),\end{aligned} \tag{1.100}$$

where u is the velocity obtained by the electron after scattering, and θ and ϕ are the angles between the worldlines of the scattered photon and electron and the x^1-axis. By 4-momentum conservation

$$p^\mu + q^\mu = \bar{p}^\mu + \bar{q}^\mu, \tag{1.101}$$

we get

$$\begin{aligned}h\nu/c + m_e c &= h\bar{\nu}/c + \gamma_u m_e c,\\ h\nu/c &= (h\bar{\nu}/c)\cos\theta + \gamma_u m_e u\cos\phi,\\ 0 &= (h\bar{\nu}/c)\sin\theta - \gamma_u m_e u\sin\phi.\end{aligned} \tag{1.102}$$

Removing u and ϕ from the above three equations, we find

$$\bar{\nu} = \nu\left(1 + \frac{h\nu}{m_e c^2}(1-\cos\theta)\right)^{-1}. \tag{1.103}$$

This gives the relationship between the frequency of the incident photon and the scattered photon with a scattering angle. The relationship can be well verified by observing photons at different angles.

1.2.3 *Dynamics in special relativity*

In this section, we examine the dynamics of special relativity. Newton's first law states that an object remains at rest or in uniform linear motion unless acted upon by an external force. The relativistic generalization of this law is straightforward: in the absence of external forces, the 4-velocity of an object remains constant over time,

$$\frac{\mathrm{d}\hat{u}}{\mathrm{d}\tau} = 0. \tag{1.104}$$

Newton's second law states that the acceleration of an object's motion is proportional to the external force,

$$\boldsymbol{F} = m\boldsymbol{a}. \tag{1.105}$$

This law lacks a first-principles derivation and must consequently be treated as an a priori formulation. What might constitute its relativistic generalization? Several fundamental requirements must be satisfied:

(1) It must satisfy the principle of relativity, i.e., it should have the same form in different reference frames;
(2) When the external force is zero, it must return to the first law, (1.104);
(3) When the velocity is much less than the speed of light, it should return to Newton's second law.

Therefore, we can guess its possible form as

$$\widehat{f} = m\frac{\mathrm{d}\widehat{u}}{\mathrm{d}\tau} = m\widehat{a}, \tag{1.106}$$

where m is the rest mass, and $\widehat{f}$ is called the 4-force vector. Note that since the 4-acceleration is orthogonal to the 4-velocity, we always have $\widehat{f} \cdot \widehat{u} = 0$.

Although Newton's second law lacks a fundamental derivation, the first law can be naturally understood from the principle of least action. Consider a free particle whose worldline extremizes (in fact, maximizes) the proper time between two timelike-separated points A and B. The proper time for any timelike curve from A to B is

$$\tau_{AB} = \int_A^B \mathrm{d}\tau. \tag{1.107}$$

The line integral is along a certain timelike curve $x^\mu(\lambda)$, integrated from $A = x^\mu(\lambda = 0)$ to $B = x^\mu(\lambda = 1)$. Therefore,

$$\tau_{AB} = \int_0^1 \mathrm{d}\lambda \left(\left(\frac{\mathrm{d}t}{\mathrm{d}\lambda}\right)^2 - \left(\frac{\mathrm{d}x}{\mathrm{d}\lambda}\right)^2 - \cdots \right)^{1/2}. \tag{1.108}$$

Consider all possible curves with fixed endpoints A and B, and find the one with the extreme value of proper time. This is equivalent to doing a variation of the above integral with respect to $\delta x^\mu(\lambda)$. The problem is similar to the principle of least action in analytical mechanics. At this time, the action is a first-order integral $I = \displaystyle\int \mathrm{d}\lambda L$. As we know, the variation of this action will give the Euler–Lagrange equation

$$-\frac{\mathrm{d}}{\mathrm{d}\lambda}\left(\frac{\partial L}{\partial(\mathrm{d}x^\mu/\mathrm{d}\lambda)}\right) + \frac{\partial L}{\partial x^\mu} = 0. \tag{1.109}$$

For the situation we are facing, this equation gives

$$\frac{\mathrm{d}^2 x^\mu}{\mathrm{d}\tau^2} = 0. \tag{1.110}$$

This is precisely the equation satisfied by a particle moving uniformly along a straight line. In essence, the principle of least action dictates that a free particle must move with constant velocity along a straight trajectory, thereby recovering Newton's first law. Later, we will explore how to generalize this framework to describe particle motion in curved spacetime.

1.3 Observers and Observations

In the framework of relativity, it is essential to rigorously define observers and their observations. The same physical phenomenon may yield different observational results for different observers—this lies at the heart of relativity. To establish a precise foundation, we first clarify the operational definition of an observation. An observation is conducted within the observer's laboratory, which constitutes their comoving reference frame. More precisely, since an observer follows a unique worldline in spacetime, their laboratory must be determined in accordance with this worldline.

(1) In the laboratory, the clock carried by the observer defines the direction of time:

$$\widehat{e}_0 = \widehat{u}_{\mathrm{obs}} = \text{the 4-velocity of the observer.}$$

(2) The other spatial directions require orthogonality with the time direction, and are also mutually orthogonal and normalized:

$$\widehat{e}_i \cdot \widehat{e}_0 = 0, \quad \widehat{e}_i \cdot \widehat{e}_j = \delta_{ij}. \tag{1.111}$$

Usually, $\widehat{e}_a$ is called the tetrad vector. Note that the lower index here does not represent components, but refers to different tetrad 4-vectors.

Example 1.7. In the previous discussion of a spaceship accelerating uniformly along the x direction, the following laboratory can be established:

$$\begin{aligned}
(\widehat{e}_0(\tau))^\mu &= (\cosh(a\tau), \sinh(a\tau), 0, 0), \\
(\widehat{e}_1(\tau))^\mu &= (\sinh(a\tau), \cosh(a\tau), 0, 0), \\
(\widehat{e}_2(\tau))^\mu &= (0, 0, 1, 0), \\
(\widehat{e}_3(\tau))^\mu &= (0, 0, 0, 1).
\end{aligned}$$

It can be seen that the time and the first spatial direction of the spaceship change with the proper time.

When the worldline of the particle under observation intersects the observer's worldline at a given event P, the particle's properties can be measured at P. Crucially, such an observation is defined with respect to the observer's laboratory—their comoving reference frame. At event P, the measured values of any physical quantities (vectors or tensors) are determined by their projections onto the observer's tetrad vectors. In other words, the observation results correspond to the scalar products of these

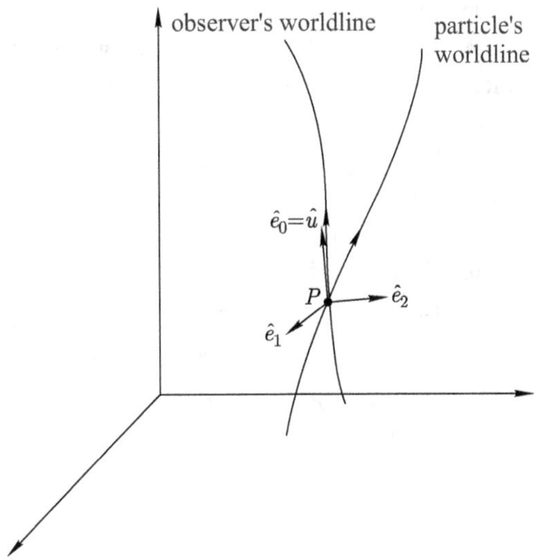

Fig. 1.10 Observation is relative to the observer's comoving reference frame.

physical quantities with the tetrad basis. If the particle's 4-momentum is $\widehat{p}$, the energy and momentum observed by the observer are

$$E_{\text{obs}} = -\widehat{p} \cdot \widehat{e}_0, \quad \boldsymbol{P}^i_{\text{obs}} = \widehat{p} \cdot \widehat{e}_i, \tag{1.112}$$

that is, the projection of the particle's 4-momentum along the tetrad 4-vectors $\{\widehat{e}_a\}$, or the scalar product with the tetrad 4-vectors. It is evident that this definition does not require the observer to move at a constant velocity, making it broadly applicable. In fact, the formulation depends solely on the observer's comoving reference frame. As we will demonstrate, this approach remains valid even for measurements in curved spacetime. Furthermore, note that the measured energy and momentum components are ultimately determined by scalar projections (i.e., contractions with the tetrad vectors). Consequently, these measurement results are inherently coordinate-independent. (*Question: How does this generalize to other tensor quantities?*)

From the above discussion, it is clear that the choice of spacelike basis vectors is not unique and possesses considerable freedom—the only constraint being orthogonality to the timelike vector. In fact, the three spacelike basis vectors can be arbitrarily rotated, corresponding to an $SO(3)$ rotational degree of freedom. These different choices have no absolute physical meaning. The crucial point, however, is that once the basis vectors are chosen at a given moment, we require their evolution to be naturally determined without further arbitrary choices. In other words, the basis vectors at subsequent moments should follow a unique transition dictated solely by the observer's motion. In such a non-rotating tetrad, the evolution of the basis vectors is governed by the rate of change of the timelike vector $\widehat{u}$, with no additional rotational effects. Specifically, we permit only the unavoidable rotation in the plane spanned by the 4-velocity $\widehat{u}$ and the 4-acceleration $\widehat{a}$, while explicitly excluding rotations purely among the spacelike vectors. Thus, the evolution of the tetrad basis vectors must satisfy the following two conditions:

(1) If $\widehat{e}_a$ is on the plane defined by $\widehat{u}$ and $\widehat{a}$, the change of the basis vector is a certain linear transformation;
(2) If $\widehat{e}_a$ is on other planes, especially any spacelike plane, the basis vector will not have any rotation.

The only possibility that satisfies the above two requirements is

$$\frac{\mathrm{d}\widehat{e}_a}{\mathrm{d}\tau} = -(\widehat{u} \cdot \widehat{e}_a)\widehat{a} + (\widehat{a} \cdot \widehat{e}_a)\widehat{u}. \tag{1.113}$$

In fact, the tetrad basis vectors in (1.113) can be replaced with any other vectors. Any vector that moves along the worldline according to the above evolution rules is said to undergo Fermi–Walker transport.[6] This relationship tells us how a group of tetrad basis vectors chosen at a certain event will change with the observer's motion along the worldline. If a tetrad basis vector $\widehat{e}_a$ is simultaneously orthogonal to $\widehat{u}$ and $\widehat{a}$, then its change rate is zero, implying that it remains unchanged. And for the timelike basis vector, its change along the worldline is $\dfrac{\mathrm{d}\widehat{e}_0}{\mathrm{d}\tau} = \widehat{a}$, determined by the 4-acceleration.

The above discussion relies fundamentally on the establishment of a proper comoving reference frame and an ideal clock within it. However, excessive acceleration of the observer may introduce complications. Consider, for instance, the measurement of a complete light wave. The required measurement time is $\Delta t \propto 1/\nu$, during which the observer's position changes by $\dfrac{1}{2}a(\Delta t)^2 \approx \dfrac{1}{2}a\dfrac{1}{\nu^2}$, where a is the observer's proper acceleration. For this displacement to be negligible compared to the photon's wavelength $\lambda \sim c/\nu$, we must require

$$a \ll 2c\nu. \tag{1.114}$$

Similar constraints apply to accelerating light sources as well.

[6]The systematic treatment of vector transport along worldlines, and the associated derivative operations, will be developed rigorously in later chapters.

Using this well-defined notion of observation, we now examine several physical processes. This will demonstrate the conceptual and computational advantages offered by our formulation.

Example 1.8. Energy of a moving particle.

Consider a stable particle, and the observer moves relative to it at a velocity $\boldsymbol{v}$. The energy of the particle measured by the observer can be obtained through the Lorentz transformation. Here we try to derive it in the formulation we developed in this subsection. We choose the coordinate system that is stationary relative to the particle, so the particle's 4-momentum is $p^\mu = (m, 0, 0, 0)$. And the time direction of the observer's laboratory is $e_0^\mu = u_{\rm obs}^\mu = (\gamma, v\gamma, 0, 0)$. From the preceding discussion, the observed energy is

$$E = -\widehat{p} \cdot \widehat{e}_0 = m\gamma, \tag{1.115}$$

which is consistent with the result obtained by assuming that the observer is stationary and making observations in the reference frame. Note that the scalar product of the 4-momentum and the basis vector is a scalar, independent of the choice of reference frame.

Example 1.9. Frequency of photons observed by a uniformly accelerated observer.

Consider a stable star located at the origin of the reference frame emitting light, and the observer is in a spaceship accelerating at a constant rate a_0. If $\omega_\star$ is the frequency of the photon in the star's reference frame, and the motion of the photon is along the positive x-axis, its 4-momentum is

$$p^\mu = (\hbar\omega_\star, \hbar\omega_\star, 0, 0). \tag{1.116}$$

The observer's worldline has been discussed before, which is $(t, x, y, z) = (a_0^{-1}\sinh(a_0\tau), a_0^{-1}\cosh(a_0\tau), 0, 0)$. From $E = -\widehat{p}\cdot\widehat{u}$, and $E = \hbar\omega$, we know that the measured photon frequency is

$$\omega(\tau) = \hbar^{-1}(p^t u^t - p^x u^x) = \omega_\star(\cosh(a_0\tau) - \sinh(a_0\tau)) = \omega_\star \mathrm{e}^{-a_0\tau}. \tag{1.117}$$

If the spaceship initially moves along the negative x-axis but accelerates along the positive$x$-axis, it first approaches the star before turning around. During this initial phase, the observed photon frequency is blue-shifted due to the relative motion toward the star. After the spaceship reverses direction, it moves progressively farther away from the star, resulting in a red-shift in the observed photon frequency. Notably, unlike conventional Doppler shift scenarios involving inertial motion, the observer here undergoes constant acceleration. Consequently, both the blue-shift and red-shift factors take exponential forms. Following the turnaround, the starlight observed by the spaceship transitions to the red spectrum. The rate of this reddening depends on the acceleration magnitude—the greater the acceleration, the faster the shift occurs. A characteristic timescale for this transition is given by $\Delta\tau \approx 1/a_0$.

For a uniformly accelerated observer, a horizon exists beyond which spacetime becomes unobservable. This is illustrated in Figure 1.11. Consider a light source (star) A fixed at the origin $x = 0$ of its inertial rest frame, emitting light steadily. Its worldline coincides with the time axis of this frame. Meanwhile, an observer B is aboard a spaceship that initially approaches the star but later undergoes constant acceleration a_0 along the positive x-axis. The spaceship executes a turnaround at $x = 1/a_0$ and thereafter moves away from the star. When the spaceship turns around, the light observed at that moment was emitted by the star at time $t_0 = -1/a_0$. After a time interval $\Delta t = 1/a_0$, the observer can no longer detect any light from the star. This is because the light emitted at $t = 0$ forms the asymptote of the hyperbolic trajectory of the spaceship's worldline and thus never intersects with the observer's path. Consequently, any light emitted by the star after $t = 0$ cannot reach the spaceship. The surface $x = t$ acts as a horizon for the observer: events occurring to its left (i.e., in the region $x < t$) become causally disconnected from the observer B. This boundary is known as the acceleration horizon (or Rindler horizon). Notably, it is a null

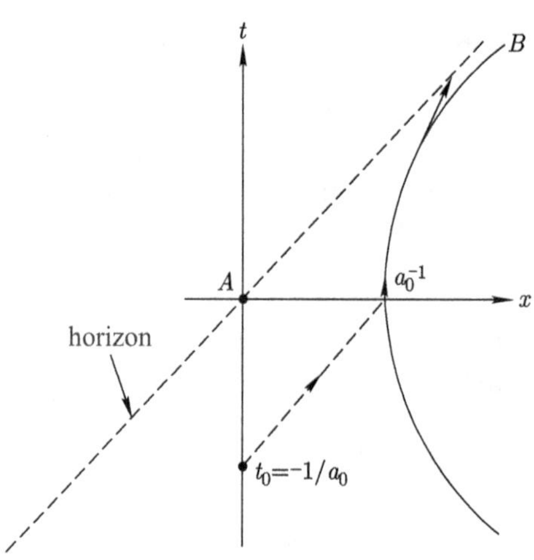

Fig. 1.11 For a uniformly accelerated observer, there exists a horizon.

surface, meaning its normal vector is a null (light-like) vector. More generally, any observer undergoing constant proper acceleration will experience such a horizon, which fundamentally limits their observable spacetime.

The existence of this horizon holds profound physical significance. Classically, an observer's state of motion fundamentally limits their access to certain regions of spacetime, rendering portions of it unobservable. This demonstrates that spacetime perception is inherently observer-dependent. When quantum mechanical effects are taken into account, a uniformly accelerated observer perceives themselves immersed in a thermal bath with a temperature directly proportional to their acceleration. This striking quantum phenomenon is known as the Unruh effect.

Example 1.10. Einstein's disk.

As illustrated in Figure 1.12, consider a photon transmitter and receiver positioned on the boundary of a disk with radius r, rotating at a constant angular velocity ω. If the transmitter emits photons that are subsequently detected by the receiver, what is the resulting redshift factor, and how does it depend on ω and r?

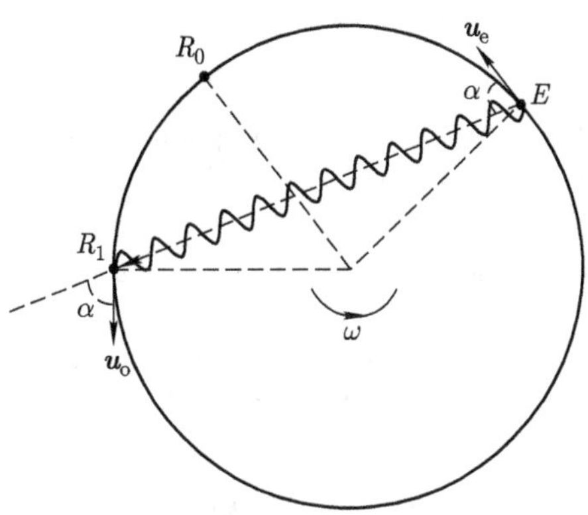

Fig. 1.12 Einstein's disk. In the figure, E represents the position of the transmitter, R_0 represents the initial position of the receiver, and R_1 represents the position of the receiver when the photon is received. From E to R_1 is the direction of the photon's 4-momentum. α is the angle between the 3-velocity of the transmitter and the direction of photon motion, and also the angle between the 3-velocity of the receiver and the direction of photon motion.

Solution Let $\widehat{u}_{\rm e}$ be the 4-velocity of the transmitter, $\widehat{u}_{\rm o}$ be the 4-velocity of the observer (receiver), and $\widehat{p}$ be the 4-momentum of the photon. We know that the energy of the emitted photon is

$$E_{\rm e} = -\widehat{p} \cdot \widehat{u}_{\rm e}, \tag{1.118}$$

and the energy of the observed photon is

$$E_{\rm o} = -\widehat{p} \cdot \widehat{u}_{\rm o}, \tag{1.119}$$

then

$$\frac{\lambda_{\rm o}}{\lambda_{\rm e}} = \frac{E_{\rm e}}{E_{\rm o}} = \frac{-\widehat{p} \cdot \widehat{u}_{\rm e}}{-\widehat{p} \cdot \widehat{u}_{\rm o}} = \frac{p^0 u^0_{\rm e} - \boldsymbol{p} \cdot \boldsymbol{u}_{\rm e}}{p^0 u^0_{\rm o} - \boldsymbol{p} \cdot \boldsymbol{u}_{\rm o}}. \tag{1.120}$$

Since the disk rotates at a constant angular velocity, the speed of the transmitter and receiver is $|v| = \omega r$, a constant. Therefore, the magnitude of the spatial component of the 4-velocity is $|\boldsymbol{u}_{\rm e}| = |\boldsymbol{u}_{\rm o}| = v\gamma$. From the normalization condition of the 4-velocity, we know $u^0_{\rm e} = u^0_{\rm o} = \gamma$. In addition, using the fact that the angle between the tangent and the chord of a circle is equal to half the arc they subtend, we find that the angle between $\boldsymbol{u}_{\rm e}$ and $\boldsymbol{p}$ is equal to the angle between $\boldsymbol{u}_{\rm o}$ and $\boldsymbol{p}$, so

$$\boldsymbol{p} \cdot \boldsymbol{u}_{\rm e} = \boldsymbol{p} \cdot \boldsymbol{u}_{\rm o}. \tag{1.121}$$

Therefore, we finally find

$$\frac{\lambda_{\rm e}}{\lambda_{\rm o}} = 1, \tag{1.122}$$

implying that there is no redshift, the frequency of the received photon is the same as the frequency of the emitted photon.

Appendix 1.1 *Further Introduction to the Lorentz Group

In the preceding discussion, we defined the general Lorentz group as the set of all coordinate transformations that leave the Minkowski spacetime line element

$$ds^2 = \eta_{\mu\nu} dx^\mu dx^\nu \tag{1.123}$$

invariant, that is

$$\Lambda^{\rm T} \eta \Lambda = \eta, \tag{1.124}$$

where $\eta_{\mu\nu} = \text{diag}(-1, 1, 1, 1)$ is the Minkowski metric, and Λ is a matrix. If further requirements are made

$$\Lambda^0{}_0 \geq 1, \quad \det(\Lambda) = 1, \tag{1.125}$$

the matrices that satisfy the above relationship constitute the $SO(1,3)$ group, called the proper Lorentz group, or simply the Lorentz group. The Lorentz group has six generators, three of which correspond to the generators of the pure three-dimensional rotation group, and the other three are related to Lorentz boosts in different directions. They satisfy the following Lie algebra commutation relations:

$$\begin{aligned}
[J_x, J_y] &= \mathrm{i}J_z \text{ (and its permutations)},\\
[K_x, K_y] &= -\mathrm{i}J_z \text{ (and its permutations)},\\
[J_x, K_y] &= \mathrm{i}K_z \text{ (and its permutations)},\\
[J_x, K_x] &= 0 \text{ (etc.)},
\end{aligned}$$

or more succinctly,

$$[J_i, J_k] = \mathrm{i}\varepsilon_{ikm}J_m, \quad [J_i, K_k] = \mathrm{i}\varepsilon_{ikm}K_m, \quad [K_i, K_k] = \mathrm{i}\varepsilon_{ikm}J_m, \tag{1.126}$$

where K_i's are the generators of the Lorentz boosts, and J_i's are the generators of the usual three-dimensional spacial rotation, ε_{ikm} is the Levi–Civita symbol in three-dimensional flat space.

In fact, the Lorentz group is isomorphic to the $SL(2, \mathbf{C})$ group. For a vector $x^\mu = (t, x, y, z)$ in four-dimensional Minkowski spacetime, we can construct a 2×2 self-adjoint (Hermitian) matrix

$$X = \begin{pmatrix} t+z & x-\mathrm{i}y \\ x+\mathrm{i}y & t-a \end{pmatrix}. \tag{1.127}$$

It is easy to see that $X^\dagger = X$, and

$$\det X = t^2 - x^2 - y^2 - z^2. \tag{1.128}$$

This matrix can also be expanded using the unit matrix and Pauli matrices:

$$X = te + x\sigma_1 + y\sigma_2 + z\sigma_3, \tag{1.129}$$

where

$$e = \begin{pmatrix} 1 & 0 \\ 0 & 1 \end{pmatrix}, \sigma_1 = \begin{pmatrix} 0 & 1 \\ 1 & 0 \end{pmatrix}, \sigma_2 = \begin{pmatrix} 0 & -\mathrm{i} \\ \mathrm{i} & 0 \end{pmatrix}, \sigma_3 = \begin{pmatrix} 1 & 0 \\ 0 & -1 \end{pmatrix}. \tag{1.130}$$

Let L be a 2×2 complex matrix, and its action on X is

$$X \to LXL^\dagger. \tag{1.131}$$

It induces a transformation $x^\mu \to \varphi(L)x^\mu$ on the vector x^μ. Since

$$\det(LXL^\dagger) = |\det(L)|^2 \det X, \tag{1.132}$$

if L is a matrix with a determinant of 1, i.e., $L \in SL(2, \mathbf{C})$, it keeps the length of the vector x^μ invariant, so the transformation $\varphi(L)$ is a Lorentz transformation.

From the definition of the Lorentz group, the determinant of the matrix Λ can be $+1$ or -1. If it is -1, these group elements cannot be connected to the unit element through a continuous curve on the group manifold. For timelike vectors, the elements of the Lorentz group always keep their size invariant. The zero (time) component of the timelike vector can be positive or negative, and the general Lorentz group element can change its sign without destroying the size of the vector. However, if this element can be connected to the unit element, it will not change the sign of the zero component of the timelike vector. These elements constitute the proper Lorentz group, their determinant is $+1$, and they keep the forward light cone of the timelike vector,

$$\Lambda^0{}_0 \geq 1, \quad \det(\Lambda) = 1. \tag{1.133}$$

Under this condition, the zero component of the timelike vector after Lorentz transformation will not change its sign. If it points to the future before the transformation, it will still point to the future after the transformation.

Space inversion (or parity transformation) in four dimensions can be represented by the following matrix:

$$I_s = \begin{pmatrix} 1 & 0 & 0 & 0 \\ 0 & -1 & 0 & 0 \\ 0 & 0 & -1 & 0 \\ 0 & 0 & 0 & -1 \end{pmatrix}. \tag{1.134}$$

It, together with time reversal, proper Lorentz group elements, and their products, can form a group, called the complete Lorentz group. Obviously, space inversion and all spacial rotation generators commute, and when it acts on the Lorentz transformation generator,

$$I_s K_i I_s^{-1} = -K_i. \tag{1.135}$$

This is easy to understand: space reflection makes the part related to space in K_i change sign.

The proper Poincaré group is formed by combining the proper Lorentz group with four translations. This group possesses a total of 10 generators: 6 for rotations and 4 for translations. Similarly, the Galilean transformation discussed earlier involves three-dimensional rotations, velocity transformations, and four translations, also comprising 10 generators. It can be mathematically demonstrated that all Galilean transformations collectively form a group, known as the Galilean group. Notably, in the non-relativistic limit, the Poincaré group reduces to the Galilean group.

Exercises

1. Prove that the Lorentz transformation is the only non-singular coordinate transformation that keeps the line element invariant.
2. How many generators does the $SO(n, m)$ group have?
3. Prove that in the non-relativistic limit, the Poincaré group can be reduced to the Galilean group.
4. Prove that a timelike vector and a null vector cannot be orthogonal.
5. Prove that $\frac{\mathrm{d}^3p}{E}$ and $\int \mathrm{d}^3x\mathrm{d}^3p$ are Lorentz invariant.
6. Consider a hypothetical particle whose motion speed exceeds the speed of light, such a particle is called a tachyon.
   (1) Consider the worldline of the tachyon, its tangent vector is $u^\mu = \mathrm{d}x^\mu/\mathrm{d}s$, where $\mathrm{d}s$ is the spacelike interval along the worldline. Prove $u^\mu u_\mu = 1$, and express u^μ using the 3-velocity $\boldsymbol{v}$.
   (2) Prove that if a tachyon moves faster than the speed of light in one reference frame, then it also moves faster than the speed of light in any inertial frame connected by a Lorentz transformation.
   (3) Define $p^\mu = mu^\mu$, give the relationship between energy and 3-momentum.
   (4) Prove that in a certain inertial frame, the energy of the tachyon is negative.
   (5) Consider the process $a \to a + t$, where a is an ordinary particle and t is a tachyon, prove that this process does not contradict energy and momentum conservation. This indicates that a world with tachyons is unstable.
7. In the LEP particle accelerator at CERN, electrons and positrons move in opposite directions around a circular ring with a radius of 10 km, each with an energy of 100 GeV.
   (1) What is the ratio of the motion speed of these particles to the speed of light?
   (2) Electrons and positrons can stay in the ring for 2 hours, how many circles can they move around the ring?

8. Carefully derive the formulas for the Doppler effect and the relativistic beaming effect.
9. Using the definition of observers and observations, prove that the frequency of photons emitted from the origin received by an observer on the edge of the disk is blue-shifted.
10. If there are two observers who meet at a certain event with 4-velocities $\widehat{u}_1$ and $\widehat{u}_2$, at this time they see each other with a 3-velocity, the magnitude of which is v, please prove

$$v^2 = \frac{(\widehat{u}_1 \cdot \widehat{u}_2)^2 - 1}{(\widehat{u}_1 \cdot \widehat{u}_2)^2}. \tag{1.136}$$

11. Consider a particle with a 4-momentum $\widehat{p}$, and the observer's 4-velocity is $\widehat{u}$. Prove that if the particle passes through the observer's laboratory, the magnitude of the measured 3-momentum is

$$|\boldsymbol{p}| = [(\widehat{p} \cdot \widehat{u})^2 + (\widehat{p} \cdot \widehat{p})]^{1/2}.$$

12. Prove that when an electron and a positron annihilate, more than one photon is emitted.
13. When a photon scatters with a particle moving at a speed close to the speed of light, inverse Compton scattering will occur. Assume that the particle has a rest mass of m_0 and a total energy of E, and it collides with a photon with an energy of E_γ.

    (1) Prove that the energy of the scattered photon is

    $$E\left(1 + \frac{m_0^2 c^4}{4EE_\gamma}\right)^{-1}. \tag{1.137}$$

    (2) The energy of extremely high-energy cosmic rays can reach 10^{20} eV, so how much energy will a proton cosmic ray with such high energy transfer to a photon in the microwave background radiation?

14. Prove that when a photon is reflected by a mirror moving parallel to its surface, the angle of incidence is equal to the angle of reflection.
15. A plane mirror moves at a constant velocity v along a direction perpendicular to the mirror surface. A beam of light with a frequency of ν_i collides with the mirror surface at an angle of incidence ϕ_i, the reflection angle is ϕ_o, and the frequency of the reflected photon is ν_o. Prove

$$\frac{\tan\frac{1}{2}\phi_\mathrm{i}}{\tan\frac{1}{2}\phi_\mathrm{o}} = \frac{c+v}{c-v}, \quad \frac{\nu_\mathrm{o}}{\nu_\mathrm{i}} = \frac{\sin\phi_\mathrm{i}}{\sin\phi_\mathrm{o}}. \tag{1.138}$$

16. A rocket obtains acceleration by ejecting fuel. Assume that in the comoving reference frame of the rocket, the ejection speed of the fuel is a constant u, and the rocket accelerates from rest to a velocity V. Using energy and momentum conservation, find the ratio of the rocket's rest mass when it reaches velocity V to its initial rest mass. (Fuel should be included in the rest mass.)
17. A spaceship moves at a variable velocity $u(t)$ in a reference frame S. An observer in the spaceship measures the proper acceleration of the spaceship to be $f(\tau)$, where τ is the proper time. Assume that at $\tau = 0$, the spaceship's velocity in S is u_0.

    (1) Prove

    $$\frac{u(\tau) - u_0}{1 - u(\tau)u_0/c^2} = c \tanh \psi(\tau), \tag{1.139}$$

    where $c\psi(\tau) = \int_0^\tau f(\tau')\mathrm{d}\tau'$.

    (2) Prove that the spaceship's velocity cannot exceed the speed of light.

18. Continue from the previous question, assume that the spaceship's proper acceleration is a constant f, and the spaceship leaves the base at $t = \tau = 0$, and moves forward in a straight line.

    (1) Prove that after $t = c/f$, the base's signal can no longer reach the spaceship.
    (2) Prove that the light signal sent by the base is increasingly red-shifted when it reaches the spaceship.
    (3) If this proper acceleration is the Earth's gravitational acceleration $g = 9.8\ \mathrm{m/s^2}$, and the spaceship is to reach a star 10 ly away from the base, how long will it take to measure with the clock on the spaceship?
    (4) In the previous question, if the spaceship decelerates with the opposite acceleration halfway to reach the star, and then returns to the base in the same way, how long will the observer on the spaceship measure? How many years has the clock on the base passed?

19. Suppose a particle with a rest mass of m_0 is subjected to a 4-force $\widehat{f}$, prove that the corresponding 3-force $\boldsymbol{f}$ satisfies

    $$\boldsymbol{f} = \gamma_u m_0 \boldsymbol{a} + \frac{\boldsymbol{f} \cdot \boldsymbol{u}}{c^2} \boldsymbol{u}, \tag{1.140}$$

where $\boldsymbol{u}$ is the particle's velocity, and γ_u is the corresponding Lorentz factor. From this, prove that when $\boldsymbol{f}$ is either parallel or perpendicular to $\boldsymbol{u}$, $\boldsymbol{a}$ is parallel to $\boldsymbol{f}$. Further prove that in these two cases, we have $\boldsymbol{f} = \gamma_u^3 m_0 \boldsymbol{a}$ and $\boldsymbol{f} = \gamma_u m_0 \boldsymbol{a}$ respectively.

Chapter 2

Tensor Calculus in Special Relativity

Tensor analysis is an indispensable tool for learning general relativity. Within this chapter, we systematically develop the formalism of tensor analysis within the context of special relativity, meticulously examining the fundamental concepts and essential operations pertaining to vectors and tensors. As a culminating demonstration of tensor analysis' practical significance, we shall explore the mathematical formulation and physical interpretation of the energy-momentum tensor, a pivotal concept in relativistic field theory.

2.1 Vectors

In flat spacetime, a vector can be conventionally defined as a directed line segment connecting two distinct points. However, this elementary definition proves inadequate for more general physical scenarios. For instance, when analyzing particle motion, the particle's trajectory, commonly referred to as its worldline, typically deviates from a straight path. Consequently, the particle's velocity at any given point does not manifest as a vector along the worldline itself, but rather aligns with the direction of the tangent vector at that specific point. To establish a more comprehensive framework for vector analysis, it becomes necessary to introduce advanced mathematical concepts. Through rigorous examination, we shall demonstrate that tangent vectors exhibit a one-to-one correspondence with directional derivatives, thereby providing a robust foundation for the generalized treatment of vectors via the formalism of directional derivatives. This sophisticated approach not only maintains its validity in flat spacetime but also extends seamlessly to the analysis of vector fields in curved spacetime.

Consider a worldline $P(\lambda)$, and at some point $\lambda = \lambda_0$, the tangent vector is

$$\widehat{t} = \left.\frac{\mathrm{d}P(\lambda)}{\mathrm{d}\lambda}\right|_{\lambda=\lambda_0}. \tag{2.1}$$

If we choose a coordinate system with the origin at O, then the points on the worldline can be described as:

$$P(\lambda) - O = x^\mu(\lambda)\widehat{e}_\mu, \tag{2.2}$$

where x^μ are coordinate functions, and $\widehat{e}_\mu$ are coordinate basis vectors. The tangent vector can be expressed as

$$\widehat{t} = \left.\frac{\mathrm{d}P(\lambda)}{\mathrm{d}\lambda}\right|_{\lambda=\lambda_0} = \left(\frac{\mathrm{d}x^\mu}{\mathrm{d}\lambda}\right)_{\lambda=\lambda_0}\widehat{e}_\mu = t^\mu(\lambda=\lambda_0)\widehat{e}_\mu, \tag{2.3}$$

where $t^\mu(\lambda=\lambda_0)$ are the components of the tangent vector at $P(\lambda_0)$.

Next, consider an arbitrary function $f(x^\mu(\lambda))$ along the worldline $x^\mu(\lambda)$. Its change along the worldline at $P(\lambda_0)$ is

$$\left.\frac{\mathrm{d}f}{\mathrm{d}\lambda}\right|_{\lambda=\lambda_0} = \lim_{\epsilon\to 0}\frac{f(x^\mu(\epsilon+\lambda_0)) - f(x^\mu(\lambda_0))}{\epsilon} = \left(\frac{\mathrm{d}x^\mu}{\mathrm{d}\lambda}\right)_{\lambda=\lambda_0}\frac{\partial f}{\partial x^\mu}. \tag{2.4}$$

Thus,

$$\left.\frac{\mathrm{d}f}{\mathrm{d}\lambda}\right|_{\lambda=\lambda_0} = t^\mu(\lambda=\lambda_0)\frac{\partial}{\partial x^\mu}f. \tag{2.5}$$

Since f is an arbitrary function, the directional derivative at a point $\lambda=\lambda_0$ on the worldline is given by $\widehat{t}$, i.e.,

$$\partial_{\widehat{t}}|_{\lambda_0} = \left.\frac{\mathrm{d}}{\mathrm{d}\lambda}\right|_{\lambda_0} = t^\mu\left.\frac{\partial}{\partial x^\mu}\right|_{\lambda_0}. \tag{2.6}$$

The directional derivative represents a fundamental differential operator that quantifies the rate of change of a given function along a specified worldline. Formally, this operator acts upon any differentiable function, providing a precise mathematical description of the function's variation in the direction tangent to the worldline.

The preceding analysis establishes a correspondence between directional derivatives and tangent vectors. Specifically, we have demonstrated that each directional derivative uniquely determines an associated tangent vector. Conversely, for any given tangent vector $\widehat{t}$, there exists a corresponding directional derivative along a parameterized curve defined by $x^\mu(\lambda) = x^\mu(\lambda_0) + t^\mu(\lambda-\lambda_0)$. It is noteworthy that in the local neighborhood of λ_0, this curve can be effectively approximated as a straight line for analytical purposes. This bijective relationship between tangent vectors and directional derivatives constitutes a cornerstone of modern differential geometry, providing a rigorous mathematical framework for the analysis of vector fields on differentiable manifolds, as we will show later.

In short, we can describe vectors from an equivalent perspective: a vector can be regarded as an operator

$$\widehat{a} = \partial_{\widehat{a}}. \tag{2.7}$$

In a certain coordinate system, this vector can be written in component form as

$$\widehat{a} = a^{\mu} \frac{\partial}{\partial x^{\mu}}. \tag{2.8}$$

The proposed definition of vectors demonstrates comprehensive compliance with the fundamental principles of vector algebra. This conceptual framework offers several significant advantages, including enhanced mathematical rigor, simplification of complex concepts (particularly in the context of vector field commutators, see below), and improved computational efficiency.

As mentioned before, vectors are independent of the choice of coordinate system. When discussing specific problems, for convenience, we choose a coordinate system $\{x^{\mu}\}$, and accordingly, there is a set of coordinate basis vectors

$$\widehat{e}_{\mu} = \frac{\partial}{\partial x^{\mu}}. \tag{2.9}$$

Each coordinate basis vector is simply the directional derivative along the corresponding coordinate axis. Under a coordinate transformation, both the coordinate basis vectors and the vector's components undergo corresponding transformations to maintain the vector's geometric and algebraic consistency. Suppose the new coordinate system is $\{x^{\nu'}\}$, then we have

$$\begin{aligned}\widehat{a} &\equiv a^{\mu} \frac{\partial}{\partial x^{\mu}} = a^{\mu} \frac{\partial x^{\nu'}}{\partial x^{\mu}} \cdot \frac{\partial}{\partial x^{\nu'}} \\ &\equiv a^{\nu'} \frac{\partial}{\partial x^{\nu'}},\end{aligned}$$

giving rise to the change of the components

$$a^{\nu'} = \frac{\partial x^{\nu'}}{\partial x^{\mu}} a^{\mu}, \tag{2.10}$$

and the change of the coordinate basis vectors

$$\widehat{e}_{\nu'} = \frac{\partial x^{\mu}}{\partial x^{\nu'}} \widehat{e}_{\mu}. \tag{2.11}$$

In the literature, a^μ is conventionally classified as a contravariant vector, whose transformation properties under coordinate changes are precisely characterized by the aforementioned description. It is crucial to reiterate that vectors possess an intrinsic coordinate-independent nature, with only their components exhibiting variation across different coordinate systems.

In the context of special relativity, the Lorentz transformation represents a specific class of coordinate transformations that govern the transition between inertial reference frames. This transformation, which corresponds to a change of observers' perspectives, is constrained by the requirement of preserving the spacetime line element invariant. Mathematically, this constraint defines the Lorentz transformation as a non-singular, linear coordinate transformation that maintains the Minkowski metric structure of spacetime:

$$x^\mu \to x^{\mu'} = \Lambda^{\mu'}{}_\nu x^\nu. \tag{2.12}$$

Therefore,

$$a^\mu \to a^{\mu'} = \Lambda^{\mu'}{}_\nu a^\nu, \tag{2.13}$$

and

$$\widehat{e}_\mu = \Lambda^{\nu'}{}_\mu \widehat{e}_{\nu'}, \quad \widehat{e}_{\nu'} = \Lambda^{\mu}{}_{\nu'} \widehat{e_\mu}, \tag{2.14}$$

where $\Lambda^{\mu}{}_{\nu'}$ is the inverse of $\Lambda^{\nu'}{}_\mu$,

$$\Lambda^{\nu'}{}_\mu \Lambda^{\mu}{}_{\sigma'} = \delta^{\nu'}_{\sigma'}, \quad \Lambda^{\mu}{}_{\nu'} \Lambda^{\nu'}{}_\rho = \delta^\mu_\rho. \tag{2.15}$$

2.2 Dual Vectors

Dual vectors, alternatively referred to as co-vectors, covariant vectors, or simply 1-forms, are formally defined as linear functionals that operate on vectors, mapping them to elements of the real number field, i.e.,

$$\widehat{\omega} : \widehat{a} \to \widehat{\omega}(\widehat{a}) \in \mathbf{R}. \tag{2.16}$$

Later, we will see that at a point p, the tangent vectors form a linear space, i.e., the tangent space T_p, and dual vectors form the cotangent space

$$T_p^* = \{\widehat{\omega} \in T_p^* | \widehat{\omega}(T_p) \in \mathbf{R}\}. \tag{2.17}$$

In fact, from the perspective of mappings, we can consider the tangent space and cotangent space to be dual to each other:

$$T_p \leftrightarrow T_p^*. \tag{2.18}$$

From the above definition, we can derive the following properties of dual vectors:

$$\begin{aligned} \widehat{\omega}(a\widehat{v}+b\widehat{w}) &= a\widehat{\omega}(\widehat{v})+b\widehat{\omega}(\widehat{w}), \\ (a\widehat{\omega}+b\widehat{\eta})(\widehat{v}) &= a\widehat{\omega}(\widehat{v})+b\widehat{\eta}(\widehat{v}), \end{aligned} \tag{2.19}$$

where $\widehat{\omega}$, $\widehat{\eta} \in T_p^*$, $\widehat{v}$, $\widehat{w} \in T_p$, and a, $b \in \mathbf{R}$. According to linear algebra, it is readily verifiable that dual vectors constitute a well-defined vector space under the standard operations of addition and scalar multiplication. However, it is crucial to distinguish that dual vectors are fundamentally distinct from vectors, because if they were, we would need an additional structure—a metric—to define the scalar product, thereby obtaining a real number. Importantly, the functional relationship between dual vectors and vectors exists independently of any metric structure, relying solely on the linear algebraic properties of the spaces involved.

Dual vectors or 1-forms possess an intuitive geometric interpretation that provides valuable insight into their nature. Locally, a 1-form $\widehat{\omega}$ can be visualized as a family of parallel hypersurfaces (codimension-1 surfaces) in the tangent space, where the magnitude of $\widehat{\omega}$ is inversely proportional to the spacing between these surfaces. The action of a 1-form $\widehat{\omega}$ on a vector $\widehat{v}$, denoted by $\widehat{\omega}(\widehat{v})$, is geometrically determined by the number of these hypersurfaces that the vector $\widehat{v}$ traverses. This interpretation is illustrated in Figure 2.1, where the evaluation of the 1-form corresponds to counting

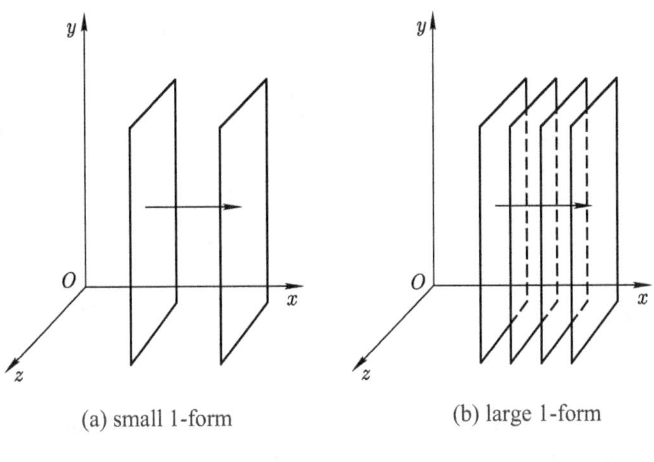

Fig. 2.1 Geometric image of a 1-form. It can be locally viewed as a series of surfaces, with the density of the surfaces related to the magnitude of the 1-form. The interaction between the 1-form and a vector is determined by how many surfaces the vector passes through.

the intersections of the vector with the characteristic hypersurface family.[1] The simplest example is the surface family $x = n$, with two surfaces separated by 1, which gives the basis vector $\mathrm{d}x$. If the surface interval widens to $x = 2n$, then the corresponding 1-form is $\frac{1}{2}\mathrm{d}x$. In four dimensions, a 1-form is a three-dimensional surface, and in n dimensions, a 1-form is an $(n-1)$-dimensional surface. Therefore, a 1-form defines a way of dividing space. Note that using codimension-1 hypersurfaces to divide space only holds locally. In other words, if we consider the interaction between a 1-form field and a vector field near a point, this image holds. The same is true for general scalar functions, whose gradients are equivalent to 1-form fields. An intuitive image is contour lines, which cut through space. When adjacent contour lines are closer, it represents a larger gradient, i.e., the value of the 1-form is larger at that point. It is worth reminding everyone that the existence of contour lines is independent of the choice of coordinate system, so the 1-form field itself is independent of the choice of coordinate system, and the number of times the vector passes through the contour lines gives the value between the vector and the 1-form.

Consider a vector at point P_0, with its endpoint at P. For a 1-form defined by the function $f =$ constant, the number of times the vector passes through the surface is given by

$$\langle \mathrm{d}f, P - P_0 \rangle = f(P) - f(P_0), \tag{2.20}$$

where $\mathrm{d}f$ represents the 1-form. The vector $P - P_0 = \lambda \widehat{V}$ corresponds to the directional derivative

$$\partial_{\widehat{V}} = \frac{\mathrm{d}}{\mathrm{d}\lambda}. \tag{2.21}$$

Thus,

$$\partial_{\widehat{V}} f(P) = \left\langle \mathrm{d}f, \frac{\mathrm{d}P}{\mathrm{d}\lambda} \right\rangle = \langle \mathrm{d}f, \widehat{V} \rangle, \tag{2.22}$$

i.e., the directional derivative of the function along $\widehat{V}$ gives the functional relationship between the 1-form $\mathrm{d}f$ and the vector $\widehat{V}$. This discussion is independent of the choice of coordinate system.

[1] Here, to provide an intuitive geometric representation, we have deliberately relaxed certain mathematical rigor. This approach involves two key simplifications: Firstly, we employ finite differences as approximations for differentials, thereby rendering the family of hypersurfaces more accessible to geometric intuition. Secondly, we interpret the intersection of a vector with these surfaces as yielding a real-valued quantity, rather than restricting it to integer values.

For dual vectors, in a certain coordinate system, their basis vectors $\widehat{\theta}^{\nu}$ can be defined as

$$\widehat{\theta}^{\nu}(\widehat{e}_{\mu}) = \delta^{\nu}_{\mu}. \tag{2.23}$$

For a dual vector (1-form), we have

$$\widehat{\omega} = \omega_{\mu}\widehat{\theta}^{\mu}. \tag{2.24}$$

Therefore,

$$\widehat{\omega}(\widehat{v}) = \omega_{\mu}\widehat{\theta}^{\mu}(v^{\nu}\widehat{e}_{\nu}) = \omega_{\mu}v^{\nu}\widehat{\theta}^{\mu}(\widehat{e}_{\nu}) = \omega_{\mu}v^{\nu}\delta^{\mu}_{\nu} = \omega_{\mu}v^{\mu}. \tag{2.25}$$

Like vectors, dual vectors are independent of the choice of coordinate system, from which we can know the transformation properties of the components and basis vectors of 1-forms under Lorentz transformations as

$$\omega_{\mu'} = \Lambda^{\nu}{}_{\mu'}\omega_{\nu}, \quad \widehat{\theta}^{\rho'} = \Lambda^{\rho'}{}_{\sigma}\widehat{\theta}^{\sigma}. \tag{2.26}$$

The above discussion may seem very abstract, so let's look at a specific example.

Example 2.1. Gradient of a scalar function.

As mentioned before, the gradient of a scalar function gives a geometric description of dual vectors. We discuss it through a certain coordinate system. Consider the gradient of a scalar function $f(x^{\mu})$.[2] The change of f along a vector $\widehat{t}$ is the directional derivative $t^{\mu}\dfrac{\partial f}{\partial x^{\mu}}$, which gives the functional relationship between the 1-form and the vector. We can rewrite it as

$$\mathrm{d}f(\widehat{t}) \equiv t^{\mu}\frac{\partial f}{\partial x^{\mu}}. \tag{2.27}$$

It can be seen that $\mathrm{d}f$ defines a mapping from vectors to real numbers, so it is a dual vector. More clearly, we know that the gradient of a scalar function in a coordinate system can be written as

$$\mathrm{d}f = \frac{\partial f}{\partial x^{\mu}}\mathrm{d}x^{\mu}. \tag{2.28}$$

This indicates the change of the function along the x^{μ} direction. Obviously, we have

$$\widehat{\theta}^{\mu} = \mathrm{d}x^{\mu}. \tag{2.29}$$

[2]Sometimes, the gradient of a scalar function is denoted as ∇f. In this book, we will use both notations, hoping it will not cause confusion.

It satisfies the properties of dual basis vectors because $\mathrm{d}x^\mu \left(\frac{\partial}{\partial x^\nu}\right) = \delta^\mu_\nu$. In other words, if we choose the scalar function to be $x^\mu =$ constant, then the gradient of this scalar function defines the dual basis vector, which only has changes along the direction of the directional derivative defined by the x^μ coordinate axis.

More generally, for a scalar function $g(x) = -t^2 + x^2 + y^2 + z^2$, its gradient has components

$$\frac{\partial g}{\partial x^\mu} = (-2t, 2x, 2y, 2z), \tag{2.30}$$

which can be written as

$$\mathrm{d}g = -2t\mathrm{d}t + 2x\mathrm{d}x + 2y\mathrm{d}y + 2z\mathrm{d}z. \tag{2.31}$$

As previously discussed, the equation "scalar function = constant" defines a level surface, and its gradient naturally gives rise to a 1-form field. This 1-form field corresponds to the (co)normal vector field of the conventional surface representation. More precisely, this 1-form field is orthogonal to all tangent vectors on the surface. Specifically, if $\widehat{t}$ is an arbitrary tangent vector to the surface defined by $f =$ constant, and $\mathrm{d}f$ denotes the 1-form associated with this surface, then the following orthogonality condition holds:

$$\langle \widehat{t}, \mathrm{d}f \rangle = \widehat{t}(\mathrm{d}f) = 0, \tag{2.32}$$

because the function is unchanged along the surface, and the change along the tangent direction is zero. This clearly shows the difference between the 1-form field and the conventional normal vector. In standard treatments, we define a normal vector as being orthogonal to tangent vectors, but this orthogonality relies on an inner product—requiring the introduction of a metric structure. In contrast, the natural pairing between a 1-form field and tangent vectors is metric-independent, making it a more fundamental geometric concept. A more intrinsic way to understand this is that the 1-form field itself defines the orthogonal complement to the surface. The gradient naturally gives what we might call the "normal 1-form" of the surface. For closed surfaces that partition space into interior and exterior regions, we can define orientation through the action of this 1-form: if its evaluation on an outward-pointing vector yields a positive value, we call it an outward normal 1-form; otherwise, it represents an inward orientation.

2.3 Tensors

A (k, l)-type tensor defines a mapping that acts on the tensor product of k cotangent spaces and l tangent spaces.

$$\widehat{T}: T_p^* \otimes \cdots \otimes T_p^* \otimes T_p \otimes \cdots \otimes T_p \to \mathbf{R}. \tag{2.33}$$

Example 2.2. Tensors:

(1) Scalar: a (0, 0)-type tensor.
(2) Vector: a (1, 0)-type tensor.
(3) Dual vector: a (0, 1)-type tensor.
(4) (1, 1)-type tensor:

$$\widehat{T}(a\widehat{\omega} + b\widehat{\eta}, c\widehat{v} + d\widehat{w}) = ac\widehat{T}(\widehat{\omega}, \widehat{v}) + ad\widehat{T}(\widehat{\omega}, \widehat{w}) + bc\widehat{T}(\widehat{\eta}, \widehat{v}) + bd\widehat{T}(\widehat{\eta}, \widehat{w}), \tag{2.34}$$

where $\widehat{T}(\widehat{\omega}, \widehat{v}) \in \mathbf{R}$, $\widehat{\omega} \in T_p^*$, $\widehat{v} \in T_p$ and a, b, c, d are all real numbers.[3]

Using the tensor product, we can define higher-rank tensors from lower-rank ones. Given a (k, l)-type tensor $\widehat{T}$, and another (m, n)-type tensor $\widehat{S}$, their tensor product is a $(k + m, l + n)$-type tensor $T \otimes S$:

$$\begin{aligned}&\widehat{T} \otimes \widehat{S}(\widehat{\omega}^{(1)}, \cdots, \widehat{\omega}^{(k)}, \cdots, \widehat{\omega}^{(k+m)}, \widehat{v}^{(1)}, \cdots, \widehat{v}^{(l)}, \cdots, \widehat{v}^{(l+n)})\\ &\equiv \widehat{T}(\widehat{\omega}^{(1)}, \cdots, \widehat{\omega}^{(k)}, \widehat{v}^{(1)}, \cdots, \widehat{v}^{(l)})\widehat{S}(\widehat{\omega}^{(k+1)}, \cdots, \widehat{\omega}^{(k+m)}, \widehat{v}^{(l+1)}, \cdots, \widehat{v}^{(l+n)}).\end{aligned} \tag{2.35}$$

Note that generally, the tensor product of two tensors cannot be arbitrarily exchanged, i.e., $\widehat{T} \otimes \widehat{S} \neq \widehat{S} \otimes \widehat{T}$, because the vector spaces they act on are different.

Like vectors and dual vectors, tensors exist independently of the coordinate system. For a (k, l)-type tensor, its basis vectors are

$$\widehat{e}_{\mu_1} \otimes \cdots \otimes \widehat{e}_{\mu_k} \otimes \widehat{\theta}^{\nu_1} \otimes \cdots \otimes \widehat{\theta}^{\nu_l}. \tag{2.36}$$

Thus, the tensor can be expanded as

$$\widehat{T} = T^{\mu_1 \cdots \mu_k}{}_{\nu_1 \cdots \nu_l} \widehat{e}_{\mu_1} \otimes \cdots \otimes \widehat{e}_{\mu_k} \otimes \widehat{\theta}^{\nu_1} \otimes \cdots \otimes \widehat{\theta}^{\nu_l} \tag{2.37}$$

and

$$\widehat{T}(\widehat{\omega}^{(1)}, \cdots, \widehat{\omega}^{(k)}, \widehat{v}^{(1)}, \cdots, \widehat{v}^{(l)}) = T^{\mu_1 \cdots \mu_k}{}_{\nu_1 \cdots \nu_l} \omega_{\mu_1} \cdots \omega_{\mu_k} v^{\nu_1 \cdots} v^{\nu_l}.$$

Under Lorentz transformations, the transformation rule for tensor components is

$$T^{\mu_1' \cdots \mu_k'}{}_{\nu_1' \cdots \nu_l'} = \Lambda^{\mu_1'}_{\mu_1} \cdots \Lambda^{\mu_k'}_{\mu_k} \Lambda^{\nu_1}_{\nu_1'} \cdots \Lambda^{\nu_l}_{\nu_l'} T^{\mu_1 \cdots \mu_k}{}_{\nu_1 \cdots \nu_l}. \tag{2.38}$$

[3]In this book, we discuss real manifolds, so when defining functionals, we only require the mapping to the real number field. Of course, this condition can be relaxed.

Let us now examine several commonly employed tensors. The first is the (0, 2)-type metric tensor, which in Minkowski spacetime is

$$\widehat{\eta} = \eta_{\mu\nu} \mathrm{d}x^\mu \otimes \mathrm{d}x^\nu. \tag{2.39}$$

From the above definition, it naturally gives the scalar product, or inner product, of two vectors:

$$\widehat{\eta}(\widehat{v}, \widehat{w}) = \eta_{\mu\nu} v^\mu w^\nu = \widehat{v} \cdot \widehat{w}. \tag{2.40}$$

If $\widehat{v} \cdot \widehat{w} = 0$, then $\widehat{v}$ and $\widehat{w}$ are orthogonal. This orthogonality is independent of the choice of coordinate system (inertial frame) because the inner product is a scalar. On the other hand, considering the inner product of a vector with itself can define the squared norm (magnitude) of the vector, $\widehat{\eta}(\widehat{v}, \widehat{v})$. The sign of the squared norm can be used to classify vectors as follows:

$$\widehat{\eta}(\widehat{v}, \widehat{v}) \begin{cases} < 0, & \text{timelike vector,} \\ = 0, & \text{null (lightlike) vector,} \\ > 0, & \text{spacelike vector.} \end{cases} \tag{2.41}$$

If the tangent vector of a curve is always timelike, then the curve is known as a timelike curve. For a particle with mass, its worldline is always timelike, indicating that the tangent vector of a timelike curve is always inside the light cone.

Below are some other common examples of tensors.

(1) The Kronecker delta function δ^μ_ν. It is a (1, 1)-type tensor that gives the identity mapping from vectors to vectors.
(2) The inverse of the metric tensor gives a (2, 0)-type tensor $\eta^{\mu\nu}$, satisfying

$$\eta^{\mu\nu} \eta_{\nu\rho} = \delta^\mu_\rho, \quad \eta_{\rho\nu} \eta^{\nu\mu} = \delta^\mu_\rho. \tag{2.42}$$

(3) The Levi–Civita "tensor" (symbol). It is a (0, 4)-tensor:

$$\varepsilon_{\mu\nu\rho\sigma} = \begin{cases} 1, & \text{if the indices are an even permutation of } (0,1,2,3), \\ -1, & \text{if the indices are an odd permutation of } (0,1,2,3), \\ 0, & \text{otherwise.} \end{cases} \tag{2.43}$$

For example,

$$\varepsilon_{1032} = 1, \quad \varepsilon_{3120} = -1.$$

The tensors $\eta_{\mu\nu}$, $\eta^{\mu\nu}$, and $\varepsilon_{\mu\nu\rho\sigma}$ are related to the structure of spacetime and have corresponding generalizations in curved spacetime.

(4) Electromagnetic field strength tensor.
The electromagnetic field strength tensor can be defined from the electromagnetic gauge potential:

$$F_{\mu\nu} = \partial_\mu A_\nu - \partial_\nu A_\mu, \tag{2.44}$$

where A_μ is a gauge potential with gauge transformation

$$A_\mu \to A_\mu - \partial_\mu \psi, \tag{2.45}$$

where ψ is an arbitrary scalar function. It is evident that under this gauge transformation, the electromagnetic field strength tensor is invariant. The components of this field strength tensor are

$$F_{\mu\nu} = \begin{pmatrix} 0 & -E_1 & -E_2 & -E_3 \\ E_1 & 0 & B_3 & -B_2 \\ E_2 & -B_3 & 0 & B_1 \\ E_3 & B_2 & -B_1 & 0 \end{pmatrix}. \tag{2.46}$$

Obviously, it is a (0, 2)-type tensor and is antisymmetric: $F_{\mu\nu} = -F_{\nu\mu}$.

2.4 Operations on Tensors

The action of tensors on other tensors induces various tensor operations. While these operations can be formulated abstractly, we will employ component notation for expository convenience.

First, we can define the contraction operation on tensors, i.e., an upper index and a lower index can be summed over and contracted:

$$S^{\mu\rho}{}_{\sigma} = T^{\mu\nu\rho}{}_{\sigma\nu}. \tag{2.47}$$

Such operations can be induced through the Kronecker delta function. It is important to note that since the indices represent vector spaces, the order is very important, and it is necessary to maintain the relative order of the indices. Generally speaking,

$$T^{\mu\nu\rho}{}_{\sigma\nu} \neq T^{\mu\rho\nu}{}_{\sigma\nu}. \tag{2.48}$$

Second, we can use the metric tensor $\eta_{\mu\nu}$ and its inverse $\eta^{\mu\nu}$ to raise or lower indices:

$$T^{\alpha\beta\mu}{}_{\delta} = \eta^{\mu\gamma} T^{\alpha\beta}{}_{\gamma\delta}, \quad T_{\mu}{}^{\beta}{}_{\gamma\delta} = \eta_{\mu\alpha} T^{\alpha\beta}{}_{\gamma\delta}. \tag{2.49}$$

When raising or lowering indices and performing contractions, the following points should be noted:

(1) The order of indices should not be altered;

(2) Free indices that are not summed over must preserve their relative positions;
(3) Summed indices must appear on the same side of the equation;
(4) A vector can be converted to its dual form, and vice versa, using the metric:

$$v_\mu = \eta_{\mu\nu} v^\nu, \quad \omega^\mu = \eta^{\mu\nu}\omega_\nu. \tag{2.50}$$

(5) In Euclidean space, there is no distinction between v_μ and v^μ, but in Minkowski spacetime, they differ:

$$\omega_\mu = (\omega_0, \omega_1, \omega_2, \omega_3) \to \omega^\mu = (-\omega_0, \omega_1, \omega_2, \omega_3). \tag{2.51}$$

In curved spacetime, the distinction becomes even more pronounced.

By using the metric field, a 1-form field can be converted into a vector field, thereby turning the scalar product between vectors into an operation between a 1-form and a vector. Similarly, using the inverse of the metric field, a scalar product between 1-form fields can be defined as:

$$\widehat{\omega} \cdot \widehat{\eta} = g^{\mu\nu}\omega_\mu \eta_\nu. \tag{2.52}$$

From this, we can compute

$$\|\widehat{\omega}\|^2 \equiv \widehat{\omega} \cdot \widehat{\omega}. \tag{2.53}$$

A 1-form is referred to as *spacelike* if this quantity is positive, *timelike* if negative, and *null* or *lightlike* if zero. A surface defined by $f =$ constant is called:

- *timelike* if its normal 1-form is spacelike,
- *spacelike* if its normal 1-form is timelike,
- *null* if its normal 1-form is null.

For non-null surfaces, the normal 1-form can be normalized as

$$\widehat{n} = \frac{\mathrm{d}f}{\|\mathrm{d}f\|}. \tag{2.54}$$

A corresponding normal vector can then be defined with components

$$n^\mu = \eta^{\mu\nu} n_\nu. \tag{2.55}$$

It is worth noting that in Minkowski spacetime, the normal vector and the normal 1-form point in opposite directions when the surface is spacelike. For example, for the surface defined by $t =$ constant, the normal 1-form is

$$n_\mu = (1, 0, 0, 0), \tag{2.56}$$

and the corresponding normal vector is

$$n^\mu = (-1, 0, 0, 0). \tag{2.57}$$

Thus, the outward-pointing normal 1-form corresponds to an inward-pointing normal vector. This illustrates that the normal 1-form carries a more intrinsic geometric meaning.

In fact, we can further examine the behavior of tensors under the permutation of their indices. A tensor is said to be symmetric with respect to two specific indices if its components remain unchanged when those two indices are swapped. For example, if $S_{\mu\nu\rho} = S_{\nu\mu\rho}$, then the tensor is symmetric in its first two indices. More generally, if a tensor remains invariant under any permutation of its indices, such as

$$S_{\mu\nu\rho} = S_{\nu\mu\rho} = S_{\rho\mu\nu} = \cdots, \tag{2.58}$$

it is referred to as a fully symmetric tensor. Conversely, if the tensor changes sign under the exchange of two indices, for instance,

$$A_{\mu\nu\rho} = -A_{\nu\mu\rho}, \tag{2.59}$$

it is antisymmetric with respect to the first two indices. If the tensor changes sign under the exchange of any pair of indices, it is called a fully antisymmetric tensor, or a skew-symmetric tensor. Naturally, for a fully antisymmetric tensor, the value remains unchanged under an even number of index swaps and changes sign under an odd number of swaps.

Example 2.3. The metric tensor is a fully symmetric tensor, while the Levi–Civita tensor and the electromagnetic field strength tensor are fully antisymmetric tensors.

Tensor operations allow us to convert any tensor into a fully symmetric or fully antisymmetric form through the processes of symmetrization and antisymmetrization. Symmetrization with respect to $\mu_1, \cdots, \mu_n$ is defined as:

$$T_{(\mu_1\cdots\mu_n)\rho}{}^{\sigma} = \frac{1}{n!}(T_{\mu_1\cdots\mu_n\rho}{}^{\sigma} + \text{sum over all permutations of } \mu_1, \cdots, \mu_n). \tag{2.60}$$

Similarly, antisymmetrization with respect to $\mu_1, \cdots, \mu_n$ is defined as:

$$T_{[\mu_1\cdots\mu_n]\rho}{}^{\sigma}$$
$$= \frac{1}{n!}(T_{\mu_1\cdots\mu_n\rho}{}^{\sigma} + \text{ alternating sum over all permutations of } \mu_1, \cdots, \mu_n),$$

where "alternating sum" means that even permutations contribute with a positive sign, while odd permutations contribute with a negative sign. By convention, symmetrization is denoted by parentheses around the indices, and antisymmetrization by square brackets. An expressions such as $T_{(\mu|\nu|\rho)}$ indicates full symmetrization over the indices μ and ρ, while the index ν is excluded from the operation but its position is preserved.

Below are several important properties related to tensor operations:

(1) For any rank-2 tensor Y, $X^{(\mu\nu)}Y_{\mu\nu} = X^{(\mu\nu)}Y_{(\mu\nu)}$, even if Y is not symmetric. This follows from the fact that any rank-2 tensor Y can be decomposed into symmetric and antisymmetric parts. The contraction of the symmetric tensor X with the antisymmetric part of Y vanishes, leaving only the symmetric part of Y. Note that this property does not generalize simply to tensors of higher rank.

(2) A tensor $T_{\mu\nu\rho\sigma}$ can be decomposed as $T_{\mu\nu\rho\sigma} = T_{(\mu\nu)\rho\sigma} + T_{[\mu\nu]\rho\sigma}$. However, such a decomposition does not generally hold over three or more indices. For example, $T_{\mu\nu\rho\sigma} \neq T_{(\mu\nu\rho)\sigma} + T_{[\mu\nu\rho]\sigma}$.

(3) Taking trace. For a (1, 1)-type tensor $X^{\mu}{}_{\nu}$, its trace is a scalar, defined as $X = X^{\mu}{}_{\mu}$. When written in matrix form, this corresponds to the usual matrix trace.

(4) For a (2, 0)-type or (0, 2)-type tensor, the trace is not the sum of its diagonal components. Instead, one must first raise or lower an index to convert it to a (1,1)-type tensor, and then take the trace:

$$Y = Y^{\mu}_{\mu} = \eta^{\mu\nu}Y_{\nu\mu}. \tag{2.61}$$

Note, in particular, that

$$\eta^{\nu\mu}\eta_{\mu\nu} = \delta^{\mu}_{\mu} = 4, \tag{2.62}$$

and not $-1+1+1+1 = 2$, due to the implicit summation over repeated indices.

2.5 Ideal Fluids in Special Relativity

This section demonstrates tensor analysis through its application to ideal fluids in special relativity. Such fluids are particularly important because: (1) they closely approximate gravitational sources in astrophysics and cosmology, and (2) they are indispensable for modeling cosmic expansion, stellar collapse, and related astrophysical phenomena.

A fluid is a distinct class of continuous medium. Unlike the mechanics of single particle or multi-particle point masses, a continuous medium comprises an immense number of particles whose motion can be highly complex.

While tracking the trajectory of each individual particle is impractical, we can employ an "averaging" approach to describe the system's macroscopic behavior. This includes quantities such as particle density, energy density, momentum, pressure, and temperature. The term averaging here refers to selecting a representative unit volume that contains a sufficiently large number of particles to ensure meaningful statistics, yet remains small enough to preserve local variations. This unit must also be spatially localized, allowing the definition of local thermodynamic properties like pressure $p(x)$ and temperature $T(x)$. For a rigorous treatment of the averaging methodology, refer to [26].

A fluid is a continuous medium capable of flow. It can be treated as such when the mean interparticle spacing and collision timescale are both negligible compared to the characteristic length and time scales of interest. While this definition primarily applies to liquids, gases—though less intuitive—also qualify as fluids under this framework. Consider two adjacent fluid elements, A and B. Due to intermolecular interactions and thermal motion, a pressure exists between them. In a rigid medium, no relative motion (sliding) occurs between adjacent elements. In contrast, a fluid permits significant tangential motion—its defining feature. An ideal fluid is a theoretical construct in which viscous forces are entirely absent. In such a fluid, adjacent layers may slide past one another without resistance, exhibiting zero rigidity.

2.5.1 *Particle number-flux 4-vector*

We begin by analyzing an idealized system consisting of non-interacting particles. In a suitably chosen reference frame, all particles remain at rest, implying the absence of both external forces and interparticle interactions. Such particles are referred to as dust, where each particle is sufficiently massive to be treated as an independent and the interactions between particles can be neglected. Suppose the rest mass of each particle is m, then in the rest reference frame of these particles, the particle number density of each unit is

$$n = N/V = N/(\Delta x \Delta y \Delta z). \tag{2.63}$$

The number of particles in the unit is a quantity that does not change with the reference frame, but the volume element is not. If we are not in the rest reference frame, due to the Lorentz contraction effect, the unit volume becomes $\Delta x \Delta y \Delta z \sqrt{1-v^2}$. Therefore, in a reference frame where all particles move at a speed v, the particle number density is $n/\sqrt{1-v^2}$.

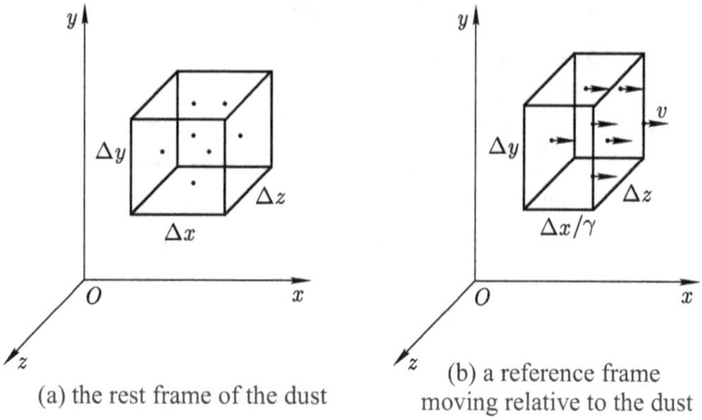

Fig. 2.2 The states of dust observed in different inertial frames are different.

On the other hand, if the particles have speed, there will be flux. The flux through a surface is the number of particles passing through a unit area surface per unit time. As shown in Figure 2.2, in the rest reference frame, all particles are stationary, and the flux is zero, while in a moving reference frame, for example, moving along the x-axis, the unit area is

$$\Delta A = \Delta y \Delta z, \tag{2.64}$$

and the number of particles passing through this area per unit time is

$$\frac{n}{\sqrt{1-v^2}} v \Delta t \Delta A. \tag{2.65}$$

Therefore, the flux along the x-axis is

$$(\text{flux})^x = \frac{nv}{\sqrt{1-v^2}}. \tag{2.66}$$

Even if the particles do not strictly move along the x-axis, we have

$$(\text{flux})^x = \frac{nv^x}{\sqrt{1-v^2}}. \tag{2.67}$$

Further, we can define a particle number-flux 4-vector

$$\widehat{N} \equiv n\widehat{u}, \tag{2.68}$$

where $\widehat{u}$ is the 4-velocity of the particles. In a reference frame where the velocity is $\boldsymbol{v}$, the components of the 4-velocity are

$$u^\mu = (\gamma, \gamma v^x, \gamma v^y, \gamma v^z), \tag{2.69}$$

so

$$N^\mu = (n\gamma, n\gamma v^x, n\gamma v^y, n\gamma v^z). \tag{2.70}$$

The first term is exactly the particle number density, and the latter three terms are the fluxes along different coordinate axes, respectively. Note that $\widehat{N}$ is a 4-vector that is independent of the reference frame, and its square of the magnitude is

$$\widehat{N} \cdot \widehat{N} = -n^2. \tag{2.71}$$

If the unit is small enough, $\widehat{N}$ can be considered as a local field $\widehat{N}(x)$, which is related to the position where it is located. This local 4-vector field satisfies the conservation law of particle number

$$\frac{\partial N}{\partial t} + \nabla \cdot \boldsymbol{N} = 0, \tag{2.72}$$

i.e.,

$$\frac{\partial N^\mu}{\partial x^\mu} = \partial_\mu N^\mu = 0. \tag{2.73}$$

This implies that in a certain local area, the change of particle number with time is equal to the change of particle flux in this local area.

Consider the motion of a system of multiple particles. These particles need not be dust particles, and their motion may be arbitrary. The particle number flux density vector N^μ can be expressed as

$$N^\mu = \sum_a \int \mathrm{d}\tau_a \frac{\mathrm{d}x_a^\mu}{\mathrm{d}\tau_a} \delta^4(x^\lambda - x_a^\lambda(\tau_a)), \tag{2.74}$$

where the index a labels distinct particles, and x_a^λ denotes the worldline of the a-th particle. In the comoving reference frame of a single particle, $N^\mu = (N^0, \mathbf{0})$, and $N^0 = \delta^3(\boldsymbol{x})$ indicates that the particle is located at the spatial origin. Using the identity

$$\begin{aligned} \delta^3(\boldsymbol{x}) &= \int \mathrm{d}x^0 \delta(x^0 - x^0(\tau))\delta^3(\boldsymbol{x}) \\ &= \int \mathrm{d}\tau \frac{\mathrm{d}x^0}{\mathrm{d}\tau} \delta^4(x^\lambda - x^\lambda(\tau)), \end{aligned} \tag{2.75}$$

we can generalize this to a four-vector expression:

$$N^\mu = \int \mathrm{d}\tau \frac{\mathrm{d}x^\mu(\tau)}{\mathrm{d}\tau} \delta^4(x^\lambda - x^\lambda(\tau)). \tag{2.76}$$

The extension to multiple particles follows naturally as in (2.74). It is straightforward to verify that N^μ is conserved, $\partial_\mu N^\mu = 0$.

2.5.2 *Energy-momentum tensor*

As previously discussed, a 1-form can be geometrically interpreted as a family of surfaces partitioning space. By employing a 1-form, we can extract the components of vectors and tensors. In this context, the contraction of a vector with a 1-form corresponds to the number of times the vector pierces the surfaces defined by the 1-form. For example, for $\mathrm{d}x^\mu$, we have

$$\langle \mathrm{d}x^\mu, \widehat{V} \rangle = V^\mu. \tag{2.77}$$

In three dimensions, the surface element is equal to the unit normal vector multiplied by the area element. Similarly, in four-dimensional spacetime, the volume element of 3-space is $\widehat{n}\Delta x^\alpha \Delta x^\beta \Delta x^\gamma$. The flux through the surface defined by the scalar function $\phi =$ constant is $\langle \widehat{n}, \widehat{N} \rangle$, where $\widehat{n}$ is the normalized dual normal vector. The notation here should be understood as $\widehat{n}(\widehat{N})$, because $\widehat{n}$ is a 1-form, and $\widehat{N}$ is a vector, so their contraction is not the scalar product of vectors, but is defined by the functional relationship. In fact, this operation precisely captures the geometric interpretation of a 1-form: its contraction with a vector is determined by the number of times the vector pierces the associated surfaces. Through appropriate normalization, $\widehat{n}(\widehat{N})$ indeed gives the flux through the surface. For example, if we take ϕ as $\{x = \text{constant}\}$, then $n_\mu = (0, 1, 0, 0)$, and $\langle \mathrm{d}x, \widehat{N} \rangle = N^x$ gives the flux through the surface $\{x = \text{constant}\}$. And if we take $\phi = t =$ constant, then $n_\mu = (1, 0, 0, 0)$, so $\langle \mathrm{d}t, \widehat{N} \rangle = N^0$ gives the particle number density $N^0 = N$. That is to say, the particle number density can be understood as a timelike flux.

Similarly, we can understand the energy and momentum of particles as $\langle \widehat{n}, \widehat{p} \rangle$. For example, the energy is $E = \langle \mathrm{d}t, \widehat{p} \rangle = p^0$. In other words, the derived energy and momentum values are intrinsically dependent on the choice of the reference surface. Since our analysis presupposes a specific coordinate system, which inherently defines these surfaces, the resulting energy and momentum quantities exhibit an apparent coordinate dependence.

Next, let us revisit the dust system discussed earlier. In the rest frame of the dust, the particles are stationary, and the energy density in a unit volume is given by

$$\rho = nm. \tag{2.78}$$

Now consider an inertial frame S, moving relative to the rest frame. Under a Lorentz transformation, the number density transforms as $n \to n\gamma$, and the energy of each particle as $m \to m\gamma$. Consequently, the energy density in frame S becomes

$$\rho|_S = \gamma^2 \rho. \tag{2.79}$$

This transformation behavior shows that the energy density is neither a scalar nor a component of a vector. Instead, it is the $(0,0)$-component of a rank-2 tensor, specifically, the energy-momentum tensor,[4] also known as the stress-energy tensor or stress tensor. The energy-momentum tensor is a (2,0)-type tensor. Its action on the tensor product of two 1-form basis vectors yields its various components:

$$\begin{aligned}\widehat{T}(\mathrm{d}x^\alpha, \mathrm{d}x^\beta) &= T^{\alpha\beta} \\ &= \{\text{4-momentum's } \alpha \text{ component through the surface } x^\beta = \text{ constant}\}.\end{aligned} \tag{2.80}$$

In other words, the 1-form $\mathrm{d}x^\alpha$ selects the component of the 4-momentum, while $\mathrm{d}x^\beta$ specifies the hypersurface across which the flux is measured. In a given reference frame, the energy of a particle is extracted via the 1-form $\mathrm{d}t$, which yields the temporal 0-component of the 4-momentum. Similarly, the particle number density is obtained by contracting the flux 4-vector with $\mathrm{d}t$ to isolate its 0-component. Combining these elements, we arrive at the energy density. The physical interpretations of the components of the energy-momentum tensor $T^{\alpha\beta}$ are as follows:

(1) $T^{00} = \widehat{T}(\mathrm{d}t, \mathrm{d}t)$ is the energy density.

(2) T^{0i} represents the energy flux through the surface $x^i = \text{constant}$. This arises from mechanisms such as heat conduction and is equal to the energy density multiplied by the flow velocity. Since the tensor is symmetric, $T^{i0} = T^{0i}$, which corresponds to the equivalence between momentum density and energy flux:

$$T^{i0} = T^{0i}. \tag{2.81}$$

(3) T^{i0} denotes the flux of the ith momentum through a surface of constant t, i.e., the ith momentum density.

(4) T^{ij} is the flux of the ith component of momentum through the $x^j = \text{constant}$ surface. This represents the force per unit area between adjacent fluid units and is referred to as stress. In general, it can be shown that the stress tensor is symmetric:

$$T^{ij} = T^{ji}. \tag{2.82}$$

For a detailed proof, see [3].

[4] Also known as the stress-energy tensor or stress tensor.

(5) If only forces perpendicular to the intersecting surface act, i.e., no shear stresses, then

$$T^{ij} = 0, \text{ when } i \neq j. \tag{2.83}$$

(6) If $T^{ij} \neq 0$ for $i \neq j$, this indicates the presence of viscosity.

For dust, in its rest reference frame, the only non-zero component of the energy-momentum tensor is $T^{00} = mn$, while all other components vanish. The energy-momentum tensor for dust can be expressed in a coordinate-free form as

$$\widehat{T} = \widehat{p} \otimes \widehat{N} = mn\widehat{u} \otimes \widehat{u}. \tag{2.84}$$

In component form, this becomes

$$T^{\alpha\beta} = \widehat{T}(\mathrm{d}x^\alpha, \mathrm{d}x^\beta) = \rho\widehat{u}(\mathrm{d}x^\alpha)\widehat{u}(\mathrm{d}x^\beta) = \rho u^\alpha u^\beta. \tag{2.85}$$

In an arbitrary reference frame, the explicit form of the 4-velocity can be used to construct the energy-momentum tensor accordingly. It is manifestly symmetric:

$$T^{\alpha\beta} = T^{\beta\alpha}. \tag{2.86}$$

This symmetry property holds not only for dust but also for general fluid systems.

For an observer, a laboratory can be established, i.e., a comoving reference frame, with basis vectors $\{\widehat{e}_a\}$. The energy density of the fluid measured by the observer is

$$T_{\mu\nu} u^\mu_{\text{obs}} u^\nu_{\text{obs}}. \tag{2.87}$$

The measured momentum density's ith component is $-T_{\mu\nu} u^\mu (\widehat{e}_i)^\nu$, and the three-dimensional stress tensor's (ij)-component is $T_{\mu\nu}(\widehat{e}_i)^\mu (\widehat{e}_j)^\nu$. The observed 4-momentum density is $W^\mu = -T^\mu_\nu u^\nu$.

For a system of multiple particles that is not necessarily dust, the energy-momentum tensor is given by

$$\begin{aligned} T^{\mu\nu}(x) &= \sum_a \int \mathrm{d}\tau_a \frac{\mathrm{d}x^\mu_a}{\mathrm{d}\tau_a} p^\nu_a(\tau_a) \delta^4(x^\lambda - x^\lambda_a(\tau_a)) \\ &= \sum_a \int \mathrm{d}\tau_a \left(m_a \frac{\mathrm{d}x^\mu_a}{\mathrm{d}\tau_a} \frac{\mathrm{d}x^\nu_a}{\mathrm{d}\tau_a} \right) \delta^4(x^\lambda - x^\lambda_a(\tau_a)), \end{aligned} \tag{2.88}$$

which effectively sums the energy-momentum contributions from all individual particles. It is straightforward to verify that this tensor satisfies the conservation law $\partial_\mu T^{\mu\nu} = 0$. Considering a fixed spatial volume V, the total 4-momentum within V at time t is

$$\int_V \mathrm{d}^3x T^{0\nu}(x) = \sum_{a\in V} \int \mathrm{d}\tau_a \frac{\mathrm{d}x_a^0}{\mathrm{d}\tau_a} p_a^\nu(\tau_a)\delta(x^0 - x_a^0(\tau_a)) = \sum_{a\in V} p_a^\nu = p_V^\nu(t). \tag{2.89}$$

Thus, T^{00} represents the energy density, and T^{0i} the momentum density. Note that in interacting systems, the energy density includes not only the rest mass of the particles but also potential energy from interactions. The time derivative of the total momentum is

$$\frac{\mathrm{d}p_V^\nu(t)}{\mathrm{d}t} = \int_V \mathrm{d}^3x \frac{\mathrm{d}}{\mathrm{d}t} T^{0\nu}(x) = -\int_V \mathrm{d}^3x \frac{\partial T^{i\nu}}{\partial x^i} = -\int_{\partial V} \mathrm{d}S_i T^{i\nu}, \tag{2.90}$$

where the last equality follows from the divergence theorem. For the spatial components $\nu = j$, this gives

$$\frac{\mathrm{d}p_V^j(t)}{\mathrm{d}t} = -\int_{\partial V} \mathrm{d}S_i T^{ij}, \tag{2.91}$$

which is the force acting on the volume V. The quantity T^{ij} represents the force per unit area in the j-th direction acting on a surface element with normal in the i-th direction, i.e., the stress tensor. Its symmetry $T^{ij} = T^{ji}$ follows from angular momentum conservation.

Using the integral identity

$$\int \mathrm{d}\tau_a \delta(t - x_a^0(\tau_a)) f(\tau_a) = \left(f(\tau_a) \left| \frac{\mathrm{d}x_a^0}{\mathrm{d}\tau_a} \right|^{-1} \right) \Bigg|_{x_a^0(\tau_a)=t}, \tag{2.92}$$

and noting that the worldline is future-directed, we have

$$\frac{\mathrm{d}x_a^\mu}{\mathrm{d}\tau_a} \bigg/ \frac{\mathrm{d}x_a^0}{\mathrm{d}\tau_a} = (1, v_a^i). \tag{2.93}$$

The particle number flux components are then

$$\begin{aligned} N^0 &= \sum_a \delta^3(\boldsymbol{x} - \boldsymbol{x}_a(\tau_a)), \\ N^i &= \sum_a v_a^i \delta^3(\boldsymbol{x} - \boldsymbol{x}_a(\tau_a)), \end{aligned} \tag{2.94}$$

and the energy-momentum tensor takes the form

$$T^{\mu\nu}(x) = \sum_a \frac{p_a^\mu(\tau_a) p_a^\nu(\tau_a)}{E_a} \delta^3(\boldsymbol{x} - \boldsymbol{x}_a(\tau_a)). \tag{2.95}$$

2.5.3 *Ideal fluids*

The dust model introduced earlier represents an idealized fluid in which particle interactions and random thermal motion are neglected. For realistic fluids, however, the situation is more complex: Beyond the bulk motion of the fluid element, individual particles exhibit random thermal motions; Moreover, there are interactions between particles that contribute to potential energy. To properly incorporate these effects into the energy-momentum tensor, we analyze the system in the fluid element's rest frame (i.e., the comoving frame), where the total spatial momentum vanishes. All local quantities, e.g., number density, energy density, temperature, are defined as measured in this frame. This frame is local, valid only for a single fluid element, since neighboring elements may have relative motion. For simplicity, we restrict our discussion to a single-component fluid. Under these assumptions, the energy-momentum tensor takes the following form:

(1) T^{00} is the total energy density, including potential and kinetic energy;
(2) T^{0i} is the energy propagated by heat conduction, so it is essentially a heat conduction term;
(3) T^{i0}, if heat conduction exists, energy will also include contributions from momentum;
(4) T^{ij}, the random thermal motion of particles gives momentum flow, so T^{ii} is the isotropic pressure along the i-th direction, and $T^{ij}(i \neq j)$ is the shear viscosity of the fluid.

An ideal fluid is a relativistic generalization of an ideal gas, characterized by the absence of viscosity and heat conduction in its local rest frame. These properties impose strict constraints on the energy-momentum tensor $T^{\mu\nu}$: Since there is no heat conduction, $T^{0i} = T^{i0} = 0$; No viscosity means $T^{ij} = 0$ unless $i = j$. Therefore, an ideal fluid only has diagonal terms,

$$T^{ij} = p\delta^{ij}. \tag{2.96}$$

This condition naturally follows from the requirement of spatial isotropy in the fluid's rest frame. In the absence of heat conduction and viscosity, the fluid must exhibit identical physical properties in all spatial directions. Therefore, in the rest reference frame, the components of the energy-momentum tensor of an ideal fluid are

$$\begin{aligned} T^{\alpha\beta} &= \begin{pmatrix} \rho & 0 & 0 & 0 \\ 0 & p & 0 & 0 \\ 0 & 0 & p & 0 \\ 0 & 0 & 0 & p \end{pmatrix} \\ &= (\rho + p) u^\alpha u^\beta + p\eta^{\alpha\beta}. \end{aligned} \tag{2.97}$$

Using tensor language, the energy-momentum tensor can be written as

$$\widehat{T} = (\rho + p)\widehat{u} \otimes \widehat{u} + p\widehat{g}. \tag{2.98}$$

In the last term, we have introduced the general metric tensor. In Minkowski spacetime, $\widehat{g} = \widehat{\eta}$. Obviously, $\widehat{T}$ is a $(2,0)$-type tensor. Note that in a moving reference frame,

$$T^{00} = (\rho + p)(u^0)^2 - p = \frac{\rho + \boldsymbol{v}^2 p}{1 - \boldsymbol{v}^2}, \tag{2.99}$$

that is, pressure also appears in the energy density.

If $p = 0$, then the ideal fluid is pressureless, which corresponds to dust. The pressureless ideal fluid indicates that the particles within it have no random motion. If the particles have random velocities, it would lead to the generation of pressure. In Newtonian gravity,

$$\nabla^2 \Phi = 4\pi G \rho, \tag{2.100}$$

where ρ is the mass density. According to special relativity, ρ should be replaced by various forms of energy. From the above discussion on the energy-momentum tensor, ρ is only one component of the energy-momentum tensor, $\rho \to T^{00}$. This implies that in the relativistic theory of gravity, $\rho \to \widehat{T}$. We will see in Einstein's general relativity that this is indeed the case. This also indicates that pressure may play an important role in general relativity.

In the energy-momentum tensor of an ideal fluid, there are two seemingly independent quantities—energy density and pressure. However, these quantities are not truly independent. Their relationship is governed by an equation of state (EOS), which varies for different types of ideal fluids,

$$p = p(\rho). \tag{2.101}$$

It characterizes the properties of the fluid. Below are some ideal fluids often encountered in cosmology:

(1) Dust, $p = 0$.

(2) Radiation, $p = \frac{1}{3}\rho$. Massless electromagnetic radiation plays a crucial role in cosmology. Unlike non-relativistic matter, radiation exerts non-zero pressure, leading to distinct characteristics in the evolution of a radiation-dominated universe.

(3) Vacuum energy, also known as the cosmological constant,

$$p = -\rho. \tag{2.102}$$

Therefore, it is also often said to have negative pressure. Its energy-momentum tensor is

$$T^{\mu\nu} = -\rho_{\text{vac}}\eta^{\mu\nu}. \tag{2.103}$$

This particular ideal fluid plays a dominant role in the late stage of the universe's evolution.

We now consider two special types of ideal fluids derived from the coarse-grained average of multi-particle systems. In the rest frame of an ideal fluid, there is no preferred direction; the three-momenta of the particles are isotropically distributed, and their vector sum within any fluid element vanishes. Averaging over all directions yields

$$\langle p_a^0 p_a^i \rangle_{\text{direction}} = 0, \tag{2.104}$$

so that $T^{0i} = 0$. Moreover, using the angular average

$$\frac{1}{4\pi}\int \mathrm{d}\theta \mathrm{d}\varphi \sin\theta p^i p^j = \frac{1}{3}\boldsymbol{p}^2\delta^{ij}, \tag{2.105}$$

we get

$$\langle p_a^i p_a^j \rangle_{\text{direction}} = \frac{1}{3}\boldsymbol{p}_a^2(\tau_a)\delta^{ij}. \tag{2.106}$$

From this, the energy density and pressure in the fluid element are given by

$$\begin{aligned} \rho &= \left\langle \sum_a E_a(\tau_a)\delta^3(\boldsymbol{x} - \boldsymbol{x}_a(\tau_a)) \right\rangle, \\ p &= \left\langle \sum_a \frac{\boldsymbol{p}_a^2(\tau_a)}{3E_a(\tau_a)}\delta^3(\boldsymbol{x} - \boldsymbol{x}_a(\tau_a)) \right\rangle. \end{aligned} \tag{2.107}$$

Since for each particle $E_a^2 = m^2 + \boldsymbol{p}_a^2 \geq \boldsymbol{p}_a^2$, it follows that

$$\rho \geq 3p \geq 0. \tag{2.108}$$

For non-relativistic particles, with $E = m + \dfrac{\boldsymbol{p}^2}{2m} + \cdots$, the energy density becomes

$$\rho = n\left(m + \frac{3}{2}k_{\rm B}T\right). \tag{2.109}$$

Where $k_{\rm B}$ is the Boltzmann constant, and T is the temperature of the fluid. Using the approximation

$$\left\langle \frac{\boldsymbol{p}_a^2(\tau_a)}{3E_a(\tau_a)} \right\rangle \approx \frac{\boldsymbol{p}^2}{3m} = k_{\rm B}T, \tag{2.110}$$

we recover the equation of state for an ideal gas $p = nk_{\rm B}T$ or $pV = Nk_{\rm B}T$. Thus, an ideal fluid can be viewed as a generalization of an ideal gas, where particle interactions are neglected and only random thermal motion is considered.

For an ultra-relativistic gas, with $E = |\boldsymbol{p}| + \cdots$, we find $\rho \approx 3p$, In particular, for a photon gas,

$$p = \frac{1}{3}\rho, \tag{2.111}$$

which is exactly the equation of state for radiation. In this case, the trace of the energy-momentum tensor satisfies $\eta_{\mu\nu}T^{\mu\nu} = 0$, consistent with the traceless property of the electromagnetic energy-momentum tensor in four dimensions.

2.5.4 *Conservation of energy-momentum*

In special relativity, the conservation of energy-momentum equation is

$$\partial_\mu T^{\mu\nu} = 0. \tag{2.112}$$

This equation sums over one index, and the remaining index indicates that this equation actually contains four equations. For $\nu = 0$, this equation gives energy conservation. And for $\nu = k \neq 0$, this equation gives

$$\partial_\mu T^{\mu k} = 0. \tag{2.113}$$

To simplify the discussion, we consider only the case of an ideal fluid. The divergence of the energy-momentum tensor is given by

$$\begin{aligned}\partial_\mu T^{\mu\nu} &= \partial_\mu[(\rho + p)u^\mu u^\nu + p\eta^{\mu\nu}] \\ &= \partial_\mu(\rho + p)u^\mu u^\nu + (\rho + p)\partial_\mu(u^\mu u^\nu) + \partial_\mu p\eta^{\mu\nu}.\end{aligned} \tag{2.114}$$

We first consider the component of this equation parallel to the 4-velocity. Contracting with u_ν yields

$$u_\nu \partial_\mu T^{\mu\nu} = -\partial_\mu(\rho u^\mu) - p\partial_\mu u^\mu, \tag{2.115}$$

where we have used $u_\nu \partial_\mu u^\nu = \frac{1}{2}\partial_\mu(u_\nu u^\nu) = 0$. This gives the relativistic energy conservation law:

$$\partial_\mu(\rho u^\mu) + p\partial_\mu u^\mu = 0. \tag{2.116}$$

In the non-relativistic limit,

$$u^\mu = (1, v^i), \quad v^i \ll 1, p \ll \rho,$$

where the relationship $p \ll \rho$ is because pressure comes from random motion, and since $v^i \ll 1$, p is very small. In this limit, Eq. (2.116) reduces to

$$\partial_t \rho + \nabla \cdot (\rho \boldsymbol{v}) = 0, \tag{2.117}$$

which is the continuity equation for energy density, i.e., the law of energy conservation.

Next, we project the conservation equation perpendicular to the 4-velocity by using the operator

$$P^\sigma_\nu = \delta^\sigma_\nu + u^\sigma u_\nu, \tag{2.118}$$

which annihilates components parallel to u^ν and preserves orthogonal components, i.e.,

$$P^\sigma_\nu V^\nu_\parallel = V^\sigma_\parallel + u^\sigma u_\nu V^\nu_\parallel = 0, \quad P^\sigma_\nu W^\nu_\perp = W^\sigma_\perp. \tag{2.119}$$

Applying this to the conservation equation of energy-momentum gives

$$P^\sigma_\nu \partial_\mu T^{\mu\nu} = (\rho + p)u^\mu \partial_\mu u^\sigma + \partial^\sigma p + u^\sigma u^\mu \partial_\mu p. \tag{2.120}$$

In the rest frame, the equation is automatically satisfied when $\sigma = 0$, and only when $\sigma = i$ is non-trivial. Since $u^i = 0$, the spatial components yields

$$0 = (\rho + p)u^\mu \partial_\mu u_i + \partial_i p = (\rho + p)a_i + \partial_i p, \tag{2.121}$$

where the 4-acceleration

$$a_\sigma = \frac{\mathrm{d}}{\mathrm{d}\tau}u_\sigma = \frac{\mathrm{d}x^\mu}{\mathrm{d}\tau}\frac{\partial}{\partial x^\mu}u^\sigma = u^\mu \partial_\mu u_\sigma. \tag{2.122}$$

To understand physical meaning of (2.121), we can compare it with the equation in the non-relativistic limit. In taking the non-relativistic limit, we only retain the first-order terms: The part of $\sigma = 0$ is trivial, and the part of $\sigma \neq 0$ gives

$$\rho[\partial_t \boldsymbol{v} + (\boldsymbol{v} \cdot \nabla)\boldsymbol{v}] = -\nabla p, \tag{2.123}$$

which is the Euler equation in fluid mechanics. Alternatively, it can be written as

$$\rho \boldsymbol{a} + \nabla p = 0. \tag{2.124}$$

The relativistic version (2.121) effectively replaces ρ with $\rho + p$, indicating that pressure contributes to inertial mass in relativity. In fact, in relativity, the role of $\rho + p$ is similar to the inertial mass density because Eq. (2.121) is essentially $\boldsymbol{F} = m\boldsymbol{a}$.

Before concluding this section, we now briefly discuss the fundamental role of the energy-momentum tensor in general relativity. In Newtonian gravity, the gravitational source is simply the mass density ρ, as in Poisson's equation. However, relativity requires that all forms of energy, encoded in T^{00}, contribute to gravity. Einstein's key insight was that the full energy-momentum tensor $\widehat{T}$ must serve as the source of the gravitational field in general relativity.

The role of pressure p in general relativity is particularly significant for several reasons. First, pressure itself is a source of the gravitational field. Second, in compact objects such as neutron stars, strong pressure gradients are essential to resist gravitational collapse. Third, as demonstrated in Eq. (2.121), relativistic effects require stronger pressure gradients for stability compared to Newtonian gravity. Finally, as will be discussed later, pressure also plays a pivotal role in cosmology.

2.6 Covariant Form of Electrodynamics

Maxwell's theory provided a unified framework for describing electric and magnetic phenomena, laying the groundwork for Einstein's development of special relativity. In turn, the formulation of special relativity necessitated a reinterpretation of electromagnetic phenomena within a relativistic framework. By employing four-dimensional tensors, Maxwell's equations can be elegantly recast in a manifestly covariant form, offering profound insights and simplifying the analysis of diverse problems in electrodynamics. Given the scope constraints of this discussion, we present only a concise discussion here. For a comprehensive and rigorous derivation, we refer the reader to Ref. [27].

2.6.1 *Maxwell's equations*

In the mid-19th century, James Clerk Maxwell demonstrated that electromagnetic phenomena could be elegantly unified through a system of fundamental equations. When the effects of polarization and magnetization in a medium are neglected, these equations take a particularly simplified and symmetric form:

$$\begin{aligned} \nabla \times \boldsymbol{B} - \partial_t \boldsymbol{E} &= 4\pi \boldsymbol{J}, \\ \nabla \cdot \boldsymbol{E} &= 4\pi\rho, \\ \nabla \times \boldsymbol{E} + \partial_t \boldsymbol{B} &= 0, \\ \nabla \cdot \boldsymbol{B} &= 0. \end{aligned} \tag{2.125}$$

Maxwell's theory is actually the source of Einstein's proposal of special relativity. In a vacuum, this set of equations becomes[5]

$$\begin{aligned} \nabla \cdot \boldsymbol{E} &= 0, \\ \nabla \cdot \boldsymbol{B} &= 0, \\ \nabla \times \boldsymbol{E} &= -\partial_t \boldsymbol{B}, \\ \nabla \times \boldsymbol{B} &= \mu_0 \epsilon_0 \partial_t \boldsymbol{E}. \end{aligned} \tag{2.126}$$

From this, we can obtain the electromagnetic wave equation

$$\begin{aligned} \nabla^2 \boldsymbol{E} &= \mu_0 \epsilon_0 \frac{\partial^2 \boldsymbol{E}}{\partial t^2}, \\ \nabla^2 \boldsymbol{B} &= \mu_0 \epsilon_0 \frac{\partial^2 \boldsymbol{B}}{\partial t^2}. \end{aligned} \tag{2.127}$$

This suggests that both the electric field and the magnetic field propagate at the speed of light $c^2 = 1/(\mu_0\epsilon_0)$. An astonishing conclusion of Maxwell's theory is that visible light is itself an electromagnetic wave, and the speed of light is determined by the two constants μ_0 and ϵ_0 in electromagnetic theory. Maxwell's theory not only unifies electricity and magnetism but also unifies optics with electromagnetism. However, if Maxwell's theory is correct, it is incompatible with the Galilean transformation. Under the Galilean transformation,

$$\boldsymbol{x}' = \boldsymbol{x} - \boldsymbol{v}t, \quad t' = t, \tag{2.128}$$

if a field Ψ satisfies the wave equation in the reference frame $(t', \boldsymbol{x}')$

$$\left(\sum_i \frac{\partial^2}{\partial x_i'^2} - \frac{1}{c^2}\frac{\partial^2}{\partial t'^2} \right) \Psi = 0, \tag{2.129}$$

then in the reference frame $(t, \boldsymbol{x})$, its equation of motion is

$$\left(\nabla^2 - \frac{1}{c^2}\frac{\partial^2}{\partial t^2} - \frac{2}{c^2}\boldsymbol{v}\cdot\nabla\frac{\partial}{\partial t} - \frac{1}{c^2}\boldsymbol{v}\cdot\nabla\boldsymbol{v}\cdot\nabla \right) \Psi = 0, \tag{2.130}$$

[5]In the discussions of this chapter, to more clearly understand relativistic effects, we try to retain physical constants in the formulas.

so the form of the wave equation is not invariant under the Galilean transformation, nor is there any kinematic redefinition of the wave function Ψ that would restore it to a standard wave equation. This stands in contrast to the non-relativistic Schrödinger equation, where Galilean invariance can be achieved by redefining the wave function—a modification that merely introduces an additional phase factor.

A fundamental question arises: Do electromagnetic waves propagate at the same speed in all inertial reference frames? If not, how do the relevant physical constants vary between frames? The concept of a vacuum—often idealized as a space devoid of matter—plays a crucial role here, as it defines the medium in which electromagnetic waves propagate without dispersion or attenuation. A. Einstein postulated that the speed of light must be invariant across all inertial reference frames, a cornerstone assumption of special relativity. Consequently, Maxwell's equations must retain their form under Lorentz transformations, ensuring that the wave equation governing electromagnetic propagation remains unchanged. To see this clearly, we use the component form to transform the above equations into

$$\begin{aligned} \varepsilon^{ijk}\partial_j B_k - \partial_0 E^i &= 4\pi J^i, \\ \partial_i E^i &= 4\pi J^0, \\ \varepsilon^{ijk}\partial_j E_k + \partial_0 B^i &= 0, \\ \partial_i B^i &= 0. \end{aligned} \tag{2.131}$$

We can combine $J^0 = \rho$ and J^i into a charge current 4-vector J^μ. Expressing the electromagnetic field with the electromagnetic field strength, $F^{0i} = E^i$, $F^{ij} = \varepsilon^{ijk}B_k$, the first two equations can be written as

$$\begin{aligned} \partial_j F^{ij} - \partial_0 F^{0i} &= 4\pi J^i, \\ \partial_i F^{0i} &= 4\pi J^0. \end{aligned} \tag{2.132}$$

More compactly, using 4-vectors and tensors, these two equations can be written as

$$\partial_\mu F^{\nu\mu} = 4\pi J^\nu. \tag{2.133}$$

The last two equations can be written as

$$\partial_{[\mu} F_{\nu\lambda]} = 0. \tag{2.134}$$

Clearly, Eqs. (2.133) and (2.134) are in covariant form and are the same in all inertial frames.

In fact, we can introduce the electromagnetic potential to simplify the above discussion. The electromagnetic potential $\widehat{A}$ is a 1-form field,

$$\widehat{A} = A_\mu \mathrm{d}x^\mu, \tag{2.135}$$

and equation (2.134) implies that the electromagnetic field strength tensor can be expressed as

$$F_{\mu\nu} = \partial_\mu A_\nu - \partial_\nu A_\mu. \tag{2.136}$$

From a field-theoretic perspective, $\widehat{A}$ is an Abelian $U(1)$ gauge potential possessing gauge symmetry. It is straightforward to verify that under a gauge transformation

$$A_\mu \to A_\mu + \partial_\mu \psi, \tag{2.137}$$

the field strength tensor remains invariant. This gauge freedom allows us to choose ψ to impose convenient conditions, such as the static gauge $A_0 = 0$, or the Lorenz gauge[6]

$$\partial^\mu A_\mu = 0. \tag{2.138}$$

Notably, while the gauge potential exhibits arbitrariness, it is more fundamental than the field strength itself, a fact confirmed experimentally by the Aharonov–Bohm effect in quantum theory.

The charge current 4-vector $\widehat{J}$ is analogous to the particle number flux 4-vector introduced earlier. For a fluid with proper charge density $\rho_0(x)$, it is defined as

$$\widehat{J} = \rho_0(x)\widehat{u}, \tag{2.139}$$

satisfying $\widehat{J} \cdot \widehat{J} = -\rho_0^2$. In a reference frame where the charged fluid moves with velocity $\boldsymbol{v}$, the components are

$$J^\mu = \rho_0 \gamma_v (1, v^i). \tag{2.140}$$

Due to the antisymmetry of the electromagnetic field strength tensor, we have

$$\partial_\nu \partial_\mu F^{\nu\mu} = 0, \tag{2.141}$$

which, via the inhomogeneous Maxwell equation, implies

$$\partial_\nu J^\nu = 0. \tag{2.142}$$

[6]Hendrik A. Lorentz (1853–1928) was a renowned Dutch theoretical physicist and mathematician, awarded the Nobel Prize in Physics in 1902 for his work on the Zeeman effect. His contributions include the Lorentz force law, the Lorentz transformation, and the development of classical electron theory. Ludvig V. Lorenz (1829–1891) was a Danish physicist who first introduced the Lorenz gauge and retarded potentials in 1867. The modern understanding of gauge transformations stems from Lorenz's foundational work in "Theory of Electrons".

This is the continuity equation expressing local charge conservation. In component form

$$\frac{\partial \rho}{\partial t} + \nabla \cdot \boldsymbol{J}^i = 0. \tag{2.143}$$

Charge conservation follows directly from particle number conservation. If each particle carries charge q, then the charge density is

$$\rho(x) = qn(x), \tag{2.144}$$

where $n(x)$ is the particle number density, and the current density is

$$\boldsymbol{J}(x) = qn(x)\boldsymbol{v}(x), \tag{2.145}$$

where $\boldsymbol{v}(x)$ is the flow velocity. The charge current 4-vector is simply $\widehat{J} = q\widehat{N}$, so charge conservation is equivalent to particle number conservation.

Maxwell's theory retains the same mathematical form across all inertial reference frames. Equations (2.133) and (2.134) are expressed in tensor form, making them manifestly covariant under Lorentz transformations while preserving their structural invariance. Furthermore, the electromagnetic field tensor $F^{\mu\nu}$ constitutes a second-rank antisymmetric tensor, with the electric and magnetic fields representing merely its components. Under Lorentz transformations, these fields do not transform as simple vectors; rather, their transformation properties must be derived from $F^{\mu\nu}$ itself. Consequently, the precise transformation rules for the electric and magnetic fields emerge naturally from the transformation law of $F^{\mu\nu}$

$$F^{\mu'\nu'} = \Lambda^{\mu'}{}_{\mu}\Lambda^{\nu'}{}_{\nu}F^{\mu\nu}. \tag{2.146}$$

Crucially, these transformation rules reveal the fundamental unity of electricity and magnetism—they are not independent phenomena but can interconvert between reference frames. What appears as a purely electric effect in one frame may manifest as a magnetic effect in another. This profound insight into electromagnetic relativity was not fully realized until Einstein's formulation of special relativity.

We begin by examining the electromagnetic field of a charged particle in uniform motion. Consider a reference frame S in which the particle moves with velocity v along the x-axis. In the particle's rest frame S', the electromagnetic field is purely electrostatic

$$\boldsymbol{E}' = \frac{q}{4\pi\epsilon_0 r'^3}\boldsymbol{r}', \quad \boldsymbol{B}' = \boldsymbol{0}, \tag{2.147}$$

where the electric field is radially symmetric and isotropic. Transforming to the lab frame S via Lorentz transformation yields

$$\begin{aligned} \boldsymbol{B} &= \frac{1}{c}\boldsymbol{v} \times \boldsymbol{E}, \\ \boldsymbol{E} &= \frac{q\boldsymbol{r}}{4\pi\epsilon_0\gamma^2 r^3\left(1-\left(\frac{v}{c}\right)^2\sin^2\theta\right)^{3/2}}, \end{aligned} \tag{2.148}$$

where $\boldsymbol{v} = (v, 0, 0)$, γ is the corresponding Lorentz factor, and θ is the angle between the position vector with the x-axis. In frame S, there is not only a magnetic field, but the electric field is no longer uniformly isotropic, with its magnitude depends on θ. At $\theta = \pi/2$, the electric field is the strongest, $E_{\text{max}} = \gamma q/(4\pi\epsilon_0 r^2)$, and at $\theta = 0$, the electric field is the weakest, $E_{\text{min}} = q/(4\pi\epsilon_0\gamma^2 r^2)$. The electric field lines, illustrated in Figure 2.3, exhibit Lorentz contraction along the direction of motion. Although the total number of field lines remains unchanged, as required by Gauss's law, their distribution is compressed in the x-direction due to relativistic effects. This compression reflects the loss of isotropy and the directional dependence of the field strength in the lab frame.

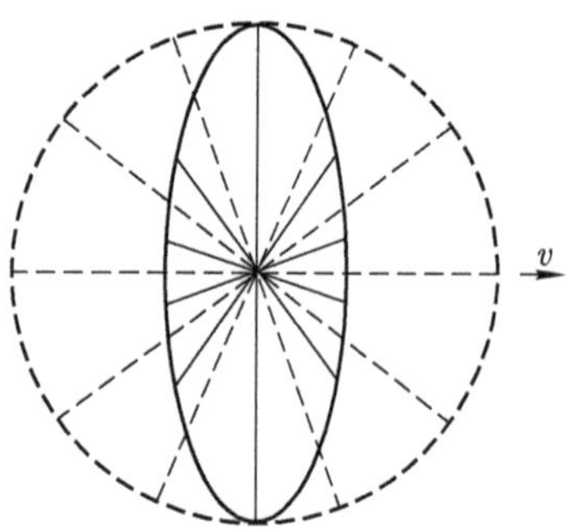

Fig. 2.3 Distribution of electric field lines in different reference frames. The dashed lines represent the spherically symmetric electric field lines in S', and the solid lines show the electric field lines in the reference frame S.

2.6.2 *Electromagnetic radiation*

Finally, we briefly discuss electromagnetic radiation, which exhibits important similarities to the gravitational radiation that will be introduced later. Starting from Maxwell's Eqs. (2.133), we derive

$$\partial_\mu(\partial^\nu A^\mu - \partial^\mu A^\nu) = 4\pi J^\nu. \tag{2.149}$$

The $U(1)$ gauge field admits a gauge transformation $A^\mu \to A^\mu + \partial^\mu \psi$, under which the field strength remains invariant. This gauge freedom allows us to impose convenient conditions. For analyzing radiation, the Lorenz gauge (a covariant condition) is particularly useful:

$$\partial_\mu A^\mu = 0. \tag{2.150}$$

In this gauge, the equation for the gauge potential becomes

$$\Box A^\mu = -4\pi J^\mu, \tag{2.151}$$

where "$\Box$" is the d'Alembert operator in flat spacetime, $\Box = \partial_\mu \partial^\mu$. In vacuum, this matches the wave equation satisfied by the electric and magnetic fields (2.127).

Consider electromagnetic waves in vacuum, where $J^\mu = 0$. The solution is a plane wave:

$$A^\mu = C^\mu e^{i\widehat{k}\cdot\widehat{x}} = C^\mu e^{ik_\nu x^\nu}, \tag{2.152}$$

where C^μ is a constant amplitude vector, and the wave 4-vector $\widehat{k}$ is null, $\widehat{k} \cdot \widehat{k} = 0$, indicating propagation at the speed of light. The Lorenz gauge implies

$$\widehat{C} \cdot \widehat{k} = 0, \tag{2.153}$$

so the amplitude is transverse to the wave vector, leaving three independent components. For a wave propagating along the x^3-direction,

$$k^\mu = (k, 0, 0, k), \tag{2.154}$$

the amplitude vector can be written as $C^\mu = (C^0, C^1, C^2, C^0)$, where C^1 and C^2 are arbitrary. Moreover, the Lorenz gauge does not fully fix the gauge; any transformation $A_\mu \to A_\mu + \partial\lambda$ with $\Box\lambda = 0$ preserves it. Choosing

$$\lambda = \varepsilon e^{i\widehat{k}\cdot\widehat{x}}, \tag{2.155}$$

the amplitude vector change as

$$C^\mu \to C'^\mu = C^\mu + i\varepsilon k^\mu. \tag{2.156}$$

For the wave along x^3-direction, this gives

$$C'^0 = C^0 + i\varepsilon k, \quad C'^1 = C^1, \quad C'^2 = C^2. \tag{2.157}$$

By choosing $\varepsilon = iC^0/k$, we can set $C'^0 = 0$, making the amplitude purely spatial and transverse. Thus, only two physical degrees of freedom remain. We can define linear polarization vectors

$$\widehat{e}_1 = (0, 1, 0, 0),$$
$$\widehat{e}_2 = (0, 0, 1, 0),$$

and the general amplitude vector is

$$C^\mu = c_1 e_1^\mu + c_2 e_2^\mu, \tag{2.158}$$

where c_1 and c_2 are complex constants representing the two independent polarization states.

To analyze the physical effects of electromagnetic waves, consider a positively charged test particle in vacuum. Its motion under the influence of the wave is determined by the Lorentz force in covariant form. As the wave passes, the force induces oscillatory motion whose nature depends on the wave's polarization: If $c_2 = 0$, the wave is linearly polarized along x^1, and the particle oscillates linearly in the x^1-direction; If $c_1 = 0$, the wave is linearly polarized along x^2, and the particle oscillates linearly in the x^2-direction; If $c_2 = \pm i c_1$, the wave is circularly polarized, and the particle undergoes circular motion in response. This demonstrates how the polarization state of the electromagnetic wave directly determines the trajectory of the test charge.

To determine what kinds of sources can produce electromagnetic radiation, we analyze the inhomogeneous electromagnetic wave equation. The Green's function method and multipole expansion provide powerful tools for solving this problem—an approach that will later prove useful in the study of gravitational radiation. For now, we highlight several key conclusions:

(1) Due to charge conservation, the monopole moment will not produce electromagnetic radiation;
(2) The leading-order electromagnetic radiation comes from the electric dipole moment;
(3) The next-to-leading-order electromagnetic radiation comes from the magnetic dipole moment, but since the magnetic field can be regarded as a relativistic effect, it is relatively suppressed by a factor of v/c compared to the electric dipole moment.

Appendix 2.1 A Brief History of Special Relativity

Special relativity was proposed by A. Einstein in 1905 and has now become a cornerstone of modern physics. Special relativity originated from the study of electromagnetism in the 19th century. It can be said that the unification of electricity, magnetism, and optics by Maxwell's theory led to the birth of special relativity. Lorentz's research on electrodynamics since 1890 laid the foundation for special relativity, and Henri Poincaré and others also made important contributions, but it was ultimately Einstein who successfully established special relativity.

The seminal paper published by Einstein in 1905 was entitled "On the Electrodynamics of Moving Bodies". When Einstein recognized that Maxwell's equations lacked relativistic invariance, he was confronted with multiple theoretical alternatives:

(1) Maxwell's theory is fundamentally incompatible with Galilean relativity because it fails to remain invariant under Galilean transformations. If one insists that the correct electromagnetic theory must satisfy Galilean invariance, then Galilean transformations would hold an absolute and privileged status in physics.
(2) The Galilean principle of relativity only applies to classical mechanics, and Maxwell's theory has a special reference frame in which the ether is at rest.
(3) There exists a principle of relativity that applies to both classical mechanics and electromagnetic theory, but it is obviously not the Galilean principle of relativity. If so, in the new theory of relativity, Maxwell's theory remains invariant in different reference frames. Of course, this means that the original mechanical system needs to be appropriately modified.

The first possibility is hard to accept because Maxwell's theory has achieved unprecedented success. With the widespread application of electromagnetic waves, it is difficult to imagine that Maxwell's theory is wrong. Therefore, the second view was popular at the end of the 19th century and the beginning of the 20th century, and the existence of the ether was widely accepted. However, the Michelson–Morley experiment confirmed that the speed of light is constant, which posed a challenge to the ether theory. George F. FitzGerald and Hendrik A. Lorentz made a proposal: objects moving in the ether would contract in the direction of motion, that is

$$L(v) = L_0\left(1 - \frac{v^2}{2c^2}\right). \tag{2.159}$$

This can explain the Michelson–Morley experiment.[7] This proposal seems not to be in Maxwell's theory. However, H. Lorentz and H. Poincaré proved that Maxwell's theory is invariant under the Lorentz transformation,[8] and

[7]This relationship is the first-order expansion of the length contraction effect in special relativity.

[8]In fact, the Lorentz transformation was first proposed by Woldemar Voigt in 1887.

the above scale contraction holds for moving charge density. Thus, Lorentz argued that length contraction was fundamentally an electrodynamic phenomenon. With the discovery of the electron, he further reasoned that since all matter is ultimately electromagnetic in nature, this contraction must also apply to macroscopic objects—an idea that appeared to salvage the aether theory. However, other experimental findings continued to challenge the aether framework. Among these was the famous experiment by Hippolyte Fizeau (1851, 1853), which measured the speed of light in moving fluids. Albert A. Michelson and Edward W. Morley later conducted analogous experiments in 1886. The results suggested that if the aether theory were correct, the aether would have to be partially dragged by the moving fluid, with the degree of drag depending on the fluid's refractive index, and the net light speed is

$$c' = \frac{c}{n} + v\left(1 - \frac{1}{n^2}\right), \tag{2.160}$$

where n is the refractive index of the liquid (assuming the liquid is non-dispersive), and v is the speed of the fluid. This formula was first predicted by Augustin-Jean Fresnel, and the factor $1 - 1/n^2$ is called the Fresnel drag coefficient. Fresnel postulated that when light propagates through the aether, it induces elastic vibrations in the medium and is consequently partially dragged along with the aether. Another crucial experimental phenomenon was the stellar aberration of light, which we have previously examined. To account for this effect, Fresnel introduced in 1818 the concept of an absolutely stationary aether. It became increasingly evident that prior to Einstein's formulation of special relativity, the aether theory framework provided unsatisfactory explanations for these experimental observations.

In formulating special relativity, Einstein was primarily motivated by optical phenomena—particularly the aberration of light and the Fizeau experiment.[9] Among the possible theoretical approaches, he regarded the third option (retaining Maxwell's equations while modifying classical kinematics) as the fundamental solution to the electrodynamic contradictions. Einstein maintained an unwavering conviction in the correctness of Maxwell's theory, and this belief became the cornerstone for his development of special relativity. Historical evidence suggests that while he was aware of the Michelson-Morley experiment, it did not serve as the primary impetus for his revolutionary theory. His familiarity with Lorentz's pre-1895 work was substantial, and the 1895 Lorentz paper particularly influenced

[9]For the historical context of Einstein's work on special relativity, see [28].

his thinking. However, Einstein had not yet encountered the full Lorentz transformation at this stage. Remarkably, he independently derived what we now call the Lorentz transformation during his construction of special relativity.

In 1921, special relativity had not yet been widely accepted. When Einstein visited Princeton, there was a rumor that someone had discovered a non-zero drift of the ether. Einstein commented, "God is subtle, but he is not malicious."

Exercises

1. Calculate in 4 dimensions

$$\varepsilon_{\alpha\beta\gamma\delta}\eta^{\alpha\mu}\eta^{\beta\nu}\eta^{\gamma\rho}\eta^{\delta\rho}\eta^{\delta\lambda}\varepsilon_{\mu\nu\rho\lambda}. \tag{2.161}$$

2. Suppose in Minkowski spacetime, a second-order tensor $X^{\mu\nu}$ and a first-order vector V^{μ} have components respectively

$$X^{\mu\nu} = \begin{pmatrix} 1 & -1 & 0 & 0 \\ -1 & 0 & 3 & 3 \\ -2 & 0 & 0 & 0 \\ 0 & 1 & 0 & 2 \end{pmatrix}, \qquad V^{\mu} = (0, 1, 1, -2),$$

find the components of the following tensors:

(1) $X^{\mu}{}_{\nu}$;
(2) $X_{\mu}{}^{\nu}$;
(3) $X^{(\mu\nu)}$;
(4) $X_{[\mu\nu]}$;
(5) $X^{\lambda}{}_{\lambda}$;
(6) $V^{\mu}V_{\mu}$;
(7) $V_{\mu}X^{\mu\nu}$.

3. Prove that any second-order tensor can always be written as the sum of its symmetrized part and antisymmetrized part, i.e., $A_{\mu\nu} = A_{(\mu\nu)} + A_{[\mu\nu]}$. If $B^{\mu\nu} = B^{\nu\mu}$ and $C^{\mu\nu} = -C^{\nu\mu}$, prove that $A_{\mu\nu}B^{\mu\nu} = A_{(\mu\nu)}B^{\mu\nu}$ and $A_{\mu\nu}C^{\mu\nu} = A_{[\mu\nu]}C^{\mu\nu}$.
4. If $t_{\mu\nu}$ are the components of a symmetric tensor and v_{μ} are the components of a vector, prove that if

$$v_{\mu}t_{\nu\rho} + v_{\rho}t_{\mu\nu} + v_{\nu}t_{\rho\mu} = 0, \tag{2.162}$$

then either $t_{\mu\nu} = 0$ or $v_{\mu} = 0$.

5. In astronomical observations, the brightness of a celestial body is determined by the radiation flux T^{0i} observed by an observer on Earth. Here we calculate how this flux depends on the relative motion between the celestial body and Earth.

   (1) In the rest inertial frame S of a star with fixed brightness L (total energy radiated per second), assume the star is at the origin, prove that at the event point $(t, x, 0, 0)$, the energy-momentum tensor of the star's radiation has components $T^{00} = T^{0x} = T^{x0} = T^{xx} = L/(4\pi x^2)$.
   (2) Let $\widehat{X}$ be the null vector that separates the emission and reception of radiation. Prove that for the radiation observed at the event point $(x, x, 0, 0)$, $\widehat{X} = (x, x, 0, 0)|_S$. Prove that the energy-momentum tensor in (1) has the reference frame-independent form

$$\widehat{T} = \frac{L}{4\pi} \frac{\widehat{X} \otimes \widehat{X}}{(\widehat{U}_{\mathrm{s}} \cdot \widehat{X})^4}, \tag{2.163}$$

   where $\widehat{U}_{\mathrm{s}}$ is the star's 4-velocity.
   (3) Let the observer O on Earth move away from the star along the x-direction at velocity v, measuring the same radiation. Let $\widehat{X} = (R, R, 0, 0)$, find the functional dependence of R on x, and calculate the energy flux observed by this observer.

6. Use the conservation of energy-momentum equation to prove that for a system of finite size (i.e., the energy-momentum tensor is non-zero only in a finite region of space):

   (1) In a certain inertial frame, $\partial_t \int T^{0\nu} \mathrm{d}^3x = 0$. This indicates the conservation of energy and momentum.
   (2) $\partial_t^2 \int T^{00} x^i x^j \mathrm{d}^3x = 2 \int T^{ij} \mathrm{d}^3x$.

7. Suppose two reference frames S and S' differ by a constant velocity along the x-direction, use the Lorentz transformation to give the relationship between the electromagnetic fields in the two reference frames.
8. If we do not assume that two reference frames S and S' are connected by a constant velocity along the x-direction, but consider S' to differ from S by a constant velocity $\boldsymbol{v}$ in an arbitrary direction, prove that the relationship between the electromagnetic fields in the two reference

frames is

$$\begin{aligned} \boldsymbol{E}' &= \gamma(\boldsymbol{E} + \boldsymbol{v} \times \boldsymbol{B}) + \frac{1-\gamma}{v^2}(\boldsymbol{v} \cdot \boldsymbol{E})\boldsymbol{v}, \\ \boldsymbol{B}' &= \gamma\left(\boldsymbol{B} - \frac{1}{c^2}\boldsymbol{v} \times \boldsymbol{E}\right) + \frac{1-\gamma}{v^2}(\boldsymbol{v} \cdot \boldsymbol{B})\boldsymbol{v}. \end{aligned} \tag{2.164}$$

9. Prove that $c^2\boldsymbol{B}^2 - \boldsymbol{E}^2$ and $\boldsymbol{E} \cdot \boldsymbol{B}$ are Lorentz invariants.
10. In a certain reference frame, prove that the 3-acceleration of a charged particle in an electromagnetic field is

$$\boldsymbol{a} = \frac{\mathrm{d}\boldsymbol{u}}{\mathrm{d}t} = \frac{q}{\gamma m_0}\left(\boldsymbol{E} + \boldsymbol{u} \times \boldsymbol{B} - \frac{1}{c^2}(\boldsymbol{u} \cdot \boldsymbol{E})\boldsymbol{u}\right). \tag{2.165}$$

11. Consider the Maxwell theory and scalar field theory of the electromagnetic field in a vacuum, their energy-momentum tensors are respectively

$$\begin{aligned} T^{\mu\nu}_{\text{EM}} &= F^{\mu\lambda}F^{\nu}{}_{\lambda} - \frac{1}{4}\eta^{\mu\nu}F^{\lambda\sigma}F_{\lambda\sigma}, \\ T^{\mu\nu}_{\text{scalar}} &= \eta^{\mu\lambda}\eta^{\nu\sigma}\partial_\lambda\phi\partial_\sigma\phi - \eta^{\mu\nu}\left(\frac{1}{2}\eta^{\lambda\sigma}\partial_\lambda\phi\partial_\sigma\phi + V(\phi)\right). \end{aligned}$$

   (1) Using divergence, gradient, curl in three-dimensional vector analysis as well as electric and magnetic fields, re-express the components of the energy-momentum tensor in each theory.
   (2) Using the equations of motion, prove the conservation of the energy-momentum tensor.
   (3) Prove that the energy-momentum tensor of the (four-dimensional) electromagnetic field is traceless.

Chapter 3

Gravity and Geometry

To date, we have identified four fundamental interactions in nature: the electromagnetic, strong, weak, and gravitational forces. Among these, gravity appears deceptively simple yet remains the most enigmatic. Let us first revisit the established properties of gravity. Firstly, gravity acts universally between all forms of energy, earning it the designation of universal gravitation. Secondly, unlike other forces, gravity cannot be shielded and is exclusively attractive. This arises because gravitational "charge" (i.e., mass-energy) is always positive, precluding the existence of negative gravitational charges. Thirdly, gravity, like electromagnetism, obeys an inverse-square law, diminishing with the square of the distance. However, while electromagnetic charges can be positive or negative—leading to attraction or repulsion—gravitational interactions are invariably attractive due to the positivity of mass-energy. From the perspective of modern quantum field theory, gravity is mediated by a massless spin-2 particle, the graviton. Yet, unlike the other three forces—which are successfully described by quantum gauge field theory—gravity resists such a framework. The quantization of gravity remains one of the most profound unsolved puzzles in theoretical physics.[1] Finally, gravity is by far the weakest of the four fundamental forces, a fact that further compounds its theoretical peculiarity. For example, in a proton-proton system, gravity is 36 orders of magnitude weaker than the electromagnetic force:

$$\frac{F_{\mathrm{g}}}{F_{\mathrm{e}}} = \frac{Gm_{\mathrm{p}}^2/r^2}{e^2/(4\pi\epsilon_0 r^2)} \sim 10^{-36}.$$

[1]The most promising candidate for a theory of quantum gravity is superstring theory, which has achieved significant milestones but still grapples with deep conceptual challenges.

Despite being the weakest of the four forces, gravity may hold unparalleled cosmic significance. It governs the universe's large-scale evolution, from the formation of galaxies and galaxy clusters to the dynamics of cosmic expansion. On smaller scales, it dictates stellar evolution, planetary formation, and orbital mechanics—processes that shape the very fabric of habitable worlds. This stark contrast between its feeble local strength and its overarching cosmic influence underscores gravity's unique role as the architect of the universe.

In the study of physics, gravity was the first interaction to be focused on. As early as the 16th century, the Italian scientist Galileo refuted the previous erroneous notion, pointing out that the motion of objects under the influence of Earth's gravity is independent of their mass. In modern terms, this means that gravitational mass is equal to inertial mass. On the other hand, Tycho Brahe meticulously compiled an extensive dataset on the motion of celestial bodies within the solar system through decades of precise astronomical observations. Based on this wealth of empirical evidence, Johannes Kepler subsequently derived and formulated his three fundamental laws of planetary motion. Nevertheless, despite these groundbreaking discoveries, the underlying physical principles governing these laws remained elusive to contemporary scholars. In the 17th century, the eminent British physicist Isaac Newton formulated the law of universal gravitation, asserting that the force governing planetary motion is fundamentally the same as Earth's gravity—both arising from the mutual attraction between masses. Newton's universal gravitation not only provided a rigorous physical explanation for the motion of planets within the solar system but also offered a compelling account of various phenomena in the Earth-Moon system, such as tidal forces. Moreover, his law was seamlessly integrated into his broader mechanical framework. Through subsequent refinements by mathematicians and physicists—including Laplace, Poisson, and Hamilton—Newtonian mechanics grew increasingly rigorous and sophisticated. By the 19th century, aided by advances in perturbation theory, Newtonian gravity achieved remarkable precision in describing and predicting planetary motion, leaving little room for doubt about its completeness. However, in the early 20th century, with the advent of Einstein's special relativity, Newtonian gravity encountered a profound challenge, ultimately necessitating a fundamental revision of gravitational theory.

In Newtonian gravity, the gravitational potential generated by a given matter distribution can be described by the Poisson equation:

$$\nabla^2 \Phi = 4\pi G_\rho,$$

where Φ is the gravitational potential, G is the Newtonian gravitational constant, and ρ is the mass density distribution. It can be seen that once the matter distribution of the system is given, the gravitational potential of the system is determined. If an object with mass m interacts with this system through universal gravitation, according to Newton's second law, the interaction is $\boldsymbol{F} = m\boldsymbol{g}$, where $\boldsymbol{g}$ is determined by the gradient of the gravitational potential:

$$\boldsymbol{g} = -\nabla\Phi,$$

that is to say, the gravitational acceleration is completely determined by the gravitational potential of the original system and is independent of the mass of the object itself.

In Newtonian gravity, the gravitational interaction arises solely from mass and appears to have no intrinsic connection with spacetime geometry. However, this framework is fundamentally incompatible with the principle of relativity. A key manifestation of this incompatibility is the gravitational redshift effect: when a photon propagates through a gravitational potential Φ, it must expend energy to overcome the potential, resulting in a measurable shift in its wavelength. This phenomenon has profound implications—it demonstrates that spacetime cannot be described by a flat geometry in the presence of gravity. Consequently, within a relativistic theory of gravity, the curvature of spacetime becomes an inescapable necessity.

This chapter begins with a review of the fundamental principles of Newtonian gravity, including the crucial observation that inertial mass and gravitational mass are experimentally indistinguishable. We will then examine precision experimental tests of the equivalence principle that validate this foundational concept. Next, we analyze the gravitational redshift effect through carefully constructed thought experiments, demonstrating its measurable consequences. A key application of this phenomenon—its critical role in the Global Positioning System (GPS)—will be discussed in detail, highlighting how relativistic gravity corrections are essential for the system's accuracy. Finally, through rigorous examination of gravitational redshift, we will establish that this effect necessarily requires a curved spacetime geometry, providing a compelling bridge from Newtonian gravity to Einstein's general theory of relativity.

3.1 Incompatibility of Newtonian Gravity with the Principle of Relativity

By the late 19th century, Newtonian gravity had reached an advanced stage of maturity after two centuries of development. Through the application of perturbation theory, it provided remarkably accurate descriptions of planetary motions within the solar system. The crowning achievement came with the successful prediction of Neptune's existence using Newtonian theory, which reinforced the prevailing belief that it represented a complete and ultimate description of gravitational phenomena in the universe. This confidence persisted despite a small but persistent anomaly in Mercury's orbit. Careful observations spanning over a century had revealed a minute discrepancy[2] between the observed precession of Mercury's perihelion and Newtonian predictions. The scientific community initially attributed this deviation—amounting to just about 10^{-7} of the total effect—to possible undiscovered asteroids or other celestial bodies rather than considering any fundamental limitations in Newtonian gravity. The deviation was so extraordinarily small that it failed to raise significant concerns about the theory's validity.

In his seminal 1687 work "Philosophiæ Naturalis Principia Mathematica", Isaac Newton first formulated the law of universal gravitation. This fundamental principle states that all massive bodies attract each other through a force that: (1) acts along the line connecting their centers, (2) is directly proportional to the product of their masses, and (3) decreases with the square of the distance between them. If two objects are approximated as point masses with masses M and m respectively, then the gravitational force between them is

$$\boldsymbol{F} = -\frac{GMm}{r^2}\boldsymbol{e}_r, \tag{3.1}$$

where G is the Newtonian gravitational constant, and $\boldsymbol{e}_r$ represents the unit vector connecting these two point masses. Using Newton's second law $\boldsymbol{F} = m\boldsymbol{g}$, we can introduce a gravitational potential

$$\Phi = -\frac{GM}{r}. \tag{3.2}$$

This is the gravitational potential of the object M, and its gradient gives the gravitational acceleration

$$\boldsymbol{g} = -\nabla\Phi = -\frac{GM}{r^2}\boldsymbol{e}_r. \tag{3.3}$$

[2]The theoretical calculation of Mercury's perihelion precession was first completed by Urbain Le Verrier in 1845 and later refined by Simon Newcomb in 1882.

In fact, the forms of $\boldsymbol{F}$ and Φ given above are only correct when the mass distribution is spherically symmetric. The previous discussion assumes that the objects are point masses. More generally, we can consider the gravitational potential of N point particles, which is the linear superposition of the gravitational potentials of single point particles:

$$\Phi(\boldsymbol{x}) = -G\sum_{i=1}^{N}\frac{m_i}{|\boldsymbol{x}-\boldsymbol{x}_i|}. \tag{3.4}$$

If it is a continuous distribution, we can turn the summation into an integral, thus obtaining

$$\Phi(\boldsymbol{x}) = -G\int\frac{\rho(\boldsymbol{x}')}{|\boldsymbol{x}-\boldsymbol{x}'|}\mathrm{d}x'. \tag{3.5}$$

From this, we can derive the Poisson equation that the gravitational potential satisfies

$$\nabla^2\Phi(\boldsymbol{x}) = 4\pi G\rho(\boldsymbol{x}). \tag{3.6}$$

This equation can be considered the core of Newtonian gravitational theory, where Φ is the gravitational potential, G is the Newtonian gravitational constant, and ρ is the mass density that characterizes the distribution of matter. Equation (3.6) indicates that once the distribution of matter is known, the gravitational potential caused by these matters is determined. Simply put, ρ is the source that forms the gravitational potential. If we consider a particle with mass m in the gravitational potential, then the force experienced by this particle is

$$\boldsymbol{f} = m\boldsymbol{g}, \tag{3.7}$$

where the gravitational acceleration is entirely determined by the gradient of the gravitational potential:

$$\boldsymbol{g} = -\nabla\Phi. \tag{3.8}$$

In fact, the discussion here involves the (weak) equivalence principle that inertial mass equals gravitational mass. Therefore, all objects in a certain gravitational potential, regardless of their material or mass, have the same gravitational acceleration.

The Poisson equation is a second-order linear differential equation, which can be solved using the principle of linear superposition. It can be compared with the electrostatic potential equation in Maxwell's equations, the only difference being that mass density is always positive, while charge density can be positive or negative. To solve this equation, additional boundary conditions for the gravitational potential are required. Usually, we require the gravitational potential to be zero at infinity. By Gauss's theorem,

$$\int_S \boldsymbol{g} \cdot \mathrm{d}\boldsymbol{S} = -\int_V \nabla \cdot (\nabla \Phi) = -\int_V 4\pi G\rho = -4\pi GM. \tag{3.9}$$

For a spherically symmetric mass distribution,

$$\boldsymbol{g} = -\frac{GM}{r^2}\boldsymbol{e}_r. \tag{3.10}$$

If the mass distribution is not spherically symmetric, we can handle it through multipole expansion when the distance is relatively far. Assuming the position of the source is represented by $\boldsymbol{x}'$, we can expand in Eq. (3.5) as follows:

$$\frac{1}{|\boldsymbol{x} - \boldsymbol{x}'|} = \frac{1}{r} + \sum_i \frac{x^i x'^i}{r^3} + \frac{1}{2}\sum_{i,j}(3x'^i x'^j - r'^2\delta^{ij})\frac{x^i x^j}{r^5} + \cdots \tag{3.11}$$

and obtain the Newtonian gravitational potential order by order:

$$\Phi(\boldsymbol{x}) = -\frac{GM}{r} - \frac{G}{r^3}\sum_i x^i D^i - \frac{G}{2r^5}\sum_{i,j} x^i x^j Q^{ij} + \cdots, \tag{3.12}$$

where

$$\begin{aligned} M &= \int \rho(\boldsymbol{x}')\mathrm{d}^3x', \\ D^i &= \int x'^i \rho(\boldsymbol{x}')\mathrm{d}^3x', \\ Q^{ij} &= \int (3x'^i x'^j - r'^2\delta^{ij})\rho(\boldsymbol{x}')\mathrm{d}^3x' \end{aligned}$$

are the total mass (monopole moment), dipole moment, and quadrupole moment of the mass distribution source, respectively. When the coordinate origin is chosen at the center of mass, the dipole moment vanishes. Consequently, the leading-order correction to the inverse-square law arises from the quadrupole moment, introducing a $1/r^4$ perturbation. This quadrupole correction has an important physical consequence: it disrupts the closed nature of elliptical orbits, leading to orbital precession. For instance, Earth's slightly oblate shape, with a polar radius being about 0.3% smaller than its equatorial radius, generates a quadrupole moment sufficient to induce measurable precession in satellite orbits. Remarkably, such precession serves as a sensitive probe of Earth's mass distribution. The Sun, though nearly spherical, with polar-equatorial radius differences of order 10^{-4}, also exhibits a small but non-zero quadrupole moment. While its gravitational influence on Mercury's orbit is minor (contributing a few arcseconds/century to precession), the Newtonian perturbations from other planets dominate, producing about 500 arcseconds/century of perihelion advance. However, after accounting for both planetary perturbations and solar oblateness, observations reveal an unresolved anomalous precession of about 43 arcseconds/century in Mercury's orbit. This discrepancy, though small, became a pivotal empirical test—later exquisitely explained by general relativity's curved spacetime description of gravity.

Since the Poisson equation is linear, the gravitational potential in Newtonian gravity admits a straightforward linear superposition. However, from a modern perspective, gravity is mediated by gravitons, and the gravitational field itself possesses energy. According to the relativistic mass-energy equivalence $E = mc^2$, mass and energy are fundamentally inseparable. Consequently, the gravitational field carries an effective "mass" and must itself generate a secondary gravitational field. This self-interaction implies that gravity cannot be fully described by a linear theory. Instead, a nonlinear formalism is required—a criterion the Poisson equation clearly fails to satisfy.

The incompatibility of Newtonian gravity with special relativity manifests in several key ways. First, Newtonian gravity assumes instantaneous propagation of gravitational effects. For instance, if the Sun's mass were to change, Earth would experience an immediate alteration in its gravitational field—a scenario that violates the fundamental relativistic constraint that no information can propagate faster than the speed of light. Such instantaneous action-at-a-distance inherently undermines causality. Second, the Newtonian gravitational equation lacks relativistic covariance. The Poisson equation, which governs Newtonian gravity, contains only spatial derivatives on its left-hand side, with no time dependence, while its source term is merely the mass density ρ—a non-tensorial quantity that represents just one component of the full energy-momentum tensor. Consequently, the equation fails to satisfy the requirements of Lorentz covariance. This structural limitation means that no straightforward modification of the Poisson equation can yield a relativistic theory of gravity.

For most people, the incompatibility of Newtonian gravity with special relativity might seem inconsequential. After all, Newtonian gravity has been rigorously tested by countless experiments, and no compelling empirical evidence demands its revision. Yet for Einstein, this inconsistency represented a profound crisis—a fracture in the fundamental harmony of physics. He was driven by a deep conviction that the universe ought to be governed by simple, elegant, and unified laws. It was this unwavering pursuit of beauty and coherence that ultimately led him to seek a relativistic theory of gravity, one that would reconcile the apparent contradictions and unveil a deeper, more harmonious description of nature. In 1907, while preparing a review article on special relativity for *Annalen der Physik*, Einstein encountered a fundamental problem: the irreconcilable conflict between Newtonian gravity and special relativity. This realization marked the beginning of an eight-year intellectual journey, one characterized by profound challenges and conceptual breakthroughs. By November 1915, his persistent efforts culminated in the formulation of general relativity, a revolutionary relativistic theory of gravity. Beyond its groundbreaking physical insights, Einstein's achievement represented a paradigm shift in scientific methodology. He demonstrated that theoretical self-consistency—not merely empirical adequacy—must serve as a foundational criterion in physical theories. Today, this principle has become indispensable in modern theoretical physics, guiding the development of new frameworks and the evaluation of their validity.

3.2 Weak Equivalence Principle

The weak equivalence principle (WEP) holds fundamental significance in the framework of Newtonian gravity. It can be stated as follows:

Weak Equivalence Principle The inertial mass is equal to the gravitational mass, i.e., $m_{\rm i} = m_{\rm g}$.

Inertial mass refers to the mass $m_{\rm i}$ in Newton's second law

$$\boldsymbol{f} = m_{\rm i}\boldsymbol{a}, \tag{3.13}$$

which characterizes the inertia exhibited by an object under any external force; the greater the mass, the greater the inertia. Gravitational mass, on the other hand, specifically refers to the mass involved in gravitational interactions when two objects attract each other. The gravitational acceleration derived earlier is related to the gradient of the gravitational potential, which actually utilizes the weak equivalence principle, because

$$m_{\rm i}\ddot{\boldsymbol{r}} = -m_{\rm g}\nabla\Phi, \tag{3.14}$$

where the left-hand side is Newton's second law and the right-hand side represents the gravitational force. Only when the inertial mass equals the gravitational mass do we have

$$g = -\nabla\Phi. \tag{3.15}$$

Equation (3.15) has significant physical implications, indicating that all matter responds to gravity in the same way: they all fall with the same acceleration in a given gravitational field. For instance, in the gravitational field of the Earth, regardless of the material composition, objects with the same initial conditions will have the same motion state.

The experimental verification of WEP has a long history, culminating in remarkably high-precision tests. One of the most iconic demonstrations dates back to Galileo's legendary experiment from the Leaning Tower of Pisa, where he dropped two iron balls of differing masses. Their identical motion conclusively refuted Aristotle's assertion that 'heavier objects fall faster,' marking a pivotal moment in the empirical validation of WEP.[3] Similar experiments can be performed using objects of different materials while minimizing air resistance, yielding consistent results. However, the precision of such tests remains limited. Galileo proposed a more refined method: using pendulums to test WEP. By comparing the oscillation periods of pendulums with bobs of different materials, he achieved a precision of approximately 10^{-3}. Following Galileo, scientists such as Isaac Newton, Friedrich Bessel, and Heinrich Potter further refined pendulum-based experiments, progressively enhancing the accuracy of WEP verification

In the late 19th century, Loránd Eötvös revolutionized the experimental verification of WEP by introducing the torsion balance technique, achieving an unprecedented precision of 10^{-9}. This breakthrough laid the foundation for further advancements in the mid-20th century. Building on Eötvös' work, Robert H. Dicke, Vladimir Braginsky, and their collaborators refined the methodology, pushing the precision to 10^{-11} by the 1960s. The quest for higher accuracy culminated in the MICROSCOPE satellite experiment, which by the end of 2017 had improved the precision to 10^{-14}. Looking ahead, the STEP (Satellite Test of the Equivalence Principle) program aims to extend these measurements to the 10^{-17} level by using advanced space-based techniques. Figure 3.1 summarizes the historical progression of WEP tests up to the early 21st century, illustrating the remarkable evolution in experimental precision.

[3]Historically, Galileo did not actually carry out experiments from the Leaning Tower of Pisa. Instead, he proposed a thought experiment: if Aristotle's view were correct—that heavier objects fall faster than lighter ones—what would happen if two iron balls of different masses were connected by a chain? Would the combined object fall faster, slower, or at some intermediate speed?

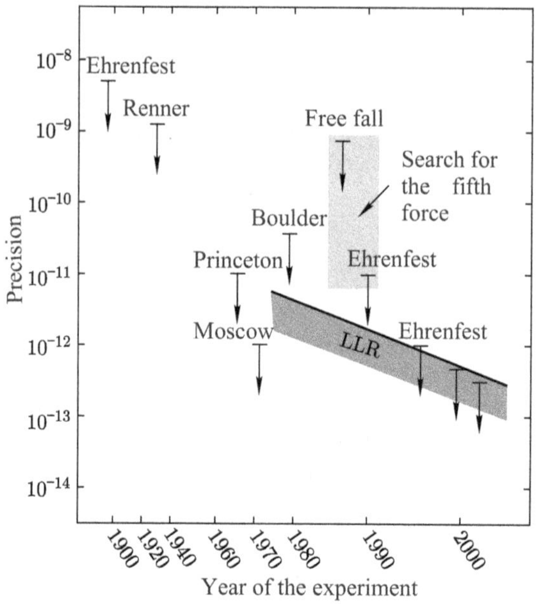

Fig. 3.1 Experimental tests of the weak equivalence principle.

The weak equivalence principle has undergone the most stringent experimental tests in the history of physics, achieving precision levels that even surpass those of electroweak standard model verifications. This remarkable accuracy has firmly established WEP as one of the most rigorously validated foundations of modern physics. We highlight two landmark experiments that demonstrate this extraordinary precision:

(1) Lunar laser ranging.

Modern tests of WEP can be conducted by analyzing the Earth-Moon system's dynamics within the Sun's gravitational field. Given the well-constrained solar gravitational potential, precise measurements of the Earth-Moon relative position enable WEP tests at the 10^{-13} level. Notably, both celestial bodies possess substantial self-gravitational energy, which may potentially violate the equivalence between inertial and gravitational mass. The crucial experimental requirement is determining their relative position with extreme accuracy. This necessitates lunar laser ranging (LLR) techniques capable of distance measurements with 10^{-10} precision. The principle is simple: spacecraft deploy arrays of retroreflectors, such as corner-cube mirrors, on the lunar surface. These mirrors efficiently reflect incident light back toward its source. Since 1969, research teams, including the McDonald Observatory in Texas and a French collaboration, have utilized these reflectors for the LLR experiments. By emitting short laser pulses (e.g., 200 ps duration with 10^{18} photons per pulse) at a rate of 10 Hz from Earth, they precisely measure the Earth-Moon distance. However, the process is remarkably inefficient due to atmospheric effects. Upon reaching the Moon, the laser beam diffracts into a 7 km radius spot, with only 10^{-10} of the photons striking the retroreflectors. The returning photons are further dispersed over a 20 km radius area on Earth, reducing the detected flux to 10^{-20} of the emitted pulse. Consequently, a 1 m aperture telescope typically captures just one reflected photon every few seconds—a testament to the extraordinary precision required for these measurements.

(2) Eötvös's torsion balance experiment.

The most renowned experimental test of the weak equivalence principle remains the torsion balance experiment developed by Eötvös in 1890. Remarkably, the experiment's exceptional design was so groundbreaking that it won Eötvös the prize money originally designated for a theoretical physics competition at the University of Göttingen. Below we outline the main idea of this pioneering experiment. Consider a torsion balance on Earth with two spheres of potentially different masses attached to its ends, initially in equilibrium. In addition to the gravitational force, the system experiences a centrifugal force due to the Earth's rotation. This centrifugal force can be decomposed into two components: a vertical component, opposing the gravitational force and a horizontal component, acting perpendicular to gravity,

$$\boldsymbol{a} = \boldsymbol{a}_z + \boldsymbol{a}_x. \tag{3.16}$$

The two weights can have different masses but are in balance, which requires the torques in different directions to be balanced. As shown in Figure 3.2, the possible rotation around the z-axis is

$$\tau = m_{A,\mathrm{i}} a_x l - m_{B,\mathrm{i}} a_x l', \tag{3.17}$$

and the balance condition around the x-axis requires

$$(m_{A,\mathrm{g}} g - m_{A,\mathrm{i}} a_z) l = (m_{B,\mathrm{g}} g - m_{B,\mathrm{i}} a_z) l'. \tag{3.18}$$

Equation (3.18) leads to

$$l' = \frac{m_{A,\mathrm{g}} g - m_{A,\mathrm{i}} a_z}{m_{B,\mathrm{g}} g - m_{B,\mathrm{i}} a_z} l. \tag{3.19}$$

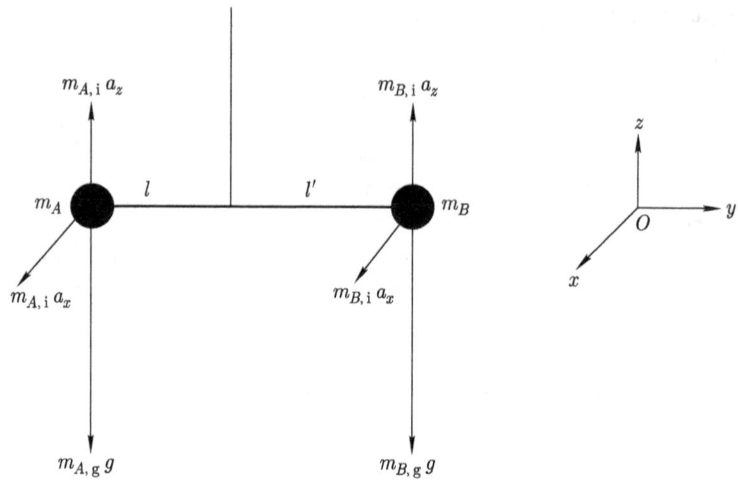

Fig. 3.2 Schematic diagram of the Eötvös experiment.

Therefore, the possible rotation can be expressed as

$$\tau = m_{A,\mathrm{i}} g l \left(\frac{\dfrac{m_{B,\mathrm{g}}}{m_{B,\mathrm{i}}} - \dfrac{m_{A,\mathrm{g}}}{m_{A,\mathrm{i}}}}{\dfrac{m_{B,\mathrm{g}}}{m_{B,\mathrm{i}}} g - a_z} \right). \tag{3.20}$$

Thus, the necessary and sufficient condition for no additional rotation is

$$\frac{m_{B,\mathrm{g}}}{m_{B,\mathrm{i}}} = \frac{m_{A,\mathrm{g}}}{m_{A,\mathrm{i}}}. \tag{3.21}$$

Only when the weak equivalence principle holds for both weights do we observe no torsion.

From the above discussion, it is evident that the Eötvös experiment, while conceptually simple, was brilliantly designed. Despite the technological limitations of the late 19th century, it achieved a remarkable precision of 10^{-9} in testing WEP. Modern iterations of this experiment have further refined the methodology, with key improvements including: replacing the two test masses with four precisely arranged mass blocks, minimizing the influence of Earth's gravitational potential gradient, mitigating magnetic and thermal disturbances, and periodically rotating the suspension arm to improve sensitivity. These advancements have pushed the experimental precision to an extraordinary level of 10^{-12}, demonstrating the enduring power of Eötvös' foundational idea.

3.3 Einstein's Equivalence Principle

Special relativity postulates the existence of inertial reference frames as a foundational assumption. However, this assumption is not strictly valid in our universe, where gravity universally influences particle motion, precluding the existence of perfectly free particles or ideal inertial frames. Nevertheless, in many physical scenarios, we can define approximately inertial frames with sufficient accuracy. The question of how to properly define an inertial frame in the presence of gravity was among the first relativistic problems Einstein investigated. The weak equivalence principle, expressed as the equality of inertial and gravitational masses $m_{\mathrm{i}} = m_{\mathrm{g}}$, implies that the acceleration of a freely falling particle in a gravitational field $\mathbf{a} = -\nabla\Phi$ is universal and independent of the particle's mass. Einstein's foundational insight was the realization that: "An observer in free fall does not perceive their own weight." This profound statement implies that the trajectory of a freely falling object defines an inertial reference frame. Importantly, Einstein's conception of free fall refers to an idealized scenario where air resistance is negligible and the observer has no visual reference to external surroundings, e.g., by closing their eyes. Under these conditions, at each point along the observer's worldline, we can construct a local inertial frame. The weak equivalence principle provides the mathematical foundation for this concept, expressed as:

$$\boldsymbol{a} = \boldsymbol{g}. \tag{3.22}$$

This equality demonstrates the complete observational equivalence between gravitational effects and acceleration. Crucially, it establishes that no local experiment can distinguish between being at rest in a gravitational field and undergoing uniform acceleration in flat spacetime. To operationalize this concept, we consider a thought experiment: an observer enclosed in a sufficiently small compartment. The compartment's dimensions are negligible compared to the scale of gravitational potential variation. Thus, the gradient of the gravitational potential is imperceptible to the compartment. Then consider the following two situations:

(1) The compartment is stationary on the surface of the Earth;
(2) The compartment is accelerating in a gravity-free space with an acceleration equal to the Earth's gravitational acceleration.

The observer inside the compartment cannot distinguish between these two situations by conducting any experiment. This leads to the profound question: if all physical phenomena are identical in these scenarios, in what meaningful sense does gravity "exist"?

In general relativity, while a global inertial frame (as in special relativity) cannot be defined, we can always establish a local inertial frame (LIF), a reference frame in which gravity locally vanishes. These LIFs correspond to the natural reference frames of freely falling objects. However, since the acceleration due to gravity cannot be uniquely defined in a general relativistic framework, the concept of gravitational acceleration itself becomes operationally ambiguous. In fact, within a sufficiently small spacetime region, gravity does not manifest as a force. Only when considering a larger region, by comparing the relative acceleration between neighboring LIFs, can the presence of gravity be detected. This observation reveals that gravity is fundamentally a tidal force. If two adjacent local inertial frames exhibit relative accelerated motion during free fall, this indicates the existence of gravity. In the language of general relativity, such relative acceleration signifies that spacetime is curved. Therefore, in short,

$$\text{Gravity} = \text{Spacetime Curvature}.$$

In fact, Einstein proposed that not only through free-fall experiments but also considering other interactions, it is impossible to distinguish between a and g. This is the so-called Einstein's Equivalence Principle (EEP).

Einstein's Equivalence Principle In a sufficiently small region, the laws of physics should be consistent with those in special relativity, and it is impossible to detect the existence of a gravitational field through any local experiment.

This principle tells us that no matter what experiment is conducted, it is impossible to distinguish between $\boldsymbol{a}$ and $\boldsymbol{g}$.

It is difficult to conceive of a theory that satisfies the weak equivalence principle while violating Einstein's equivalence principle. To illustrate this, consider a hydrogen atom consisting of a proton and an electron. The binding energy $E_{\mathrm{b}} < 0$ implies that the hydrogen atom's mass m_{H} is less than the sum of its constituent masses: $m_{\mathrm{H}} < m_{\mathrm{p}} + m_{\mathrm{e}}$. Because gravity couples universally to all forms of energy—including binding energy—the hydrogen atom's gravitational mass m_{g} remains consistent with the equivalence principle. This example demonstrates that gravitational interaction is not limited to rest mass alone, but rather couples to the full energy-momentum content of a system.

3.4 Gravitational Redshift

In Newtonian gravity, only objects with mass experience gravitational effects, leaving massless particles seemingly unaffected. However, special relativity's mass-energy equivalence ($E = mc^2$) reveals that energy and mass are fundamentally interchangeable. For photons—though massless—their energy $E = \hbar\omega$ implies they too carry an effective "mass" and should thus interact with gravity. While special relativity dictates that photons travel along straight paths, gravity bends spacetime, altering particle trajectories. This suggests that light should deflect in gravitational fields, a phenomenon that later became a cornerstone of general relativity. Here, however, we focus not on light deflection but on a simpler gravitational effect: the frequency shift of photons in a gravitational potential. Remarkably, this can be derived using only Einstein's equivalence principle or energy conservation, without invoking full general relativity.

(1) Energy conservation and gravitational redshift.

Below, we analyze the phenomenon of gravitational redshift using a thought experiment based on energy conservation. Consider an idealized scenario illustrated in Figure 3.3, where two points A and B are located at different heights in Earth's gravitational field, with A above B and a vertical separation h. For small values of h, the variation in gravitational acceleration can be neglected, allowing us to treat g as constant. The gravitational potential is lower at B than at A. Suppose a particle of mass m is released from rest at A and falls freely to B. Its energy at the two positions is given by

$$E_A = m, \quad E_B = m + mgh. \tag{3.23}$$

Now, assume that upon reaching B, the particle annihilates completely

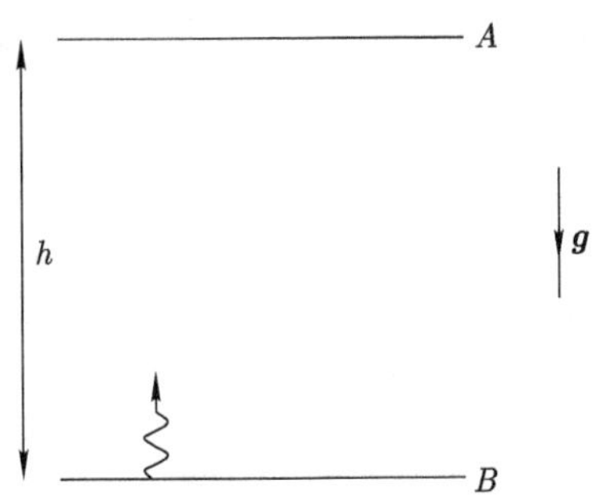

Fig. 3.3 Thought experiment for gravitational redshift.

and produces a single photon, which is then emitted upward toward A. The energy of this photon at emission is

$$\hbar\omega = E_B. \tag{3.24}$$

If the photon were unaffected by gravity during its propagation, then upon being reconverted into a particle at A, the resulting energy would be $m + mgh$, exceeding the original energy $E_A = m$ by mgh. Such an energy gain would violate conservation of energy. To avoid this inconsistency and uphold energy conservation, the photon must undergo a frequency shift as it moves through the gravitational field. Since

$$E_A = m = \hbar\omega_A, \tag{3.25}$$

$$E_B = m + mgh = \hbar\omega_B, \tag{3.26}$$

and assuming the Planck constant $\hbar$ remains unchanged, it follows that $\omega_B > \omega_A$. We define the redshift factor as

$$z \equiv \frac{\Delta\lambda}{\lambda_\mathrm{i}} = \frac{\lambda_\mathrm{f} - \lambda_\mathrm{i}}{\lambda_\mathrm{i}}, \tag{3.27}$$

where λ_i and λ_f denote the initial (emission) and final (reception) wavelengths of the photon, respectively. In this case

$$1 + z = \frac{\lambda_\mathrm{i} + \Delta\lambda}{\lambda_\mathrm{i}} = \frac{\lambda_A}{\lambda_B} = \frac{\omega_B}{\omega_A} = 1 + gh, \tag{3.28}$$

which implies $z = gh$. Strictly speaking, the gravitational acceleration is not a constant, $\boldsymbol{g} = -\nabla\Phi$. By considering the trajectory as a series of small segments within each of which g is approximately constant, the total redshift can be expressed as the integral

$$z = \frac{\Delta\lambda}{\lambda} = \frac{1}{c}\int \nabla\Phi \mathrm{d}z = \Delta\Phi, \tag{3.29}$$

where $\Delta\Phi$ is the difference in gravitational potential between the two points. Thus, the redshift is directly related to the change in the gravitational potential.

(2) EEP and gravitational redshift.

The above discussion can be approached from another angle, starting from Einstein's equivalence principle. As shown in Figure 3.4, consider an elevator accelerating uniformly with an acceleration $\boldsymbol{a} = -\boldsymbol{g}$ and a height h. Assume the speed of the elevator $v \ll c$, so the effects of special relativity can be neglected. Also assume that h is small, so other effects of general relativity can be neglected. Under these two assumptions, second-order small quantities such as $(v/c)^2$, $(gh/c^2)^2$ can be neglected, while

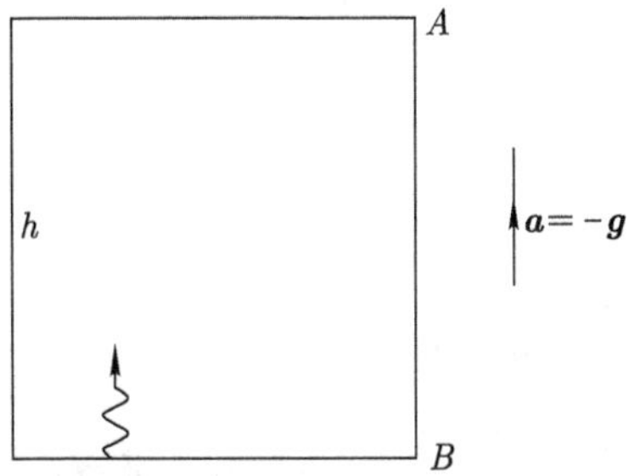

Fig. 3.4 Using Einstein's equivalence principle to study gravitational redshift.

first-order quantities (v/c), (gh/c^2) are important and need to be retained. The bottom of the elevator is denoted as B, and the top as A, with their vertical positions at heights

$$z_B(t) = \frac{1}{2}gt^2, \quad z_A(t) = h + \frac{1}{2}gt^2. \tag{3.30}$$

Consider two pulses emitted continuously from the bottom B,

$$t = 0 \to t_1, \quad \text{the first pulse is received,} \tag{3.31}$$

$$t = \Delta\tau_B \to t_1 + \Delta\tau_A, \quad \text{the second pulse is received.} \tag{3.32}$$

For the first pulse, the distance it travels is

$$z_A(t_1) - z_B(0) = ct_1. \tag{3.33}$$

For the second pulse, since the elevator is accelerating, the distance it travels to reach A is longer,

$$z_A(t_1 + \Delta\tau_A) - z_B(\Delta\tau_B) = c(t_1 + \Delta\tau_A - \Delta\tau_B). \tag{3.34}$$

Assuming the interval $\Delta\tau_B$ between the two pulses at emission is small, we expand the above quantities, keeping only the linear terms of $\Delta\tau_A$ and $\Delta\tau_B$,

$$h + \frac{1}{2}gt_1^2 = ct_1,$$

$$h + \frac{1}{2}g((t_1 + \Delta\tau_A)^2 - (\Delta\tau_B)^2) = c(t_1 + \Delta\tau_A - \Delta\tau_B),$$

thus we obtain

$$gt_1\Delta\tau_A = c(\Delta\tau_A - \Delta\tau_B). \tag{3.35}$$

Since $t_1 \approx h/c$, we have

$$\Delta\tau_A = \frac{\Delta\tau_B}{1 - gh/c^2}. \tag{3.36}$$

The frequency of the photon $\omega \propto \dfrac{1}{\Delta\tau}$, so we have

$$\omega_A = \left(1 - \frac{gh}{c^2}\right)\omega_B, \quad \lambda_A = \left(1 + \frac{gh}{c^2}\right)\lambda_B. \tag{3.37}$$

This is exactly the relationship derived earlier through energy conservation. The above discussion is only valid in the weak gravitational limit, where the Newtonian gravitational potential is not large. For strong gravitational fields, we must discuss within the framework of general relativity. We will discuss this issue in Chapter 8.

The gravitational redshift effect arises as photons expend energy to overcome a gravitational potential. Conversely, when photons travel from a region of higher gravitational potential to one of lower potential, they gain energy, resulting in a gravitational blueshift. Traditionally, the term gravitational redshift refers specifically to the process in which photons lose energy while climbing out of a gravitational well. This phenomenon has been experimentally verified, most notably in the pioneering work of Robert V. Pound and Glen A. Rebka (1960), later refined by R. V. Pound and J. L. Snider (1964). Their experiments, conducted at Harvard University's Jefferson Laboratory, measured the gravitational redshift across a 22.5-meter tower. Given the minuscule gravitational potential difference between the top and bottom of the tower, $gh/c^2 \sim 10^{-15}$, extremely precise detection methods were required. The team employed the Mössbauer effect to achieve the necessary resolution. The results confirmed the existence of gravitational redshift, with a mere 1% discrepancy between theoretical predictions and experimental observations.

3.5 Relativistic Corrections in the Global Positioning System

The Global Positioning System (GPS) consists of 24 satellites, each completing an orbit around the Earth every 12 hours. These satellites are distributed across 6 orbital planes, with 4 satellites per plane. Each satellite is equipped with a highly precise atomic clock, maintaining an accuracy of 10^{-13} seconds even over weeks of operation. From the perspective of Earth's approximately inertial frame, the satellites continuously broadcast their spatiotemporal coordinates. By analyzing signals from three satellites, a receiver can determine its position via the intersection of three spheres. A fourth satellite is required to resolve timing discrepancies, enabling full four-dimensional spacetime localization. This can be clearly seen in a two-dimensional scenario. In two dimensions, only two satellites are needed:

$$ct_{\rm p} = \frac{1}{2}(c(t_A + t_B) + (x_B - x_A)),$$

$$x_{\rm p} = \frac{1}{2}(c(t_A - t_B) + (x_B + x_A)).$$

Knowing the time and spatial information of two satellites, the time and spatial information of the receiver are also determined. This is the basic working principle of the GPS.

The operation of the GPS requires accounting for numerous physical factors. In this discussion, we focus specifically on two key relativistic effects. The first effect arises from special relativity: time dilation due to the satellites' orbital motion. As the satellites move at high velocities relative to Earth's surface, their onboard clocks experience measurable time dilation compared to ground-based reference frames. By Newtonian gravity, we know

$$\frac{v_{\rm s}^2}{R_{\rm s}} = \frac{GM_\oplus}{R_{\rm s}^2}. \tag{3.38}$$

The orbital radius of the satellite is

$$R_{\rm s} \approx 2.7 \times 10^4 \text{ km } \approx 4.2 R_\oplus, \quad R_\oplus \approx 6.4 \times 10^3 \text{ km}. \tag{3.39}$$

Thus, its velocity is $v_{\rm s} \approx 3.9$ km/s, and $v_{\rm s}/c \approx 1.3 \times 10^{-5}$. The special relativistic effect is

$$\frac{1}{2}\left(\frac{v_{\rm s}}{c}\right)^2 \approx 0.84 \times 10^{-10}. \tag{3.40}$$

The other relativistic effect is the gravitational redshift of photons. The influence of this effect is

$$\frac{GM_\oplus}{R_{\rm s}c^2} \approx 1.6 \times 10^{-10}. \tag{3.41}$$

While these relativistic effects may appear small (on the order of 10^{-10}), they are critically important for the GPS's precision. Given the speed of light, a timing error of 10^{-10} seconds translates to a positional deviation of approximately 30 cm. To maintain the system's required 2-meter accuracy, timing must be precise to within 6 nanoseconds. This stringent requirement means both special relativistic time dilation and general relativistic gravitational redshift must be meticulously accounted for in GPS operations. Remarkably, without these corrections, the system would become unusable within mere tens of minutes. The GPS thus stands as one of the most practical, everyday demonstrations of both special and general relativity in modern technology.

3.6 Gravity and Geometry

The gravitational redshift effect discussed above actually implies that spacetime must be curved. As Alfred Schild (1960) demonstrated, the presence of this effect proves spacetime cannot be described by flat Minkowski geometry. As shown in Figure 3.5, assume there is an observer at a height z_1 on the surface of the Earth, and another observer at a height $z_2 = z_1 + h$. They remain stationary relative to each other and also relative to the Earth's inertial frame. In the spacetime diagram, light must travel along a 45° line. The interval of photon pulses at z_1, Δt_1, must be the same as the interval of photon pulses at z_2, Δt_2, which contradicts the gravitational redshift. In the above discussion, it is not necessary for light to travel in a straight line; it just needs to be a null geodesic. If the null geodesics of two light pulses were perfectly parallel, gravitational redshift could not occur. This implies that gravitational redshift is fundamentally incompatible with the parallel postulate in Euclidean geometry. Consequently, the observed phenomenon of gravitational redshift necessitates that spacetime cannot be flat—it must be intrinsically curved. This conclusion aligns with Einstein's general theory of relativity, where gravity is not a force but a manifestation of spacetime curvature. The very existence of gravitational redshift thus serves as empirical evidence that spacetime is non-Euclidean on scales where gravitational fields are significant.

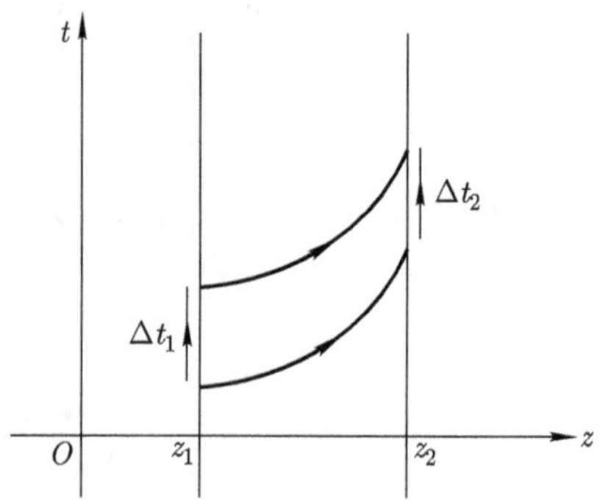

Fig. 3.5 Considering the motion of photons in a gravitational field shows that spacetime cannot be flat.

Exercises

1. Prove that the gravitational potential inside a uniform spherical shell is a constant.
2. Prove that a spherically symmetric distribution of mass has no mass quadrupole moment.
3. Suppose a planet has a uniform density and a shape like an oblate spheroid, with the polar radius $r_{\rm p}$ being less than the equatorial radius $r_{\rm e}$, calculate the mass quadrupole moment of this planet.

Chapter 4

Foundations of Curved Spacetime — Manifolds

Our discussion of gravitational redshift has demonstrated that the spacetime governing photon dynamics in a gravitational field cannot be described by Minkowski spacetime. This implies that a relativistic theory of gravity must be formulated within the framework of curved spacetime.

In this chapter, we introduce the foundational concepts of curved spacetime, beginning with an intrinsic geometric description. While our treatment prioritizes physical intuition over mathematical rigor, readers seeking a more formal exposition may consult the references [30–34].

4.1 Manifolds

Euclidean geometry stands as a paradigm of the axiomatic method. From just five postulates, all its theorems can be rigorously derived. For over two millennia after Euclid's *Elements*, mathematicians sought to refine this system—questioning whether its postulates could be simplified or altered while preserving the theory's consistency and completeness.

A focal point of these efforts was Euclid's fifth postulate, which (in one equivalent formulation) asserts that if a line intersects one of two parallel lines, it must intersect the other.[1] As early as 1824, Carl Friedrich Gauss—the "Prince of Mathematics"—hinted in correspondence at the possibility of revising this postulate. By 1832, the Hungarian mathematician János Bolyai had publicly proposed such a modification, while Nikolai Lobachevsky, working independently, had already introduced the hyperbolic plane (1826)—a two-dimensional space of constant negative curvature that defies embedding in ordinary Euclidean space.

[1] This differs from Euclid's original phrasing but is logically equivalent; another version, due to Legendre, states that the sum of a triangle's interior angles equals 180°.

These developments marked the dawn of curved geometry, a field later revolutionized by Bernhard Riemann. By the time Einstein formulated general relativity, the mathematics of curved spaces had matured over half a century, providing the very tools he needed. Yet, to most physicists of Einstein's era, Riemannian geometry remained little more than a mathematician's abstraction—devoid of apparent physical meaning.

The geometry of curved spaces is typically described by Riemannian geometry, while pseudo-Riemannian geometry extends this framework to curved spacetime. For now, we will not distinguish between space and time, focusing instead on general curved spaces—particularly those that are mathematically well-behaved.

The spaces we will examine are smooth manifolds, which can be intuitively characterized by two key properties:

(1) **Local flatness:** Every point on the manifold has a neighborhood that resembles a flat Euclidean space of the same dimension. For an n-dimensional manifold, this local approximation is n-dimensional.
(2) **Smooth compatibility:** Adjacent local regions can be seamlessly "stitched together" in a smooth manner.

Property (1) ensures that the manifold is locally indistinguishable from Euclidean space, with a consistent dimension throughout. Property (2) guarantees global smoothness, meaning there are no abrupt discontinuities in how these local patches connect.

To clarify these ideas, we will first illustrate examples of manifolds and non-manifolds, followed by precise mathematical definitions of the two properties above.

Example 4.1. Flat Euclidean space $\mathbf{R}^n$, including one-dimensional lines, two-dimensional planes, etc.

Example 4.2. The n-dimensional sphere $\mathbf{S}^n$. The surface of the Earth is approximately a two-dimensional sphere. The sphere presents a non-trivial example of a smooth manifold.

Example 4.3. The n-dimensional torus $\mathbf{T}^n$. It can be obtained by periodically compactifying each direction of the n-dimensional Euclidean space:

$$x^i \sim x^i + a^i, \quad i = 1, \cdots, n. \tag{4.1}$$

From the construction, it can be seen that there is not much difference between the n-dimensional torus and the n-dimensional flat space, both should be flat, but topologically they are completely different.

Example 4.4. Genus-g two-dimensional Riemann surfaces, as shown in Figure 4.1. Riemann surfaces can have algebraic definitions as well as geometric definitions. We only give a few obvious examples here:

(1) $g = 0$: The two-dimensional sphere $\mathbf{S}^2$;
(2) $g = 1$: The two-dimensional torus $\mathbf{T}^2$, geometrically it has a handle;
(3) $g > 1$: It has g handles.

Two-dimensional Riemann surfaces are rich in beautiful mathematics, related to complex analysis, algebraic geometry, index theorem, moduli space, etc., in mathematics, and have many important applications in physics, such as in string theory, supersymmetric Yang–Mills theory, etc.

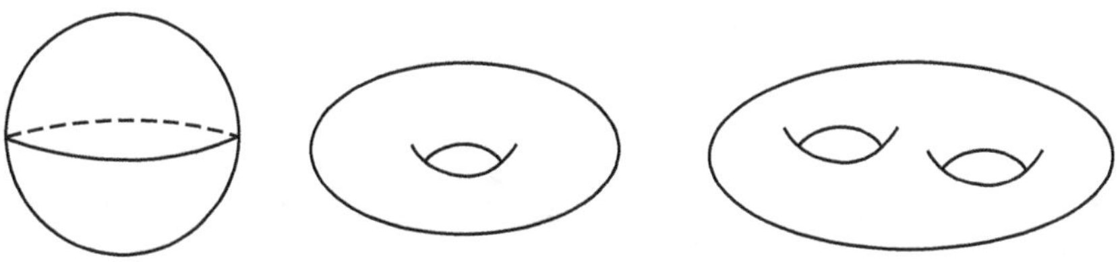

(a) sphere of genus 0 (b) torus of genus 1 (c) double torus of genus 2

Fig. 4.1 Riemann surfaces of various genus.

Example 4.5. More abstract examples: A collection of continuous transformations can define a Lie group, such as the rotation group and the Lorentz group discussed earlier.

Manifolds can be effectively studied through their embeddings in higher-dimensional spaces. In multivariable calculus, this approach is commonly used to describe curves and surfaces: For a one-dimensional curve, it possesses a single degree of freedom, parametrized by a variable λ. When embedded in a flat spacetime, its trajectory is described by a set of coordinates: $x^\mu = x^\mu(\lambda)$, where each value of λ corresponds to a distinct point along the curve. More generally, an M-dimensional surface embedded in an N-dimensional flat space ($M \leq N$) requires M independent parameters. Its embedding is specified by the coordinate functions:

$$x^\mu = x^\mu(u^1, \cdots, u^M), \quad \mu = 1, \cdots, N. \tag{4.2}$$

where $u^1, \cdots u^M$ are the intrinsic coordinates parametrizing the surface.

An alternative approach to describing embedded manifolds is through constraint functions. In an N-dimensional flat space, a hypersurface of codimension one (i.e., dimension $N-1$) can be defined by a smooth function:

$$f(x^1, \cdots, x^N) = 0, \tag{4.3}$$

where the equation defines an implicit relationship between the coordinates. Here, the function is defined on the coordinate functions $\{x^\mu\}$ of the ambient space. This single constraint equation removes one degree of freedom, resulting in the desired $(N-1)$-dimensional hypersurface. This construction naturally generalizes to lower-dimensional surfaces. An M-dimensional surface (where $M = N - k$) can be defined through k independent constraint equations:

$$\begin{cases} f_1(x^1, \cdots, x^N) = 0 \\ \quad \cdots\cdots \\ f_{N-M}(x^1, \cdots, x^N) = 0, \end{cases} \tag{4.4}$$

with the requirement that the gradients ∇f_i are linearly independent everywhere on the surface.

While the constraint function approach provides an elegant way to define surfaces, this method has important limitations that must be considered. First of all the constraint equations may have no solution, or have solutions that are non-manifolds, e.g., self-intersecting surfaces. Secondly, even when solutions exist, the resulting surface may contain singularities. Finally, this implicit definition may have global structure issue, e.g. multiple disconnected components.

Whitney's embedding theorem guarantees that any smooth N-dimensional manifold can always be embedded in a $2N$-dimensional Euclidean space. However, this embedding is primarily a mathematical convenience rather than a fundamental requirement. The manifold's intrinsic geometric structure exists independently of any embedding and its properties are defined without reference to an ambient space. We will now develop the proper mathematical framework for describing manifolds intrinsically.

For any point on an N-dimensional manifold, its position can be locally specified by N independent real coordinates $(x^1, \cdots, x^N)$.[2] However, the choice of coordinates is arbitrary, and even if the manifold itself is trivial, the choice of coordinates sometimes makes some points appear singular.

[2]In this book, we only discuss real manifolds.

For example, using polar coordinates (r, φ) to describe two-dimensional Euclidean space,

$$x = r\cos\varphi, \quad y = r\sin\varphi, \tag{4.5}$$

there is a degeneracy at the origin $r = 0$, because φ is undetermined at this time. But there is no such problem with Cartesian coordinates. Of course, this does not mean that Cartesian coordinates are better than polar coordinates. In some cases, using polar coordinates can bring a lot of convenience. The singular behavior at $r = 0$ is simply a coordinate artifact.

Generally speaking, it is impossible to cover the entire spacetime with only one non-degenerate coordinate system. Let us consider a two-dimensional sphere.

Example 4.6. The two-dimensional sphere $\mathbf{S}^2$.

We can parametrize this sphere using spherical coordinates (θ, ϕ) where $\theta \in [0, \pi]$ represents the polar angle (colatitude) and $\phi \in [0, 2\pi)$ denotes the azimuthal angle (longitude). However, this coordinate system exhibits a degeneracy at the North and South Poles ($\theta = 0$ and $\theta = \pi$), where the azimuthal angle ϕ becomes undefined. This reflects a fundamental topological constraint: unlike the two-dimensional Euclidean space, the two-dimensional sphere does not admit a single, globally defined, non-degenerate coordinate chart that covers its entire surface. Consequently, at least two overlapping coordinate patches are required to construct a complete atlas for the two-dimensional sphere, highlighting its intrinsic topological distinction from Euclidean space.

As we have seen, even for flat spaces, coordinates can be chosen arbitrarily. Locally, a manifold is homeomorphic to R^n, meaning that in a neighborhood of any point p on the manifold, we can define a coordinate chart—a smooth mapping $\phi(p)$ to $\mathbf{R}^n$. Similarly, for an adjacent point q, we may choose another coordinate chart $\psi(q)$. These two coordinate systems are not defined merely at individual points, but rather over open neighborhoods, leading to overlapping regions where both ϕ and ψ are valid. In such overlaps, a single point can be described by two different coordinate systems, and these descriptions must be compatible. The transition between coordinate systems is governed by a coordinate transformations: $x^\mu \to x'^\mu = x'^\mu(x^1, \cdots, x^N)$. From this transformation, we can compute the Jacobian determinant: $J = \left|\dfrac{\partial x'^\mu}{\partial x^\nu}\right|$. If $J \neq 0$, the transformation is invertible, ensuring a well-defined relationship between the two coordinate systems. The precise meaning of "compatibility" will be explored later, but first, we must establish some foundational concepts about mappings.

Given two sets A and B, a mapping $\varphi : A \to B$ is a rule that assigns to each element A exactly one element of B. This means that every $x \in A$ has a unique image $\varphi(x) \in B$, and multiple elements of A may map to the same element in B (i.e. φ need not be one-to-one), and there may exist elements in B that are not images of any element in A (i.e. φ need not be onto). Given two mappings $\varphi : A \to B$ and $\psi : B \to C$, we can define a composite mapping $\psi \circ \varphi : A \to C$ through the operation $(\psi \circ \varphi)(a) = \psi(\varphi(a))$, where $a \in A$, $\varphi(a) \in B$, so $(\psi \circ \varphi)(a) \in C$, as shown in Figure 4.2.

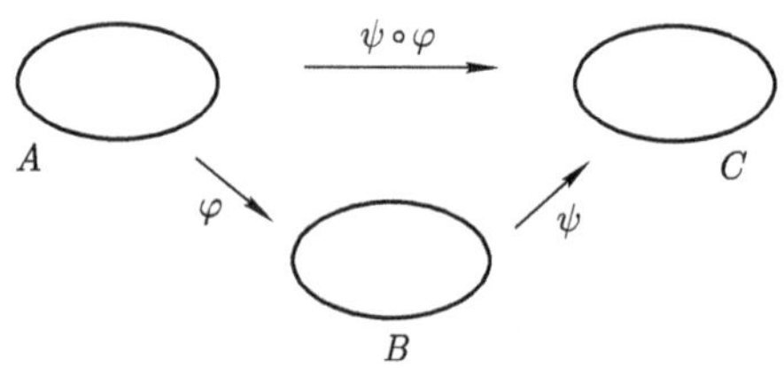

Fig. 4.2 Composite mapping.

Example 4.7. Suppose we have mappings $f : \mathbf{R}^m \to \mathbf{R}^n$ and $g : \mathbf{R}^n \to \mathbf{R}^l$, then we have a composite mapping $(g \circ f) : \mathbf{R}^m \to \mathbf{R}^l$.

Mappings between Euclidean spaces and composite mappings are very useful. They allow us to use the knowledge of multivariable calculus, one of which is the chain rule. This rule relates the partial derivatives of the composite mapping to the partial derivatives of the individual mappings:

$$\frac{\partial}{\partial x^a}(g \circ f)^c = \sum_b \frac{\partial f^b}{\partial x^a}\frac{\partial g^c}{\partial y^b}, \tag{4.6}$$

which can usually be simplified to

$$\frac{\partial}{\partial x^a} = \sum_b \frac{\partial y^b}{\partial x^a}\frac{\partial}{\partial y^b}. \tag{4.7}$$

Since manifolds are locally Euclidean, their local structure can be effectively studied using mappings between Euclidean spaces. This property allows us to extend tools from multivariable calculus—such as composite mappings and the chain rule—to the study of manifolds. To formalize this idea, we first establish that every smooth manifold admits a local homeomorphism to Euclidean space. This is achieved through the concept of a coordinate chart (or simply a chart), which provides a way to "flatten" a small region of the manifold for analysis. A coordinate chart includes an open set U of an n-dimensional manifold M and a 1-1 mapping $\varphi : U \to \mathbf{R}^n$, such that $\varphi(U)$ is also an open set in $\mathbf{R}^n$, as shown in Figure 4.3.

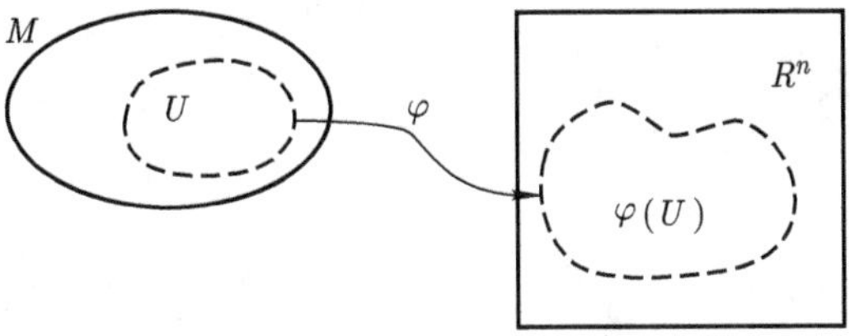

Fig. 4.3 Manifolds are described using coordinate charts. Locally, a manifold is homeomorphic to an open set of Euclidean space.

The definition of a smooth manifold M is an N-dimensional smooth ($\mathbf{C}^\infty$ differentiable) manifold if

(1) M is a topological manifold,
(2) There is a family of coordinate charts $\{(U_i, \varphi_i)\}$ on M, where $\cup_i U_i = M$ and each U_i is an open set, φ_i is a homeomorphism from U_i to an open set U_i' in $\mathbf{R}^n$,
(3) If $U_i \cap U_j \neq \phi$, then the mapping from the open set U_i' of Euclidean space to U_j' and its inverse are smooth (infinitely differentiable).

The above definition is shown in Figure 4.4. To avoid introducing excessive formalism, we briefly summarize the key ideas behind coordinate charts while acknowledging some relaxed rigor: Condition (1) means that the manifold M retains a meaningful topology, ruling out pathological or overly abstract sets. Condition (2) essentially means that each local part of the manifold can be regarded as Euclidean space. Condition (3) means that where multiple charts overlap, their descriptions must agree smoothly. This ensures seamless "gluing" of local coordinates to reconstruct the global manifold. It is remarkable that while M resembles $\mathbf{R}^n$ locally, its global structure may differ entirely. For example, the two-dimensional sphere is different from the two-dimensional Euclidean space: $\mathbf{S}^2 \not\approx \mathbf{R}^2$.

In the above definition, the collection $\{(U_i, \varphi_i)\}$ is called an atlas.[3] In this framework, each U_i is called the coordinate neighborhood (or

[3]The terminology originates from nautical navigation, where multiple overlapping maps were historically needed to chart the Earth's surface for precise voyage planning.

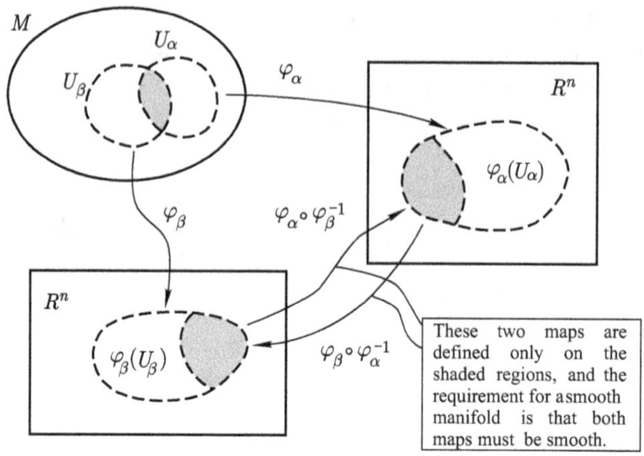

Fig. 4.4 A smooth manifold is defined through an collection of open covers and smooth mappings between the open covers.

coordinate patch), and each φ_i is a coordinate function (or local coordinate system). The transition function ψ_{ij} describes how coordinates transform between overlapping charts U_i and U_j. By requiring these transitions to be k-times differentiable, we obtain a $\mathbf{C}^k$ manifold. In the above definition, the mappings and transition functions are assumed to be real. However, if they are instead defined over the complex field (with holomorphic transition functions), the resulting structure is a complex manifold. In this book, we only discuss smooth real manifolds.

Example 4.8. Coordinate charts of $\mathbf{S}^1$.

Obviously, we can describe $\mathbf{S}^1$ with the coordinate $\theta \in [0, 2\pi)$, but this set is not open, so we need at least two coordinate charts, as shown in Figure 4.5.

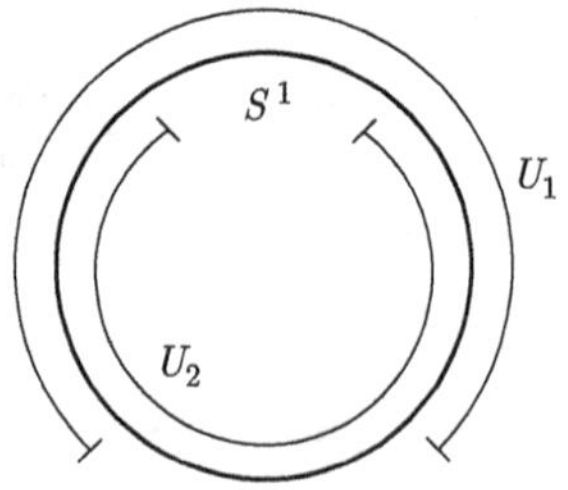

Fig. 4.5 For a unit circle, we need at least two open sets U_1 and U_2 to cover it.

Example 4.9. Coordinate charts of $\mathbf{S}^2$.

The two-dimensional unit sphere can be defined in three-dimensional Euclidean space as

$$(x^1)^2 + (x^2)^2 + (x^3)^2 = 1. \tag{4.8}$$

The first coordinate chart (U_1, φ_1) can be defined as U_1 covering the sphere except for the North Pole, and φ_1 is the stereographic projection, as shown in Figure 4.6: projecting from the North Pole, the points on the sphere are projected onto the two-dimensional plane where $x^3 = -1$,

$$\varphi_1(x^1, x^2, x^3) \equiv (y^1, y^2) = \left(\frac{2x^1}{1 - x^3}, \frac{2x^2}{1 - x^3}\right). \tag{4.9}$$

It is easy to see that in this projection mapping, the North Pole is mapped to infinity, and its value is undefined, that is to say, the North Pole must be excluded from the above mapping.

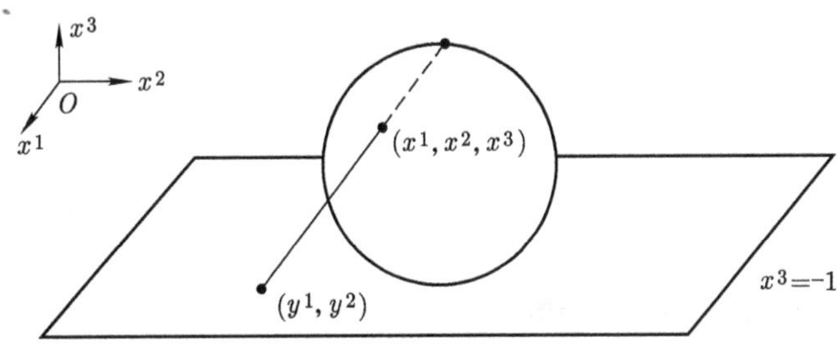

Fig. 4.6 The stereographic projection from the North Pole provides a mapping from the sphere to the plane, but the projection of the North Pole itself is undefined.

The other coordinate chart (U_2, φ_2) is defined as U_2 covering the sphere except for the South Pole, and φ_2 is the stereographic projection from the South Pole to the plane where $x^3 = 1$,

$$\varphi_2(x^1, x^2, x^3) \equiv (z^1, z^2) = \left(\frac{2x^1}{1 + x^3}, \frac{2x^2}{1 + x^3}\right). \tag{4.10}$$

In the overlapping area, the transition function $\varphi_2 \circ \varphi_1^{-1}$ is

$$z^i = \frac{4y^i}{[(y^1)^2 + (y^2)^2]}, \tag{4.11}$$

which is a smooth function.

Mappings can be defined between manifolds. Given two smooth manifolds: M (dimension m) with chart $(U, \varphi : U \to \mathbf{R}^m)$, and N (dimension n)

with chart $(V, \psi : V \to \mathbf{R}^n)$, a mapping $f : M \to N$ is said to be differentiable at $p \in M$ if $p \in U, f(p) \in V$, and the mapping $\psi \circ f \circ \varphi^{-1} : \mathbf{R}^m \to \mathbf{R}^n$ is differentiable at $\varphi(p)$, as shown in Figure 4.7. This is a mapping between two Euclidean spaces: $\mathbf{R}^m \to \mathbf{R}^n$, so the function f is differentiable with respect to the coordinates x^μ in $\mathbf{R}^m$,

$$\frac{\partial f}{\partial x^\mu} \equiv \frac{\partial}{\partial x^\mu}(\psi \circ f \circ \varphi^{-1})(x^\mu). \tag{4.12}$$

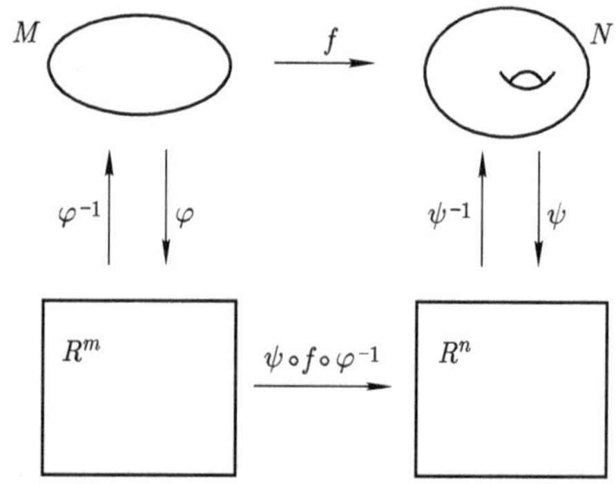

Fig. 4.7 Mappings between two manifolds can be defined through mappings between coordinate charts.

The coordinate chart formalism allows us to study mappings between manifolds by reducing them to mappings between Euclidean spaces. If a mapping and its inverse are both smooth, then the two manifolds are called diffeomorphic. In particular, a manifold typically have infinitely many atlases. We can consider mappings between the same manifold with different atlases, thereby discussing the compatibility of the two atlases. If two atlases are compatible, then their union is also an atlas. In this way, we may define an equivalence relation. Compatible atlases can be summarized into an equivalence class, while incompatible atlases belong to different equivalence classes. This equivalence relation defines the differential structure of the manifold. An equivalence class corresponds to a differential structure. Some manifolds admit multiple distinct differential structures.

Example 4.10. Differential structures of manifolds.

(1) $\mathbf{S}^7$ has 28 differential structures.[4]

[4]American mathematician John Milnor discovered that $\mathbf{S}^7$ has 28 differentiable structures. He was awarded the Fields Medal in 1962.

(2) $\mathbf{S}^n$ has only one differential structure when $n < 7$, and the number of differential structures grows rapidly when $n > 7$.
(3) $\mathbf{R}^4$ has infinitely many differential structures.[5]

4.2 Tangent and Cotangent Spaces

We shall now examine the tangent space of a manifold from an intrinsic perspective. Let us consider the collection of all smooth parameterized curves $\gamma : \mathbf{R} \to M$ passing through a point p on the manifold M, where γ is a non-degenerate mapping such that $p \in \gamma(\mathbf{R})$. As previously established, the tangent vector to a curve at p can be characterized in terms of directional derivatives. Consequently, the tangent space T_p at p can be constructed as the space of all directional derivatives induced by smooth curves passing through p. The dimension of the tangent space T_p, determined by the maximal number of linearly independent tangent vectors at p, coincides with the dimension of the manifold itself, as illustrated in Figure 4.8.

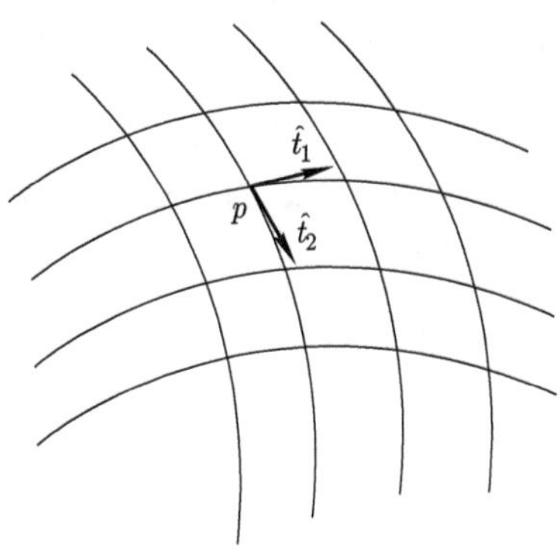

Fig. 4.8 The tangent space at point p is composed of the directional derivatives of all curves passing through p.

Consider an n-dimensional smooth manifold M. A curve on M is defined by a smooth map $\gamma : \mathbf{R} \to M$, where $\lambda \in \mathbf{R}$ parametrizes the curve. Now, let $f \in \mathcal{F}(M)$ be a smooth real-valued function on M. The directional derivative of f along γ at a point $p = \gamma(\lambda_0)$ describes how f changes as

[5]Through the study of the moduli space of Yang–Mills theory on four-dimensional manifolds, Simon Donaldson and Michael Freedman found that flat four-dimensional space actually has infinitely many differentiable structures. They were awarded the Fields Medal in 1986.

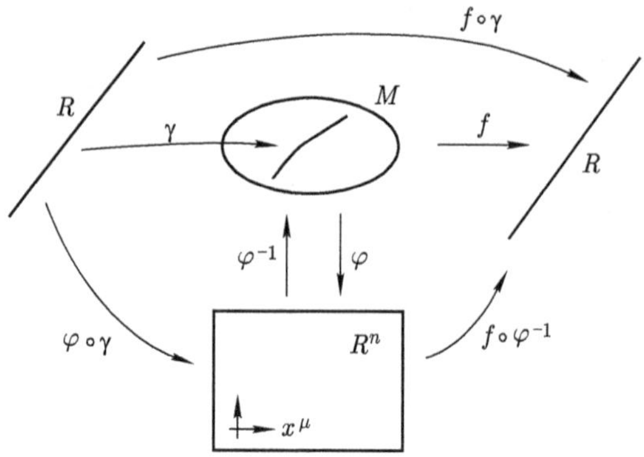

Fig. 4.9 The directional derivative along a curve on a manifold can be locally defined using coordinate charts.

we move along the curve. In a neighborhood of p, we have a coordinate chart (U, φ). This allows us to express the directional derivative in terms of coordinates. First, the function itself defines a mapping $f : M \to \mathbf{R}$. As shown in Figure 4.9, the composite mapping $f \circ \gamma : \mathbf{R} \to \mathbf{R}$ defines a function $f(\lambda)$ depending on the affine parameter. On the other hand, we can use the coordinate chart at p to explicitly express this mapping: $(f \circ \varphi^{-1}) \circ (\varphi \circ \gamma)$, that is, $f(x^\mu(\lambda))$. Therefore, the directional derivative of the function along the curve is

$$\begin{aligned}\frac{\mathrm{d}}{\mathrm{d}\lambda}(f \circ \gamma) &= \frac{\mathrm{d}}{\mathrm{d}\lambda}[(f \circ \varphi^{-1}) \circ (\varphi \circ \gamma)] \\ &= \frac{\mathrm{d}(\varphi \circ \gamma)^\mu}{\mathrm{d}\lambda} \frac{\partial (f \circ \varphi^{-1})}{\partial x^\mu} \\ &= \frac{\mathrm{d}x^\mu}{\mathrm{d}\lambda} \partial_\mu f.\end{aligned} \tag{4.13}$$

Since the function f is arbitrary, the directional derivative operator $\frac{\mathrm{d}}{\mathrm{d}\lambda}$ can be expanded in terms of the basis vectors ∂_μ:

$$\frac{\mathrm{d}}{\mathrm{d}\lambda} = \frac{\mathrm{d}x^\mu}{\mathrm{d}\lambda} \partial_\mu. \tag{4.14}$$

In this coordinate chart, the basis of the vector is

$$\widehat{e}_\mu = \partial_\mu, \quad \mu = 1, \cdots, n. \tag{4.15}$$

It is usually called the coordinate basis of T_p. Obviously, the dimension of the tangent space is the same as the dimension of the manifold.

Since the choice of coordinate chart near point p is not unique, we may express the directional derivative using different coordinate systems. Consider a coordinate transformation $x^\mu \to x^{\mu'}$, the basis vectors transform as:

$$\partial_{\mu'} = \frac{\partial x^\mu}{\partial x^{\mu'}} \partial_\mu. \tag{4.16}$$

As the vector itself is a geometric object, and remains invariant under coordinate transformation,

$$\begin{aligned} V^\mu \partial_\mu &= V^{\mu'} \partial_{\mu'} \\ &= V^{\mu'} \frac{\partial x^\mu}{\partial x^{\mu'}} \partial_\mu, \end{aligned} \tag{4.17}$$

the transformation of the vector components is

$$V^{\mu'} = \frac{\partial x^{\mu'}}{\partial x^\mu} V^\mu. \tag{4.18}$$

From the preceding analysis, we observe that coordinate transformations on a manifold naturally generalize the concept of Lorentz transformations in Minkowski spacetime. In the framework of general relativity, such transformations between coordinate charts are referred to as general coordinate transformations or diffeomorphisms.

Note that the Lorentz transformations in Minkowski space are global linear transformations, i.e., constant matrix transformation. This is possible because Minkowski spacetime admits a global coordinate chart, allowing inertial frames to be related by constant linear mappings. By contrast, general coordinate transformations. However, not all transformations are like this are typically position-dependent. For instance, the transformation between Cartesian coordinates and polar coordinates in $\mathbf{R}^2$ is nonlinear, not a constant transformation. On a general manifold, even if two coordinate charts both use Cartesian-like coordinates, the transition functions between them cannot generally be constant transformations unless the manifold is globally flat. Consequently, while Lorentz transformations form a finite-dimensional Lie group, general coordinate transformations constitute an infinite-dimensional diffeomorphism group, reflecting the local flexibility of spacetime structure in general relativity.

The cotangent space T_p^*M at a point p on the manifold M is the dual space to the tangent space T_pM. Formally, it consists of all linear functionals (covectors or 1-forms) on T_pM: $T_p^*M = \{\text{linear mappings } \omega : T_pM \to \mathbf{R}\}$. A canonical example of a 1-form is the gradient of a smooth function

f, denoted as $\mathrm{d}f$. Its action on a tangent vector is precisely the directional derivative of f along the vector :

$$\mathrm{d}f\left(\frac{\mathrm{d}}{\mathrm{d}\lambda}\right) = \frac{\mathrm{d}f}{\mathrm{d}\lambda}. \tag{4.19}$$

Given a coordinate chart with local coordinates x^μ near $p \in M$, the gradient of the coordinate function x^μ provides the natural basis of the cotangent space:

$$\widehat{\theta}^\mu = \mathrm{d}x^\mu, \quad \mu = 1, \cdots, n, \tag{4.20}$$

satisfying

$$\widehat{\theta}^\mu(\widehat{e}_\nu) = \mathrm{d}x^\mu(\partial_\nu) = \frac{\partial x^\mu}{\partial x^\nu} = \delta^\mu_\nu. \tag{4.21}$$

An arbitrary 1-form can be expanded in this basis: $\widehat{\omega} = \omega_\mu \mathrm{d}x^\mu$. Under coordinate transformation, the transformation law of the basis and the components of the 1-form is

$$\mathrm{d}x^{\mu'} = \frac{\partial x^{\mu'}}{\partial x^\mu}\mathrm{d}x^\mu, \quad \omega_{\mu'} = \frac{\partial x^\mu}{\partial x^{\mu'}}\omega_\mu. \tag{4.22}$$

The tangent and cotangent spaces introduced previously are defined pointwise on the manifold M. To study global geometric structures, we now consider smooth assignments of tangent and cotangent vectors across the manifold. A vector field $\widehat{V}(x)$ on the manifold specifies a vector at each point p in the tangent space at that point. The vector field defines a mapping

$$\widehat{V} : f \to \widehat{V}(f) \in \mathbf{C}^\infty(M), \quad f \in \mathcal{F}(M), \tag{4.23}$$

where $\mathcal{F}(M)$ is the set of functions on the manifold. If $\widehat{V}(f)$ is smooth for all f, then the vector field is smooth. Locally, the vector field can be expanded in terms of the basis vectors as

$$\widehat{V}(x) = V^\mu(x)\widehat{e}_\mu = V^\mu(x)\partial_\mu. \tag{4.24}$$

Similarly, there are dual vector fields, or cotangent vector fields, 1-form fields, which locally have

$$\widehat{\omega}(x) = \omega_\mu(x)\widehat{\theta}^\mu = \omega_\mu(x)\mathrm{d}x^\mu. \tag{4.25}$$

A vector field can be viewed as a collection of vectors defined on a manifold, or equivalently, as a smooth assignment of "arrows" (vectors) to each point. For a smooth vector field, these "arrows" can be seamlessly connected to form a family of curves that foliate the entire manifold. More formally, given a smooth vector field $\widehat{V}(x)$, its integral curve is defined as a curve whose tangent vector at every point coincides with the vector field at that point. Parametrizing this curve as $x^\mu(\lambda)$, we have

$$\begin{aligned}\frac{\mathrm{d}x^\mu(\lambda)}{\mathrm{d}\lambda} &= V^\mu(x(\lambda)),\\ x^\mu(\lambda=0) &= x_0.\end{aligned} \tag{4.26}$$

By the existence and uniqueness theorem of ordinary differential equations, the above equation admits a unique local solution for given initial conditions. Furthermore, the solution can be extended across overlapping coordinate patches, thereby prolonging the integral curve. In the case of a compact manifold, it can be shown that the integral curve is globally defined for all $\lambda \in \mathbf{R}$.

Example 4.11. Vector fields on the two-dimensional plane $\mathbf{R}^2$.

(1) Rotational vector field

$$\widehat{V} = -x^2\frac{\partial}{\partial x^1} + x^1\frac{\partial}{\partial x^2}. \tag{4.27}$$

Consider the integral curve passing through the point $p = (x_0^1, x_0^2)$, we obtain

$$x^\mu(\lambda) = (x_0^1\cos\lambda - x_0^2\sin\lambda, x_0^1\sin\lambda + x_0^2\cos\lambda). \tag{4.28}$$

It is a circle with the center at the origin and passing through the point p.

(2) Radially expanded vector field

$$\widehat{V} = x^1\frac{\partial}{\partial x^1} + x^2\frac{\partial}{\partial x^2}. \tag{4.29}$$

The integral curve passing through point $p = (x_0^1, x_0^2)$ is

$$x^\mu(\lambda) = (x_0^1\mathrm{e}^\lambda, x_0^2\mathrm{e}^\lambda). \tag{4.30}$$

The integral curve passing through the point p is a ray starting from the origin (but not including the origin) and passing through p.

In the above two cases, if the point p is the origin, then the integral curve is the origin.

For two vector fields $\widehat{X}$ and $\widehat{Y}$, their commutator is still a vector field:

$$[\widehat{X}, \widehat{Y}](f) \equiv \widehat{X}(\widehat{Y}(f)) - \widehat{Y}(\widehat{X}(f)). \tag{4.31}$$

The commutator $[\widehat{X}, \widehat{Y}]$ is known as the Lie bracket. This operation is a fundamental example of a Lie derivative, a concept we will introduce later. In a certain coordinate system,

$$[\widehat{X}, \widehat{Y}] = [\widehat{X}, \widehat{Y}]^\alpha\partial_\alpha = (X^\beta\partial_\beta Y^\alpha - Y^\beta\partial_\beta X^\alpha)\partial_\alpha. \tag{4.32}$$

The commutator of a vector field has a natural geometric interpretation, illustrated in Figure 4.10. Intuitively, the failure of the integral curves of two vector fields to form a closed quadrilateral locally is precisely measured by their Lie bracket. For simplicity, we consider these vector fields on a plane. The vectors between points t and r are

$$\begin{aligned}\vec{rt} &= (\widehat{U}(p) + \widehat{V}(s)) - (\widehat{V}(p) + \widehat{U}(q)) \\ &= (\widehat{V}(s) - \widehat{V}(p)) - (\widehat{U}(q) - \widehat{U}(p)) \\ &= (\partial_\alpha V^\beta U^\alpha \widehat{e}_\beta)(p) - (\partial_\alpha U^\beta V^\alpha \widehat{e}_\beta)(p) + \text{ higher order terms} \\ &= [\widehat{U}, \widehat{V}](p) + \text{ higher order terms,}\end{aligned} \tag{4.33}$$

where possible higher order terms such as $\partial_{\mu\nu} V^\beta U^\mu U^\nu \widehat{e}_\beta$. If we halve the vector $\widehat{V}$, then the commutator is reduced to 1/4, while the higher order correction terms are at least reduced to 1/8. Therefore, taking the limit as $\widehat{U}$ and $\widehat{V}$ approach zero, the higher-order terms can be neglected, and in this way, we obtain the pentagon diagram.

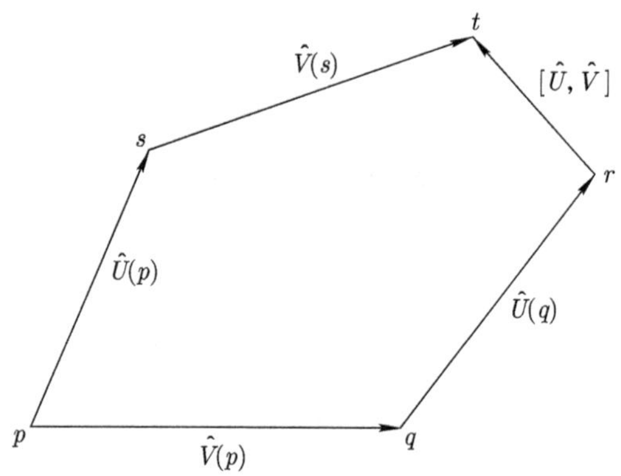

Fig. 4.10 The geometric meaning of the Lie bracket of vector fields.

4.3 Tensors

The definition and algebraic operations of tensors on a general manifold follow the same principles as those in Minkowski spacetime. Since the fundamental tensor formalism remains unchanged, we shall not reiterate these concepts here. A (k, l)-type tensor $\widehat{T}$ at point p can be expanded as

$$\widehat{T} = T^{\mu_1 \cdots \mu_k}{}_{\nu_1 \cdots \nu_l} \partial_{\mu_1} \otimes \cdots \otimes \partial_{\mu_k} \otimes \mathrm{d}x^{\nu_1} \otimes \cdots \otimes \mathrm{d}x^{\nu_l}, \tag{4.34}$$

at this time, the transformation law of the tensor components is

$$T^{\mu'_1\cdots\mu'_k}{}_{\nu'_1\cdots\nu'_l} = \frac{\partial x^{\mu'_1}}{\partial x^{\mu_1}}\cdots\frac{\partial x^{\mu'_k}}{\partial x^{\mu_k}}\frac{\partial x^{\nu_1}}{\partial x^{\nu'_1}}\cdots\frac{\partial x^{\nu_l}}{\partial x^{\nu'_l}}T^{\mu_1\cdots\mu_k}{}_{\nu_1\cdots\nu_l}. \tag{4.35}$$

Tensor transformation rules are easily remembered through index position: Upper indices transform with $\dfrac{\partial x^{\mu'_1}}{\partial x^{\mu_1}}$, while lower indices transform with the inverse. These rules extend naturally to define tensor fields on arbitrary manifolds, generalizing the vector/covector cases.

Most tensor operations defined in flat spacetime extend naturally to tensor fields on a manifold, including index contraction, symmetrization and antisymmetrization, index raising and lowering, trace, etc. However, three crucial distinctions arise when working with tensors on general manifolds:

(1) Derivative operations;
(2) Metric fields;
(3) Levi–Civita tensors.

First, let us examine derivative operations in the context of tensor calculus. In Minkowski spacetime with Cartesian coordinates, the partial derivative of a tensor field yields another tensor. However, this is not the case in curved space. For scalar functions, their gradients are legitimate $(0,1)$-tensors. However, for general tensor fields, coordinate derivatives fail to preserve tensorial character. For example, the partial derivative of a 1-form is no longer a tensor. This can easily be seen from the transformation of components under coordinate transformation:

$$\begin{aligned}\frac{\partial}{\partial x^{\mu'}}W_{\nu'} &= \frac{\partial x^{\mu}}{\partial x^{\mu'}}\frac{\partial}{\partial x^{\mu}}\left(\frac{\partial x^{\nu}}{\partial x^{\nu'}}W_{\nu}\right)\\ &= \frac{\partial x^{\mu}}{\partial x^{\mu'}}\frac{\partial x^{\nu}}{\partial x^{\nu'}}\left(\frac{\partial}{\partial x^{\mu}}W_{\nu}\right) + W_{\nu}\frac{\partial x^{\mu}}{\partial x^{\mu'}}\frac{\partial}{\partial x^{\mu}}\frac{\partial x^{\nu}}{\partial x^{\nu'}}.\end{aligned}$$

The first term on the right-hand side conforms to the transformation law of a tensor, but the extra second term destroys the covariance under coordinate transformation. How to modify the partial derivative to obtain a derivative that is covariant under coordinate transformation is an interesting question. We will learn how to define derivatives consistently in the next chapter.

4.4 Metric Tensor Fields

The metric tensor is a symmetric $(0,2)$-tensor, denoted either as $\widehat{g}$ or its component form $g_{\mu\nu}$. Since the dimension of the manifold is invariant, the metric tensor field must be non-degenerate: $g = \det(g_{\mu\nu}) \neq 0$. Its inverse, denoted $g^{\mu\nu}$ is likewise symmetric and constitutes the components of a $(2, 0)$-tensor, satisfying

$$g^{\mu\nu} g_{\nu\sigma} = \delta^{\mu}_{\sigma}. \tag{4.36}$$

Analogous to the role of the metric in flat spacetime, we can use the metric field $g_{\mu\nu}$ and $g^{\mu\nu}$ to raise and lower the indices of tensors in curved space, and as well as to define the inner product between vectors or 1-forms. Furthermore, the symmetry of $g_{\mu\nu}$ implies that it possesses $\dfrac{n(n+1)}{2}$ independent components in an n-dimensional space.

The metric tensor field plays a fundamental role in general relativity, with profound physical and geometric implications. First and foremost, it encodes the local spacetime geometry, enabling the computation of essential physical quantities, such as the proper length of spacelike curves and the proper time of timelike curves, etc. Furthermore, it helps us determine the "shortest" (geodesic) distance between two spacetime events. Physically, it governs the dynamics of test particles in curved spacetime, as their trajectories are determined by geodesic motion. In addition, it defines the inner product between two vectors, thereby classifying them as timelike, spacelike, or null, which in turn determines the attribute of the curve in the spacetime manifold. This classification establishes the causal structure of spacetime, rigorously distinguishing past, future, and causally separated events. Einstein's equivalence principle requires that the spacetime manifold must be locally flat, meaning the metric tensor field should locally take the form of the Minkowski metric $\eta_{\mu\nu}$ at any given point. Thus, the metric tensor field naturally defines local inertial frames, where special relativity holds. Remarkably, in the subsequent studies, we will also see that in the Newtonian approximation, the deviation of the metric field from the Minkowski metric reproduces the Newtonian gravitational potential, directly linking the metric to classical gravity. This correspondence underscores the metric tensor's dual role as both a geometric and a gravitational field..

Let us first see how the metric helps us understand local geometry. Consider two infinitesimally separated points P and Q. They can be described in the same coordinate chart,

$$P : x^{\mu}, \quad Q : x^{\mu} + \mathrm{d}x^{\mu}. \tag{4.37}$$

Their "distance" or "interval" line element is a certain function

$$\mathrm{d}s^2 = f(x^a, \mathrm{d}x^a). \tag{4.38}$$

This function determines the local geometry. For example, consider 2-dimensional Finsler geometry,

$$\mathrm{d}s^2 = (\mathrm{d}\xi^4 + \mathrm{d}\zeta^4)^{1/2}. \tag{4.39}$$

In (pseudo-)Riemannian geometry, the function f is given by the metric field,

$$\mathrm{d}s^2 = g_{\mu\nu}\mathrm{d}x^\mu \mathrm{d}x^\nu. \tag{4.40}$$

From P to Q, approximately we have an infinitesimal vector $\mathrm{d}\widehat{s}$:

$$\mathrm{d}\widehat{s} = \mathrm{d}x^\mu \widehat{e}_\mu. \tag{4.41}$$

Note that here $\mathrm{d}x^\mu$ represents the components of the infinitesimal vector, not a 1-form, then

$$\mathrm{d}s^2 = \mathrm{d}\widehat{s} \cdot \mathrm{d}\widehat{s} = (\widehat{e}_\mu \cdot \widehat{e}_\nu)\mathrm{d}x^\mu \mathrm{d}x^\nu = g_{\mu\nu}\mathrm{d}x^\mu \mathrm{d}x^\nu. \tag{4.42}$$

Here we have already used the fact that the inner product of two basis vectors is defined by the metric field:

$$\begin{aligned} \widehat{e}_\mu \cdot \widehat{e}_\nu &= \widehat{g}(\widehat{e}_\mu, \widehat{e}_\nu) = g_{\sigma\rho}\widehat{\theta}^\sigma(\widehat{e}_\mu)\widehat{\theta}^\rho(\widehat{e}_\nu) \\ &= g_{\sigma\rho}\delta^\sigma_\mu \delta^\rho_\nu = g_{\mu\nu}. \end{aligned} \tag{4.43}$$

The metric tensor field can be locally expressed in the standard diagonal form through an appropriate coordinate transformation:

$$g_{\mu\nu} = \mathrm{diag}(-1, -1, \cdots, -1, +1, +1, \cdots, +1, 0, 0, \cdots, 0),$$

where we do not impose the non-degeneracy condition on the metric. If the metric is further required to be non-degenerate, the diagonal matrix contains no zero elements. In this case, denoting the number of "+1" entries as s, and the number of "−1" entries as t, the quantity $s - t$ constitute a topological invariant, known as the signature. Meanwhile, $s+t$ corresponds to the dimension of spacetime, which is likewise a topological invariant. Depending on the signature, the metric is classified as follows: If $t = 0$, then the metric is termed Euclidean (globally flat) or Riemannian; If $t = 1$, it is referred to as Lorentzian or pseudo-Riemannian. For a continuous metric field, both its rank and signature remain constant everywhere.

4.4.1 *Riemann normal coordinates*

If we believe that a relativistic theory of gravity should be formulated in terms of curved spacetime, then the first crucial step is to verify Einstein's assertion regarding the existence of a local inertial frame (LIF)—namely, that in the neighborhood of any spacetime point, such a frame should exist. In this inertial frame, the metric must reduce to the Minkowski form, or more precisely, the standard diagonal form previously discussed. What does this requirement impose on the metric field? Moreover, can such a standard form always be achieved? We will demonstrate that it is always feasible by choosing an appropriate coordinate system in the vicinity of a given point $p \in M$. In fact, a stronger result holds: not only can the metric be brought into standard Minkowski form at p, but its first-order derivative can also be made to vanish at that point. However, it is generally impossible to eliminate all second-order derivatives of the metric through coordinate transformations.

Proposition In the neighborhood of a point p on the manifold, there always exists a coordinate system such that the metric and its derivatives satisfy

$$\begin{aligned} g_{\mu\nu}|_p &= \eta_{\mu\nu}, \\ \partial_\sigma g_{\mu\nu}|_p &= 0, \end{aligned} \tag{4.44}$$

and $\partial_\sigma \partial_\rho g_{\mu\nu}|_p$ is not necessarily all zero.

Proof Under a coordinate transformation, the metric transforms according to

$$g_{\mu'\nu'} = \frac{\partial x^\mu}{\partial x^{\mu'}} \frac{\partial x^\nu}{\partial x^{\nu'}} g_{\mu\nu}. \tag{4.45}$$

Without loss of generality, we take the point p as the origin in both coordinate systems, i.e., $x^\mu(p)$, $x^{\mu'}(p) = 0$. The coordinates in a neighborhood of p can be expanded as

$$\begin{aligned} x^\mu = &\left(\frac{\partial x^\mu}{\partial x^{\mu'}}\right)_p x^{\mu'} + \frac{1}{2}\left(\frac{\partial^2 x^\mu}{\partial x^{\mu'_1}\partial x^{\mu'_2}}\right)_p x^{\mu'_1} x^{\mu'_2} \\ &+ \frac{1}{6}\left(\frac{\partial^3 x^\mu}{\partial x^{\mu'_1}\partial x^{\mu'_2}\partial x^{\mu'_3}}\right)_p x^{\mu'_1} x^{\mu'_2} x^{\mu'_3} + \cdots, \end{aligned} \tag{4.46}$$

and the metric field can be similarly expanded as

$$\begin{aligned} g_{\mu\nu} = &(g_{\mu\nu})_p + (\partial g)_p \left(\frac{\partial x}{\partial x'}\right)_p x' + (\partial g)_p \left(\frac{\partial^2 x}{\partial x' \partial x'}\right)_p x' x' \\ &+ (\partial^2 g)_p \left(\frac{\partial x}{\partial x'} \frac{\partial x}{\partial x'}\right)_p x' x' + \cdots. \end{aligned} \tag{4.47}$$

For brevity, index structures and numerical factors are partially omitted. Suppose the original coordinates are x^μ, and we seek new coordinates $x^{\mu'}$. Expanding both sides of Eq. (4.45) to second order in x', we have

$$\begin{aligned}
&(g')_p + (\partial' g')_p x' + (\partial'\partial' g')_p x' x' \\
&\quad \approx \left(\frac{\partial x}{\partial x'}\frac{\partial x}{\partial x'} g\right)_p + \left(\frac{\partial x}{\partial x'}\frac{\partial^2 x}{\partial x' \partial x'} g + \frac{\partial x}{\partial x'}\frac{\partial x}{\partial x'}\partial' g\right)_p x' \\
&\qquad + \left(\frac{\partial x}{\partial x'}\frac{\partial^3 x}{\partial x'\partial x'\partial x'} g + \frac{\partial^2 x}{\partial x'\partial x'}\frac{\partial^2 x}{\partial x'\partial x'} g\right. \\
&\qquad \left. + \frac{\partial x}{\partial x'}\frac{\partial^2 x}{\partial x'\partial x'}\partial' g + \frac{\partial x}{\partial x'}\frac{\partial x}{\partial x'}\partial'\partial' g\right)_p x' x'.
\end{aligned} \tag{4.48}$$

Indices are omitted in some terms for simplicity, but can be restored straightforwardly. We now analyze Eq. (4.48) order by order in the context of an n-dimensional manifold.

(1) Zeroth order: The left-hand side of Eq. (4.48) g' has $\frac{n(n+1)}{2}$ degrees of freedom. The right-hand side is controlled by the Jacobian matrix $(\partial x^\mu / \partial x^{\mu'})_p$, which has $n \times n = n^2$ degrees of freedom. Thus, there are sufficient degrees of freedom to transform $g_{\mu'\nu'}(p)$ into a standard form, e.g., Minkowski metric. The remaining $\frac{n(n-1)}{2}$ degrees of freedom correspond to the generators of the Lorentz group in the local inertial frame.

(2) First order: The left-hand side $\partial'_\sigma g'_{\mu\nu}$ of Eq. (4.48) has $n \times \frac{n(n+1)}{2} = \frac{n^2(n+1)}{2}$ degrees of freedom. The additional freedom on the right-hand side comes from the second derivative matrix $(\partial^2 x^\mu / \partial x^{\mu_1'} \partial x^{\mu_2'})_p$, which has $n \times \frac{n(n+1)}{2}$ (since μ_1', μ_2' are symmetric) degrees of freedom. This is exactly the number required to set $\partial'_\sigma g'_{\mu\nu} = 0$.

(3) Second order: The left-hand side $\partial'_\rho \partial'_\sigma g'_{\mu\nu}$ of Eq. (4.48) has $\frac{n(n+1)}{2} \times \frac{n(n+1)}{2} = \left(\frac{n(n+1)}{2}\right)^2$ degrees of freedom. The remaining freedom on the right is governed by the third derivative object $(\partial^3 x^\mu / \partial x^{\mu_1'} \partial x^{\mu_2'} \partial x^{\mu_3'})_p$, which has $n \times \frac{n(n+1)(n+2)}{3 \times 2 \times 1} = \frac{n^2(n+1)(n+2)}{6}$ (since μ_1', μ_2', μ_3') are fully symmetric, so there are $\frac{(n+2) \times (n+1) \times n}{3 \times 2 \times 1}$ of them) degrees of freedom. This is insufficient to force all second derivatives of the metric to vanish. The deficit, $\frac{n^2(n-1)^2}{12}$,

matches the number of independent components of the Riemann curvature tensor in n-dimensions. Hence, the curvature generally cannot be eliminated, and the manifold is non-flat.

The coordinates that realize condition (4.44) are referred to as Riemann normal coordinates (RNC), and the associated basis vectors constitute a local Lorentz inertial frame. In RNC, the metric at point p coincides with the flat-space Minkowski metric to first order.

The introduction of a metric field helps us investigate the local geometry of the manifold intrinsically, without reference to any embedding space. This local geometry is an inherent property of the manifold itself. Consider, for example, a flat sheet of paper. In Cartesian coordinates (x, y), its metric is given by $\mathrm{d}s^2 = \mathrm{d}x^2 + \mathrm{d}y^2$. Now suppose there is a bug confined to this paper. If the paper is rolled into a cylinder, the bug—lacking any perception of the embedding three-dimensional space[6]—cannot detect any change in the local geometric properties, as illustrated in Figure 4.11. This is because the cylinder can be smoothly unrolled back into a flat plane without introducing wrinkles, tears, or distortions. Mathematically, this equivalence is reflected in the metric of the cylindrical surface, which retains the same local form

$$\mathrm{d}s^2 = \mathrm{d}z^2 + a^2\mathrm{d}\phi^2. \tag{4.49}$$

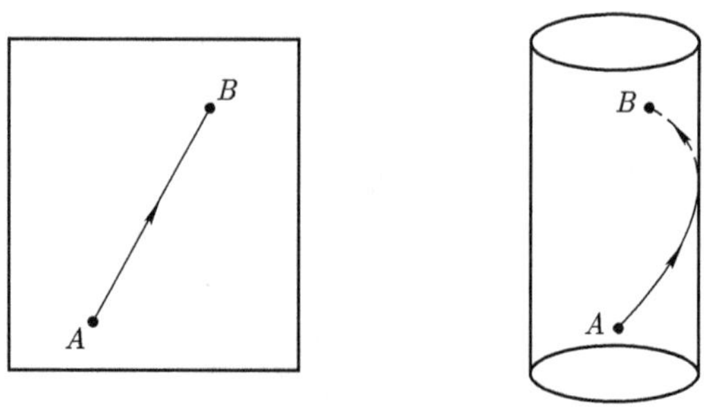

Fig. 4.11 On a flat sheet of paper, a bug crawls from A to B along a straight line. After rolling the paper into a cylinder, the bug's trajectory appears to be a curve.

Let $x = z$, $y = a\phi$, the metric reduces to that of a flat plane $ds^2 = dz^2 + dy^2$. This demonstrates that the cylindrical surface is intrinsically flat—its curvature arises solely from its embedding in three-dimensional space and is thus purely extrinsic. More generally, if there exists a globally

[6]We assume the bug has no external spatial awareness beyond the surface.

defined coordinate transformation that reduces a manifold's metric to the Euclidean (or Minkowski) form everywhere, then the manifold must be intrinsically flat.

4.4.2 *Geometry on the sphere*

Let us now examine a non-trivial example: the two-dimensional sphere $\mathbf{S}^2$. Unlike the cylinder, a sphere cannot be formed from a flat sheet without tearing or distorting the material, reflecting its fundamentally different intrinsic geometry. The metric of $\mathbf{S}^2$ in standard spherical coordinates is given by:

$$\mathrm{d}s^2 = a^2(\mathrm{d}\theta^2 + \sin^2\theta \mathrm{d}\phi^2), \tag{4.50}$$

where a is the radius of the sphere. Crucially, there exists no coordinate transformation that can reduce this metric to the Euclidean form $\mathrm{d}s^2 = \mathrm{d}x^2 + \mathrm{d}y^2$ everywhere on the sphere. This obstruction stems from the sphere's intrinsic curvature, which we will later show to be non-zero and inversely proportional to a^2. A direct geometric consequence is that the sum of interior angles of any triangle on the sphere exceeds π, in contrast to the flat Euclidean case.

We can describe this sphere using embedding in three-dimensional Euclidean space $\mathbf{R}^3$. Using Cartesian coordinates, the metric of three-dimensional Euclidean space $\mathbf{R}^3$ takes the form

$$\mathrm{d}s^2 = \mathrm{d}x^2 + \mathrm{d}y^2 + \mathrm{d}z^2. \tag{4.51}$$

In spherical coordinates,

$$\begin{aligned} x &= r\sin\theta\cos\phi, \\ y &= r\sin\theta\sin\phi, \\ z &= r\cos\theta, \end{aligned} \tag{4.52}$$

so the metric becomes

$$\mathrm{d}s^2 = \mathrm{d}r^2 + r^2\mathrm{d}\theta^2 + r^2\sin^2\theta \mathrm{d}\phi^2. \tag{4.53}$$

Although the underlying space remains unchanged, the metric tensor varies across different coordinate systems. Notably, in spherical coordinates, the metric exhibits a singularity at $r = 0$. However, this singularity is merely a coordinate artifact, as the point $r = 0$ is geometrically indistinguishable from any other point in the space.

In three-dimensional Euclidean space, the two-dimensional sphere is defined by the following equation:

$$x^2 + y^2 + z^2 = a^2, \tag{4.54}$$

from which we have

$$\mathrm{d}z = -\frac{x\mathrm{d}x + y\mathrm{d}y}{z}, \tag{4.55}$$

so the induced metric on the sphere (i.e., the metric induced through the Euclidean space metric) is

$$\mathrm{d}s^2 = \mathrm{d}x^2 + \mathrm{d}y^2 + \frac{(x\mathrm{d}x + y\mathrm{d}y)^2}{a^2 - x^2 - y^2}. \tag{4.56}$$

Consider the North Pole $A : x = y = 0$. In its vicinity, the metric takes the form $\mathrm{d}s^2 = \mathrm{d}x^2 + \mathrm{d}y^2$, indicating that the local geometry can be approximated as a two-dimensional Euclidean plane. Owing to spherical symmetry, this reasoning extends to any arbitrary point on the sphere. Using polar coordinates $x = \rho\cos\phi$, $y = \rho\sin\phi$, the metric can be written as

$$\mathrm{d}s^2 = \frac{a^2\mathrm{d}\rho^2}{a^2 - \rho^2} + \rho^2\mathrm{d}\phi^2. \tag{4.57}$$

In this coordinate system, the coordinate ρ represents the perpendicular distance from a point on the sphere to the central axis passing through the North and South Poles, with its values ranging from 0 to a. The angular coordinate ϕ corresponds to the azimuthal angle along the circle formed by the intersection of the sphere with the plane perpendicular to the central axis, spanning from 0 to 2π. The above metric appears to exhibit a singularity at $\rho = a$ or equivalently when $\sqrt{x^2 + y^2} = a$, which corresponds to the equator of the sphere. However, we recognize that all points on the sphere are equivalent and the equator is no different from any other region. This suggests that the singularity arises solely from the choice of coordinates, making it a coordinate singularity. Such singularities are not intrinsic to the geometry and can be eliminated by selecting a more suitable coordinate system.

In a curved space, the distance between two points along a path is

$$L_{AB} = \int_A^B \mathrm{d}s = \int_A^B \sqrt{g_{\mu\nu}\mathrm{d}x^\mu\mathrm{d}x^\nu}. \tag{4.58}$$

Obviously, the distance depends on the path connecting points A and B. If this path is parameterized by $x^\mu(\lambda)$, then the distance is

$$L_{AB} = \int_{\lambda_A}^{\lambda_B} \sqrt{g_{\mu\nu}\frac{\mathrm{d}x^\mu}{\mathrm{d}\lambda}\frac{\mathrm{d}x^\nu}{\mathrm{d}\lambda}}\mathrm{d}\lambda. \tag{4.59}$$

In curved space, the area element of a surface can be determined as follows. For simplicity, assume that the metric takes a diagonal form, i.e., $g_{\mu\nu}(x) = 0$

for $\mu \neq \nu$. We can rescale the coordinate so that the metric takes a standard form, effectively defining the proper length along each coordinate direction: $\sqrt{g_{11}}\mathrm{d}x^1$, $\sqrt{g_{22}}\mathrm{d}x^2, \cdots$. Using these proper length, the area element of the surface defined by fixing the coordinates $x^3, \cdots, x^N$ is given by:

$$\mathrm{d}A = \sqrt{|g_{11}g_{22}|}\mathrm{d}x^1\mathrm{d}x^2. \tag{4.60}$$

This means that in flat space, where the metric reduces to $g_{\mu\nu} = \delta_{\mu\nu}$, the area element simplifies to the product of two perpendicular infinitesimal displacements, dx^1dx^2. For higher-dimensional surface in curved space, the proper volume elements can be defined in an analogous manner.

We use the two-dimensional sphere introduced earlier to illustrate the calculation of length and area. Taking the metric (4.57), i.e., $g_{\rho\rho} = \dfrac{a^2}{a^2 - \rho^2}$, $g_{\phi\phi} = \rho^2$. Consider a circle on the sphere, as shown in Figure 4.12,

$$\rho = R. \tag{4.61}$$

Just as a circle on a plane is defined as the set of points equidistant from a fixed center, we can analogously define circles on a sphere. If we take the North Pole as the reference point, the set of points at a constant distance from it traces out a latitude line (a parallel of constant latitude). However, unlike planar circles where the radius is measured along a straight line, the radius of such a spherical circle is the arc length measured along a meridian from the North Pole to the latitude line.

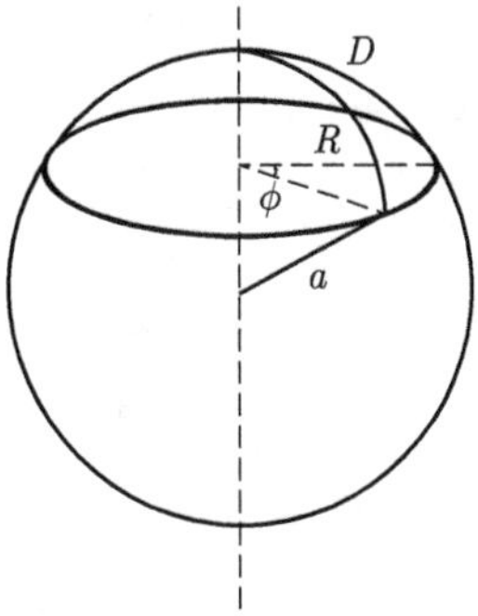

Fig. 4.12 A circle on the two-dimensional sphere with radius D originating from the North Pole. The radius of the sphere is a.

(1) The distance from the North Pole to the position of the circle is obtained by using a meridian, which has $\phi =$ constant, the distance on the sphere

is

$$D = \int_0^R \frac{a}{\sqrt{a^2 - \rho^2}} \mathrm{d}\rho = a \sin^{-1}(R/a). \tag{4.62}$$

This actually gives the radius of the circle on the sphere.

(2) The circumference of the circle is $C = \int_0^{2\pi} R\mathrm{d}\phi = 2\pi R$.

(3) The area of the cap formed from the circle to the pole is

$$A = \int_0^{2\pi} \int_0^R \frac{a}{\sqrt{a^2 - \rho^2}} \rho \mathrm{d}\rho \mathrm{d}\phi = 2\pi a^2 \left(1 - \left(1 - \frac{R^2}{a^2}\right)^{1/2}\right). \tag{4.63}$$

We should use the radius D of the circle on the sphere to express the circumference and area of the circle on the sphere, thereby obtaining

$$C = 2\pi a \sin(D/a), \quad A = 2\pi a^2(1 - \cos(D/a)). \tag{4.64}$$

As D increases, C and A also increase simultaneously, until when $D = \frac{\pi a}{2}$, i.e., at the equator. The ratio of the circumference of the circle on the sphere to its radius is

$$\frac{C}{D} = 2\pi \frac{\sin(D/a)}{D/a} < 2\pi, \tag{4.65}$$

which equals 2π only when $D \ll a$. Note that since this set of coordinates is singular at the equator, we cannot cross the equator, and the total area of the sphere should be

$$A_{\text{tot}} = 2A(R = a) = 4\pi a^2. \tag{4.66}$$

Similar to the circular arc on a plane, we can consider a circular arc relative to a very small angle ϕ, whose length is equal to the radius multiplied by the angle,

$$\eta = R\phi = a \sin(D/a)\phi. \tag{4.67}$$

When the radius D on the sphere is very small, it can be expanded

$$\eta \approx \phi \left(D - \frac{1}{6} K D^3 + \cdots \right), \tag{4.68}$$

where K is a constant

$$K = \frac{1}{a^2}. \tag{4.69}$$

Furthermore, we can analyze how the arc length changes relative to the radius, obtaining

$$\frac{\mathrm{d}^2\eta}{\mathrm{d}D^2} = -K\eta. \tag{4.70}$$

This fundamental relation describes how neighboring geodesics emanating from a common point diverge, with their relative separation rate being governed by the curvature, as illustrated in Figure 4.13. The quantity K is known as the Gaussian curvature. For a two-dimensional sphere, being a maximally symmetric space, the Gaussian curvature remains constant everywhere. However, for arbitrary two-dimensional manifolds, the Gaussian curvature typically varies from point to point. Remarkably, the expansion in Eq. (4.68) remains valid for any surface, provided we consider sufficiently close geodesics (the surface's "straight lines") passing through a given point.

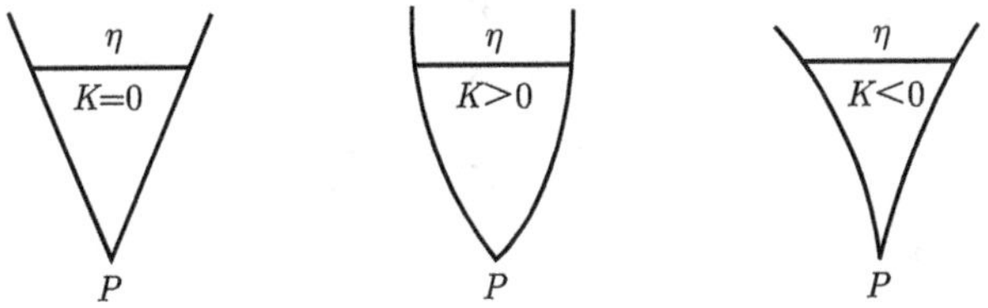

Fig. 4.13 The relative change of adjacent geodesics under different Gaussian curvatures.

The Gaussian curvature at a point can be determined locally by examining the circumference and area of infinitesimal circles centered at that point. From our previous analysis, for sufficiently small circles of geodesic radius D, we obtain the expansions:

$$C = 2\pi\left(D - \frac{1}{6}KD^3 + \cdots\right), \quad A = \pi\left(D^2 - \frac{1}{12}KD^4 + \cdots\right), \tag{4.71}$$

where K represents the Gaussian curvature. These expansions lead to two equivalent operational definitions of the curvature:

$$\begin{aligned} K &= \frac{3}{\pi}\lim_{D\to 0}\frac{2\pi D - C}{D^3} \\ &= \frac{12}{\pi}\lim_{D\to 0}\frac{\pi D^2 - A}{D^4}. \end{aligned}$$

The sign of K carries important geometric meaning: In locally convex region (where the surface bends outward like a sphere), both C and A are smaller than their Euclidean counterparts, yielding $K > 0$; In locally convex region (where the surface bends outward like a sphere), C and A exceed their Euclidean values, resulting in $K < 0$. For the Gaussian curvature, more generally, we have the Gauss–Bonnet theorem: for any two-dimensional compact manifold Σ without boundary,

$$\int_\Sigma K\mathrm{d}S = 2\pi\chi(\Sigma), \tag{4.72}$$

where $\chi(\Sigma)$ is the Euler characteristic of the two-dimensional manifold, which is a topological invariant,[7]

$$\chi(\Sigma) = 2 - 2g, \tag{4.73}$$

g is the genus of the surface. Therefore, for a sphere, $\int_{\mathbf{S}^2} K\mathrm{d}S = 4\pi$, for a torus with genus 1, $\int_{\mathbf{T}^2} K\mathrm{d}S = 0$, and for a more general Riemann surface with genus n, $\int_{\Sigma} K\mathrm{d}S = -(n-1)4\pi$.

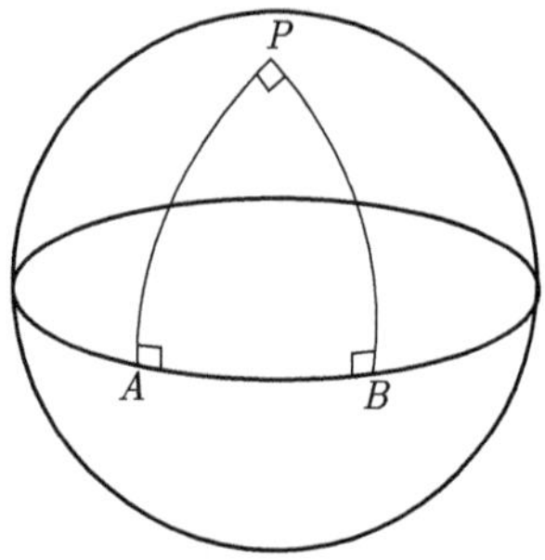

Fig. 4.14 A triangle on the sphere formed by two 90°-separated meridians and the equator.

In Euclidean geometry, the circumference-to-radius ratio of a circle is exactly 2π, and the sum of a triangle's interior angles is π. However, these fundamental relationships take different forms in spherical geometry. On a spherical surface, the sum of a triangle's interior angles is given by $\pi + \dfrac{A}{a^2}$, where A denotes the triangle's area and a is the sphere's radius. Here, the triangle on the sphere is formed by three points connected by geodesics—the shortest paths between pairs of points. Each angle of the triangle is determined by the angle between the tangent vectors of the two intersecting geodesics at a vertex. As shown in Figure 4.14, consider a spherical triangle bounded by two 90°-separated meridians and the equator. At each vertex, the tangent vectors of the intersecting geodesics are mutually orthogonal, resulting in interior angles of $\dfrac{\pi}{2}$ at all three vertices. Consequently, the angle sum for this triangle is $\dfrac{3\pi}{2}$. This geometric configuration covers

[7]The Gauss-Bonnet theorem mentioned here is in the context of a two-dimensional manifold without boundary. If there is a boundary, the contribution of the boundary must be taken into account.

one-eighth of the sphere's total surface area, which can be verified by direct computation.

4.4.3 *Volume elements*

The preceding discussion on surface area can be generalized naturally to other high-dimensional manifolds. Consider a metric tensor in a coordinate system where it is diagonal: $g_{\mu\nu} = \mathrm{diag}(g_{11}, \cdots, g_{nn})$, then the volume element takes the form

$$\mathrm{d}^n V = \sqrt{|g_{11} g_{22} \cdots g_{nn}|} \mathrm{d}x^1 \cdots \mathrm{d}x^n. \tag{4.74}$$

Observe that the determinant of the metric satisfies $|g| = |g_{11} g_{22} \cdots g_{nn}| = |\det(g_{\mu\nu})|$, allowing us to rewrite the volume element compactly as $\mathrm{d}^n V = \sqrt{|g|} \mathrm{d}^n x$. Crucially, this expression remains valid even for non-diagonal metrics

$$\mathrm{d}^n V = |\det(g_{\mu\nu})|^{1/2} \mathrm{d}x^1 \cdots \mathrm{d}x^n. \tag{4.75}$$

Under coordinate transformation $x \to x'$, we know that

$$\begin{aligned}\mathrm{d}^n V &= \sqrt{|g'|} \mathrm{d}x'^1 \cdots dx'^n \\ &= \sqrt{|g'|} J \mathrm{d}x^1 \cdots dx^n,\end{aligned}$$

where J is the Jacobian determinant of the coordinate transformation,

$$J = \left| \frac{\partial x'^a}{\partial x^b} \right|, \tag{4.76}$$

under such a coordinate transformation, the metric tensor transforms at a point p as

$$g'_{\mu\nu} = \left(\frac{\partial x^\sigma}{\partial x'^\mu} \right)_p \left(\frac{\partial x^\rho}{\partial x'^\nu} \right)_p g_{\sigma\rho}(p). \tag{4.77}$$

This implies that the determinant of the metric obeys the transformation law $|g'| = \dfrac{1}{J^2} |g|$. Consequently, the volume element retains its form:

$$\mathrm{d}^n V = \sqrt{|g'|} \mathrm{d}x'^1 \cdots \mathrm{d}x'^n = \sqrt{|g|} \mathrm{d}x^1 \cdots \mathrm{d}x^n, \tag{4.78}$$

demonstrating its invariance under coordinate transformations. This invariant volume element is closely related to the Levi–Civita tensor, which we will explore later.

For submanifolds defined by embedding, their volume elements is determined by the induced metric. Consider an m-dimensional submanifold defined by embedding $x^a = x^a(u^1, \cdots, u^m)$. The differential of the embedding coordinates is given by $\mathrm{d}x^a = \frac{\partial x^a}{\partial u^i}\mathrm{d}u^i$, which leads to the induced metric on the submanifold:

$$\begin{aligned} \mathrm{d}s^2 &= g_{ab}\mathrm{d}x^a\mathrm{d}x^b = g_{ab}\frac{\partial x^a}{\partial u^i}\frac{\partial x^b}{\partial u^j}\mathrm{d}u^i\mathrm{d}u^j \\ &= h_{ij}\mathrm{d}u^i\mathrm{d}u^j, \end{aligned} \tag{4.79}$$

where g_{ab} is the metric of the ambient space and h_{ij} is the induced metric on the submanifold,

$$h_{ij} = g_{ab}\frac{\partial x^a}{\partial u^i}\frac{\partial x^b}{\partial u^j}. \tag{4.80}$$

The corresponding volume element of the submanifold is then:

$$\mathrm{d}^m V = \sqrt{|h|}\mathrm{d}u^1 \cdots \mathrm{d}u^m, \tag{4.81}$$

where $|h| = \det(h_{ij})$. This volume element explicitly depends on the embedding through the induced metric. Remarkably, these results hold generally, regardless of whether the embedded submanifold or the ambient space is curved. The construction depends only on the local differential structure of the embedding.

As a concrete illustration, consider calculating the length of a curve on a manifold. Let a curve be parameterized in local coordinates as $x^a(\lambda)$, where λ serves as the curve parameter. The infinitesimal line element along the curve is given by:

$$\mathrm{d}s^2 = g_{ab}\mathrm{d}x^a\mathrm{d}x^b. \tag{4.82}$$

The total length of the curve between points A (at λ_A) and B (at λ_B) is obtained by integrating the line element:

$$\Delta s = \int_{\lambda_A}^{\lambda_B} \mathrm{d}s = \int_{\lambda_A}^{\lambda_B} \sqrt{g_{ab}\frac{\partial x^a}{\partial \lambda}\frac{\partial x^b}{\partial \lambda}}\mathrm{d}\lambda. \tag{4.83}$$

This result can be understood through the lens of induced metrics. For the 1-dimensional curve, the induced metric has a single component:

$$h_{11} = g_{ab}\frac{\partial x^a}{\partial \lambda}\frac{\partial x^b}{\partial \lambda}, \tag{4.84}$$

which leads to the curve's "volume element" (in this case, the line element):

$$\mathrm{d}s = \sqrt{h_{11}}\mathrm{d}\lambda. \tag{4.85}$$

This formulation is completely consistent with our general treatment of submanifolds, where the induced metric determines the intrinsic geometric quantities. For a curve, the "volume" is simply its length, and the induced metric approach correctly reproduces the standard arc length formula.

4.5 Non-coordinate Basis

In the previous discussion of tangent spaces, we introduced the coordinate basis for the tangent space T_pM associated with a chart $\{x^\mu\}$ as $\left\{\widehat{e}_\mu = \dfrac{\partial}{\partial x^\mu}\right\}$, with the corresponding dual basis for the cotangent space T_p^*M: $\{\widehat{\theta}^\mu = \mathrm{d}x^\mu\}$. In this coordinate basis, the metric tensor field admits the representation:

$$\widehat{g} = g_{\mu\nu}\theta^\mu \otimes \theta^\nu. \tag{4.86}$$

We may alternatively consider a general linear transformation of the coordinate basis:

$$\widehat{e}_m = e^\mu_m \widehat{e}_\mu, \tag{4.87}$$

where e^μ_m constitutes a non-degenerate $n \times n$ matrix with $\det(e^\mu_m) > 0$, whose components are smooth functions of the coordinates. This transformation yields a non-coordinate basis (or vielbein) that preserves the manifold's orientation. We can impose an additional orthonormality condition on the transformed basis:

$$\widehat{g}(\widehat{e}_m, \widehat{e}_n) = \eta_{mn}, \tag{4.88}$$

where η_{mn} is the canonical form of the metric (e.g., $\mathrm{diag}(-1, 1, \cdots, 1)$ in Lorentzian signature). This orthonormality condition relates the frame field e^μ_m to the original metric $g_{\mu\nu}$ via:

$$g_{\mu\nu} = e^m_\mu e^n_\nu \eta_{mn}, \tag{4.89}$$

where e^m_μ is the inverse of e^μ_m, satisfying

$$e^m_\mu e^\mu_n = \delta^m_n, \quad e^m_\mu e^\nu_m = \delta^\nu_\mu. \tag{4.90}$$

The dual basis $\{\widehat{\theta}^m\}$ is similarly transformed from the coordinate dual basis:

$$\widehat{\theta}^m = e^m_\mu \widehat{\theta}^\mu, \tag{4.91}$$

and maintains the defining duality relation:

$$\widehat{\theta}^m(\widehat{e}_n) = \delta^m_n. \tag{4.92}$$

The basis $\{\widehat{e}_m, \widehat{\theta}^m\}$ is called the non-coordinate basis, or the orthonormal Lorentz frame. In this basis, the metric tensor field take the canonical form

$$\widehat{g} = \eta_{mn}\widehat{\theta}^m \otimes \widehat{\theta}^n. \tag{4.93}$$

The existence of such orthonormal frames is consistent with Riemannian normal coordinates, where the metric can be locally transformed to η_{mn} at a point (though not necessarily in a finite neighborhood). This reflects the equivalence principle in general relativity. Unlike coordinate bases, non-coordinate bases generally do not commute. Their commutator defines the structure constants C^p_{mn} of the frame:

$$[\widehat{e}_m, \widehat{e}_n] = C^p_{mn}\widehat{e}_p. \tag{4.94}$$

These structure constants encode the failure of the basis to arise from a coordinate system and are related to the spin connection in the Cartan formalism.

Example 4.12. Three-dimensional flat space.

In three-dimensional flat Euclidean space, we can choose spherical coordinates (r, θ, ϕ), in which the metric can be written as

$$\mathrm{d}s^2 = \mathrm{d}r^2 + r^2\mathrm{d}\theta^2 + r^2\sin^2\theta\mathrm{d}\phi^2, \tag{4.95}$$

the coordinate basis is

$$\widehat{e}_r = \frac{\partial}{\partial r}, \quad \widehat{e}_\theta = \frac{\partial}{\partial \theta}, \quad \widehat{e}_\phi = \frac{\partial}{\partial \phi}, \tag{4.96}$$

$$\widehat{\theta}^r = \mathrm{d}r, \quad \widehat{\theta}^\theta = \mathrm{d}\theta, \quad \widehat{\theta}^\phi = \mathrm{d}\phi, \tag{4.97}$$

and the metric coefficients take the diagonal form $g_{ij} = \mathrm{diag}(1, r^2, r^2\sin^2\theta)$. The orthogonal normalized non-coordinate basis is

$$\widehat{e}_1 = \frac{\partial}{\partial r}, \quad \widehat{e}_2 = \frac{1}{r}\frac{\partial}{\partial \theta}, \quad \widehat{e}_3 = \frac{1}{r\sin\theta}\frac{\partial}{\partial \phi}, \tag{4.98}$$

$$\widehat{\theta}^1 = \mathrm{d}r, \quad \widehat{\theta}^2 = r\mathrm{d}\theta, \quad \widehat{\theta}^3 = r\sin\theta\mathrm{d}\phi, \tag{4.99}$$

and at this time the metric coefficients are $g_{ij} = \delta_{ij}$. It is easy to see that

$$[\widehat{e}_1, \widehat{e}_2] = -\frac{1}{r}\widehat{e}_2. \tag{4.100}$$

Note that the choice of a non-coordinate basis is not unique. Much like Riemann normal coordinates, there exists a local gauge freedom in selecting an orthonormal frame. Specifically, two orthonormal frames $\{\widehat{e}_m\}$ and $\{\widehat{e}'_m\}$ at a point p are related by a local Lorentz transformation: $\widehat{e}'_m = \Lambda^n{}_m\widehat{e}_n$, where $\Lambda^n{}_m$ preserves the metric η_{mn}. This gauge symmetry reflects the equivalence of all local inertial frames under Lorentz transformations.

In addition, for a particular observer, we can always choose an orthonormal frame field accompanying him to define a local reference frame. This will be convenient when discussing the laboratory reference frame relative to this observer. In addition, non-coordinate bases have important significance in discussing various mathematical and physical problems: they are indispensable in theories with spinors, e.g., Dirac fields in curved spacetime, where the canonical form η_{mn} simplifies the coupling of matter fields to gravity; they also underlie the tetrad formalism, which relates spacetime curvature to local gauge-like structures.

4.6 Time Orientation

It is well-known that two-dimensional surfaces can be classified as either orientable or non-orientable. A canonical example of a non-orientable surface is the Möbius strip. This distinction extends to Riemannian manifolds, where orientability is determined by the sign of the Jacobian determinant for coordinate transformations. Specifically, a manifold is said to be orientable if the Jacobian determinant of the transition functions between any pair of overlapping coordinate charts is strictly positive. In the context of non-coordinate bases discussed earlier, the preservation of orientation necessitates the condition $\det(e^{\mu}_{m}) > 0$. For pseudo-Riemannian manifolds, the situation becomes more subtle due to the presence of a temporal dimension. In such cases, one must additionally consider the notion of temporal orientation to properly characterize the manifold's structure.

First, consider the tangent space $T_p(M)$ at a point p on the manifold. A vector $\widehat{u}$ is called timelike if $\widehat{g}(\widehat{u}, \widehat{u}) < 0$; it is called null or lightlike if $\widehat{g}(\widehat{u}, \widehat{u}) = 0$; it is called spacelike if $\widehat{g}(\widehat{u}, \widehat{u}) > 0$. A vector $\widehat{u}$ is called causal if it is either timelike or null, i.e., $\widehat{g}(\widehat{u}, \widehat{u}) \leq 0$.

Proposition Let $\widehat{u}$, $\widehat{v} \in T_p(M)$.

(1) If $\widehat{u}$ is timelike and $\widehat{v}$ is orthogonal to it, then either $\widehat{v} = 0$ or $\widehat{v}$ is a spacelike vector;
(2) If $\widehat{u}$ and $\widehat{v}$ are both null vectors, then they are orthogonal if and only if they are proportional to each other.

We say that two timelike vectors $\widehat{u}$ and $\widehat{v}$ are co-oriented if their inner product satisfies $\widehat{u} \cdot \widehat{v} < 0$. It can be shown that co-orientation defines an equivalence relation on the set of timelike vectors at a point. Specifically: every timelike vector is co-oriented with itself; if $\widehat{u}$ is co-oriented with $\widehat{v}$, then $\widehat{v}$ is co-oriented with $\widehat{u}$; if $\widehat{u}$ and $\widehat{v}$ are co-oriented, and $\widehat{u}$ and $\widehat{w}$ are co-oriented, then $\widehat{u}$ and $\widehat{w}$ must be co-oriented. This equivalence relation partitions the timelike vectors at each point into two distinct equivalence classes, representing the two possible time orientations on the manifold. Notably, a timelike vector $\widehat{u}$ and its negative $-\widehat{u}$ are never co-oriented; they necessarily belong to different classes.

At a given event p on the manifold, the tangent space $T_p(M)$ captures the local causal structure. By introducing a non-coordinate basis, we can locally define an orthogonal Lorentz frame, which allows us to characterize the causal cone:

$$\mathcal{C} = \{\widehat{g}(\widehat{u}, \widehat{u}) \leq 0, \forall \widehat{u} \in T_p(M)\}, \tag{4.101}$$

where the boundary $\widehat{g}(\widehat{u}, \widehat{u}) = 0$ defines the light cone, separating time-like and spacelike vectors. The causal cone naturally decomposes into two disjoint components:

$$\mathcal{C} = \mathcal{C}_+ \cup \mathcal{C}_-, \tag{4.102}$$

where

$$\mathcal{C}_+ : \widehat{\theta}^0(\widehat{u}) = u^0 > 0, \quad \mathcal{C}_- : u^0 < 0. \tag{4.103}$$

This partitioning reflects the local Minkowski structure, with $\mathcal{C}_+$ and $\mathcal{C}_-$ representing the future and past causal cones, respectively. Notably, for any timelike vector $\widehat{u} \in \mathcal{C}_+$, its negative $-\widehat{u}$ lies in $\mathcal{C}_-$, and vice versa. Furthermore, as established earlier, the co-orientation condition $\widehat{u} \cdot \widehat{v} < 0$ ensures that all timelike vectors within $\mathcal{C}_+$ (or $\mathcal{C}_-$) are mutually co-oriented, while vectors from opposite cones are not. Thus, the choice of $\mathcal{C}_+$ or $\mathcal{C}_-$ at each point defines a time orientation on the manifold.

A pseudo-Riemannian manifold $(M, \widehat{g})$ is said to be time-orientable if it admits a continuous timelike vector field $\widehat{\tau}$ globally defined on M. Given a time-orientable manifold, two timelike vector fields $\widehat{\tau}$ and $\widehat{\tau}'$ are called co-oriented if they are co-oriented at every point on the manifold. This relation partitions the set of all timelike vector fields into two global equivalence classes, representing the two possible consistent time orientations on M. One of these classes is designated as the future direction, while the other corresponds to the past. In the context of the local Minkowski structure discussed earlier, the causal cone $\mathcal{C}_+$provides a local representation of the future cone. The time-orientability condition ensures that this local choice of future/past can be extended continuously to the entire manifold, eliminating topological obstructions (e.g., "time-reversing" loops like those on a Möbius strip adapted to Lorentzian geometry). In general relativity, the physical interpretation of spacetime requires the exclusion of non-time-orientable manifolds. Thus, the term spacetime is reserved for time-orientable Lorentzian manifolds.

Physically, in the comoving reference frame of an observer, the 4-velocity $\widehat{u}_0$ defines the local time direction, pointing toward the future. A non-zero causal vector $\widehat{\xi}$ in this frame is classified as future-directed or past-directed based on its alignment with $\widehat{u}_0$: if $\widehat{u}_0 \cdot \widehat{\xi} < 0$, the vector is future-directed; if $\widehat{u}_0 \cdot \widehat{\xi} > 0$, the vector is past-directed. This classification extends to smooth causal curves, which are correspondingly termed future-directed or past-directed based on their tangent vectors.

Given two points p and q in a spacetime manifold M, we write $p \ll q$ (or $p < q$), if there exists a smooth future-directed timelike (or causal) curve $\gamma : [a, b] \to M$, such that $\gamma(a) = p$, $\gamma(b) = q$. Note that for all $p \in M$, the relation $p < p$ holds trivially (via a constant curve), but $p \ll p$ is generally false—it holds if and only if there exists a smooth closed timelike curve (CTC) through p. Such a CTC violates causality, leading to pathological scenarios (e.g., time travel).

4.7 Tensor Densities

We now examine the definition of the Levi–Civita tensor in curved space. At first glance, one might attempt to define it in the same way as in four-dimensional Minkowski spacetime:

$$\widetilde{\varepsilon}_{\mu_1\mu_2\cdots\mu_n} = \begin{cases} +1, \text{ if } (\mu_1\mu_2\cdots\mu_n) \text{ is an even permutation of } (01\cdots(n-1)), \\ -1, \text{ if } (\mu_1\mu_2\cdots\mu_n) \text{ is an odd permutation of } (01\cdots(n-1)), \\ 0, \quad \text{otherwise.} \end{cases} \tag{4.104}$$

However, as we shall demonstrate, this object does not transform as a tensor under general coordinate transformations. For this reason, it is more appropriate to refer to it as the Levi–Civita symbol. By definition, the symbol remains invariant under coordinate changes—a property that a genuine tensor cannot satisfy. To understand this distinction, consider an $n \times n$ matrix M with determinant $|M|$. The Levi–Civita symbol obeys the relation:

$$\widetilde{\varepsilon}_{\mu_1'\mu_2'\cdots\mu_n'}|M| = \widetilde{\varepsilon}_{\mu_1\mu_2\cdots\mu_n} M^{\mu_1}_{\mu_1'} M^{\mu_2}_{\mu_2'} \cdots M^{\mu_n}_{\mu_n'}. \tag{4.105}$$

Now, let $M^{\mu}_{\mu'} = \partial x^{\mu}/\partial x^{\mu'}$ represent the Jacobian matrix of a coordinate transformation. Substituting this into the above equation yields:

$$\widetilde{\varepsilon}_{\mu_1'\mu_2'\cdots\mu_n'} = \left|\frac{\partial x^{\mu'}}{\partial x^{\mu}}\right| \widetilde{\varepsilon}_{\mu_1\mu_2\cdots\mu_n} \frac{\partial x^{\mu_1}}{\partial x^{\mu_1'}} \frac{\partial x^{\mu_2}}{\partial x^{\mu_2'}} \cdots \frac{\partial x^{\mu_n}}{\partial x^{\mu_n'}}. \tag{4.106}$$

This transformation rule closely resembles that of a tensor, except for the presence of the Jacobian determinant factor. This discrepancy confirms that the Levi–Civita symbol does not qualify as a true tensor in curved spacetime.

A quantity that transforms like a tensor but includes an additional factor of the Jacobian determinant is referred to as a tensor density. The exponent of the Jacobian determinant in the transformation rule defines the weight of the density. From the transformation law derived earlier, we see that the Levi–Civita symbol picks up a single power of the Jacobian determinant. Thus, it constitutes a tensor density of weight 1. Another fundamental example of a tensor density is the determinant of the metric field $g = |g_{\mu\nu}|$, which transforms as

$$g(x^{\mu'}) = \left| \frac{\partial x^{\mu'}}{\partial x^{\mu}} \right|^{-2} g(x^{\mu}). \tag{4.107}$$

This shows that g is a (scalar) density of weight -2.

The existence of tensor densities of varying weights allows us to systematically construct true tensors. Given a tensor density of weight w, the product with $|g|^{w/2}$ transforms as a proper tensor under coordinate transformations. For the Levi–Civita symbol, we may define the Levi–Civita tensor

$$\varepsilon_{\mu_1\mu_2\cdots\mu_n} = \sqrt{|g|}\tilde{\varepsilon}_{\mu_1\mu_2\cdots\mu_n}. \tag{4.108}$$

In certain contexts, it is necessary to work with the Levi–Civita tensor expressed with upper indices. This raised-index version is obtained through the following relation:

$$\varepsilon^{\mu_1\mu_2\cdots\mu_n} = \mathrm{sgn}(g)\frac{1}{\sqrt{|g|}}\tilde{\varepsilon}^{\mu_1\mu_2\cdots\mu_n}, \tag{4.109}$$

where $\mathrm{sgn}(g)$ denotes the sign of the determinant of the metric, positive for Riemannian manifolds and negative for pseudo-Riemannian manifolds. By construction, both the raised- and lowered-index forms of the Levi–Civita tensor are completely antisymmetric. An important topological restriction arises here: the Levi–Civita tensor can only be globally defined on orientable manifolds, as its existence requires the vanishing of the second Stiefel–Whitney class.[8] For our purposes, we implicitly assume all manifolds under consideration are orientable, ensuring the Levi–Civita tensor is well-defined throughout.

As previously discussed, the Levi–Civita tensor reduces to the Levi–Civita symbol in flat spacetime. To generalize this to curved spacetime, we first define the tensor in a local orthogonal Lorentz frame—where the metric takes its Minkowski form—and then extend it covariantly to arbitrary

[8]This aligns with the requirement that the manifold admits a nowhere-vanishing volume form.

coordinates. We have

$$\begin{aligned}\widehat{\varepsilon} &= \widetilde{\varepsilon}_{mnpq}\widehat{\theta}^m \otimes \widehat{\theta}^n \otimes \widehat{\theta}^p \otimes \widehat{\theta}^q \\ &= \widetilde{\varepsilon}_{mnpq}e^m_\mu e^n_\nu e^p_\sigma e^q_\rho \widehat{\theta}^\mu \otimes \widehat{\theta}^\nu \otimes \widehat{\theta}^\sigma \otimes \widehat{\theta}^\rho \\ &= \widetilde{\varepsilon}_{\mu\nu\sigma\rho}|e|\widehat{\theta}^\mu \otimes \widehat{\theta}^\nu \otimes \widehat{\theta}^\sigma \otimes \widehat{\theta}^\rho \\ &= \varepsilon_{\mu\nu\sigma\rho}\widehat{\theta}^\mu \otimes \widehat{\theta}^\nu \otimes \widehat{\theta}^\sigma \otimes \widehat{\theta}^\rho.\end{aligned} \tag{4.110}$$

In fact, the Levi–Civita tensor is related to the volume element of the manifold. Before discussing this relationship, we first introduce some knowledge of exterior differential forms and differential forms.

4.8 Differential Forms

A differential p-form is a completely antisymmetric tensor of type $(0, p)$. For instance, a scalar is a 0-form, a dual (co-)vector is a 1-form, the electromagnetic field strength tensor corresponds to a 2-form, and the Levi–Civita tensor is a 4-form (on a 4-dimensional manifold). The space of all p-forms is denoted by Λ^p, while the space of p-form fields on a manifold M is written as $\Lambda^p(M)$. In an n-dimensional vector space, the number of linearly independent p-forms is given by $n!/(p!(n-p)!)$, which also gives the dimension of $\Lambda^p(M)$. Due to antisymmetry, there are no differential forms with $p > n$ on an n-dimensional manifold. One major advantage of differential forms is that they allow for intrinsic definitions of differentiation and integration without relying on additional geometric structures, such as connections or a metric.

Given a p-form A and a q-form B, their exterior product (or wedge product) $A \wedge B$ yields a $(p+q)$-form. The components of this product are defined as:

$$(A \wedge B)_{\mu_1 \cdots \mu_{p+q}} = \frac{(p+q)!}{p!q!} A_{[\mu_1 \cdots \mu_p} B_{\mu_{p+1} \cdots \mu_{p+q}]}, \tag{4.111}$$

where the brackets $[\cdots]$ denote antisymmetrization over all indices. For example, for two 1-forms, their exterior product is a 2-form given by

$$(A \wedge B)_{\mu\nu} = 2A_{[\mu}B_{\nu]} = A_\mu B_\nu - A_\nu B_\mu. \tag{4.112}$$

Note that the exterior product is not commutative in general, but instead satisfies:

$$A \wedge B = (-1)^{pq} B \wedge A. \tag{4.113}$$

Any p-form A_p can be expanded in terms of the coordinate basis $\{\mathrm{d}x^\mu\}$ as

$$A_p = \frac{1}{p!} A_{\mu_1\mu_2\cdots\mu_p} \mathrm{d}x^{\mu_1} \wedge \mathrm{d}x^{\mu_2} \wedge \cdots \wedge \mathrm{d}x^{\mu_p}. \tag{4.114}$$

A fundamental operation on differential form fields is the exterior derivative, denoted by the symbol "d". This operator maps a p-form field to a $p+1$-form field, expressed as:

$$\mathrm{d} : \Lambda^p \to \Lambda^{p+1}. \tag{4.115}$$

The explicit definition of the exterior derivative is given by:

$$(\mathrm{d}A)_{\mu_1\cdots\mu_{p+1}} \equiv (p+1)\partial_{[\mu_1}A_{\mu_2\cdots\mu_{p+1}]}. \tag{4.116}$$

For a scalar field, $(\mathrm{d}\phi)_\mu = \partial_\mu\phi$. For a 1-form (such as a gauge potential) $A = A_\mu \mathrm{d}x^\mu$,

$$(\mathrm{d}A)_{\mu\nu} = \partial_\mu A_\nu - \partial_\nu A_\mu. \tag{4.117}$$

An essential property of the exterior derivative is that it yields a well-defined tensor even in curved spacetime, in contrast to the ordinary partial derivative. This robustness arises because the antisymmetrization in its definition ensures that the non-tensorial terms introduced by coordinate transformations cancel out, leaving only components that transform correctly under a change of coordinates.

By definition, the exterior derivative satisfies nilpotency, meaning that applying it twice to any differential form A yields zero:

$$\mathrm{d}(\mathrm{d}A) = 0. \tag{4.118}$$

This property, $\mathrm{d}^2 = 0$, plays a crucial role in the algebraic and geometric structure of differential forms, particularly in homology and cohomology theory. Additionally, the exterior derivative obeys a graded Leibniz rule when acting on the wedge product of a p-form A and a q-form B,

$$\mathrm{d}(A \wedge B) = \mathrm{d}A \wedge B + (-1)^p A \wedge dB. \tag{4.119}$$

This rule reflects the anticommutative nature of differential forms, where the sign factor $(-1)^p$ arises from permuting d past the p-form A.

Another fundamental operation on differential forms is the Hodge duality. The Hodge star operator, defined on an n-dimensional manifold, is a linear map $* : \Lambda^p \to \Lambda^{n-p}$, that takes a p-form to an $(n-p)$-form. Explicitly, its action on a p-form A_p is given by

$$(*A)_{\mu_1\cdots\mu_{n-p}} = \frac{1}{p!}\varepsilon^{\nu_1\cdots\nu_p}{}_{\mu_1\cdots\mu_{n-p}} A_{\nu_1\cdots\nu_p}, \tag{4.120}$$

where ε denotes the Levi–Civita tensor. A key property of the Hodge star operator is its behavior under repeated application. Specifically, acting twice on a p-form A yields

$$** A = (-1)^{s+p(n-p)} A, \tag{4.121}$$

where s counts the number of negative eigenvalues (i.e., "-1" terms) in the canonical form of the metric tensor.

The Hodge dual is distinct from the standard duality between vectors and covectors, despite the fact that the space of p-form Λ^p has the same dimension as the space of $n-p$ form Λ^{n-p}. Moreover, the Hodge star operator satisfies

$$*(A^{(n-p)} \wedge B^{(p)}) \in \mathbf{R}, \tag{4.122}$$

meaning that the Hodge dual of the wedge product of an $(n-p)$-form and a p-form yields a scalar function. A particularly important case arises in three-dimensional Euclidean space, where the Hodge dual of the exterior product of two 1-forms U and V is itself a 1-form:

$$*(U \wedge V)_i = \varepsilon_i{}^{jk} U_j V_k. \tag{4.123}$$

This operation is equivalent to the cross product of two vectors in three-dimensional vector calculus. However, such a direct correspondence does not generalize to higher dimensions, where no analogous vector product exists.

Example 4.13. Electromagnetic fields and U(1) gauge fields.

As mentioned earlier, a better language for describing electromagnetic theory is to use Abelian gauge fields. First, we have the 1-form gauge potential $\widehat{A} = A_\mu \mathrm{d}x^\mu$, and the electromagnetic field strength is the 2-form, $\widehat{F} = \mathrm{d}\widehat{A}$, which remains invariant under the gauge transformation $\widehat{A} \to \widehat{A} + \mathrm{d}\chi$ due to $\mathrm{d}^2 = 0$. Using form fields, the Maxwell equations can be written very concisely. One set of equations is trivial,

$$\mathrm{d}\widehat{F} = 0, \tag{4.124}$$

which comes from the nilpotency $\mathrm{d}^2 = 0$ of the exterior derivative operator. In other words, the two source-free equations indicate that the electromagnetic field strength can be expressed by the gauge field, with gauge invariance. As for the two equations related to the flow, we can write the flow as a 1-form, thereby obtaining

$$\mathrm{d}(*\widehat{F}) = 4\pi(*\widehat{J}). \tag{4.125}$$

In vacuum, since there is no flow, the above two equations have the following Hodge dual invariance:

$$\begin{aligned} \widehat{F} &\to *\widehat{F}, \\ *\widehat{F} &\to -\widehat{F}. \end{aligned} \tag{4.126}$$

If written in component form, we will see that this Hodge dual actually exchanges the electric and magnetic fields, so it is also commonly referred to as electromagnetic duality. If there is a source present, it seems that we can introduce a "magnetic" current $\widehat{J}_M$, and then add $\widehat{J} \leftrightarrow \widehat{J}_M$, and it appears that the duality still holds. However, if there is a magnetic source, it means that the gauge field cannot be globally defined, and we must introduce a magnetic monopole with singularities. The concept of magnetic monopoles was first proposed by Dirac. If there are magnetic monopoles, considering the wave function of a charged particle, quantum mechanics gives the quantization condition of the charge

$$eg = 2\pi n, \tag{4.127}$$

where e is the charge of the charged particle, and g is the magnetic charge. Dirac's quantization condition can naturally explain why the charges of fundamental particles in nature are integer multiples of the electron charge. Furthermore, this condition indicates that electromagnetic duality is a strong-weak duality, e small then g large, and vice versa, so electromagnetic duality is also commonly referred to as S-duality. However, in electromagnetic theory, the equation $\mathrm{d}\widehat{F} = \widehat{J}_M$ cannot find a globally defined non-singular gauge potential, and it is necessary to introduce the Dirac string to eliminate the singularity. But magnetic monopoles are naturally realized in quantum field theory. In particular, in supersymmetric quantum field theory and superstring theory, electromagnetic duality has important physical significance. In addition, in recent years, it has been found that S-duality is closely related to the Langlands program in mathematics.

To conclude, we revisit the concept of integration on manifolds. On an n-dimensional manifold M, the integration of a differential form can be viewed as a map that takes an n-form to a real number. Specifically, for a domain of integration $\Sigma \subset M$, we define

$$\Sigma \in M, \quad \int_\Sigma : \widehat{\omega} \to \mathbf{R}, \tag{4.128}$$

where $\widehat{\omega}$ is an n-form. For example, for a line integral, the integrand is a 1-form $\widehat{\omega} = \omega(x)\mathrm{d}x$, yielding the familiar expression $\displaystyle\int \omega(x)\mathrm{d}x$. For high-dimensional integrals, the integrand is constructed from the volume element associated with the manifold. Consider a three-dimensional manifold as an illustrative example. An infinitesimal volume element can be described as a parallelepiped spanned by three tangent vectors $\boldsymbol{U}$, $\boldsymbol{V}$, $\boldsymbol{W}$. More rigorously, this volume element is defined by the exterior product of these vectors, resulting in a 3-vector. The corresponding integration element is a real-valued functional:

$$\mathrm{d}\widehat{\mu}(\boldsymbol{U},\,\boldsymbol{V},\,\boldsymbol{W})\in\mathbf{R},\tag{4.129}$$

which satisfies multilinearity and antisymmetry: $\mathrm{d}\widehat{\mu}(a\boldsymbol{U},b\boldsymbol{V},c\boldsymbol{W}) = (abc)\mathrm{d}\widehat{\mu}(\boldsymbol{U},\,\boldsymbol{V}.\boldsymbol{W})$. This implies that $\mathrm{d}\widehat{\mu}$ is a totally antisymmetric (0, 3)-type tensor, i.e., a 3-form. The antisymmetry reflects the oriented nature of the volume element—swapping any two vectors introduces a sign change. On a general n-dimensional manifold, the integration element is an n-form, encoding both the infinitesimal volume and the orientation of the domain of integration. For further details on the geometric interpretation of p-forms and their integrals, see [5].

As mentioned earlier, the integration element of the form $\mathrm{d}^n x = \mathrm{d}x^0\cdots\mathrm{d}x^{n-1}$ is not a tensor, but a tensor density,

$$\begin{aligned}\mathrm{d}x^0\cdots\mathrm{d}x^{n-1} &= \frac{1}{n!}\widetilde{\varepsilon}_{\mu_1\cdots\mu_n}\mathrm{d}x^{\mu_1}\wedge\cdots\wedge\mathrm{d}x^{\mu_n}\\ &\to \frac{1}{n!}\widetilde{\varepsilon}_{\mu_1\cdots\mu_n}\frac{\partial x^{\mu_1}}{\partial x^{\mu_1'}}\cdots\frac{\partial x^{\mu_n}}{\partial x^{\mu_n'}}\mathrm{d}x^{\mu_1'}\wedge\cdots\wedge\mathrm{d}x^{\mu_n'}\\ &= \left|\frac{\partial x^{\mu}}{\partial x'^{\mu}}\right|\widetilde{\varepsilon}_{\mu_1'\cdots\mu_n'}\mathrm{d}x^{\mu_1'}\wedge\cdots\wedge\mathrm{d}x^{\mu_n'}.\end{aligned}\tag{4.130}$$

We can insert a factor $\sqrt{|g|}$ to obtain a tensor:

$$\sqrt{|g|}\mathrm{d}^n x = \sqrt{|g|}\mathrm{d}x^0\cdots\mathrm{d}x^{n-1}.\tag{4.131}$$

This is actually the Levi–Civita tensor

$$\begin{aligned}\widehat{\varepsilon} &= \varepsilon_{\mu_1\cdots\mu_n}\mathrm{d}x^{\mu_1}\otimes\cdots\otimes\mathrm{d}x^{\mu_n}\\ &= \frac{\sqrt{|g|}}{n!}\widetilde{\varepsilon}_{\mu_1\cdots\mu_n}\mathrm{d}x^{\mu_1}\wedge\cdots\wedge\mathrm{d}x^{\mu_n} = \sqrt{|g|}\mathrm{d}^n x.\end{aligned}\tag{4.132}$$

In short, the Levi–Civita tensor is the volume element.

The general form of integration on an n-dimensional manifold M is given by

$$I = \int_M \phi(x)\sqrt{|g|}\mathrm{d}^n x,\tag{4.133}$$

where $\phi(x)$ is the scalar integrand and $\sqrt{|g|}\mathrm{d}^n x$ represents the invariant volume element induced by the metric. Suppose the manifold M has a boundary ∂M, and there is an $n-1$ form field $\widehat{\omega}$ on M, then $\mathrm{d}\widehat{\omega}$ is an n-form field that can be integrated on the manifold, and $\widehat{\omega}$ itself can be integrated on ∂M, thus we have Stokes' theorem

$$\int_M \mathrm{d}\widehat{\omega} = \int_{\partial M}\widehat{\omega}.\tag{4.134}$$

This elegant result unifies several classical theorems in vector calculus: the Newton–Leibniz formula for definite integrals, as well as the Green's theorem, Gauss's divergence theorem, and classical Stokes' theorem for surface integrals.

Exercises

1. Prove that the commutator of two vector fields is still a vector field.
2. The surface of the Earth can be regarded as a two-dimensional sphere with radius a. Using the standard spherical coordinates (θ, ϕ), Greenwich is taken as $\phi = 0$, and east is taken as the positive direction of ϕ. The longitude of a point on the Earth is simply ϕ, and the latitude is $\lambda = \pi/2 - \theta$. Prove that using longitude and latitude, the metric of the sphere can be written as
$$\mathrm{d}s^2 = a^2(\mathrm{d}\lambda^2 + \cos^2\lambda \mathrm{d}\phi^2). \tag{4.135}$$
Furthermore, we can introduce functions $x = x(\lambda, \phi)$, $y = y(\lambda, \phi)$ as Cartesian coordinates to draw maps. The Mercator projection[9] is a commonly used way to draw maps, which is defined as
$$x = \frac{W\phi}{2\pi}, \quad y = \frac{H}{2\pi}\ln\left(\tan\left(\frac{\pi}{4} + \frac{\lambda}{2}\right)\right), \tag{4.136}$$
where W and H are the width and length of the map, respectively. Please give the line element of this projection.
3. In the choice of drawing maps in the previous exercise, prove that if the choice of functions x and y satisfies
$$\Omega(x, y)(\mathrm{d}x^2 + \mathrm{d}y^2) = a^2(\mathrm{d}\lambda^2 + \cos^2\lambda \mathrm{d}\phi^2), \tag{4.137}$$
where $\Omega(x, y)$ is some function, then the angle between two directions at a point on the sphere is preserved in the drawn map. Prove that the Mercator projection drawing of the map satisfies this relationship.

[9]The Mercator projection, developed by the Flemish cartographer Gerardus Mercator in 1569, is a conformal (angle-preserving) cylindrical projection that revolutionized navigation. Its geometric construction and mathematical properties are as follows. Imagine the Earth (modeled as a sphere) enclosed in a tangent cylinder, where the cylinder touches the Earth along the equator and the Earth's rotational axis aligns with the cylinder's axis. Project the figures on the sphere from the center of the Earth onto the cylinder, and then unroll the cylinder. This is how the map is drawn using the "Mercator projection." The Mercator projection has no angular distortion, with the scale of length in all directions from each point being equal. Its meridians and parallels are all parallel straight lines, intersecting at right angles. The intervals between meridians are equal, while the intervals between parallels gradually increase from the reference latitude towards the poles. On the map drawn with the Mercator projection, there is significant distortion in length and area, but the correctness of direction and angle is maintained.

4. In the previous exercises, if we require the drawn map to preserve the area, what are the requirements for x and y? Is there a map that preserves both angles and area at the same time?
5. In four-dimensional spacetime, for the convenience of discussing problems, we often divide the metric components into three groups: g_{00}, g_{0i}, g_{ij}, where the 0-coordinate represents the time direction. This so-called $1+3$ decomposition can be done in two ways. The first way is to write the metric as

$$\mathrm{d}s^2 = -M^2(\mathrm{d}t - M_i \mathrm{d}x^i)^2 + \gamma_{ij}\mathrm{d}x^i\mathrm{d}x^j, \tag{4.138}$$

where M, M_i, γ_{ij} are all real functions. The second way is to write the metric as

$$\mathrm{d}s^2 = -N^2\mathrm{d}t^2 + g_{ij}(\mathrm{d}x^i + N^i\mathrm{d}t)(\mathrm{d}x^j + N^j\mathrm{d}t), \tag{4.139}$$

where N, N_i, g_{ij} are real functions. Please find the components of these two metrics and the components of the inverse metric.
6. The metric of a two-dimensional manifold can be written as

$$\mathrm{d}s^2 = \mathrm{d}r^2 + f^2(r)\mathrm{d}\theta^2, \tag{4.140}$$

prove that its Gaussian curvature is $K = -\dfrac{f''}{f}$.
7. The metric of a three-dimensional sphere with radius a can be written as

$$\mathrm{d}s^2 = a^2(\mathrm{d}\xi^2 + \sin^2\xi(\mathrm{d}\theta^2 + \sin^2\theta\mathrm{d}\phi^2)). \tag{4.141}$$

(1) Calculate the area of the two-dimensional sphere defined by $\xi = \xi_0$;
(2) Calculate the total volume of this three-dimensional sphere.

8. If the line element of a three-dimensional space is

$$\mathrm{d}s^2 = \frac{\mathrm{d}r^2}{1 - 2GM/r} + r^2(\mathrm{d}\theta^2 + \sin^2\theta\mathrm{d}\phi^2), \tag{4.142}$$

calculate the following quantities:

(1) The surface area of the sphere with coordinate radius $r = R$;
(2) The volume of the sphere with coordinate radius $r = R$;
(3) The radial distance from the sphere with radius $r = 2GM$ to the sphere with $r = 3GM$;
(4) The volume between the two spheres in the previous question.

9. Prove that in an n-dimensional spacetime, we can always find Riemann normal coordinates such that

$$g_{\mu\nu}|_P = \eta_{\mu\nu}, \quad \partial_\sigma g_{\mu\nu}|_P = 0, \tag{4.143}$$

but we cannot make $\partial_\sigma \partial_\rho g_{\mu\nu}|_P$ all zero. Moreover, we can always ensure the existence of the local Lorentz symmetry of the inertial frame, and the number of $\partial_\sigma \partial_\rho g_{\mu\nu}|_P \neq 0$ is $\frac{1}{2}n^2(n^2-1)$.

10. Prove the proposition in Section 4.6.
11. For a p-form $\widehat{\omega}$ and a q-form $\widehat{\eta}$, prove

(1) Their exterior product satisfies $\widehat{\omega} \wedge \widehat{\eta} = (-1)^{pq} \widehat{\eta} \wedge \widehat{\omega}$;

(2) The exterior derivative of their exterior product satisfies

$$\mathrm{d}(\widehat{\omega} \wedge \widehat{\eta}) = (\mathrm{d}\widehat{\omega}) \wedge \widehat{\eta} + (-1)^p \widehat{\omega} \wedge (\mathrm{d}\widehat{\eta}). \tag{4.144}$$

12. For a p-form $\widehat{A}$, prove

$$* * (\widehat{A}) = (-1)^{s+p(n-p)} \widehat{A}, \tag{4.145}$$

where $*$ is the Hodge star operator, s is the number of negative eigenvalues in the eigenvalues of the metric, and n is the dimension of the manifold.

Chapter 5

Connections and Geodesics

We have observed that the partial derivative of a general tensor field fails to produce a new tensor. This necessitates careful consideration of both tensor field comparison in curved spaces and the proper definition of derivative operations. In this chapter, we present the fundamental concept of an affine connection, define the covariant derivative, and discuss the comparison of vectors and tensors across different spacetime points via parallel transport. We establish the fundamental theorem of Riemannian geometry and characterize the dynamics of free point particles in curved spacetime through geodesic motion. The chapter concludes with an introduction to two additional derivative operations of particular utility in differential geometry.

5.1 Covariant Derivatives and Affine Connections

In flat (Euclidean or Minkowski) space, vectors are naturally embedded within the space, each possessing both an initial point and a terminal point. Comparing two vectors is straightforward: one need only translate their initial points to coincide. In contrast, curved spaces present a fundamental complication—vectors are intrinsically local objects. Consequently, comparing vector fields at distinct points becomes a nontrivial task. To perform such a comparison, we must establish a consistent method to transport vectors from one point to another. A natural approach is to connect the two points with a smooth path and perform parallel transport of the vectors along this path. However, this introduces a path-dependence: the result of the comparison depends on the specific curve chosen for the transport.

Let us consider a familiar example: a particle moving in a two-dimensional plane. Its velocity field is tangent to the trajectory and can be expressed in polar coordinates as: $\boldsymbol{v} = v^r \boldsymbol{e}_r + v^\phi \boldsymbol{e}_\phi$, where $\boldsymbol{e}_r$ and $\boldsymbol{e}_\phi$

denote the unit basis vectors in the radial and angular directions, respectively. To determine the particle's acceleration, we must examine how this velocity field changes along the trajectory. This requires computing the time derivative of $\boldsymbol{v}$, which involves not only the derivatives of the components v^r and v^ϕ but also accounting for the changing basis vectors $\boldsymbol{e}_r$ and $\boldsymbol{e}_\phi$ as the particle moves. The acceleration vector is given by

$$\boldsymbol{a} = \frac{\mathrm{d}\boldsymbol{v}}{\mathrm{d}t} = \frac{\mathrm{d}v^r}{\mathrm{d}t}\boldsymbol{e}_r + \frac{\mathrm{d}v^\phi}{\mathrm{d}t}\boldsymbol{e}_\phi + v^r\frac{\mathrm{d}\boldsymbol{e}_r}{\mathrm{d}t} + v^\phi\frac{\mathrm{d}\boldsymbol{e}_\phi}{\mathrm{d}t}. \tag{5.1}$$

Since $\mathrm{d}\boldsymbol{e}_r/\mathrm{d}t = \omega\boldsymbol{e}_\phi = (\mathrm{d}\phi/\mathrm{d}t)\boldsymbol{e}_\phi$ and $\mathrm{d}\boldsymbol{e}_\phi/\mathrm{d}t = -\omega\boldsymbol{e}_r = -(\mathrm{d}\phi/\mathrm{d}t)\boldsymbol{e}_r$, the acceleration vector has components

$$\begin{aligned} a^r &= \frac{\mathrm{d}v^r}{\mathrm{d}t} - v^\phi\frac{\mathrm{d}\phi}{\mathrm{d}t} = \frac{\mathrm{d}^2 r}{\mathrm{d}t^2} - r\left(\frac{\mathrm{d}\phi}{\mathrm{d}t}\right)^2, \\ a^\phi &= \frac{\mathrm{d}v^\phi}{\mathrm{d}t} + v^r\frac{\mathrm{d}\phi}{\mathrm{d}t} = \frac{\mathrm{d}}{\mathrm{d}t}\left(r\frac{\mathrm{d}\phi}{\mathrm{d}t}\right). \end{aligned}$$

From this analysis, we observe that the evolution of a vector field is inherently path-dependent. More importantly, the basis vectors themselves undergo variation along the curve, as illustrated in Figure 5.1.

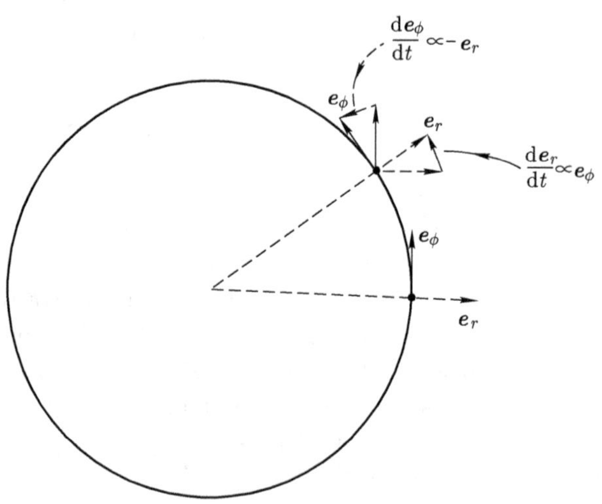

Fig. 5.1 The variation of radial basis vectors and angular basis vectors along the circle.

Specifying a path, which locally be characterized by its tangent vector $\hat{t}$, allows us to approximate the curve in an infinitesimal neighborhood by its linear approximation along $\hat{t}$. The derivative of a vector field $\hat{V}$ along this path is then defined through parallel transport,

$$\nabla_{\hat{t}}\hat{V}(x^\mu) = \lim_{\epsilon\to 0}\frac{[\hat{V}(x^\mu + t^\mu\epsilon)]_{\text{parallel transported to } x^\mu} - \hat{V}(x^\mu)}{\epsilon}, \tag{5.2}$$

where the fundamental requirement is that vectors must be compared at the same spacetime point. Here, parallel transport along the infinitesimal displacement $t^\mu\epsilon$ effectively "slides" the vector from $x^\mu + t^\mu\epsilon$ back to x^μ while maintaining its intrinsic direction in a locally flat approximation. In a local Cartesian coordinate, this reduces to the partial derivative $(\nabla_{\hat{t}}\hat{V})^\alpha = t^\beta\dfrac{\partial V^\alpha}{\partial x^\beta}$, i.e., $\nabla_\beta V^\alpha = \dfrac{\partial V^\alpha}{\partial x^\beta}$. However, in curvilinear coordinates (even in flat space), the basis vectors themselves vary, requiring additional care. This was explicitly demonstrated in our earlier polar coordinates example. While the general definition (5.2) remains valid, the practical computation of the covariant derivative must account for the derivatives of the basis vectors. Thus, the covariant derivative of a vector (or tensor) depends not only on the field's components but also on the tangent vector that characterizes the path.

5.1.1 *Affine connections*

We begin by considering the covariant derivative of a vector field. An affine connection can be formally defined as a bilinear mapping

$$\nabla : \chi(M)\otimes\chi(M)\to\chi(M), \tag{5.3}$$

where $\chi(M)$ denotes the space of all smooth vector fields on the manifold M. This mapping takes two vector fields and assigns to them a third vector field, interpreted as the covariant derivative of one field with respect to the other

$$(\hat{X},\hat{Y})\to\nabla_{\hat{X}}\hat{Y}, \quad \hat{X},\hat{Y}\in\chi(M). \tag{5.4}$$

This mapping should satisfy the following conditions:

(1) $\nabla_{\hat{X}}(\hat{Y}+\hat{Z}) = \nabla_{\hat{X}}\hat{Y} + \nabla_{\hat{X}}\hat{Z}$,
(2) $\nabla_{(\hat{X}+\hat{Y})}\hat{Z} = \nabla_{\hat{X}}\hat{Z} + \nabla_{\hat{Y}}\hat{Z}$,
(3) $\nabla_{f\hat{X}}\hat{Y} = f\nabla_{\hat{X}}\hat{Y}$,
(4) $\nabla_{\hat{X}}(f\hat{Y}) = \hat{X}[f]\hat{Y} + f\nabla_{\hat{X}}\hat{Y}$,

where $f\in\mathcal{F}(M)$ is any function on M, and $\hat{X}$, $\hat{Y}$, $\hat{Z}\in\chi(M)$.

We can interpret $\nabla_{\widehat{X}}\widehat{Y}$ as the change of the vector field $\widehat{Y}$ along the direction $\widehat{X}$, resulting in a new vector. The first two properties follow naturally from the superposition principle of vectors. The third property highlights that the direction of $\widehat{X}$, rather than its magnitude, determines the connection. The final property emphasizes the role of the directional derivative of a scalar function f. Specifically, $\widehat{X}[f] = X^\mu \partial_\mu f$ represents the action of the tangent vector field $\widehat{X}$ on f, which is precisely the directional derivative of f along $\widehat{X}$.

A smooth manifold (M, ∇) endowed with a fixed affine connection ∇ is referred to as an affine connection space, where the notion of a covariant derivative is intrinsically defined. Given vector fields $\widehat{X}, \widehat{Y} \in \chi(M)$, the expression $\nabla_{\widehat{X}}\widehat{Y}$ denotes the covariant derivative (or covariant differential) of $\widehat{Y}$ with respect to $\widehat{X}$. This operation measures the intrinsic change of $\widehat{Y}$ along the direction specified by $\widehat{X}$.

From the above definition, it is clear that the key aspect of an affine connection lies in how basis vectors vary along different directions. Consider a coordinate chart (U, φ) on the manifold, with local coordinates $x = \varphi(p)$. We define the covariant derivative of a basis vector $\widehat{e}_\nu$ along $\widehat{e}_\mu$ as

$$\nabla_\mu \widehat{e}_\nu \equiv \nabla_{\widehat{e}_\mu} \widehat{e}_\nu = \widehat{e}_\lambda \Gamma^\lambda_{\mu\nu}, \tag{5.5}$$

where $\Gamma^\lambda_{\mu\nu}$ are the connection coefficients.[1] These coefficients encode how basis vectors transform from point to point, capturing their twisting, rotating, expansion, or contraction along various directions. Obviously, the connection coefficients depend on the choice of coordinates—that is, the choice of basis vectors. As we will see later, they do not transform as tensors.

Example 5.1. Connection coefficients in two-dimensional plane in polar coordinates.

Consider a two-dimensional flat space with the metric in polar coordinates given by

$$\mathrm{d}s^2 = \mathrm{d}\rho^2 + \rho^2 \mathrm{d}\phi^2. \tag{5.6}$$

The basis vectors in polar coordinates are

$$\widehat{e}_\rho = \cos\phi \widehat{e}_x + \sin\phi \widehat{e}_y, \quad \widehat{e}_\phi = -\rho \sin\phi \widehat{e}_x + \rho \cos\phi \widehat{e}_y, \tag{5.7}$$

[1]Some mathematical texts adopt the convention $\nabla_{\widehat{e}_\mu} \widehat{e}_\nu = \widehat{e}_\lambda \Gamma^\lambda_{\nu\mu}$. However, for torsion-free connections, where $\Gamma^\lambda_{\mu\nu} = \Gamma^\lambda_{\nu\mu}$, the distinction vanishes.

and the changes in the basis vectors are

$$\frac{\partial \widehat{e}_\rho}{\partial \rho} = 0, \quad \frac{\partial \widehat{e}_\rho}{\partial \phi} = -\sin\phi \widehat{e}_x + \cos\phi \widehat{e}_y = \frac{1}{\rho}\widehat{e}_\phi. \tag{5.8}$$

Similarly,

$$\frac{\partial \widehat{e}_\phi}{\partial \rho} = \frac{1}{\rho}\widehat{e}_\phi, \quad \frac{\partial \widehat{e}_\phi}{\partial \phi} = -\rho\widehat{e}_\rho. \tag{5.9}$$

Thus, the connection coefficients are

$$\Gamma^\rho_{\rho\rho} = \Gamma^\phi_{\rho\rho} = \Gamma^\rho_{\rho\phi} = \Gamma^\rho_{\phi\rho} = \Gamma^\phi_{\phi\phi} = 0, \tag{5.10}$$

$$\Gamma^\phi_{\rho\phi} = \Gamma^\phi_{\phi\rho} = \frac{1}{\rho}, \quad \Gamma^\rho_{\phi\phi} = -\rho. \tag{5.11}$$

In this simple example, we can easily determine the basis vectors and their changes, but for general curved spaces, such calculations are not easy to achieve.

5.1.2 *Covariant derivatives of vectors*

Using the connection coefficients, we can determine how a general vector field $\widehat{W}$ changes along the direction of another vector field $\widehat{V}$. Expanding the covariant derivative $\nabla_{\widehat{V}}\widehat{W}$ in terms of local coordinates, we obtain:

$$\begin{aligned}
\nabla_{\widehat{V}}\widehat{W} &= V^\mu \nabla_{\widehat{e}_\mu}(W^\nu \widehat{e}_\nu) \\
&= V^\mu(\widehat{e}_\mu[W^\nu]\widehat{e}_\nu + W^\nu \nabla_{\widehat{e}_\mu}(\widehat{e}_\nu)) \\
&= V^\mu\left(\frac{\partial W^\lambda}{\partial x^\mu} + W^\nu \Gamma^\lambda_{\mu\nu}\right)\widehat{e}_\lambda \\
&= V^\mu \nabla_\mu W^\lambda \widehat{e}_\lambda,
\end{aligned} \tag{5.12}$$

where

$$\nabla_\mu W^\lambda = (\nabla_\mu \widehat{W})^\lambda = \frac{\partial W^\lambda}{\partial x^\mu} + W^\nu \Gamma^\lambda_{\mu\nu} \tag{5.13}$$

denotes the λ-th component of the vector $\nabla_\mu \widehat{W}$. Unlike the partial derivative $\partial_\mu W^\lambda$, which fails to transform as a tensor under coordinate changes, the covariant derivative $\nabla_\mu W^\lambda$ does constitute a well-defined tensor.[2]

[2]Strictly speaking, $\nabla_\mu W^\lambda$ represent the components of a (1, 1)-type tensor. For brevity, however, we will often refer to the components themselves as the tensor, following standard convention. In subsequent discussions, we will generally not distinguish between a tensor and its components unless necessary.

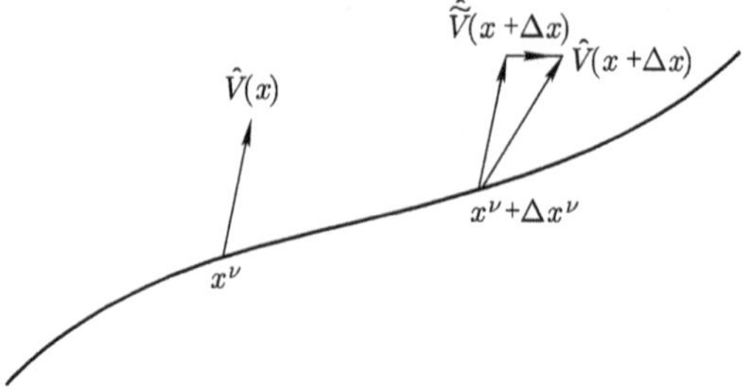

Fig. 5.2 Defining the covariant derivative using parallel transport.

Returning to the geometric definition of the covariant derivative via parallel transport, as shown in Figure 5.2, consider a vector field $V^\mu(x)$ in a given coordinate system. At two infinitesimally separated points x^ν and $x^\nu + \Delta x^\nu$, we have distinct vectors $V^\mu(x)$ and $V^\mu(x + \Delta x)$. To compare them meaningfully, we must transport $V^\mu(x)$ to $x^\nu + \Delta x^\nu$ along a specified path—a process called parallel transport. The resulting vector at $x^\nu + \Delta x^\nu$ is denoted as $\widetilde{V}(x + \Delta x)$. For sufficiently small Δx^ν, the neighborhood can be treated as locally flat, and the natural path for parallel transport is the straight line (geodesic) connecting the two points, with tangent vector Δx^ν. The parallel transport condition requires that:

(1) $\widetilde{V}^\mu(x + \Delta x) - V^\mu(x) \propto \Delta x$, i.e., the change is small;
(2) $(\widetilde{V}^\mu + \widetilde{W}^\mu)(x + \Delta x) = \widetilde{V}^\mu(x + \Delta x) + \widetilde{W}^\mu(x + \Delta x)$.

To properly compare the vector components $V^\mu(x)$ at neighboring points, we must account for the change in basis vectors under parallel transport. For an infinitesimal displacement Δx^ν, the parallel-transported vector components satisfy:

$$\widetilde{V}^\mu(x + \Delta x) = V^\mu(x) - V^\lambda(x)\Gamma^\mu_{\nu\lambda}(x)\Delta x^\nu, \tag{5.14}$$

where the second term precisely compensates for basis vector changes. This definition ensures consistency with our earlier defined affine connection, and linear dependence on Δx^ν reflects the local nature of the affine connection. The covariant derivative is then obtained by comparing the actual vector change to the parallel-transported one:

$$\begin{aligned}\frac{\mathrm{D}\widehat{V}}{\mathrm{d}x^\nu} &= \lim_{\Delta x^\nu \to 0} \frac{V^\mu(x + \Delta x) - \widetilde{V}^\mu(x + \Delta x)}{\Delta x^\nu} \frac{\partial}{\partial x^\mu} \\ &= \left(\frac{\partial V^\mu}{\partial x^\nu} + V^\lambda \Gamma^\mu_{\nu\lambda} \right) \frac{\partial}{\partial x^\mu},\end{aligned}$$

i.e., $\nabla_\nu V^\mu = \dfrac{\partial V^\mu}{\partial x^\nu} + \Gamma^\mu_{\nu\lambda} V^\lambda$, in complete agreement with the previous discussion.

The connection coefficients fundamentally depend on the choice of coordinate system and can be shown to transform non-tensorially. To demonstrate this explicitly, consider two overlapping coordinate charts: one with coordinates $(x^\mu, \widehat{e}_\mu)$ denoted as (U, φ), and another with coordinates $(y^\alpha, \widehat{f}_\alpha)$ denoted as (V, ψ). In the y coordinates, the connection coefficients $\widetilde{\Gamma}^\gamma_{\alpha\beta}$ are given by $\nabla_{\widehat{f}_\alpha} \widehat{f}_\beta = \widetilde{\Gamma}^\gamma_{\alpha\beta} \widehat{f}_\gamma$. Since $\widehat{f}_\alpha = \left(\dfrac{\partial x^\mu}{\partial y^\alpha}\right) \widehat{e}_\mu$, we have

$$
\begin{aligned}
\nabla_{\left(\frac{\partial x^\mu}{\partial y^\alpha}\widehat{e}_\mu\right)} \left(\frac{\partial x^\nu}{\partial y^\beta}\widehat{e}_\nu\right) &= \frac{\partial x^\mu}{\partial y^\alpha}\left(\widehat{e}_\mu \left[\frac{\partial x^\nu}{\partial y^\beta}\right] \widehat{e}_\nu + \frac{\partial x^\nu}{\partial y^\beta} \nabla_{\widehat{e}_\mu} \widehat{e}_\nu\right) \\
&= \frac{\partial x^\mu}{\partial y^\alpha}\left(\frac{\partial^2 x^\nu}{\partial x^\mu \partial y^\beta}\widehat{e}_\nu + \frac{\partial x^\nu}{\partial y^\beta}\Gamma^\lambda_{\mu\nu}\widehat{e}_\lambda\right) \\
&= \left(\frac{\partial^2 x^\lambda}{\partial y^\alpha \partial y^\beta} + \frac{\partial x^\mu}{\partial y^\alpha}\frac{\partial x^\nu}{\partial y^\beta}\Gamma^\lambda_{\mu\nu}\right)\widehat{e}_\lambda,
\end{aligned}
$$

so the transformation of the connection coefficients is

$$
\widetilde{\Gamma}^\gamma_{\alpha\beta} = \frac{\partial^2 x^\nu}{\partial y^\alpha \partial y^\beta}\frac{\partial y^\gamma}{\partial x^\nu} + \frac{\partial x^\lambda}{\partial y^\alpha}\frac{\partial x^\mu}{\partial y^\beta}\frac{\partial y^\gamma}{\partial x^\nu}\Gamma^\nu_{\lambda\mu}. \tag{5.15}
$$

The presence of the first term in the transformation distinguishes connection coefficients from true tensors.

5.1.3 *Covariant derivatives of tensors*

The previously established definitions and properties of affine connections can be naturally generalized to arbitrary tensor fields. Formally, the covariant derivative defines a mapping $\nabla : \chi(M) \otimes (k,l)$-tensor $\to (k,l)$-tensor, satisfying the following properties:

(1) Linearity: $\nabla_{\widehat{X}}(\widehat{T} + \widehat{S}) = \nabla_{\widehat{X}}\widehat{T} + \nabla_{\widehat{X}}\widehat{S}$, where $\widehat{X} \in \chi(M), \widehat{T}, \widehat{S} \in T^{k,l}(M)$;
(2) Leibniz rule: $\nabla_{\widehat{X}}(\widehat{T} \otimes \widehat{S}) = \nabla_{\widehat{X}}\widehat{T} \otimes \widehat{S} + \widehat{T} \otimes \nabla_{\widehat{X}}\widehat{S}$.

For a scalar function f, the covariant derivative reduces to the ordinary partial derivative:$\nabla_\mu f = \partial_\mu f$. Furthermore, the covariant derivative is required to commute with tensor contraction, which implies: $\nabla_\mu \delta^\alpha_\beta = 0$. Using these properties, we can derive the covariant derivative of a 1-form ω. Consider the scalar function $\langle \widehat{\omega}, \widehat{Y} \rangle$, which is the contraction of $\widehat{\omega}$ with a vector field $\widehat{Y}$. Applying the covariant derivative $\nabla_{\widehat{X}}$ and using the Leibniz rule, we obtain:

$$\begin{aligned}\widehat{X}[\langle\widehat{\omega},\widehat{Y}\rangle] &= \nabla_{\widehat{X}}\langle\widehat{\omega},\widehat{Y}\rangle \\ &= \langle\nabla_{\widehat{X}}\widehat{\omega},\widehat{Y}\rangle + \langle\widehat{\omega},\nabla_{\widehat{X}}\widehat{Y}\rangle.\end{aligned}$$

For simplicity, let $\widehat{X} = \widehat{e}_\mu$ and $\widehat{Y} = \widehat{e}_\nu$. Then

$$\begin{aligned}\partial_\mu\omega_\nu &= \langle\nabla_\mu\widehat{\omega},\widehat{e}_\nu\rangle + \langle\widehat{\omega},\Gamma^\sigma_{\mu\nu}\widehat{e}_\sigma\rangle \\ &= (\nabla_\mu\widehat{\omega})_\nu + \omega_\sigma\Gamma^\sigma_{\mu\nu},\end{aligned} \tag{5.16}$$

so

$$(\nabla_\mu\widehat{\omega})_\nu = \partial_\mu\omega_\nu - \omega_\sigma\Gamma^\sigma_{\mu\nu}. \tag{5.17}$$

If $\widehat{\omega} = \mathrm{d}x^\nu$, then we can define $\nabla_\mu \mathrm{d}x^\nu = \widetilde{\Gamma}^\nu_{\mu\lambda}\mathrm{d}x^\lambda$, where $\widetilde{\Gamma}^\nu_{\mu\lambda}$ gives the connection coefficients for the 1-form basis, i.e., it tells us how the 1-form basis changes along different directions. Considering the relation

$$\nabla_\mu(\langle\mathrm{d}x^\nu,\widehat{e}_\sigma\rangle) = \nabla_\mu\delta^\nu_\sigma = 0, \tag{5.18}$$

the left-hand side of the relation can be covariantly differentiated separately, yielding

$$\begin{aligned}0 &= \langle\nabla_\mu\mathrm{d}x^\nu,\widehat{e}_\sigma\rangle + \langle\mathrm{d}x^\nu,\nabla_\mu\widehat{e}_\sigma\rangle \\ &= \widetilde{\Gamma}^\nu_{\mu\lambda}\delta^\lambda_\sigma + \Gamma^\lambda_{\mu\sigma}\delta^\nu_\lambda \\ &= \widetilde{\Gamma}^\nu_{\mu\sigma} + \Gamma^\nu_{\mu\sigma},\end{aligned}$$

thus, we have $\widetilde{\Gamma}^\nu_{\mu\sigma} = -\Gamma^\nu_{\mu\sigma}$, so $\nabla_\mu\mathrm{d}x^\nu = -\Gamma^\nu_{\mu\lambda}\mathrm{d}x^\lambda$.

For a general tensor, using the definition of the affine connection and the connection coefficients, we can easily obtain its covariant derivative. For a (k,l)-tensor $\widehat{T} = T^{\mu_1\cdots\mu_k}{}_{\nu_1\cdots\nu_l}\widehat{e}_{\mu_1}\otimes\cdots\otimes\widehat{e}_{\mu_k}\otimes\widehat{\theta}^{\nu_1}\otimes\cdots\otimes\widehat{\theta}^{\nu_l}$, when the covariant derivative acts on it, it first acts on the component functions and then acts on each basis vector separately. For each vector basis $\widehat{e}$, a positive Γ is obtained, and for each 1-form basis $\widehat{\theta}$, a negative Γ is obtained. Considering all contributions, we finally get

$$\begin{aligned}\nabla_\sigma T^{\mu_1\mu_2\cdots\mu_k}{}_{\nu_1\nu_2\cdots\nu_l} &= \partial_\sigma T^{\mu_1\mu_2\cdots\mu_k}{}_{\nu_1\nu_2\cdots\nu_l} \\ &+ \Gamma^{\mu_1}_{\sigma\lambda}T^{\lambda\mu_2\cdots\mu_k}{}_{\nu_1\nu_2\cdots\nu_l} + \Gamma^{\mu_2}_{\sigma\lambda}T^{\mu_1\lambda\cdots\mu_k}{}_{\nu_1\nu_2\cdots\nu_l} + \cdots \\ &- \Gamma^\lambda_{\sigma\nu_1}T^{\mu_1\mu_2\cdots\mu_k}{}_{\lambda\nu_2\cdots\nu_l} - \Gamma^\lambda_{\sigma\nu_2}T^{\mu_1\mu_2\cdots\mu_k}{}_{\nu_1\lambda\cdots\nu_l} - \cdots.\end{aligned} \tag{5.19}$$

In some literature, an alternative convention is employed: a comma denotes a partial derivative, whereas a semicolon indicates a covariant derivative:

$$\nabla_\sigma T^{\mu_1\mu_2\cdots\mu_k}{}_{\nu_1\nu_2\cdots\nu_l} \equiv T^{\mu_1\mu_2\cdots\mu_k}{}_{\nu_1\nu_2\cdots\nu_l;\sigma}. \tag{5.20}$$

In four-dimensional spacetime, the connection coefficients $\Gamma^{\lambda}_{\mu\nu}$ appear to have $4^3 = 64$ independent components. However, these coefficients are not uniquely determined—multiple possible connections exist, though we will not explore this ambiguity further here. Crucially, in general relativity, we will demonstrate that a given metric tensor uniquely specifies a single connection. First, we observe that if there are two connections, then the difference in their connection coefficients is a (1, 2)-type tensor. That is, if we have two sets of connection coefficients $\Gamma^{\lambda}_{\mu\nu}$ and $\widehat{\Gamma}^{\lambda}_{\mu\nu}$, then their difference $T^{\lambda}_{\mu\nu} = \Gamma^{\lambda}_{\mu\nu} - \widehat{\Gamma}^{\lambda}_{\mu\nu}$ transforms under a coordinate transformation as

$$\begin{aligned} T^{\lambda'}_{\mu'\nu'} &= \Gamma^{\lambda'}_{\mu'\nu'} - \widehat{\Gamma}^{\lambda'}_{\mu'\nu'} \\ &= \frac{\partial x^{\mu}}{\partial x^{\mu'}}\frac{\partial x^{\nu}}{\partial x^{\nu'}}\frac{\partial x^{\lambda'}}{\partial x^{\lambda}}\Gamma^{\lambda}_{\mu\nu} - \frac{\partial x^{\mu}}{\partial x^{\mu'}}\frac{\partial x^{\nu}}{\partial x^{\nu'}}\frac{\partial^2 x^{\lambda'}}{\partial x^{\mu}\partial x^{\nu}} \\ &\quad - \frac{\partial x^{\mu}}{\partial x^{\mu'}}\frac{\partial x^{\nu}}{\partial x^{\nu'}}\frac{\partial x^{\lambda'}}{\partial x^{\lambda}}\widehat{\Gamma}^{\lambda}_{\mu\nu} + \frac{\partial x^{\mu}}{\partial x^{\mu'}}\frac{\partial x^{\nu}}{\partial x^{\nu'}}\frac{\partial^2 x^{\lambda'}}{\partial x^{\mu}\partial x^{\nu}} \\ &= \frac{\partial x^{\mu}}{\partial x^{\mu'}}\frac{\partial x^{\nu}}{\partial x^{\nu'}}\frac{\partial x^{\lambda'}}{\partial x^{\lambda}}(\Gamma^{\lambda}_{\mu\nu} - \widehat{\Gamma}^{\lambda}_{\mu\nu}) \\ &= \frac{\partial x^{\mu}}{\partial x^{\mu'}}\frac{\partial x^{\nu}}{\partial x^{\nu'}}\frac{\partial x^{\lambda'}}{\partial x^{\lambda}}\Gamma^{\lambda}_{\mu\nu}, \end{aligned} \tag{5.21}$$

thus, $T^{\lambda}_{\mu\nu}$ is a (1, 2)-type tensor. In fact, any connection can be expressed as the sum of a given connection and a tensor correction term. Furthermore, if $\Gamma^{\lambda}_{\mu\nu}$ represents a connection coefficient, then $\Gamma^{\lambda}_{\nu\mu}$ also defines a valid connection. Their difference yields the torsion tensor:

$$T^{\lambda}_{\mu\nu} = \Gamma^{\lambda}_{\mu\nu} - \Gamma^{\lambda}_{\nu\mu} = 2\Gamma^{\lambda}_{[\mu\nu]}. \tag{5.22}$$

The above expression is strictly valid only in a coordinate basis. A more general definition of the torsion tensor will be introduced later.

5.1.4 *Fundamental theorem of Riemannian geometry*

Given a metric tensor $g_{\mu\nu}$, a unique affine connection can be determined by imposing the following two conditions:

(1) Torsion-free condition: $T^{\lambda}_{\mu\nu} = 0$ or $\Gamma^{\lambda}_{\mu\nu} = \Gamma^{\lambda}_{\nu\mu}$;
(2) Metric compatibility condition:

$$\nabla_{\rho} g_{\mu\nu} = 0. \tag{5.23}$$

In a local inertial frame (LIF), the metric compatibility condition reduces to $\partial_{\rho} g_{\mu\nu} = 0$, which is identical to the condition for establishing Riemannian normal coordinates. This reveals that the metric compatibility condition represents the covariant generalization of the Riemannian normal coordinate condition, with the replacement: $\partial \to \nabla$. From the metric compatibility condition, we can derive several important consequences:

(1) $\nabla_\rho g^{\mu\nu} = 0$;
(2) $g_{\mu\lambda}\nabla_\rho V^\lambda = \nabla_\rho(g_{\mu\lambda}V^\lambda) = \nabla_\rho V_\mu$.

The metric compatibility condition arises naturally from requiring that the connection preserves inner products under parallel transport. Consider two vector fields $\widehat{X}$ and $\widehat{Y}$ defined along a parameterized curve with tangent vector $\partial/\partial t$. The Leibniz rule for the covariant derivative implies that their inner product evolves as

$$\frac{\mathrm{d}}{\mathrm{d}t}\langle\widehat{X},\widehat{Y}\rangle = \langle\nabla_{\partial/\partial t}\widehat{X},\widehat{Y}\rangle + \langle\widehat{X},\nabla_{\partial/\partial t}\widehat{Y}\rangle. \tag{5.24}$$

This condition is equivalent to metric compatibility. More generally, for any vector fields $\widehat{X}$, $\widehat{Y}$, and direction $\widehat{T}$, then the directional derivative of the function $\langle\widehat{X},\widehat{Y}\rangle$ along $\widehat{T}$ is

$$\widehat{T}\langle\widehat{X},\widehat{Y}\rangle = \langle\nabla_{\widehat{T}}\widehat{X},\widehat{Y}\rangle + \langle\widehat{X},\nabla_{\widehat{T}}\widehat{Y}\rangle. \tag{5.25}$$

The fundamental theorem of Riemannian geometry: On any given Riemannian manifold, there is a unique torsion-free connection compatible with the given metric on the manifold.

Proof We prove the existence and uniqueness by explicitly constructing the connection coefficients that satisfy the above two conditions. First, we expand the metric compatibility condition and permute its indices:

$$\begin{aligned}
\nabla_\rho g_{\mu\nu} &= \partial_\rho g_{\mu\nu} - \Gamma^\lambda_{\rho\mu}g_{\lambda\nu} - \Gamma^\lambda_{\rho\nu}g_{\mu\lambda} = 0,\\
\nabla_\mu g_{\nu\rho} &= \partial_\mu g_{\nu\rho} - \Gamma^\lambda_{\mu\nu}g_{\lambda\rho} - \Gamma^\lambda_{\mu\rho}g_{\nu\lambda} = 0,\\
\nabla_\nu g_{\rho\mu} &= \partial_\nu g_{\rho\mu} - \Gamma^\lambda_{\nu\rho}g_{\lambda\mu} - \Gamma^\lambda_{\nu\mu}g_{\rho\lambda} = 0.
\end{aligned} \tag{5.26}$$

Subtracting the sum of the second and third equations from the first equation in (5.26), we get

$$\partial_\rho g_{\mu\nu} - \partial_\mu g_{\nu\rho} - \partial_\nu g_{\rho\mu} + 2\Gamma^\lambda_{\mu\nu}g_{\lambda\rho} = 0, \tag{5.27}$$

where we have used the torsion-free condition and the symmetry of the metric. Multiplying (5.27) by $g^{\sigma\rho}$, we obtain

$$\Gamma^\sigma_{\mu\nu} = \frac{1}{2}g^{\sigma\rho}(\partial_\mu g_{\nu\rho} + \partial_\nu g_{\rho\mu} - \partial_\rho g_{\mu\nu}). \tag{5.28}$$

Since the metric is non-degenerate, the connection coefficients are uniquely determined. This expression for the connection coefficients is perhaps the most important relation in general relativity.

The affine connection determined by these coefficients is called the Christoffel connection, also known as the Levi–Civita connection or the Riemann connection. Its connection coefficients, called Christoffel symbols, are often denoted in the literature by the compact notation, $\left\{ \begin{matrix} \sigma \\ \mu\nu \end{matrix} \right\}$.

Connections need not be metric-compatible or derived from a metric structure. For instance, a spin connection arises naturally when describing fermionic fields in curved spacetime through the formalism of frame fields (vierbeins). The non-vanishing of connection coefficients does not, by itself, indicate spacetime curvature. Even in flat spacetime, non-zero connection coefficients may appear in non-Cartesian coordinate systems. A canonical example is the use of polar coordinates in a two-dimensional Euclidean space, where the non-zero Christoffel symbols reflect the coordinate choice rather than intrinsic curvature. Conversely, in curved spacetime, the Riemannian normal coordinates can always be constructed to make $\Gamma^\lambda_{\mu\nu} = 0$ at a single point. However, this vanishing cannot generally be extended to a finite neighborhood unless the spacetime is locally flat.

In Riemannian normal coordinates, at a point P, we have

$$g'_{\mu\nu}|_P = \eta_{\mu\nu}, \tag{5.29}$$

$$\partial_\sigma g'_{\mu\nu}|_P = 0. \tag{5.30}$$

Obviously, if (5.30) holds, then the connection coefficients $\Gamma'^\mu_{\nu\sigma}|_P = 0$. Conversely, if the connection coefficients are zero, then by the metric compatibility condition, (5.30) holds. Coordinates that make (5.30) true are conventionally referred to as geodesic coordinates, although their construction does not explicitly involve geodesic curves. Suppose in some coordinate system $\{x^\mu\}$, the metric at point P does not take the form satisfying (5.29) and (5.30), and the coordinates of point P are x^μ_P, then we can define a new coordinate system

$$x'^\mu = x^\mu - x^\mu_P + \frac{1}{2}\Gamma^\mu_{\nu\sigma}|_P(x^\nu - x^\nu_P)(x^\sigma - x^\sigma_P). \tag{5.31}$$

Its first and second derivatives are

$$\begin{aligned} \frac{\partial x'^\mu}{\partial x^\nu} &= \delta^\mu_\nu + \Gamma^\mu_{\nu\sigma}|_P(x^\sigma - x^\sigma_P), \\ \frac{\partial^2 x'^\mu}{\partial x^\nu \partial x^\sigma} &= \Gamma^\mu_{\nu\sigma}|_P. \end{aligned} \tag{5.32}$$

From the behavior of the connection coefficients under coordinate transformations, we find

$$\Gamma'^\mu_{\nu\sigma}|_P = 0, \tag{5.33}$$

namely, in the new coordinate system, the metric field is constant. The coordinate system at point P is a geodesic coordinate system. Furthermore, we can use the coordinate transformation

$$X'^{\mu} = T^{\mu}_{\nu} x'^{\nu} \tag{5.34}$$

to diagonalize the metric field and make the diagonal elements ± 1, where the transformation matrix T^{μ}_{ν} is a constant matrix that does not damage the previous discussion on derivatives. The final coordinates $\{X'\}$ are Riemannian normal coordinates.

Example 5.2. Plane in polar coordinates

$$\mathrm{d}s^2 = \mathrm{d}r^2 + r^2 \mathrm{d}\theta^2. \tag{5.35}$$

The non-zero components of the inverse metric are $g^{rr} = 1$ and $g^{\theta\theta} = r^{-2}$, and among the connection coefficients,

$$\begin{aligned}
\Gamma^{r}_{rr} &= \frac{1}{2} g^{r\rho}(\partial_r g_{r\rho} + \partial_r g_{\rho r} - \partial_\rho g_{rr}) \\
&= \frac{1}{2} g^{rr}(\partial_r g_{rr} + \partial_r g_{rr} - \partial_r g_{rr}) \\
&\quad + \frac{1}{2} g^{r\theta}(\partial_r g_{r\theta} + \partial_r g_{\theta r} - \partial_\theta g_{rr}) \\
&= \frac{1}{2}(1)(0+0-0) + \frac{1}{2}(0)(0+0-0) \\
&= 0,
\end{aligned} \tag{5.36}$$

while

$$\begin{aligned}
\Gamma^{r}_{\theta\theta} &= \frac{1}{2} g^{r\rho}(\partial_\theta g_{\theta\rho} + \partial_\theta g_{\rho\theta} - \partial_\rho g_{\theta\theta}) \\
&= \frac{1}{2} g^{rr}(\partial_\theta g_{\theta r} + \partial_\theta g_{r\theta} - \partial_r g_{\theta\theta}) \\
&= \frac{1}{2}(1)(0+0-2r) \\
&= -r.
\end{aligned} \tag{5.37}$$

Finally, we find

$$\begin{aligned}
\Gamma^{r}_{\theta r} &= \Gamma^{r}_{r\theta} = 0, \\
\Gamma^{\theta}_{rr} &= 0, \\
\Gamma^{\theta}_{r\theta} &= \Gamma^{\theta}_{\theta r} = \frac{1}{r}, \\
\Gamma^{\theta}_{\theta\theta} &= 0.
\end{aligned} \tag{5.38}$$

Example 5.3. Divergence in curved spacetime

$$\nabla_\mu V^\mu = \frac{1}{\sqrt{|g|}} \partial_\mu (\sqrt{|g|} V^\mu). \tag{5.39}$$

Proof The left-hand side of (5.39) has

$$\nabla_\mu V^\mu = \partial_\mu V^\mu + \Gamma^\mu_{\mu\lambda} V^\lambda. \tag{5.40}$$

We need a key formula

$$\Gamma^\mu_{\mu\lambda} = \frac{1}{\sqrt{|g|}} \partial_\lambda \sqrt{|g|}. \tag{5.41}$$

We first prove (5.41). From the definition of the Christoffel symbols, we know

$$\Gamma^\mu_{\mu\lambda} = \frac{1}{2} g^{\mu\rho}(\partial_\mu g_{\rho\lambda} + \partial_\lambda g_{\mu\rho} - \partial_\rho g_{\mu\lambda}) = \frac{1}{2} g^{\mu\rho} \partial_\lambda g_{\mu\rho}. \tag{5.42}$$

The first and third terms in the parentheses on the right-hand side of the first equality in (5.42) are antisymmetric with respect to (μ, ρ), while the inverse metric outside the parentheses is fully symmetric with respect to (μ, ρ), so their sum is zero after summation.

Lemma For a general matrix $M(x)$,

$$\mathrm{Tr}(M^{-1}\partial_\lambda M) = \partial_\lambda \ln(\mathrm{Det} M). \tag{5.43}$$

Proof of the Lemma Consider a general variation

$$\begin{aligned}
\delta \ln \mathrm{Det} M &= \ln \mathrm{Det}(M + \delta M) - \ln \mathrm{Det} M \\
&= \ln \mathrm{Det}(M^{-1}(M + \delta M) \\
&= \ln \mathrm{Det}(1 + M^{-1}\delta M) \\
&\sim \ln(1 + \mathrm{Tr}(M^{-1}\delta M)) \text{ (keeping the leading order)} \\
&\sim \mathrm{Tr}(M^{-1}\delta M).
\end{aligned} \tag{5.44}$$

If such a variation is with respect to x^λ, then taking the derivative with respect to x^λ on both sides of (5.44), i.e., $\delta \to \frac{\delta}{\delta x^\lambda}$, we obtain the equation for matrices. The lemma is proved.

In (5.43), let $M = g_{\mu\nu}$, then we have

$$g^{\mu\rho} \frac{\partial g_{\mu\rho}}{\partial x^\lambda} = \frac{\partial}{\partial x^\lambda} = \ln g = \frac{2}{\sqrt{|g|}} \partial_\lambda \sqrt{|g|}. \tag{5.45}$$

This proves the key relation (5.41), thereby proving the divergence formula (5.39).

From (5.39), it is easy to obtain the action of the Laplace operator on a scalar function

$$\begin{aligned}
\nabla^2 \Phi &= \nabla^\mu \nabla_\mu \Phi \\
&= \frac{1}{\sqrt{|g|}} \partial_\mu (\sqrt{|g|} \partial^\mu \Phi).
\end{aligned} \tag{5.46}$$

This discussion applies to any dimension and any signature of the metric, as long as the $|g|$ in the above is understood as the absolute value of the determinant of the metric field. In fact, this relation has already been used in multivariable calculus in vector analysis. For example, consider the Laplace operator in spherical coordinates in three-dimensional Euclidean space. The metric in spherical coordinates is

$$\mathrm{d}s^2 = \mathrm{d}r^2 + r^2(\mathrm{d}\theta^2 + \sin^2\theta \mathrm{d}\phi^2), \tag{5.47}$$

then the action of the Laplace operator on a scalar function is

$$\begin{aligned} \nabla^2\Phi &= \nabla^i\nabla_i\Phi \\ &= \frac{1}{r^2}\partial_r(r^2\partial_r\Phi) + \frac{1}{r^2\sin\theta}\partial_\theta(\sin\theta\partial_\theta\Phi) + \frac{1}{r^2\sin^2\theta}\partial_\phi^2\Phi. \end{aligned} \tag{5.48}$$

Using the definition of divergence, we can directly extend the divergence theorem to curved spacetime. Suppose we are in a 4-dimensional spacetime with the proper volume element $\sqrt{|g|}\mathrm{d}^4x$. Consider a spacetime region Σ with boundary $\partial\Sigma$, then we have

$$\int_\Sigma \sqrt{|g|}\mathrm{d}^4x(\nabla_\mu J^\mu) = \int_{\partial\Sigma} |\gamma|^{1/2}\mathrm{d}^3y(n_\mu J^\mu), \tag{5.49}$$

where γ is the induced metric on the boundary $\partial\Sigma$, y^i are the coordinates on the boundary, and n_μ are the components of the outward, normalized normal vector to the boundary. Note that in the above integral, J^μ is not required to be a 4-vector component; it can be any set of four functions. If the manifold is boundary-less, then the above integral is zero.

For the usual vector analysis of curl calculations,

$$\begin{aligned} (\text{curl}\,\widehat{V})_{\mu\nu} &= \nabla_\mu V_\nu - \nabla_\nu V_\mu \\ &= \partial_\mu V_\nu - \partial_\nu V_\mu, \end{aligned} \tag{5.50}$$

it appears that the terms related to the connection coefficients disappear, so it seems unrelated to the metric. In fact, the last line of (5.50) is exactly the component of the exterior derivative 2-form $\mathrm{d}\widehat{V}$ of the vector, i.e.,

$$(\mathrm{d}\widehat{V})_{\mu\nu} = \partial_\mu V_\nu - \partial_\nu V_\mu = \nabla_\mu V_\nu - \nabla_\nu V_\mu. \tag{5.51}$$

The exterior derivative is an operation independent of the choice of connection; even without any connection, the exterior derivative operation can define a well-behaved tensor.

5.2 Parallel Transport

We begin by examining the intrinsic derivative of a vector field along a curve. This concept is particularly important for certain physical quantities, such as the spin 4-vector of a gyroscope, which are inherently particle-dependent and consequently defined only along the particle's worldline rather than throughout the entire manifold. In such cases, the conventional covariant derivative defined on the manifold proves inadequate for describing the evolution of these quantities along the worldline. Consider a vector field $\widehat{V}(\lambda)$ defined along a curve $\mathcal{C}$. In a local coordinate system, this vector can be written as $\widehat{V}(\lambda) = V^\mu \widehat{e}_\mu(\lambda)$, where $\widehat{e}_\mu(\lambda)$ represent the coordinate basis vectors that vary along the curve. The intrinsic derivative characterizes how this vector field changes as we move along $\mathcal{C}$,

$$\begin{aligned}\frac{\mathrm{D}\widehat{V}}{\mathrm{d}\lambda} &= \frac{\mathrm{d}V^\mu}{\mathrm{d}\lambda}\widehat{e}_\mu + V^\mu \frac{\mathrm{D}\widehat{e}_\mu}{\mathrm{d}\lambda} = \frac{\mathrm{d}V^\mu}{\mathrm{d}\lambda}\widehat{e}_\mu + V^\mu \frac{\mathrm{d}x^\nu}{\mathrm{d}\lambda}\nabla_\nu \widehat{e}_\mu \\ &= \frac{\mathrm{d}V^\mu}{\mathrm{d}\lambda}\widehat{e}_\mu + \Gamma^\sigma_{\nu\mu} V^\mu \widehat{e}_\sigma \frac{\mathrm{d}x^\nu}{\mathrm{d}\lambda} \\ &= \left(\frac{\mathrm{d}V^\mu}{\mathrm{d}\lambda} + \Gamma^\mu_{\nu\sigma} V^\sigma \frac{\mathrm{d}x^\nu}{\mathrm{d}\lambda}\right)\widehat{e}_\mu \\ &= \frac{\mathrm{D}V^\mu}{\mathrm{d}\lambda}\widehat{e}_\mu,\end{aligned} \tag{5.52}$$

where

$$\begin{aligned}\frac{\mathrm{D}V^\mu}{\mathrm{d}\lambda} &= \frac{\mathrm{d}V^\mu}{\mathrm{d}\lambda} + \Gamma^\mu_{\nu\sigma} V^\sigma \frac{\mathrm{d}x^\nu}{\mathrm{d}\lambda} \\ &= \frac{\mathrm{d}x^\nu}{\mathrm{d}\lambda}(\nabla_\nu V^\mu),\end{aligned} \tag{5.53}$$

and in the last equality of (5.53), we used $\frac{\mathrm{d}V^\mu}{\mathrm{d}\lambda} = \frac{\partial V^\mu}{\partial x^\nu}\frac{\mathrm{d}x^\nu}{\mathrm{d}\lambda}$. Note that if the vector field $\widehat{V}$ is only defined along the curve $\mathcal{C}$, the last term in (5.53) involving the covariant derivative loses its meaning, as it requires the vector field to be defined in an open neighborhood of the curve. In such cases, we must instead rely on the previous definition (i.e., the intrinsic derivative) to properly describe the evolution of $\widehat{V}$ along $\mathcal{C}$. The same approach naturally extends to define the intrinsic derivative of a 1-form along a curve

$$\frac{\mathrm{D}V_\mu}{\mathrm{d}\lambda} = \frac{\mathrm{d}V_\mu}{\mathrm{d}\lambda} - \Gamma^\sigma_{\nu\mu} V_\sigma \frac{\mathrm{d}x^\nu}{\mathrm{d}\lambda}. \tag{5.54}$$

This construction naturally generalizes to arbitrary tensor fields along $\mathcal{C}$. For example, for a $(2,0)$-type tensor field

$$\widehat{T}(\lambda) = T^{\mu\nu}(\lambda)\widehat{e}_\mu(\lambda) \otimes \widehat{e}_\nu(\lambda) \tag{5.55}$$

its change along the curve is

$$\begin{aligned}\frac{\mathrm{D}\widehat{T}}{\mathrm{d}\lambda} &= \frac{\mathrm{d}T^{\mu\nu}}{\mathrm{d}\lambda}\widehat{e}_\mu \otimes \widehat{e}_\nu + T^{\mu\nu}\frac{\mathrm{D}\widehat{e}_\mu}{\mathrm{d}\lambda} \otimes \widehat{e}_\nu + T^{\mu\nu}\widehat{e}_\mu \otimes \frac{\mathrm{D}\widehat{e}_\nu}{\mathrm{d}\lambda} \\ &= \frac{\mathrm{d}T^{\mu\nu}}{\mathrm{d}\lambda}\widehat{e}_\mu \otimes \widehat{e}_\nu + T^{\mu\nu}\Gamma^\sigma_{\mu\rho}\widehat{e}_\sigma \frac{\mathrm{d}x^\rho}{\mathrm{d}\lambda} \otimes \widehat{e}_\nu + T^{\mu\nu}\Gamma^\sigma_{\nu\rho}\widehat{e}_\mu \otimes \widehat{e}_\sigma \frac{\mathrm{d}x^\rho}{\mathrm{d}\lambda} \\ &= \left(\frac{\mathrm{d}T^{\mu\nu}}{\mathrm{d}\lambda} + \Gamma^\mu_{\sigma\rho}T^{\sigma\nu}\frac{\mathrm{d}x^\rho}{\mathrm{d}\lambda} + \Gamma^\nu_{\sigma\rho}T^{\mu\sigma}\frac{\mathrm{d}x^\rho}{\mathrm{d}\lambda}\right)\widehat{e}_\mu \otimes \widehat{e}_\nu .\end{aligned} \tag{5.56}$$

In component form, we obtain the intrinsic derivative of the second-order tensor components along the curve

$$\frac{\mathrm{D}T^{\mu\nu}}{\mathrm{d}\lambda} = \frac{\mathrm{d}T^{\mu\nu}}{\mathrm{d}\lambda} + \Gamma^\mu_{\sigma\rho}\Gamma^{\sigma\nu}\frac{\mathrm{d}x^\rho}{\mathrm{d}\lambda} + \Gamma^\nu_{\sigma\rho}T^{\mu\sigma}\frac{\mathrm{d}x^\rho}{\mathrm{d}\lambda}. \tag{5.57}$$

With a connection defined, we can now introduce the concept of parallel transport. In flat spacetime, parallel transport implicitly relies on the Christoffel connection, which vanishes in inertial coordinates. A key property of flat spacetime is that transporting a vector around any closed loop returns it to its original orientation—the vector remains globally parallel to itself at all points along the curve, as illustrated in Figure 5.3(a). In curved spacetime, however, parallel transport becomes path-dependent. Unlike in flat space, there is no canonical way to identify vectors in different tangent spaces without specifying a connecting path. This intrinsic curvature effect is demonstrated in Figure 5.3(b) for transport on a sphere: first parallel transporting the vector $\widehat{V}_A$ at point A along the curve AB to point B, and

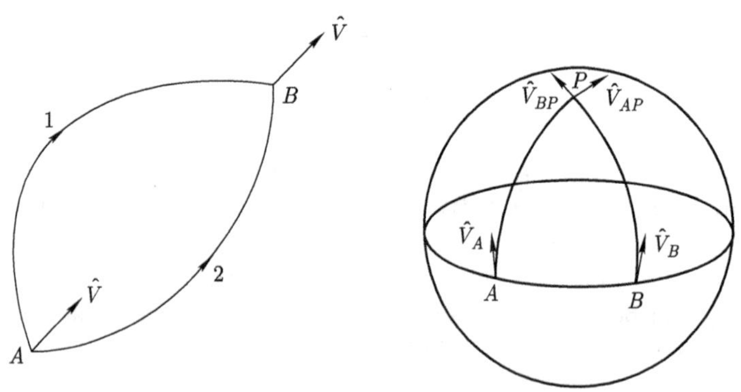

Left: (a) Parallel transport of vectors on a plane. Right: (b) Parallel transport of vectors on a sphere.

Fig. 5.3 On a plane, parallel transport of a vector along different paths results in the same vector, while on a sphere, transporting a vector along different paths leads to different outcomes.

then transporting it along BP to point P to obtain the vector $\widehat{V}_{BP}$; directly transporting the vector along the curve AP to point P to obtain the vector $\widehat{V}_{AP}$. The resulting vectors $\widehat{V}_{BP}$ and $\widehat{V}_{AP}$ differ—a direct manifestation of spacetime curvature. This path dependence is a fundamental feature of parallel transport in curved geometries.

The notion of parallel transport on a sphere—or any curved manifold—requires careful definition. At its core, parallel transport generalizes the intuitive idea of "keeping a vector constant" to curved geometries. Given an initial vector at some point along a curve, we define its parallel transport (PT) along the curve as a process that preserves the vector's direction relative to the local geometry. This is illustrated in Figure 5.4, where we maintain the vector's orientation while moving it infinitesimally along the path.

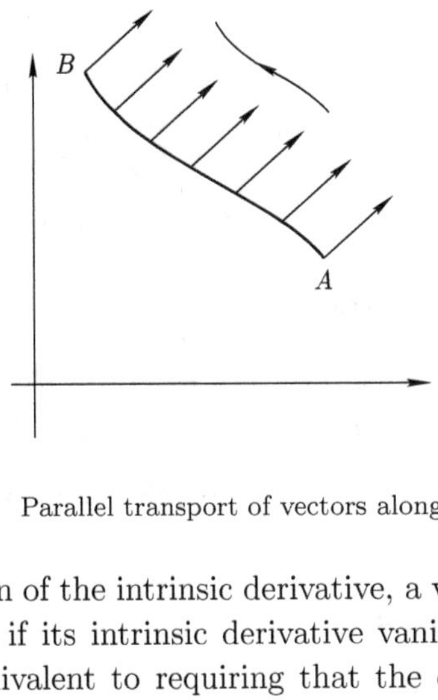

Fig. 5.4 Parallel transport of vectors along a curve.

From the definition of the intrinsic derivative, a vector is parallel transported along a curve if its intrinsic derivative vanishes along that curve. This condition is equivalent to requiring that the covariant derivative of the vector with respect to the curve's tangent vector is zero, i.e.,

$$\frac{\mathrm{D}\widehat{V}}{\mathrm{d}\lambda} = 0 \quad \Rightarrow \quad t^{\mu}\nabla_{\mu}V^{\nu} = 0, \tag{5.58}$$

or more compactly, $\nabla_{\hat{t}}\widehat{V} = 0$, where $\hat{t}$ is the tangent vector to the curve. This concept naturally extends to the parallel transport of arbitrary tensors. Consider a curve $x^{\mu}(\lambda)$. The requirement that a tensor $\widehat{T}$ remain unchanged along the curve may initially be written as $\frac{\mathrm{D}\widehat{T}}{\mathrm{d}\lambda} = \frac{\mathrm{d}x^{\mu}}{\mathrm{d}\lambda}\frac{\partial\widehat{T}}{\partial x^{\mu}} = 0$. However, this expression is not tensorial due to the use of partial derivatives. To formulate a covariant condition, we replace the partial derivative with the covariant derivative, $\partial_\mu \to \nabla_\mu$, leading to the definition of the covariant derivative along the curve:

$$\frac{\mathrm{D}}{\mathrm{d}\lambda} = \frac{\mathrm{d}x^\mu}{\mathrm{d}\lambda}\nabla_\mu. \tag{5.59}$$

Consequently, the parallel transport of a tensor along the curve must satisfy the parallel transport equation

$$\left(\frac{\mathrm{D}}{\mathrm{d}\lambda}T\right)^{\mu_1\mu_2\cdots\mu_k}{}_{\nu_1\nu_2\cdots\nu_l} \equiv \frac{\mathrm{d}x^\sigma}{\mathrm{d}\lambda}\nabla_\sigma T^{\mu_1\mu_2\cdots\mu_k}{}_{\nu_1\nu_2\cdots\nu_l} = 0, \tag{5.60}$$

or succinctly, $\nabla_{\hat{t}}\hat{T} = 0$, where $\hat{t} = \mathrm{d}/\mathrm{d}\lambda$ is the tangent vector of the curve.

For a vector field V^μ, its parallel transport equation along a curve $x^\mu(\lambda)$ is given by

$$\frac{\mathrm{d}}{\mathrm{d}\lambda}V^\mu + \Gamma^\mu_{\sigma\rho}\frac{\mathrm{d}x^\sigma}{\mathrm{d}\lambda}V^\rho = 0. \tag{5.61}$$

This is a first-order linear ordinary differential equation (ODE) with a well-defined initial value problem: given a curve $x^\mu(\lambda)$ and an initial vector $\hat{V}_0$ at a point λ_0, the equation uniquely determines a vector field $\hat{V}(\lambda)$ along the curve $\mathcal{C}$. The existence and uniqueness of the solution follow from standard ODE theory.

If the connection is metric-compatible, then the metric field is always parallel transported along any curve. This is because

$$\frac{\mathrm{d}}{\mathrm{d}\lambda}g_{\mu\nu} = \frac{\mathrm{d}x^\sigma}{\mathrm{d}\lambda}\nabla_\sigma g_{\mu\nu} = 0. \tag{5.62}$$

Moreover, the inner product of two parallel transported vectors remains unchanged:

$$\frac{\mathrm{D}}{\mathrm{d}\lambda}(\hat{V}\cdot\hat{W}) = \left(\frac{\mathrm{D}}{\mathrm{d}\lambda}g_{\mu\nu}\right)V^\mu W^\nu + g_{\mu\nu}\frac{\mathrm{D}}{\mathrm{d}\lambda}V^\mu W^\nu + g_{\mu\nu}V^\mu\frac{\mathrm{D}}{\mathrm{d}\lambda}W^\nu = 0.$$

Therefore, for a metric-compatible connection, parallel transport preserves the magnitude of vectors and the orthogonality between vectors.

5.3 Geodesics

Geodesics generalize the notion of "straight lines" in Euclidean space to curved spaces. In Euclidean geometry, the shortest path between two points is a straight line segment. Extending this idea, a geodesic in a curved space is defined as the "shortest" path connecting two points—this constitutes the first definition of a geodesic. Consequently, geodesics are often referred to as shortest paths. Here, "shortest" refers to the minimal distance between two points in Riemannian space or between points connected by a spacelike curve in a pseudo-Riemannian manifold. However, in the case of timelike curves in a pseudo-Riemannian manifold, such as spacetime in general relativity, the geodesic maximizes the spacetime distance, known as the proper time. For null curves, the spacetime interval is always zero, rendering the notions of "longest" or "shortest" meaningless.

In Euclidean space, a straight line can alternatively be characterized by the property that its tangent vector remains constantly aligned along the curve. Generalizing this idea to curved spaces, we define a geodesic as a curve whose tangent vector is parallel-transported along itself.

Consider a curve $x^\mu(\lambda)$ parameterized by λ, with tangent vector components $t^\mu = \mathrm{d}x^\mu/\mathrm{d}\lambda$. The condition of parallel transport of $\widehat{t}$ along the curve is given by

$$\frac{\mathrm{D}}{\mathrm{d}\lambda}\widehat{t} = 0, \tag{5.63}$$

where $\frac{\mathrm{D}}{\mathrm{d}\lambda}$ denotes the covariant derivative along the curve. Equivalently, this can be expressed as:

$$\nabla_{\widehat{t}}\widehat{t} = 0. \tag{5.64}$$

In a coordinate system $\{x^\mu\}$, this condition leads to the geodesic equation.

Geodesic Equation

$$\frac{\mathrm{d}^2x^\mu}{\mathrm{d}\lambda^2} + \Gamma^\mu_{\rho\sigma}\frac{\mathrm{d}x^\rho}{\mathrm{d}\lambda}\frac{\mathrm{d}x^\sigma}{\mathrm{d}\lambda} = 0. \tag{5.65}$$

The geodesic equation governs the motion of free particles in curved spacetime. In flat spacetime using Cartesian coordinates, it reduces to $\mathrm{d}^2x^\mu/\mathrm{d}\lambda^2 = 0$, which describes a straight-line motion—exactly as expected from Newton's first law. This reflects the fact that free particles follow inertial paths in the absence of forces. In curved spacetime, it is locally flat. At an given point, we can choose coordinates (a local inertial frame) where the Christoffel symbols vanish $\Gamma^\mu_{\sigma\rho} = 0$, making the geodesic equation locally identical to its flat-space form, $\mathrm{d}^2x^\mu/\mathrm{d}\lambda^2 = 0$. This means that in a sufficiently small region, geodesics appear straight—a manifestation of the equivalence principle in general relativity. Remarkably, the geodesic equation is universal: it applies not only to timelike and spacelike curves but also to null curves. This universality reflects the geometric nature of Einstein's theory, where spacetime curvature dictates the motion of all particles, regardless of their mass or causal character.

The above two definitions of geodesics are equivalent only when using the Christoffel connection. We now demonstrate this equivalence explicitly for a timelike curve. For a timelike curve parameterized by λ, the proper time between points A and B is $\tau = \int_{\lambda_A}^{\lambda_B} \mathcal{L} \mathrm{d}\lambda$, where

$$\mathcal{L} = \sqrt{-g_{\mu\nu}\frac{\mathrm{d}x^\mu}{\mathrm{d}\lambda}\frac{\mathrm{d}x^\nu}{\mathrm{d}\lambda}}. \tag{5.66}$$

The condition that geodesics extremize proper time is a variational principle where we fix the two endpoints and consider all possible paths to make τ take an extremum, i.e., $\delta\tau = 0$. This leads to the Euler–Lagrange equations (if no special boundary conditions are added)

$$\frac{\partial \mathcal{L}}{\partial x^\mu} - \frac{\mathrm{d}}{\mathrm{d}\lambda}\left(\frac{\partial \mathcal{L}}{\partial \dot{x}^\mu}\right) = 0. \tag{5.67}$$

Equation (5.67) is equivalent to

$$\frac{\mathrm{d}}{\mathrm{d}\lambda}\left(\frac{\partial \mathcal{L}^2}{\partial \dot{x}^\mu}\right) - \frac{\partial \mathcal{L}^2}{\partial x^\mu} = 2\frac{\partial \mathcal{L}}{\partial \dot{x}^\mu}\frac{\partial \mathcal{L}}{\partial \lambda}. \tag{5.68}$$

The left-hand side of Eq. (5.68) is

$$\begin{aligned}
&\frac{\mathrm{d}}{\mathrm{d}\lambda}\left(\frac{\partial}{\partial \dot{x}^\mu}(-g_{\sigma\rho}\dot{x}^\sigma\dot{x}^\rho)\right) - \frac{\partial}{\partial x^\mu}(g_{\sigma\rho}\dot{x}^\sigma\dot{x}^\rho) \\
&= -\frac{\mathrm{d}}{\mathrm{d}\lambda}(2g_{\sigma\mu}\dot{x}^\sigma) + (\partial_\mu g_{\sigma\rho})\dot{x}^\sigma\dot{x}^\rho \\
&= (\partial_\mu g_{\sigma\rho})\dot{x}^\sigma\dot{x}^\rho - 2g_{\sigma\mu}\ddot{x}^\sigma - 2\partial_\rho g_{\sigma\mu}\dot{x}^\sigma\dot{x}^\rho \\
&= -2g_{\sigma\mu}\ddot{x}^\sigma - 2\dot{x}^\sigma\dot{x}^\rho\Gamma_{\mu\sigma\rho} \quad (\Gamma_{\mu\sigma\rho} = g_{\mu\nu}\Gamma^\nu_{\ \sigma\rho}),
\end{aligned} \tag{5.69}$$

where the definition of the Christoffel symbols has been used. The right-hand side of Eq. (5.68) is

$$\begin{aligned}
2\frac{\partial \mathcal{L}}{\partial \dot{x}^\mu}\frac{\partial \mathcal{L}}{\partial \lambda} &= 2\frac{\partial}{\partial \dot{x}^\mu}(-g_{\sigma\rho}\dot{x}^\sigma\dot{x}^\rho)^{1/2}\frac{\mathrm{d}}{\mathrm{d}\lambda}\left(\frac{\mathrm{d}\tau}{\mathrm{d}\lambda}\right) \\
&= -2(-g_{\sigma\rho}\dot{x}^\sigma\dot{x}^\rho)^{1/2} g_{\mu\nu}\dot{x}^\nu\frac{\mathrm{d}^2\tau}{\mathrm{d}\lambda^2} \\
&= -2\left(\frac{\mathrm{d}^2\tau}{\mathrm{d}\lambda^2}\Big/\frac{\mathrm{d}\tau}{\mathrm{d}\lambda}\right) g_{\mu\nu}\dot{x}^\nu.
\end{aligned}$$

Finally, we obtain

$$\ddot{x}^\mu + \Gamma^\mu_{\sigma\rho}\dot{x}^\sigma\dot{x}^\rho = \frac{\ddot{\tau}}{\dot{\tau}}\dot{x}^\mu. \tag{5.70}$$

If we take $\lambda = \tau$, then $\ddot{\tau} = 0$, and we get the geodesic equation

$$\ddot{x}^\mu + \Gamma^\mu_{\sigma\rho}\dot{x}^\sigma\dot{x}^\rho = 0. \tag{5.71}$$

More generally, we may take $\lambda = \alpha\tau + \beta$ and still preserve the form of the geodesic equation. Such a parameterization is referred to as an affine parameterization, with λ termed the affine parameter. Even if other parameterizations are employed, any curve satisfying (5.70) remains a geodesic. In such case, a suitable reparameterization λ can always be chosen to recast the equation into the standard geodesic form.

For a null geodesic, the interval between any two points on it vanishes, i.e., $\mathcal{L} \equiv 0$. In this case, proper time is undefined, and τ is no longer a valid affine parameter. Nevertheless, we may still derive the geodesic equation via variation, yielding

$$\ddot{x}^\mu + \Gamma^\mu_{\sigma\rho}\dot{x}^\sigma\dot{x}^\rho = f(\lambda)\dot{x}^\mu, \tag{5.72}$$

subject to the null condition $g_{\mu\nu}\dot{x}^\mu\dot{x}^\nu = 0$. If $f(\lambda) = 0$, the parameter λ is called an affine parameter, and any linear rescaling $\bar{\lambda} = \alpha\lambda + \beta$ also qualifies as one. Notably, within this two-parameter family of affine parameters, no single choice is privileged over the others.

The preceding discussion can also be framed in terms of the action principle for a point particle. For a particle with mass m following a timelike curve, its action can be taken as

$$I = -m\int_A^B \mathrm{d}\tau = -m\int_A^B \sqrt{-g_{\mu\nu}\frac{\mathrm{d}x^\mu}{\mathrm{d}\lambda}\frac{\mathrm{d}x^\nu}{\mathrm{d}\lambda}}\mathrm{d}\lambda, \tag{5.73}$$

where τ is the proper time. However, this action is cumbersome for variational purposes. To address this, we introduce an alternative formulation using an auxiliary frame field (or einbein) $e(\lambda)$, defining the new action:

$$I = \frac{1}{2}\int_{\lambda_A}^{\lambda_B} \mathrm{d}\lambda \left(e^{-1}(\lambda)g_{\mu\nu}\frac{\mathrm{d}x^\mu}{\mathrm{d}\lambda}\frac{\mathrm{d}x^\nu}{\mathrm{d}\lambda} - m^2 e(\lambda)\right). \tag{5.74}$$

Here $e(\lambda)$ is a non-dynamical auxiliary field. Integrating it out reproduces the original action (5.73). The freedom to reparameterize the curve is equivalent to choosing $e(\lambda)$. Varying the action (5.74) with respect to $e(\lambda)$ and $x^\mu(\lambda)$ yields

$$\frac{\delta I}{\delta e} = 0 \Rightarrow e = \frac{1}{m}\frac{\mathrm{d}\tau}{\mathrm{d}\lambda}, \tag{5.75}$$

$$\frac{\delta I}{\delta x^\mu} = 0 \Rightarrow \frac{\mathrm{D}^2 x^\mu}{\mathrm{d}\lambda^2} = \left(e^{-1}\frac{\mathrm{d}e}{\mathrm{d}\lambda}\right)\frac{\mathrm{d}x^\mu}{\mathrm{d}\lambda}. \tag{5.76}$$

Substituting (5.75) into (5.76) recovers the geodesic equation obtained earlier by extremizing the proper time. Therefore, if we consider the geodesic as the curve that makes the action take an extremum, then generally we have

$$\frac{\mathrm{D}t^\mu}{\mathrm{d}\lambda} = f(\lambda)t^\mu, \tag{5.77}$$

where t^μ is the component of the tangent vector of the geodesic, and $f(\lambda) = e^{-1}\frac{\mathrm{d}e}{\mathrm{d}\lambda}\frac{\mathrm{d}^2\tau}{\mathrm{d}\lambda^2}\Big/\frac{\mathrm{d}\tau}{\mathrm{d}\lambda}$ is an arbitrary function. The arbitrariness of $f(\lambda)$ comes from the arbitrariness of the choice of the function $e(\lambda)$. The most natural parameterization is the affine parameterization, which eliminates the inhomogeneous term:

$$\frac{\mathrm{D}t^\mu}{\mathrm{d}\lambda} = 0. \tag{5.78}$$

For timelike geodesics, this corresponds to $e(\lambda) = \text{constant}$, or equivalently,

$$\lambda \propto \tau + \text{constant}. \tag{5.79}$$

A key benefit of the frame-field action (5.74) is its smooth behavior in the massless limit $m \to 0$, making it applicable to both massive and massless particles.

Consequently, when adopting an affine parameterization and setting $e(\lambda) = \text{constant}$, then the action (5.74) gets simplied. We can define

$$2K = g_{\mu\nu}\dot{x}^\mu\dot{x}^\nu \tag{5.80}$$

to obtain a new action

$$I_K = \int_{\lambda_A}^{\lambda_B} K. \tag{5.81}$$

If λ is taken as an affine parameter, then the geodesic equation is

$$\frac{\partial K}{\partial x^\mu} - \frac{\mathrm{d}}{\mathrm{d}\lambda}\left(\frac{\partial K}{\partial \dot{x}^\mu}\right) = 0, \tag{5.82}$$

where

$$2K = \begin{cases} 0, & \text{null curve,} \\ 1, & \text{spacelike curve,} \\ -1, & \text{timelike curve.} \end{cases} \tag{5.83}$$

In the latter two cases, we take λ as the distance parameter s and the proper time τ.

Example 5.4. Two-dimensional plane.

If we take Cartesian coordinates, then the metric is diagonal, and all Christoffel symbols are zero, so the geodesic equation is

$$\frac{\mathrm{d}^2x}{\mathrm{d}s^2} = 0, \quad \frac{\mathrm{d}^2y}{\mathrm{d}s^2} = 0. \tag{5.84}$$

The solutions are

$$\begin{aligned} x &= a_1 s + b_1, \\ y &= a_2 s + b_2. \end{aligned} \tag{5.85}$$

Eliminating s, we get $y = ax + b$, which is a straight line. Alternatively, we can also discuss this problem in polar coordinates. From the previously calculated connection coefficients, we have the geodesic equations

$$\frac{\mathrm{d}^2r}{\mathrm{d}s^2} - r\left(\frac{\mathrm{d}\theta}{\mathrm{d}s}\right)^2 = 0, \tag{5.86}$$

$$\frac{\mathrm{d}^2\theta}{\mathrm{d}s^2} + \frac{2}{r}\frac{\mathrm{d}r}{\mathrm{d}s}\frac{\mathrm{d}\theta}{\mathrm{d}s} = 0. \tag{5.87}$$

If $\frac{\mathrm{d}\theta}{\mathrm{d}s} = 0$, then $\theta =$ constant, and $r = as + b$ gives a straight line passing through the origin. If $\frac{\mathrm{d}\theta}{\mathrm{d}s} \neq 0$, then (5.87) can be transformed into

$$\frac{\theta''}{\theta'} + \frac{2}{r}r' = 0, \tag{5.88}$$

where the prime denotes differentiation with respect to s. Integrating (5.88), we get

$$\ln|\theta'| + \ln r^2 = \text{constant} \quad \Rightarrow \quad r^2\theta' = h = \text{constant}. \tag{5.89}$$

From the spacelike curve $2K = 1$, we have

$$r' = \pm\frac{1}{r}\sqrt{r^2 - h^2}, \tag{5.90}$$

and further obtain

$$\frac{\mathrm{d}\theta}{\mathrm{d}r} = \pm\frac{h}{r\sqrt{r^2 - h^2}}. \tag{5.91}$$

Therefore, we finally get

$$r = \frac{h}{\cos(\theta - \theta_0)}, \tag{5.92}$$

which is a straight line not passing through the origin in polar coordinates.

If spacetime is static, i.e., there exist coordinates such that $g_{0i} = 0$ and all metric components are independent of $x^0 = t$, a variational principle analogous to Fermat's principle in optics can be formulated. This principle generalizes Fermat's principle to curved spacetime. Consider the set of all null curves connecting two events P and Q. These curves satisfy

$$0 = \mathrm{d}t^2 + \left(\frac{g_{ij}}{g_{00}}\right) \mathrm{d}x^i \mathrm{d}x^j. \tag{5.93}$$

Each null curve is parametrized by three functions $x^i(t)$, and the coordinate time elapsed between P to Q is Δt. We shall demonstrate that null geodesics extremize this coordinate time. Starting from (5.93), the geodesic equation can be rewritten as

$$g_{ik}\frac{\mathrm{d}^2x^k}{\mathrm{d}t^2} + \Gamma_{ijk}\frac{\mathrm{d}x^j}{\mathrm{d}t}\frac{\mathrm{d}x^k}{\mathrm{d}t} - \Gamma_{i00}\frac{g_{jk}}{g_{00}}\frac{\mathrm{d}x^j}{\mathrm{d}t}\frac{\mathrm{d}x^k}{\mathrm{d}t} + \frac{\mathrm{d}^2t/\mathrm{d}\lambda^2}{(\mathrm{d}t/\mathrm{d}\lambda)^2}g_{ik}\frac{\mathrm{d}x^k}{\mathrm{d}t} = 0. \tag{5.94}$$

The time component of the geodesic yields

$$\frac{\mathrm{d}^2t/\mathrm{d}\lambda^2}{(\mathrm{d}t/\mathrm{d}\lambda)^2} = -2\Gamma_{0k0}\frac{\mathrm{d}x^k/\mathrm{d}t}{g_{00}}. \tag{5.95}$$

Substituting (5.95) into (5.94) and evaluating the Christoffel symbols explicitly, we get

$$h_{ik}\frac{\mathrm{d}^2x^k}{\mathrm{d}t^2} + \frac{1}{2}(\partial_j h_{ik} + \partial_k h_{ji} - \partial_i h_{jk})\frac{\mathrm{d}x^j}{\mathrm{d}t}\frac{\mathrm{d}x^k}{\mathrm{d}t} = 0, \tag{5.96}$$

where

$$h_{ij} = -\frac{g_{ij}}{g_{00}}. \tag{5.97}$$

Remarkably, this is the geodesic equation for a three-dimensional space with metric h_{ij}, where t serves as the affine parameter. Thus, in a static spacetime, null geodesic extremize the coordinate time, as expressed by the variational principle

$$\delta \int \mathrm{d}t = 0. \tag{5.98}$$

This is the precise analogue of Fermat's principle in optics. It implies that, in the presence of a gravitational field, light does not follow the shortest spatial path but rather the path of extremal coordinate time. The gravitational field effectively acts as a spatially varying optical medium, with an effective refractive index modifying the propagation of light. For instance, for a spherically symmetric spacetime $g_{ij} = f^2(x^i)\delta_{ij}$, i.e., a conformally flat space, we have

$$\mathrm{d}t = \frac{f}{\sqrt{|g_{00}|}}\mathrm{d}l, \tag{5.99}$$

where

$$\mathrm{d}l^2 = \delta_{ij}\mathrm{d}x^i\mathrm{d}x^j \tag{5.100}$$

is the Euclidean line element. Fermat's principle then identifies the refractive index induced by the gravitational field as

$$n(x) = \frac{f(x)}{\sqrt{|g_{00}(x)|}}. \tag{5.101}$$

This refractive effect leads to both the deflection of light and a time delay in its propagation.

Geodesics play a fundamental role in general relativity, describing the motion of free particles in curved spacetime. A key property of geodesics is that their character—timelike, null, or spacelike—remains invariant along the curve, as parallel transport preserves the inner product of tangent vectors. Physically, geodesics represent the trajectories of freely moving particles that experience no proper acceleration. Mathematically, this is expressed by the vanishing of the covariant derivative of the particle's 4-velocity $\widehat{u}$

$$\widehat{a} = \frac{\mathrm{D}}{\mathrm{d}\tau}\widehat{u} = 0, \tag{5.102}$$

which serves as the defining property of a geodesic. When additional forces act on the particle, the right-hand side of (5.102) must be modified to include the appropriate 4-force terms. These terms can be derived systematically from the variation of the particle's action. For example, a charged particle moving under the influence of an electromagnetic field obeys the equation of motion

$$\ddot{x}^\mu + \Gamma^\mu_{\sigma\rho}\dot{x}^\sigma\dot{x}^\rho = \frac{q}{m}F^\mu{}_\nu\dot{x}^\nu, \tag{5.103}$$

where the right-hand side represents the relativistic Lorentz force.

Timelike geodesics are characterized by maximizing proper time. This fundamental property arises from the fact that any timelike curve (whether geodesic or not) can be arbitrarily approximated by null curves with vanishing proper time. Consequently, since timelike curves can asymptotically approach zero proper time, their geodesic counterparts must necessarily correspond to maxima rather than minima of proper time. The twin paradox provides a concrete illustration of this principle: the stationary twin follows a geodesic worldline and consequently experiences greater proper time compared to the traveling twin. This result remains fully consistent with the conclusion in flat spacetime.

The extremal property of geodesics is a local concept that depends on initial conditions. For instance, on a two-dimensional sphere, two distinct geodesics (extremal paths) can connect the same pair of points—one along the shorter arc and the other along the longer arc. Only comparisons between paths with the same initial and final conditions are physically meaningful. Thus, while the shorter arc minimizes proper time between the points, the longer arc represents a saddle point rather than a true maximum due to the compactness of the space.

The geodesic equation is a second-order differential equation, and given initial values, geodesics can be locally determined. We have the following theorem:

Theorem Geodesics $\gamma(t)$ near a point p are determined by their behavior near p, which includes:

(1) $\gamma(0) = p$,
(2) The tangent vector $v^\mu(p)$ of $\gamma(t)$ at point p.

In other words, given the initial position $x^\mu|_p$ and the initial tangent vector $\left.\frac{\mathrm{d}x^\mu}{\mathrm{d}\lambda}\right|_p$, the geodesic is determined in the coordinate chart. Given $\frac{\mathrm{d}x^\mu}{\mathrm{d}\lambda}(\lambda = 0) = v^\mu$, when $\lambda = 1$, there is a unique point on the manifold M connected to the starting point by a geodesic. In fact, we can locally define an exponential map $\exp_p : T_p \to M$ at point p, specifically implemented as

$$\exp_p(v^\mu) = x^\mu(\lambda = 1), \tag{5.104}$$

where $x^\mu(\lambda)$ is determined by the geodesic equation satisfying the initial conditions. Intuitively, at point p, we can take Riemannian normal coordinates to make the Christoffel symbols zero, thus the geodesic is a straight line determined by the tangent vector. As shown in Figure 5.5.

In a neighborhood of a point p, a geodesic passing through p is uniquely determined by its tangent vector v^μ, with the local expression $x^\mu(\lambda) = \lambda v^\mu$. However, depending on the global geometry, geodesics emanating from the same point in different directions may later intersect. A classic example is the two-dimensional sphere, where all geodesics (great circles) originating from the North Pole reconverge at the South Pole. Consequently, the exponential map $\exp_p : T_p \to M$ is generally not injective (one-to-one), as distinct tangent vectors can map to the same point. Furthermore, the image of the exponential map need not cover the entire manifold, as there may be

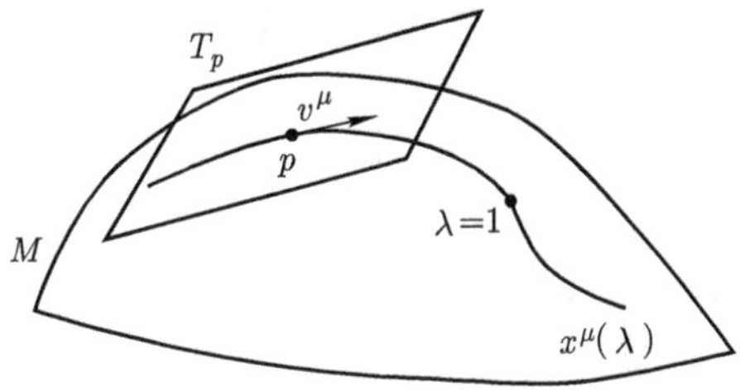

Fig. 5.5 Geodesics are locally straight lines determined by initial position and initial tangent vector.

points on the manifold that cannot be connected by geodesics. For instance, in anti-de Sitter (AdS) spacetime, certain points cannot be reached by any geodesic from p. On the other hand, the domain of exponential map may also be restricted. If a geodesic encounters a singularity during extension, the exponential map cannot be defined beyond that point. A spacetime is called geodesically complete if all geodesics can be extended indefinitely. Conversely, if some geodesics terminate at finite affine parameter values, the spacetime is geodesically incomplete, which serves as a criterion for the existence of singularities.

In the late 1960s, Roger Penrose and Stephen Hawking demonstrated that under physically reasonable energy conditions (excluding exotic matter with negative energy), spacetime in general relativity is generically geodesically incomplete. Their proof, now known as the singularity theorem, employs sophisticated techniques from global differential geometry and highlights the inevitability of singularities in classical gravity. Notable examples of geodesically incomplete spacetimes include the usual black hole spacetime, where timelike and null geodesics terminate at the singularity, and the Friedman–Lemaître–Robertson–Walker (FLRW) spacetime, which describes our expanding universe and contains an initial Big Bang singularity. Faced with the unavoidability of singularities, Penrose formulated the cosmic censorship hypothesis, which posits that all naked singularities (those visible to distant observers) must be prohibited and any physical singularity must be concealed behind an event horizon, shielding external observers from its pathological effects. This hypothesis, while still unproven, serves as a foundational principle in black hole physics and remains a central topic in theoretical relativity.

By using the exponential map, we can construct local Riemannian normal coordinates in the neighborhood of a point p_0 on the manifold. Let

$\widehat{e}_\mu(p_0)$ denote an orthonormal basis for the tangent space at p_0. Each geodesics emanating from p_0 is uniquely determined by its initial tangent vector. A nearby point p can be expressed as

$$p = \mathcal{G}(\lambda; \widehat{V}), \tag{5.105}$$

where λ is an affine parameter and $\widehat{V}$ is the tangent vector at p_0. By rescaling $\widehat{V}$, we may set $\lambda = 1$ without loss of generality, and varying $\widehat{V}$ over all possible directions covers a neighborhood of p_0. Thus, any point p near p_0 can be expressed as

$$p = \mathcal{G}(1; x^\mu \widehat{e}_\mu), \tag{5.106}$$

where the coefficients $\{x^\mu\}$ define the Riemannian normal coordinates centered at p_0. By construction, the basis vectors satisfy

$$\widehat{e}_\mu = \left.\frac{\partial}{\partial x^\mu}\right|_{p_0}, \tag{5.107}$$

and the curve $x^\mu = V^\mu \lambda$ (with V^μ constant) correspond to geodesic. Substituting into the geodesic equation yields $\Gamma^\mu_{\sigma\rho}|_{p_0} = 0$. Moreover, from the geodesic deviation equation, which will be introduced later, we deduce that the first derivatives of the connection coefficients $\Gamma^\mu_{\sigma\rho}$ at p_0 are related to the Riemann curvature tensor.

5.3.1 *Geodesics in FLRW universe*

Below, we examine the geodesics in a flat FLRW spacetime as an illustrative example. The flat FLRW universe describes a homogeneous and isotropic expanding universe consistent with the cosmological principle. Its metric is given by

$$\mathrm{d}s^2 = -\mathrm{d}t^2 + a^2(t)(\mathrm{d}x^2 + \mathrm{d}y^2 + \mathrm{d}z^2), \tag{5.108}$$

where we have assumed spatial flatness for simplicity—this is a special case of the more general FRW spacetime.[3] The scale factor $a(t)$ in the metric characterizes the expansion of the universe. Its evolution depends on the matter content, with power-law solutions of the form

$$a(t) = t^q, 0 < q < 1, \begin{cases} q = \dfrac{2}{3}, \\ q = \dfrac{1}{2}, \end{cases} \tag{5.109}$$

[3]The FLRW metric, proposed by Alexander Friedmann, Georges Lemaître, Howard P. Robertson, and Arthur G. Walker, embodies the cosmological principle. For a detailed discussion, see the final chapter of the book [64].

where $0 < q < 1$. This describes a universe that expands indefinitely. In the early-time limit ($t \to 0$), the scale factor vanishes ($a(t) \to 0$), corresponding to the Big Bang singularity. Thus, the cosmic time t spans the range $0 < t < \infty$.

To analyze the causal structure, we examine light cones defined by $\mathrm{d}s^2 = 0$. For simplicity, we consider motion along the x-direction (fixing y and z due to isotropy), reducing the metric to

$$\begin{aligned} 0 &= -\mathrm{d}t^2 + t^{2q}\mathrm{d}x^2 \\ \Rightarrow \frac{\mathrm{d}x}{\mathrm{d}t} &= \pm t^{-q}. \end{aligned} \tag{5.110}$$

This is the correct result, but we need a more convincing derivation. Let $\widehat{V} = \left(\frac{\mathrm{d}x^\mu}{\mathrm{d}\lambda}\right)\partial_\mu$ be the tangent vector of the curve $x^\mu(\lambda)$. To obtain a null path, we must require $\widehat{g}(\widehat{V}, \widehat{V}) = 0$. By the interaction between 1-forms and vectors,

$$\mathrm{d}t(\widehat{V}) = \mathrm{d}t\left(\frac{\mathrm{d}x^\mu}{\mathrm{d}\lambda}\partial_\mu\right) = \frac{\mathrm{d}x^\mu}{\mathrm{d}\lambda}\mathrm{d}t(\partial_\mu) = \frac{\mathrm{d}x^\mu}{\mathrm{d}\lambda}\frac{\partial t}{\partial x^\mu} = \frac{\mathrm{d}t}{\mathrm{d}\lambda}, \tag{5.111}$$

thus we have

$$\mathrm{d}t^2(\widehat{V}, \widehat{V}) = (\mathrm{d}t \otimes \mathrm{d}t)(\widehat{V}, \widehat{V}) = \mathrm{d}t(\widehat{V})\mathrm{d}t(\widehat{V}) = \left(\frac{\mathrm{d}t}{\mathrm{d}\lambda}\right)^2. \tag{5.112}$$

Similarly,

$$\mathrm{d}x^2(\widehat{V}, \widehat{V}) = \left(\frac{\mathrm{d}x}{\mathrm{d}t}\right)^2. \tag{5.113}$$

Therefore, requiring the tangent vector to be a null vector implies

$$\widehat{g}(\widehat{V}, \widehat{V}) = 0 \Rightarrow 0 = -\left(\frac{\mathrm{d}t}{\mathrm{d}\lambda}\right)^2 + t^{2q}\left(\frac{\mathrm{d}x}{\mathrm{d}\lambda}\right)^2 \Rightarrow \frac{\mathrm{d}x}{\mathrm{d}t} = \pm t^{-q}, \tag{5.114}$$

after integration, we get

$$t = (1-q)^{\frac{1}{1-q}}(\pm x - x_0)^{\frac{1}{1-q}}. \tag{5.115}$$

This is the equation satisfied by the null curve. Since $0 < q < 1$, the light cone equation exhibits a singularity at $t = 0$, and the light cone becomes parallel to the x-axis at $t = 0$. This behavior differs fundamentally from the one in Minkowski spacetime, where light cones always maintain a fixed 45° opening angle. The "tipping" of light cones near $t = 0$ in the FLRW case means that not all pairs of events share a common past light cone and there exist events that were never in causal contact.

In the previous discussion on geodesics, we saw that the geodesic equation can be obtained by varying the action. This actually provides an effective method for calculating the connection coefficients. Taking the FLRW metric as an example,

$$\mathrm{d}s^2 = -\mathrm{d}t^2 + a^2(t)\delta_{ij}\mathrm{d}x^i\mathrm{d}x^j, \tag{5.116}$$

the action is

$$I = -\frac{1}{2}\int\left[-\left(\frac{\mathrm{d}t}{\mathrm{d}\tau}\right)^2 + a^2(t)\delta_{ij}\frac{\mathrm{d}x^i}{\mathrm{d}\tau}\frac{\mathrm{d}x^j}{\mathrm{d}\tau}\right]\mathrm{d}\tau. \tag{5.117}$$

Varying the action, under $x^\mu \to x^\mu + \delta x^\mu$ with $\delta I = 0$, we obtain the geodesic equation

$$\ddot{x}^\mu + \Gamma^\mu_{\sigma\rho}\dot{x}^\sigma\dot{x}^\rho = 0. \tag{5.118}$$

First, consider $t \to t + \delta t$. Since $a(t+\delta t) = a(t) + (a')\delta t$, where $a' = \frac{\mathrm{d}a}{\mathrm{d}t}$, we have

$$\begin{aligned}\delta I &= \frac{1}{2}\int\left[-2\frac{\mathrm{d}t}{\mathrm{d}\tau}\frac{\mathrm{d}\delta t}{\mathrm{d}\tau} + 2aa'\delta_{ij}\frac{\mathrm{d}x^i}{\mathrm{d}\tau}\frac{\mathrm{d}x^j}{\mathrm{d}\tau}\delta t\right]\mathrm{d}\tau \\ &= \int\left[\frac{\mathrm{d}^2 t}{\mathrm{d}\tau^2} + aa'\delta_{ij}\frac{\mathrm{d}x^i}{\mathrm{d}\tau}\frac{\mathrm{d}x^j}{\mathrm{d}\tau}\right]\delta t\mathrm{d}\tau.\end{aligned} \tag{5.119}$$

To ensure $\delta I = 0$ for any δt, the term in the integral must be zero. Comparing with the geodesic equation and taking $x^\mu = t$, we can read out the non-zero connection coefficients

$$\Gamma^0_{00} = \Gamma^0_{i0} = 0, \quad \Gamma^i_{ij} = aa'\delta_{ij}. \tag{5.120}$$

Next, consider $x^i \to x^i + \delta x^i$. This leads to

$$\begin{aligned}\delta I &= \frac{1}{2}\int\left(2a^2\delta_{ij}\frac{\mathrm{d}x^i}{\mathrm{d}\tau}\frac{\mathrm{d}(\delta x^j)}{\mathrm{d}\tau}\right)\mathrm{d}\tau \\ &= \int\left(a^2\frac{\mathrm{d}^2x^i}{\mathrm{d}\tau^2} + 2a\frac{\mathrm{d}a}{\mathrm{d}\tau}\frac{\mathrm{d}x^i}{\mathrm{d}\tau}\right)\delta_{ij}\delta x^j\mathrm{d}\tau.\end{aligned} \tag{5.121}$$

By $\frac{\mathrm{d}a}{\mathrm{d}\tau} = (a')\frac{\mathrm{d}t}{\mathrm{d}\tau}$, we need

$$\frac{\mathrm{d}^2x^i}{\mathrm{d}\tau^2} + 2\frac{a'}{a}\frac{\mathrm{d}t}{\mathrm{d}\tau}\frac{\mathrm{d}x^i}{\mathrm{d}\tau} = 0. \tag{5.122}$$

Comparing with the geodesic equation, we get

$$\Gamma^i_{00} = 0, \quad \Gamma^i_{j0} = \Gamma^i_{0j} = \frac{a'}{a}\delta^i_j, \quad \Gamma^i_{jk} = 0. \tag{5.123}$$

Thus, we have obtained the connection coefficients for the flat FLRW spacetime. Using the method of action variation, the connection coefficients can be conveniently obtained when the spacetime has good symmetry.

For geodesics in FLRW spacetime, we are particularly interested in null geodesics. From the preceding discussion, we can derive the equation that any null path must satisfy:

$$0 = -\left(\frac{\mathrm{d}t}{\mathrm{d}\lambda}\right)^2 + a^2(t)\delta_{ij}\frac{\mathrm{d}x^i}{\mathrm{d}\lambda}\frac{\mathrm{d}x^j}{\mathrm{d}\lambda}. \tag{5.124}$$

Note that this is not a geodesic equation but rather a constraint that all null path must obey. From the geodesic equation, we know

$$\begin{aligned} &\frac{\mathrm{d}^2x^0}{\mathrm{d}\lambda^2} + \Gamma^0_{\rho\delta}\frac{\mathrm{d}x^\rho}{\mathrm{d}\lambda}\frac{\mathrm{d}x^\delta}{\mathrm{d}\lambda} = 0 \\ \Rightarrow &\frac{\mathrm{d}^2t}{\mathrm{d}\lambda^2} + \frac{a'}{a}\left(\frac{\mathrm{d}t}{\mathrm{d}\lambda}\right)^2 = 0 \\ \Rightarrow &\frac{\mathrm{d}t}{\mathrm{d}\lambda} = \frac{\omega_0}{a}, \end{aligned} \tag{5.125}$$

where ω_0 is an integration constant. Given the scale factor $a(t)$, we can determine the relationship between t and the affine parameter λ. For the power-law expansion, where $a(t) = t^q$, the solution takes the form

$$\frac{t^{q+1} - t_0^{q+1}}{q+1} = \omega_0(\lambda - \lambda_0). \tag{5.126}$$

Starting from an initial position at λ_0 and t_0, a finite affine parameter interval $\Delta\lambda = \lambda - \lambda_0$ leads to the singularity at $t = 0$. This implies that the Big Bang singularity is reached within finite affine parameter time, making it impossible to extend the geodesic further. Thus, the null geodesic is incomplete, and $t = 0$ constitutes a genuine spacetime singularity.

5.3.2 *Energy-momentum conservation in FLRW spacetime*

In flat spacetime, the conservation of energy and momentum is expressed as $\partial_\mu T^{\mu\nu} = 0$. In curved spacetime, this condition is naturally generalized by replacing partial derivatives with covariant derivatives, yielding

$$\nabla_\mu T^{\mu\nu} = 0. \tag{5.127}$$

For an ideal fluid, the stress-energy tensor takes the form $T^{\mu\nu} = (\rho + p)u^\mu u^\nu + pg^{\mu\nu}$, where ρ is the energy density, p is the pressure, and u^μ is the fluid's four-velocity. In the comoving coordinate system of the fluid, where the fluid is at rest, the 4-velocity simplifies to $u^\mu = (1, 0, 0, 0)$, and the stress-energy tensor becomes diagonal $T^{\mu\nu} = \text{diag}(\rho, a^{-2}p, a^{-2}p, a^{-2}p)$. The energy-momentum conservation equation can be expanded explicitly as

$$\nabla_\mu T^{\mu\nu} = \partial_\mu T^{\mu\nu} + \Gamma^\mu_{\mu\lambda} T^{\lambda\nu} + \Gamma^\nu_{\mu\lambda} T^{\mu\lambda} = 0. \tag{5.128}$$

First, consider the case $\nu = 0$. Since

$$\begin{aligned} \partial_\mu T^{\mu 0} &= \partial_0 T^{00} = \dot{\rho}, \\ \Gamma^\mu_{\mu\lambda} T^{\lambda 0} &= \Gamma^\mu_{\mu 0} T^{00} = 3\frac{a'}{a}\rho, \\ \Gamma^0_{\mu\lambda} T^{\mu\lambda} &= \Gamma^0_{00} T^{00} + \Gamma^0_{11} T^{11} + \Gamma^0_{22} T^{22} + \Gamma^0_{33} T^{33} \\ &= 3aa'(a^{-2}p) = 3\frac{a'}{a}p, \end{aligned} \tag{5.129}$$

we obtain

$$\dot{\rho} = -3\frac{a'}{a}(\rho + p). \tag{5.130}$$

For the case $\nu = 1$ ($\nu = 2, 3$ is similar),

$$\begin{aligned} \partial_\mu T^{\mu 1} &= \partial_1 T^{11} = a^{-2}\partial_x p, \\ \Gamma^\mu_{\mu\lambda} T^{\lambda 1} &= \Gamma^\mu_{\mu 1} T^{11} = 0, \\ \Gamma^1_{\mu\lambda} T^{\mu\lambda} &= \Gamma^1_{00} T^{00} + \Gamma^1_{11} T^{11} + \Gamma^1_{22} T^{22} + \Gamma^1_{33} T^{33} = 0, \end{aligned} \tag{5.131}$$

implying that

$$\partial_1 p = 0. \tag{5.132}$$

Therefore, energy-momentum conservation requires $\partial_i p = 0$, $i = 1, 2, 3$. In Minkowski spacetime, $a = 1$, $a' = 0$, and both energy density and pressure are conserved. However, in FLRW universe, pressure does not change with spatial curvature. For an observer moving with the fluid, the fluid is at rest, and the pressure remains constant at all spatial locations at a fixed moment.

Nevertheless, in FLRW spacetime, energy is not conserved, $\dot{\rho} = -3\frac{a'}{a}(\rho + p)$. By the equation of state $p = \omega\rho$, where ω is a constant,

$$\begin{aligned} \frac{\dot{\rho}}{\rho} &= -3(1+\omega)\frac{a'}{a} \\ &\Rightarrow \rho = a^{-3(1+\omega)}. \end{aligned} \tag{5.133}$$

For different cases:

$$\Rightarrow \begin{cases} \text{Matter domination:} & p = 0, \quad \rho_m \propto a^{-3}; \\ \text{Radiation domination:} & p = \frac{\rho}{3}, \quad \rho_r \propto a^{-4}; \\ \text{Vacuum energy:} & p = -\rho,\ \rho_\Lambda = \text{constant}. \end{cases} \tag{5.134}$$

Thus, the energy density changes with the scale factor. In the matter-dominated era, the energy density ρ is primarily determined by non-relativistic matter (dust), where $\rho = nm$ (with n being the number density and m the particle mass). Since the number density scales as $n \propto a^{-3}$ due to cosmological expansion, the energy density changes with the scale factor as $\rho_m \propto a^{-3}$. During the radiation-dominated stage, in addition to the volume change due to expansion, there is also the gravitational redshift effect, $E \sim a^{-1}$. Thus, the radiation energy density scales as $\rho_r \propto a^{-4}$. In contrast, vacuum energy (or a cosmological constant) remains constant over time. From this scaling behavior, we can glimpse the evolution history of the universe's matter composition: in the early universe, radiation and matter dominate due to their high initial energy densities; as the universe expands, radiation density decays faster than matter, leading to the matter-dominated epoch; eventually, even matter density becomes negligible compared to vacuum energy, which remains constant. Thus, even if ρ_Λ was initially tiny, its relative importance grows over time, leading to the late-time accelerated expansion observed today.

In the context of FLRW cosmology, the total energy associated with a given matter component can be estimated by integrating its energy density over a comoving volume, $E = \displaystyle\int \rho a^3 \mathrm{d}^3x$, where $a^3\mathrm{d}^3x$ represents the proper volume element at a given time t. The choice of coordinates here is tied to the scale factor $a(t)$, which determines the physical boundaries of the integration region. For matter, energy is constant because matter is composed of dust particles, and energy is the sum of their rest masses. For radiation, energy decreases due to the gravitational redshift effect. For vacuum energy, energy increases due to the expansion of the universe. It is important to note that energy conservation in general relativity is subtle. In FLRW spacetime, energy is not globally conserved because the metric lacks time-translation symmetry.

However, we still refer to $\nabla_\mu T^{\mu\nu} = 0$ as the energy-momentum conservation equation. What is its significance? First, this equation is fundamentally local in nature. For the energy-momentum tensor of matter, it represents a strict dynamical law—even in cases where global energy conservation does not hold. Second, the appearance of Christoffel symbols in this equation indicates an energy-momentum exchange between the matter field and the gravitational field. Thus, the energy-momentum conservation equation implicitly encodes the interaction between matter and gravity. It is important to note, however, that defining a local energy density for the gravitational field remains an unresolved challenge in general relativity. Furthermore, as we will discuss later, the energy-momentum conservation equation arises as a direct consequence of spacetime diffeomorphism invariance. In summary, there are profound differences between flat and curved spacetime. Many familiar concepts from flat spacetime require careful re-examination in the context of general relativity.

5.4 *Lie Derivative

In the definition of the covariant derivative, the key idea is to employ parallel transport to bring tangent vectors from different points to a common location, enabling their comparison. This operation inherently depends on the path connecting the two points. However, parallel transport is not the only method for comparing vectors or tensors at different points. Alternative approaches exist for moving and contrasting such objects, leading to different notions of differentiation. Among these, the Lie derivative provides a particularly useful framework, which we now introduce.

5.4.1 *Pullback and pushforward*

To define the Lie derivative, we first introduce the concepts of pullback and pushforward. Consider two smooth manifolds M and N, which may differ in dimensions, equipped with local coordinate charts x^μ and y^α, respectively. Let $\phi : M \to N$ be a smooth map between them. Given a function $f : N \to \mathbf{R}$, we can naturally construct a corresponding function on M via composition with ϕ, yielding $f \circ \phi : M \to \mathbf{R}$. This operation defines the pullback of f by ϕ, denoted as

$$\phi_* f = f \circ \phi. \tag{5.135}$$

Here, ϕ_* "pulls back" the function f from N to M, as illustrated in Figure 5.6.

While the pullback of a function $\phi_* f = f \circ \phi$ is well-defined, the pushforward of a function $g : M \to \mathbf{R}$ is generally not meaningful, since composition $g \circ \phi^{-1}$ would require ϕ to be invertible. However, for vectors, the situation is different: a vector field on M can be "pushed forward" to N via the differential of ϕ. A vector $\widehat{V}(p)$ at $p \in M$ acts as a directional derivative on functions. We define its pushforward vector $\phi^* \widehat{V}$ at $\phi(p)$ on N by requiring that its action on any $f \in \mathcal{F}(\mathcal{N})$ agrees with $\widehat{V}$ applied to the pullback of f:

$$(\phi^* \widehat{V})(f) = \widehat{V}(\phi_* f) = \widehat{V}(f \circ \phi). \tag{5.136}$$

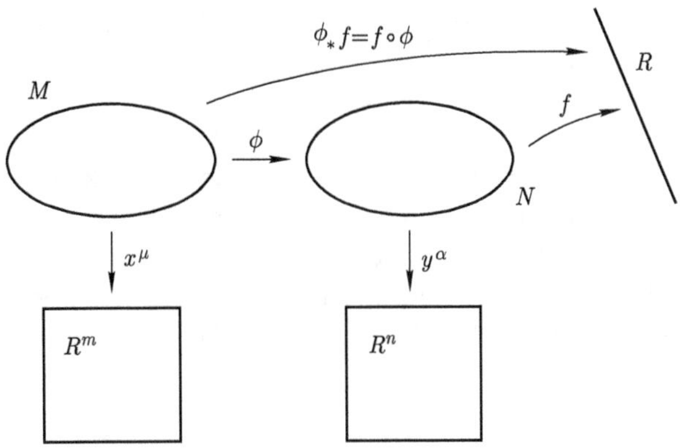

Fig. 5.6 Pullback of a function.

This means $\phi^*\widehat{V}$ is the unique vector on N whose directional derivative matches $\widehat{V}$'s action on pulled-back functions. More specifically, in coordinate charts $\{x^\mu\}$, $\widehat{V} = V^\mu\partial_\mu$, and $\phi^*\widehat{V} = (\phi^*\widehat{V})^\alpha\partial_\alpha$ in coordinates $\{y^\alpha\}$ on N. For any test function $f \in \mathcal{F}(\mathcal{N})$, applying the chain rule to (5.136) yields

$$\begin{aligned}(\phi^*\widehat{V})^\alpha\partial_\alpha f &= V^\mu\partial_\mu(\phi_* f) = V^\mu\partial_\mu(f\circ\phi)\\ &= V^\mu\frac{\partial y^\alpha}{\partial x^\mu}\partial_\alpha f.\end{aligned} \tag{5.137}$$

Since this holds for all f, we identify the components:

$$(\phi^*\widehat{V})^\alpha = (\phi^*)^\alpha_\mu V^\mu = \frac{\partial y^\alpha}{\partial x^\mu}V^\mu, \tag{5.138}$$

where $(\phi^*)^\alpha_\mu = \dfrac{\partial y^\alpha}{\partial x^\mu} = J^\alpha_\mu$ is the Jacobian matrix of ϕ. This can be seen as a generalization of the transformation law of vectors under coordinate changes, but now ϕ need not be a diffeomorphism, since M and N may even have different dimensions. The Jacobian J^α_μ is typically a rectangular matrix, so ϕ_* cannot generally be inverted. Consequently, while vectors push forward, they do not pull back unless ϕ is a diffeomorphism. If we consider the special case where M and N are the same manifold, this reduces to the standard transformation of vector fields under coordinate reparametrizations.

For a 1-form, its pullback can be well defined. A 1-form is a linear map that takes a vector to a real number. Given a smooth map $\phi : M \mapsto N$, the pullback of a 1-form $\widehat{\omega}$ on N is defined as follows: for any vector $\widehat{V}$ on M, the pullback $\phi_*\widehat{\omega}$ acts on $\widehat{V}$ by

$$(\phi_*\widehat{\omega})(\widehat{V}) = \widehat{\omega}(\phi^*\widehat{V}), \tag{5.139}$$

where $\phi^*\widehat{V}$ is the pushforward of $\widehat{V}$ under ϕ. The pullback operation admits a simple matrix representation in local coordinates. If $\widehat{\omega}$ has components ω_α on N, then the components of $\phi_*\widehat{\omega}$ on M are given by $(\phi_*\widehat{\omega})_\mu = (\phi^*)^\alpha_\mu\omega_\alpha$, where $(\phi^*)^\alpha_\mu$ is the Jacobian matrix of ϕ, in match with the pushforward matrix for vectors—the only difference being the index contraction.

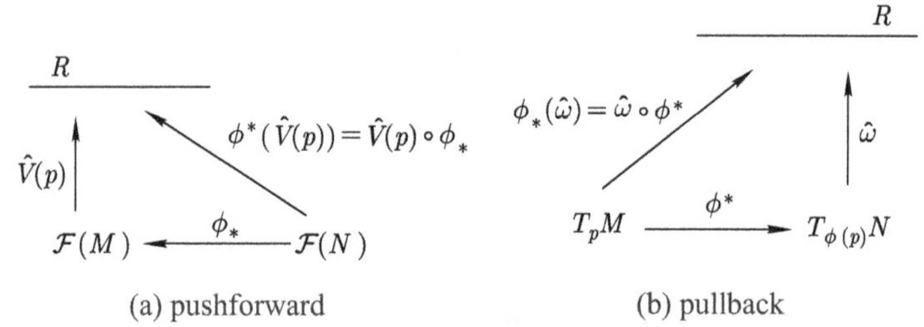

Fig. 5.7 Pushforward and pullback.

We can illustrate the concepts of pushforwards and pullbacks clearly using diagrams. Let $\mathcal{F}(M)$ and $\mathcal{F}(N)$ denote the sets of smooth functions on M and N, respectively. A vector $\widehat{V}(p)$ at a point p on M is an operator: $\mathcal{F}(M) \mapsto \mathbf{R}$. The pullback induces a mapping: $\phi^* : \mathcal{F}(N) \mapsto \mathcal{F}(M)$, allowing us to define the pushforward ϕ^* via composition of mappings, as depicted in Figure 5.7(a). Similarly, if T_qN represents the tangent space at point q on N, then a 1-form $\widehat{\omega}$ at q can be viewed as a linear functional $\widehat{\omega} : T_qN \mapsto \mathbf{R}$, i.e., an element of the cotangent space T^*_qN. Given that the pushforward ϕ^* maps T_pM to $T_{\phi(p)}N$, the pullback of a 1-form naturally arises as the composition shown in Figure 5.7(b).

The preceding discussion can be naturally extended to arbitrary tensor fields. A $(0,l)$-type tensor on N is a multi-linear map from the tensor product of l vectors to $\mathbf{R}$. Its pullback under a smooth map ϕ is defined as

$$(\phi_*\widehat{T})(\widehat{V}^{(1)}, \widehat{V}^{(2)}, \cdots, V^{(l)}) = \widehat{T}(\phi^*\widehat{V}^{(1)}, \phi^*\widehat{V}^{(2)}, \cdots, \phi^*V^{(l)}),$$

where $\widehat{T}$ is an $(0,l)$-type tensor field on the target manifold N. In local coordinates, this operation takes the form

$$(\phi_*\widehat{T})_{\mu_1\cdots\mu_l} = \frac{\partial y^{\alpha_1}}{\partial x^{\mu_1}}\cdots\frac{\partial y^{\alpha_l}}{\partial x^{\mu_l}}T_{\alpha_1\cdots\alpha_l}. \tag{5.140}$$

Dually, the pushforward of a $(k, 0)$-type tensor $\widehat{S}$ is defined as

$$(\phi^*\widehat{S})(\widehat{\omega}^{(1)}, \widehat{\omega}^{(2)}, \cdots, \widehat{\omega}^{(k)}) = \widehat{S}(\phi_*\widehat{\omega}^{(1)}, \phi_*\widehat{\omega}^{(2)}, \cdots, \phi_*\widehat{\omega}^{(k)}),$$

which, in component form, becomes

$$(\phi^*\widehat{S})^{\alpha_1\cdots\alpha_K} = \frac{\partial y^{\alpha_1}}{\partial x^{\mu_1}} \cdots \frac{\partial y^{\alpha_k}}{\partial x^{\mu_k}} S^{\mu_1\cdots\mu_k}.$$

Example 5.5. Pullback of the metric tensor.

Consider the case where M is an embedded submanifold of N, with the canonical inclusion map $\phi : M \mapsto N$ that maps each point in M to its corresponding point in N. A classic example is the embedding of the two-dimensional sphere $M = \mathbf{S}^2$ into three-dimensional Euclidean space $N = \mathbf{R}^3$. Let $x^\mu = (\theta, \phi)$ be spherical coordinates on $\mathbf{S}^2$, and let $y^\alpha = (x, y, z)$ be Cartesian coordinates on $\mathbf{R}^3$. The embedding ϕ is explicitly given by

$$y^\alpha(\theta, \phi) = (\sin\theta\cos\phi, \sin\theta\sin\phi, \cos\theta). \tag{5.141}$$

The Euclidean metric on $\mathbf{R}^3$ is $\mathrm{d}s^2 = \mathrm{d}x^2 + \mathrm{d}y^2 + \mathrm{d}z^2$. The induced metric on $\mathbf{S}^2$ is obtained by pulling back the Euclidean metric via ϕ. The Jacobian matrix of ϕ is now

$$\frac{\partial y^\alpha}{\partial x^\mu} = \begin{pmatrix} \cos\theta\cos\phi & \cos\theta\sin\phi & -\sin\theta \\ -\sin\theta\sin\theta & \sin\theta\cos\phi & 0 \end{pmatrix}. \tag{5.142}$$

The induced metric on $\mathbf{S}^2$ is simply obtained by pulling back the metric on $\mathbf{R}^3$:

$$\begin{aligned} (\phi_*\widehat{g})_{\mu\nu} &= \frac{\partial y^\alpha}{\partial x^\mu}\frac{\partial y^\beta}{ax\nu} g_{\alpha\beta} \\ &= \begin{pmatrix} 1 & 0 \\ 0 & \sin^2\theta \end{pmatrix}. \end{aligned} \tag{5.143}$$

A tensor with both contravariant and covariant indices (i.e. a (k, l)-type tensor where $k \neq 0$ and $l \neq 0$) generally cannot be pulled back or pushed forward under a smooth map ϕ unless ϕ is invertible. In the case that M and N are the same manifold and ϕ is invertible, ϕ is called a diffeomorphism. In this case, ϕ and its inverse ϕ^{-1} allow us to define both pullback and pushforward operations for arbitrary tensor fields. Specifically, given a (k, l)-type tensor field $T^{\mu_1\cdots\mu_k}{}_{\nu_1\cdots\nu_l}$ on M, its pushforward $\phi^*\widehat{T}$ is defined by

$$\begin{aligned} &(\phi^*\widehat{T})(\omega^{(1)}, \cdots, \omega^{(k)}, V^{(1)}, \cdots, V^{(l)}) \\ &= \widehat{T}(\phi_*\omega^{(1)}, \cdots, \phi_*\omega^{(k)}, [\phi^{-1}]_* V^{(1)}, \cdots, [\phi_{-1}]^* V^{(l)}). \end{aligned}$$

Here, $\omega^{(i)}$ are 1-forms on N, and $V^{(i)}$ are vector fields on N. In local coordinates, this takes the form

$$(\phi^* \widehat{T})^{\alpha_1 \cdots \alpha_k}{}_{\beta_1 \cdots \beta_l} = \frac{\partial y^{\alpha_1}}{\partial x^{\mu_1}} \cdots \frac{\partial y^{\alpha_k}}{\partial x^{\mu_k}} \frac{\partial x^{\nu_1}}{\partial y^{\beta_1}} \cdots \frac{\partial x^{\nu_l}}{\partial y^{\beta_l}} T^{\mu_1 \cdots \mu_k}{}_{\nu_1 \cdots \nu_l}, \tag{5.144}$$

which resembles a standard coordinate transformation.

In fact, there are two equivalent perspectives on coordinate transformations: consider an n-dimensional manifold M with coordinate functions $x^\mu : M \to \mathbf{R}^n$.

- Active transformation: Apply a diffeomorphism $\phi : M \to M$, mapping points and inducing new coordinates via pullback, $(\phi^* y)^\mu = y^\mu \circ \phi$.
- Passive transformation: Keep M fixed but switch to a new coordinate chart $y^\mu : M \to \mathbf{R}^n$.

These correspond, respectively, to "moving points on the manifold" and "changing the coordinate system while keeping points fixed," as illustrated in Figure 5.8.

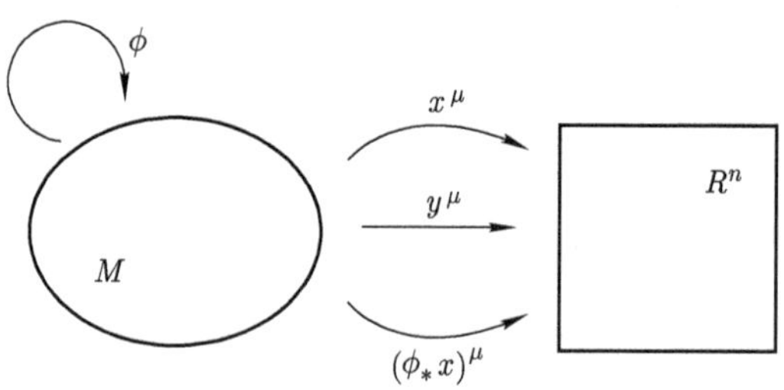

Fig. 5.8 Diffeomorphism and coordinate transformation.

5.4.2 *Lie derivative of tensors*

By employing diffeomorphisms, we establish a framework for comparing tensor fields at distinct points on a manifold, which ultimately leads to the definition of the Lie derivative. Consider a diffeomorphism $\phi : M \to M$ and a tensor field $T^{\mu_1 \cdots \mu_k}{}_{\nu_1 \cdots \nu_l}(x)$. The comparison between the tensor field at a point $p \in M$ and its pullback under ϕ is achieved by evaluating the difference between the tensor field $\widehat{T}(p)$ at p and the pullback tensor field $\phi^*[\widehat{T}(p)]$ at point $\phi(p)$. To construct a derivative operation, we introduce a one-parameter family of diffeomorphisms ϕ_t, defined as a smooth map:

$$\phi_t : \mathbf{R} \times M \to M, \quad \text{with} \quad \phi_s \circ \phi_t = \phi_{s+t}, \tag{5.145}$$

with ϕ_0 being the identity map. For a fixed point $p \in M$, the map $\phi_t : p \to \mathbf{R}$ defines an integral curve on the manifold M. This one-parameter family of diffeomorphisms is generated by a vector field $\widehat{V}$, and conversely, the tangent vectors to the curves induced by ϕ_t reconstruct this vector field. Specifically, $\phi_t(p)$ traces a curve whose tangent at $t = 0$ determines $\widehat{V}(x)$. Extending this to all points on M, the diffeomorphism ϕ_t generates a flow that fills the manifold with non-intersecting curves (unless fixed points exist, in which case degeneracy may occur). Formally, the vector field $\widehat{V}(x)$ is defined as the collection of tangent vectors to these curves at $t = 0$.

Example 5.6. A diffeomorphism on $\mathbf{S}^2$ given by $\phi_t(\theta, \phi) = (\theta, \phi + t)$ is shown in Figure 5.9.

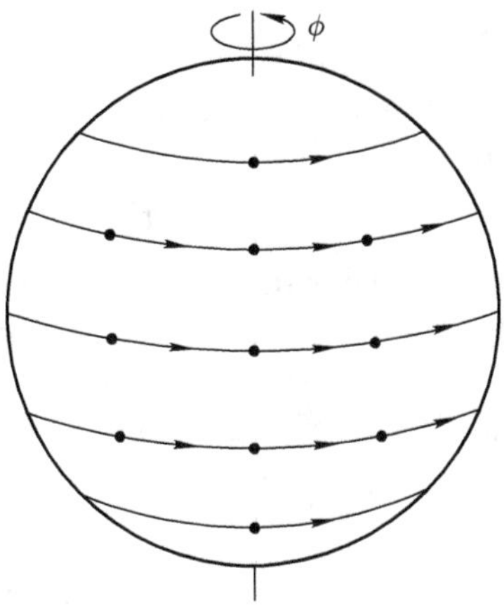

Fig. 5.9 Diffeomorphism on a sphere given by translation along the ϕ direction.

Given a vector field $V^\mu(x)$, we can define its associated integral curve $x^\mu(t)$ as the solution to the differential equation:

$$\frac{\mathrm{d}x^\mu}{\mathrm{d}t} = V^\mu. \tag{5.146}$$

Such a curve is referred to as the orbit (or flow line) of the vector field V^μ. By the theory of ordinary differential equations, the existence and uniqueness of solutions to (5.146) are guaranteed locally, provided the curve does not encounter the boundary of the manifold. A classic physical example is the magnetic field lines of a magnet, which correspond to the integral curves of the magnetic field vector $\boldsymbol{B}$. The integral curve naturally induces a one-parameter diffeomorphism, where the coordinate representation of ϕ_t is given by $x^\mu(t)$. Specifically, for a point p and the vector field $\widehat{V}$ at p, we define $\phi(p,t)$ as the solution to the initial value problem:

$$\begin{aligned} \frac{\partial \phi}{\partial t} &= V_{\phi(p,t)}, \\ \phi(p,0) &= p. \end{aligned} \tag{5.147}$$

Here $\phi(p,t)$ traces the integral curve originating from p, satisfying:

$$\phi(p,t) = \phi_t(p), \tag{5.148}$$

with $\widehat{V}(p)$ as its tangent vector at p. Geometrically, the mapping $\phi_t(p)$ transports each point along the flow generated by $\widehat{V}$, representing the local flow of the vector field. For infinitesimal parameter t, the flow approximates a linear displacement:

$$x'^\mu \approx x^\mu + tV^\mu. \tag{5.149}$$

Formally, the solution to the above differential equation can be expressed as the exponential map:

$$x'^\mu = \exp_p(t\widehat{V})x^\mu. \tag{5.150}$$

The diffeomorphism $\{\phi_t\}$ generated by $\widehat{V}$ forms a one-parameter group under composition. At $t = 0$, ϕ_0 reduces to the identity map, $\phi_0(p) = p$. If $\phi_t(p) = q$, then applying the flow for parameter s starting from q yields $\phi_s(q) = r$. This composition satisfies the group property:

$$\phi_s(\phi_t(p)) = (\phi_s \circ \phi_t)(p) = \phi_{s+t}(p) = (\phi_t \circ \phi_s)(p) = \phi_t(\phi_s(p)), \tag{5.151}$$

demonstrating the commutativity of the flow.

Consider a tensor field $T^{\mu_1\cdots\mu_k}{}_{\nu_1\cdots\nu_l}(x)$ on a manifold and a vector field $\widehat{V}$ generating a flow ϕ_t. To study how $\widehat{T}$ changes along the integral curves of $\widehat{V}$, we compare the tensor at a point p with its pullback from a nearby point $\phi_t(p)$, as illustrated in Figure 5.10. The infinitesimal change of $\widehat{T}$ at p under the flow is given by:

$$\Delta_t T^{\mu_1\cdots\mu_k}{}_{\nu_1\cdots\nu_l}(p) = \phi_{t_*}\left[T^{\mu_1\cdots\mu_k}{}_{\nu_1\cdots\nu_l}(\phi_t(p))\right] - T^{\mu_1\cdots\mu_k}{}_{\nu_1\cdots\nu_l}(p),$$

where $\phi_t(p)$ is the point obtained by flowing p along $\widehat{V}$ for parameter t, ϕ_{t_*} denotes the pullback of the tensor from $\phi_t(p)$ to p, ensuring both terms in the difference are evaluated at the same point p. The Lie derivative $\pounds_{\widehat{V}}\widehat{T}$ measures the instantaneous rate of change of $\widehat{T}$ along the vector field $\widehat{V}$. It is defined as the limit:

$$\pounds_{\widehat{V}} T^{\mu_1\cdots\mu_k}{}_{\nu_1\cdots\nu_l} = \lim_{t\to 0}\left(\frac{\Delta_t T^{\mu_1\cdots\mu_k}{}_{\nu_1\cdots\nu_l}}{t}\right). \tag{5.152}$$

Below are some key properties of the Lie derivative:

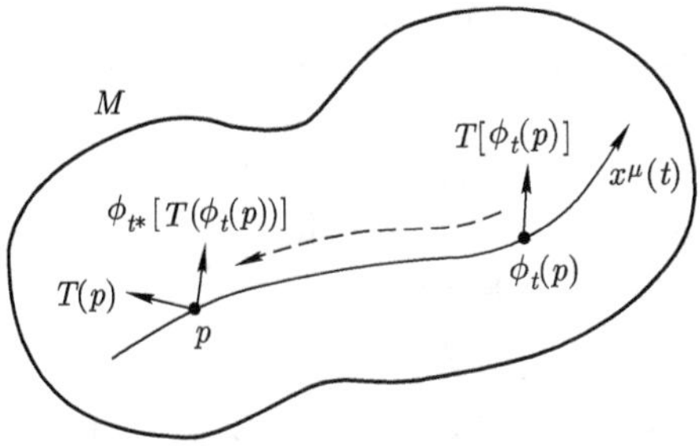

Fig. 5.10 The Lie derivative of a tensor field can be defined using a one-parameter family of diffeomorphisms.

(1) Type preservation: The Lie derivative $\pounds_{\widehat{V}}$ preserves the tensor type—a (k,l)-type tensor remains a (k,l)-type tensor under its action;
(2) Coordinate independence: The operation is intrinsically defined and does not depend on the choice of coordinates.
(3) Linearity: The Lie derivative is linear in both its argument and its vector field

$$\pounds_{\widehat{V}}(a\widehat{T} + b\widehat{S}) = a\pounds_{\widehat{V}}\widehat{T} + b\pounds_{\widehat{V}}\widehat{S}, \tag{5.153}$$

and

$$\pounds_{a\widehat{V}+b\widehat{W}}\widehat{T} = a\pounds_{\widehat{V}}\widehat{T} + b\pounds_{\widehat{W}}\widehat{T}, \tag{5.154}$$

where a and b are constants;
(4) Leibniz rule (product rule): The Lie derivative satisfies the Leibniz rule for tensor products

$$\pounds_{\widehat{V}}(\widehat{T} \otimes \widehat{S}) = (\pounds_{\widehat{V}}\widehat{T}) \otimes \widehat{S} + \widehat{T} \otimes (\pounds_{\widehat{V}}\widehat{S}), \tag{5.155}$$

where $\widehat{S}$ and $\widehat{T}$ are arbitrary tensors;
(5) Connection-free operation: Unlike the covariant derivative, the Lie derivative does not require the introduction of an affine connection.
(6) When applied to a scalar function f, the Lie derivative reduces to the ordinary directional derivative:

$$\pounds_{\widehat{V}} f = \widehat{V}(f) = V^\mu \partial_\mu f. \tag{5.156}$$

To derive a coordinate expression for the Lie derivative, we can exploit a special coordinate system adapted to the vector field $\widehat{V}$. At a point p, we can always choose local coordinates x^μ such that the integral curve of $\widehat{V}$ is aligned with the x^1-direction. In these coordinates, the vector field simplifies to $\widehat{V} = V^\mu \partial_\mu \approx \partial_1$, where $V^\mu \sim \delta_1^\mu = (1, 0, \ldots, 0)$, and the integral curves are lines of constant $(x^2, \ldots, x^n)$, with x^1 parameterizing the flow. The magic of this coordinate system is that the diffeomorphism is a coordinate transformation, transforming x^μ to $y^\mu = (x^1 + t, x^2, \cdots, x^n)$, which is a translation along x^1. The pullback of the diffeomorphism is trivial in these coordinates, with Jacobian matrix $(\phi_{t*})^\mu_\nu = \delta^\nu_\mu$. The components of the tensor field pulled back from $\phi_t(p)$ to p are given by

$$\phi_{t*}[T^{\mu_1\cdots\mu_k}{}_{\nu_1\cdots\mu_l}(\phi_t(p))] = T^{\mu_1\cdots\mu_k}{}_{\nu_1\cdots\mu_l}(x^1 + t, x^2, \cdots, x^n). \tag{5.157}$$

In this coordinate system, the Lie derivative reduces to the partial derivative with respect to x^1:

$$\pounds_{\widehat{V}} T^{\mu_1\cdots\mu_k}{}_{\nu_1\cdots\mu_l} = \frac{\partial}{\partial x^1} T^{\mu_1\cdots\mu_k}{}_{\nu_1\cdots\mu_l}. \tag{5.158}$$

For a (1, 0)-type vector field $\widehat{U} = U^\mu \partial_\nu$, its Lie derivative with respect to $\widehat{V}$ is given by

$$\pounds_{\widehat{V}} U^\mu = \frac{\partial U^\mu}{\partial x^1}, \tag{5.159}$$

which is manifestly non-covariant. However, the commutator of two vector fields yields

$$[\widehat{V}, \widehat{U}]^\mu = V^\nu \partial_\nu U^\mu - U^\mu \partial_\nu V^\mu = \frac{\partial U^\mu}{\partial x^1}, \tag{5.160}$$

coinciding with the Lie derivative in this coordinate system. Both the Lie derivative of a vector field and the commutator of two vector fields are well-defined vector quantities. The equality of their components in this particular coordinate system implies that they must be equal in all coordinate systems, i.e.,

$$\pounds_{\widehat{V}} \widehat{U} = [\widehat{V}, \widehat{U}]. \tag{5.161}$$

Consequently, the commutator of two vector fields is often referred to as the Lie bracket. For the Lie derivative of a vector field, there are two fundamental relations hold:

$$\pounds_{f\widehat{X}} \widehat{Y} = f[\widehat{X}, \widehat{Y}] - \widehat{Y}(f)\widehat{X}, \tag{5.162}$$

$$\pounds_{\widehat{X}}(f\widehat{Y}) = f[\widehat{X}, \widehat{Y}] + \widehat{X}(f)\widehat{Y}. \tag{5.163}$$

For the Lie derivative of a 1-form $\widehat{\omega}$, we first consider the contraction of $\widehat{\omega}$ with an arbitrary vector field $\widehat{U}$, yielding a scalar function $\omega_\mu U^\mu$. The Lie derivative of this scalar with respect to a vector field $\widehat{V}$ is given by

$$\begin{aligned}\pounds_{\widehat{V}}(\omega_\mu U^\mu) &= \widehat{V}(\omega_\mu U^\mu) \\ &= V^\nu \partial_\nu(\omega_\mu U^\mu) \\ &= V^\nu(\partial_\nu \omega_\mu)U^\mu + V^\nu \omega_\mu(\partial_\nu U^\mu).\end{aligned} \tag{5.164}$$

Alternatively, applying the Leibniz rule for the Lie derivative, we obtain

$$\begin{aligned}\pounds_{\widehat{V}}(\omega_\mu U^\mu) &= (\pounds_{\widehat{V}}\widehat{\omega})_\mu U^\mu + \omega_\mu(\pounds_{\widehat{V}}\widehat{U})^\mu \\ &= (\pounds_{\widehat{V}}\widehat{\omega})_\mu U^\mu + \omega_\mu V^\nu \partial_\nu U^\mu - \omega_\mu U^\nu \partial_\nu V^\mu .\end{aligned}$$

Equating the two expressions and isolating the Lie derivative term, we derive the Lie derivative of the 1-form

$$\pounds_{\widehat{V}}\omega_\mu = V^\nu \partial_\nu \omega_\mu + (\partial_\mu V^\nu)\omega_\nu . \tag{5.165}$$

Despite its appearance, this object is indeed a well-defined tensor, though its tensorial nature may not be immediately obvious.

Having established the Lie derivatives for scalar fields (0-forms), 1-forms, and vector fields, we can extend the definition to arbitrary (k,l)-tensors step by step using an iterative method. While the detailed construction follows from the Leibniz rule and linearity, we state the final result directly:

$$\begin{aligned}\pounds_{\widehat{V}} T^{\mu_1\mu_2\cdots\mu_k}{}_{\nu_1\nu_2\cdots\nu_l} &= V^\sigma \partial_\sigma T^{\mu_1\mu_2\cdots\mu_k}{}_{\nu_1\nu_2\cdots\nu_l} \\ &- (\partial_\lambda V^{\mu_1}) T^{\lambda\mu_2\cdots\mu_k}{}_{\nu_1\nu_2\cdots\nu_l} - (\partial_\lambda V^{\mu_2}) T^{\mu_2\lambda\cdots\mu_k}{}_{\nu_1\nu_2\cdots\nu_l} - \cdots \\ &+ (\partial_{\nu_1} V^\lambda) T^{\mu_1\mu_2\cdots\mu_k}{}_{\lambda\nu_2\cdots\nu_l} + (\partial_{\nu_2} V^\lambda) T^{\mu_1\mu_2\cdots\mu_k}{}_{\nu_1\lambda\cdots\nu_l} + \cdots .\end{aligned}$$

At first glance, this expression does not manifestly appear to be a tensor due to its dependence on ordinary partial derivatives. However, remarkably, this coordinate-dependent formula is equivalent to the following manifestly covariant expression:

$$\begin{aligned}\pounds_{\widehat{V}} T^{\mu_1\mu_2\cdots\mu_k}{}_{\nu_1\nu_2\cdots\nu_l} &= V^\sigma \nabla_\sigma T^{\mu_1\mu_2\cdots\mu_k}{}_{\nu_1\nu_2\cdots\nu_l} \\ &- (\nabla_\lambda V^{\mu_1}) T^{\lambda\mu_2\cdots\mu_k}{}_{\nu_1\nu_2\cdots\nu_l} - (\nabla_\lambda V^{\mu_2}) T^{\mu_1\lambda\cdots\mu_k}{}_{\nu_1\nu_2\cdots\nu_l} - \cdots \\ &+ (\nabla_{\nu_1} V^\lambda) T^{\mu_1\mu_2\cdots\mu_k}{}_{\lambda\nu_2\cdots\nu_l} + (\nabla_{\nu_2} V^\lambda) T^{\mu_1\mu_2\cdots\mu_k}{}_{\nu_1\lambda\cdots\nu_l} + \cdots ,\end{aligned}$$

where ∇_μ represents any torsion-free covariant derivative (e.g., the Levi–Civita connection in Riemannian geometry). This equivalent form makes the tensorial nature of the Lie derivative explicit, as all terms are now properly covariant under coordinate transformations.

Example 5.7. Lie bracket of vector fields.

Let $\widehat{U}$ and $\widehat{V}$ be two vector fields generating integral curves parameterized by λ and μ, respectively. Locally, these integral curves can be expressed as straight lines in suitable coordinates, e.g., $x^\mu = x^\mu(0) + \lambda U^\mu$, so that the vector fields act as directional derivatives: $\widehat{U} = \dfrac{\mathrm{d}}{\mathrm{d}\lambda}$ and $\widehat{V} = \dfrac{\mathrm{d}}{\mathrm{d}\mu}$. As shown in Figure 5.11, consider a small neighborhood of a point p, where

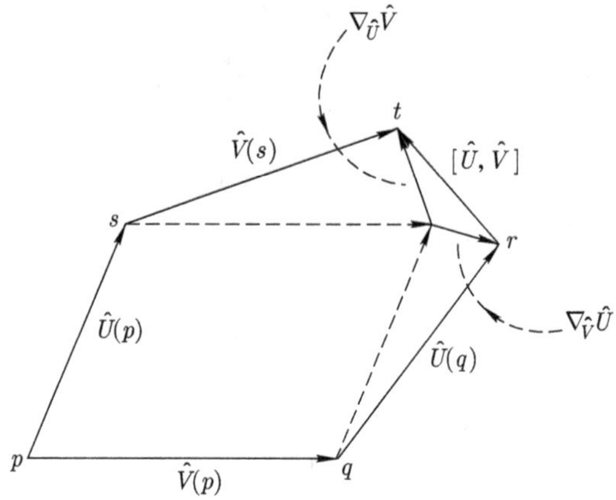

Fig. 5.11 The commutator of two vector fields can be illustrated as follows. We first move along the integral curve of $\widehat{V}$ to point q, and then along the curve of $\widehat{U}$ to point r. On the other hand, we can also first move along the integral curve of $\widehat{U}$ to point s, and then along the integral curve of $\widehat{V}$ to point t. Usually, we cannot obtain a closed curve. The vector connecting points r and t is the commutator of the two vector fields.

the integral curves defined by different vector fields do not form a closed loop. Let q, r, s, and t be points obtained by flowing along the vector fields as follows: from p flow along $\widehat{U}$ to reach q; from q flow along $\widehat{V}$ to reach r; from p flow along $\widehat{V}$ to reach s; from s flow along $\widehat{U}$ to reach t. The difference in the value of a function f at points r and t can be expressed as:

$$\begin{aligned} f(t) - f(r) = [f(t) - f(s)] + [f(s) - f(p)] \\ - [f(r) - f(q)] - [f(q) - f(p)]. \end{aligned} \tag{5.166}$$

Each term can be approximated using a Taylor expansion. For example:

$$f(t) - f(s) \approx V^\mu \partial_\mu f(s) + \frac{1}{2} V^\mu V^\nu \partial_\mu \partial_\nu f(s), \tag{5.167}$$

and similarly for the other terms. Collecting all contributions and keeping only the first-order terms, we find:

$$f(t) - f(r) = (U^\mu \partial_\mu V^\nu - V^\mu \partial_\mu U^\nu)\partial_\nu f = (\pounds_{\widehat{U}} \widehat{V}) f = [\widehat{U}, \widehat{V}] f. \tag{5.168}$$

Thus, the difference between the two points t and r gives the Lie bracket commutator of the vector fields.

Example 5.8. The Lie derivative of a metric tensor field.

For a metric tensor field, its Lie derivative with respect to $\widehat{V}$ is given by

$$\begin{aligned}\pounds_{\widehat{V}} g_{\mu\nu} &= V^\sigma \nabla_\sigma g_{\mu\nu} + (\nabla_\mu V^\lambda) g_{\lambda\nu} + (\nabla_\nu V^\lambda) g_{\mu\lambda} \\ &= \nabla_\mu V_\nu + \nabla_\nu V_\mu \\ &= 2\nabla_{(\mu} V_{\nu)},\end{aligned} \tag{5.169}$$

where ∇_μ is the covariant derivative constructed from the metric tensor field. Here we have already used the metric compatibility condition, and together with the torsion-free condition mentioned earlier, it ensures that the covariant derivative in the last line is with respect to the Christoffel symbols. From Eq. (5.169), when $\nabla_{(\mu}V_{\nu)} = 0$, the metric $g_{\mu\nu}$ is invariant along the flow generated by $\widehat{V}$. Such vector fields are called Killing vector fields, and they satisfy the Killing equation $\nabla_{(\mu}V_{\nu)} = 0$. We will explore the profound physical implications of Killing vectors in subsequent sections, including their role in defining energy, angular momentum, and other conserved quantities in curved spacetimes.

Next, we explore the relationship between the Lie derivative, exterior differentiation, as well as differential forms. Consider a 1-form $\widehat{\omega}$ and a vector field $\widehat{V}$. The exterior derivative of their contraction is given by

$$\mathrm{d}(\widehat{\omega}(\widehat{V})) = \mathrm{d}(\omega_\mu V^\mu) = (V^\mu \partial_\nu \omega_\mu + \omega_\mu \partial_\nu V^\mu)\mathrm{d}x^\nu. \tag{5.170}$$

Additionally, the action of $\mathrm{d}\widehat{\omega}$ on $\widehat{V}$ yields

$$(\mathrm{d}\widehat{\omega})(\widehat{V}) = V^\mu(\partial_\mu \omega_\nu - \partial_\nu \omega_\mu)\mathrm{d}x^\nu. \tag{5.171}$$

Combining these with the expression for the Lie derivative of a 1-form, we obtain the identity

$$\pounds_{\widehat{V}} \widehat{\omega} = \mathrm{d}(\widehat{\omega}(\widehat{V})) + (\mathrm{d}\widehat{\omega})(\widehat{V}). \tag{5.172}$$

Remarkably, this relationship generalizes to arbitrary p-forms. Specifically, the Lie derivative along $\widehat{V}$ satisfies

$$\pounds_{\widehat{V}} = \mathrm{d} \circ i_{\widehat{V}} + i_{\widehat{V}} \circ \mathrm{d}, \tag{5.173}$$

where $i_{\widehat{V}}$ denotes the interior product (contraction) of $\widehat{V}$ with the forms. Equation (5.173) is known as the Cartan formula. Finally, exploiting the nilpotency of the exterior derivative $\mathrm{d}^2 = 0$, we derive the commutation relation

$$\pounds_{\widehat{V}} \circ \mathrm{d} = \mathrm{d} \circ \pounds_{\widehat{V}}. \tag{5.174}$$

Example 5.9. The Lie derivative of the Levi–Civita tensor.

We have seen before that the Levi–Civita tensor actually gives the volume element of the manifold. In local coordinates,

$$\widehat{\varepsilon} = \sqrt{|g|}\mathrm{d}x^1 \wedge \cdots \wedge \mathrm{d}x^n. \tag{5.175}$$

Proposition The Levi–Civita tensor obeys the following identity under the Lie derivative along a vector field $\widehat{V}$:

$$\pounds_{\widehat{V}}\widehat{\varepsilon} = (\nabla \cdot \widehat{V})\widehat{\varepsilon}. \tag{5.176}$$

Proof Using the Cartan formula and considering the fact that the Levi–Civita tensor is already a top-degree form (so $\mathrm{d}\widehat{\varepsilon} = 0$), the Lie derivative simplifies to

$$\begin{aligned}\pounds_{\widehat{V}}\widehat{\varepsilon} &= \mathrm{d}(i_{\widehat{V}}\widehat{\varepsilon}) = \mathrm{d}\sum_{\mu}(-1)^{\mu-1}\sqrt{|g|}\mathrm{d}x^1 \wedge \cdots \wedge i_{\widehat{V}}(\mathrm{d}x^\mu) \wedge \cdots \wedge \mathrm{d}x^n,\\ &= \mathrm{d}\sum_{\mu}(-1)^{\mu-1}(\sqrt{|g|}V^\mu)\mathrm{d}x^1 \wedge \cdots \wedge \widehat{\mathrm{d}x^\mu} \wedge \cdots \wedge \mathrm{d}x^n,\end{aligned} \tag{5.177}$$

where $\widehat{\mathrm{d}x^\mu}$ indicates that $\mathrm{d}x^\mu$ does not appear in the wedge product because it has already been contracted with the vector $\widehat{V}$. The summation on the right-hand side of the above equation is subjected to an exterior differentiation again, and the only non-zero term comes from the exterior derivative that produces $\mathrm{d}x^\mu$, that is,

$$\begin{aligned}\pounds_{\widehat{V}}\widehat{\varepsilon} &= \sum_{\mu}(-1)^{\mu-1}(\partial_\mu(\sqrt{|g|}V^\mu))\mathrm{d}x^\mu \wedge \mathrm{d}x^1 \wedge \cdots \wedge \widehat{\mathrm{d}x^\mu} \wedge \cdots \wedge \mathrm{d}x^n\\ &= \sum_{\mu}(\partial_\mu(\sqrt{|g|}V^\mu))\mathrm{d}x^1 \wedge \cdots \wedge \mathrm{d}x^\mu \wedge \cdots \wedge \mathrm{d}x^n\\ &= (\nabla \cdot \widehat{V})\widehat{\varepsilon},\end{aligned} \tag{5.178}$$

where in the last step we used the divergence formula of a vector

$$\nabla \cdot \widehat{V} = \frac{1}{\sqrt{|g|}}\frac{\partial}{\partial x^\mu}(\sqrt{|g|}V^\mu). \tag{5.179}$$

As previously discussed, parallel transport is naturally defined through the covariant derivative. In a similar spirit, the Lie derivative provides a notion of Lie transport along a curve. A tensor field $\widehat{T}$ is said to be Lie transported along a curve $\mathcal{C}$ with tangent vector $\widehat{u}$ if its Lie derivative vanishes along $\widehat{u}$

$$\pounds_{\widehat{u}}\widehat{T} = 0. \tag{5.180}$$

This condition implies that the tensor field $\widehat{T}$ remains unchanged under an infinitesimal Lie transport along $\mathcal{C}$. In coordinates adapted to the curve—where the tangent vector components are $u^\mu = \delta^\mu_0$—the Lie derivative reduces to a partial derivative:

$$\pounds_{\widehat{u}}T^{\mu\cdots}{}_{\nu\ldots} = u^\alpha\partial_\alpha T^{\mu\cdots}{}_{\nu\ldots} = \frac{\partial}{\partial x^0}T^{\mu\cdots}{}_{\nu\ldots}. \tag{5.181}$$

Thus, $\widehat{T}$ is Lie transported if and only if its components are independent of x^0 in these coordinates. Note that for the tangent vector of the curve, it is always Lie transported along $\mathcal{C}$

$$\pounds_{\widehat{u}}\widehat{u} = 0. \tag{5.182}$$

5.5 Fermi–Walker Transport

We have already explored two well-defined methods for transporting tensor fields in curved spacetime. In this section, we introduce a third approach for vector transport, known as Fermi–Walker transport. This method will prove essential in constructing an observer's local laboratory framework.

We have previously defined the observer's local laboratory frame—or tetrad field—by the orthonormal condition:

$$\widehat{e}_a(\tau) \cdot \widehat{e}_b(\tau) = g_{\mu\nu}\widehat{e}^{\mu}_a(\tau)\widehat{e}^{\nu}_b(\tau) = \eta_{ab}, \quad \widehat{e}_0(\tau) = \widehat{u}(\tau), \tag{5.183}$$

where the key feature is that the time direction is aligned with the tangent vector of the observer's worldline. In this framework, measuring physical quantities corresponds to projecting the relevant 4-vectors or 4-tensors onto this orthonormal tetrad.

Consider an observer moving along a worldline $x^{\mu}(\lambda)$, which need not to be a geodesic, meaning the observer may be subject to external forces. Let the observer's 4-velocity be $\widehat{u}$, and their 4-acceleration be $\widehat{a} = \dfrac{\mathrm{D}\widehat{u}}{\mathrm{d}\tau}$. We now investigate how the tetrad basis vectors $\widehat{e}_a$ evolve along the worldline. The covariant derivative of the tetrad along $\widehat{u}$ is given by

$$\nabla_{\widehat{u}}\widehat{e}_a = -\Omega^b_a(\tau)\widehat{e}_b. \tag{5.184}$$

where, due to the orthonormality condition $\nabla_{\widehat{u}}(\widehat{e}_a \cdot \widehat{e}_b) = 0$, the connection coefficients are antisymmetric, $\Omega_{ab} = \Omega_{ba}$. This evolution can be expressed compactly as

$$\nabla_{\widehat{u}}\widehat{e}_a = -\Omega^b_a(\tau)\widehat{e}_b = -(\Omega^{bc}\widehat{e}_b \otimes \widehat{e}_c) \cdot \widehat{e}_a = -\widehat{\Omega} \cdot \widehat{e}_a, \tag{5.185}$$

where the inner product is taken with respect to the first index of Ω^{ab}. Geometrically, Ω^{ab} acts as a generalized rotation matrix, governing the infinitesimal change of the tetrad along the observer's trajectory. Since $\widehat{e}_0 = \widehat{u}$, we can decompose Ω^{ab} into kinematic components:

$$\Omega^{ab} = v^a u^b - u^a v^b + \omega^{ab}, \tag{5.186}$$

where ω^{ab} is antisymmetric and spatial $\omega^{ab}u_b = 0$, and v^a is orthogonal to u^a, $v^a u_a = 0$, representing a boost contribution. This decomposition separates the tetrad's evolution into rotational (ω^{ab}) and boost (v^a) effects relative to the observer's frame.

In the observer's comoving reference frame, the quantities ω^{ab} and v^a are purely spatial, each possessing only three independent components. This reflects the fact that the time-related contributions have been explicitly separated out. From the evolution of the time-like tetrad vector,

$$\nabla_{\widehat{u}}\widehat{e}_0 = \nabla_{\widehat{u}}\widehat{u} = \widehat{a}, \tag{5.187}$$

we have $\widehat{\Omega} \cdot \widehat{u} = -\widehat{a}$. Since $\widehat{\Omega} \cdot \widehat{u} = \widehat{v}$, this immediately yields

$$\widehat{v} = -\widehat{a}. \tag{5.188}$$

Substituting this back into the expression for $\widehat{\Omega}$, we obtain

$$\widehat{\Omega} = -\widehat{a} \otimes \widehat{u} + \widehat{u} \otimes \widehat{a} + \widehat{\omega}. \tag{5.189}$$

Inserting this result into the tetrad evolution equation gives

$$\frac{\mathrm{D}\widehat{e}_a}{\mathrm{d}\tau} = -[\widehat{a}(\widehat{u} \cdot \widehat{e}_a) - \widehat{u}(\widehat{a} \cdot \widehat{e}_a) + \widehat{\omega} \cdot \widehat{e}_a], \tag{5.190}$$

where ω represents the purely spatial rotation of the tetrad. If we further impose that the spatial axes remain non-rotating along the observer's worldline, i.e.,

$$\widehat{\omega} = 0, \tag{5.191}$$

then the tetrad evolution reduces to Fermi–Walker transport:

$$\frac{\mathrm{D}\widehat{e}_a}{\mathrm{d}\tau} = [\widehat{u}(\widehat{a} \cdot \widehat{e}_a) - \widehat{a}(\widehat{u} \cdot \widehat{e}_a)]. \tag{5.192}$$

This choice is physically natural: it ensures that the tetrad's spatial orientation does not rotate artificially between infinitesimally separated moments. The only change arises from the observer's 4-acceleration $\widehat{a}(\tau) = \frac{\mathrm{D}\widehat{u}}{\mathrm{d}\tau}$, which accounts for the inevitable adjustment of the local frame due to non-inertial motion. In summary, for an observer with 4-acceleration but no additional spatial rotation, the tetrad basis vectors undergo Fermi–Walker transport along the worldline.

A vector $\widehat{v}$ is said to undergo Fermi–Walker transport along a curve if it satisfies the equation

$$\frac{\mathrm{D}\widehat{v}}{\mathrm{d}\tau} = [\widehat{u}(\widehat{a} \cdot \widehat{v}) - \widehat{a}(\widehat{u} \cdot \widehat{v})], \tag{5.193}$$

where $\mathrm{D}/\mathrm{d}\tau$ denotes the covariant derivative along the curve. In component form, this condition can be expressed as

$$\frac{\mathrm{D}v^\mu}{\mathrm{d}\tau} = u^\nu \nabla_\nu v^\mu = (u^\mu a^\nu - u^\nu a^\mu)v_\nu. \tag{5.194}$$

Using Fermi–Walker transport, we can define a derivative operation. The Fermi–Walker derivative along a timelike curve $x^\mu(\tau)$ is a differential operation, $\frac{\mathrm{D_F}}{\mathrm{d}\tau}:\widehat{T}\to\widehat{T}$ defined on a tensor field $\widehat{T}$ such that it satisfies the following properties:

(1) Linearity: for any tensors $\widehat{S},\widehat{T}$ and scalars a, b,

$$\frac{\mathrm{D_F}}{\mathrm{d}\tau}(a\widehat{S}+b\widehat{T}) = a\frac{\mathrm{D_F}}{\mathrm{d}\tau}\widehat{S}+b\frac{\mathrm{D_F}}{\mathrm{d}\tau}\widehat{T}. \tag{5.195}$$

(2) Leibniz rule:

$$\frac{\mathrm{D_F}}{\mathrm{d}\tau}(\widehat{S}\otimes\widehat{T}) = \left(\frac{\mathrm{D_F}}{\mathrm{d}\tau}\widehat{S}\right)\otimes\widehat{T}+\widehat{S}\otimes\left(\frac{\mathrm{D_F}}{\mathrm{d}\tau}\widehat{T}\right). \tag{5.196}$$

(3) Agreement with scalar derivatives: $\frac{\mathrm{D_F}f}{\mathrm{d}\tau}=\frac{\mathrm{d}f}{\mathrm{d}\tau}$, $f\in\mathcal{F}(M)$.

(4) Commutation with contraction.

(5) Expression for Vectors: for a vector field $\widehat{V}$, the Fermi–Walker derivative takes the form

$$\frac{\mathrm{D_F}V^\mu}{\mathrm{d}\tau} = \frac{\mathrm{D}V^\mu}{\mathrm{d}\tau} + (a^\mu u^\nu - u^\mu a^\nu)V_\nu, \tag{5.197}$$

where $\widehat{u}$ is the tangent vector to the curve, $\widehat{a}$ is the proper 4-acceleration vector, and $\frac{\mathrm{D}V^\mu}{\mathrm{d}\tau}=u^\nu\nabla_\nu V^\mu$ is the intrinsic derivative along the curve.

The Fermi–Walker derivative operation satisfies the following key properties:

(1) If $x^\mu(\tau)$ is a geodesic, then $\frac{\mathrm{D_F}V^\mu}{\mathrm{d}\tau}=\frac{\mathrm{D}V^\mu}{\mathrm{d}\tau}$, meaning the Fermi–Walker derivative coincides with the standard covariant derivative.

(2) For the tangent vector of the curve, $\frac{\mathrm{D_F}u^\mu}{\mathrm{d}\tau}=0$, indicating that the curve's tangent is Fermi–Walker transported along itself.

(3) For any spacelike vector W^μ orthogonal to u^μ ($W^\mu\cdot u_\mu=0$),

$$\frac{\mathrm{D_F}W^\mu}{\mathrm{d}\tau} = h^\mu{}_\nu\frac{\mathrm{D_F}W^\nu}{\mathrm{d}\tau}, \tag{5.198}$$

where $h_{\mu\nu}=g_{\mu\nu}+u_\mu u_\nu$ is the spatial projection tensor. This ensures the derivative maintains the spatial nature of W^μ.

(4) When the metric field is compatible with the Fermi–Walker derivative, i.e., $\frac{\mathrm{D_F}g_{\mu\nu}}{\mathrm{d}\tau}=0$, the derivative satisfies the product rule:

$$\frac{\mathrm{D_F}(g_{\mu\nu}V^\mu W^\nu)}{\mathrm{d}\tau} = g_{\mu\nu}\frac{\mathrm{D_F}V^\mu}{\mathrm{d}\tau}W^\nu + g_{\mu\nu}V^\mu\frac{\mathrm{D_F}W^\nu}{\mathrm{d}\tau}. \tag{5.199}$$

This property ensures the preservation of inner products under Fermi–Walker transport.

These properties demonstrate that if along a geodesic, Fermi–Walker transport coincides with parallel transport, and the 4-velocity field of the curve is always Fermi–Walker transported. In addition, if two vectors are Fermi–Walker transported, then their inner product remains invariant under this transport. From the Fermi–Walker transport equation, it follows that once a point p and a vector $\widehat{V}$ at p are specified, the Fermi–Walker transported vector $\widehat{V}$ along the curve is uniquely determined.

Fermi–Walker transport defines the most natural coordinate basis for an observer moving along an arbitrary worldline in curved spacetime. It preserves the orthogonality of the tetrad basis vectors while ensuring that the time axis remains aligned with the observer's 4-velocity. The choice of local inertial frame along a worldline is not unique, Lorentz transformations can be regarded as rotations of the basis vectors. Fermi–Walker transport eliminates any extraneous spatial rotation, providing a non-rotating reference frame tied to the observer's motion. For a freely falling observer, the worldline is a geodesic, and the tetrad field $\widehat{e}_a(\tau)$ defines a freely falling reference frame. In this case, the time-like basis vector $\widehat{e}_0$ satisfies

$$\frac{\mathrm{D}\widehat{e}_0}{\mathrm{d}\tau} = 0, \tag{5.200}$$

meaning it is parallel-transported along the geodesic. Along a geodesic, the proper 4-acceleration vanishes, and Fermi–Walker transport reduces to parallel transport. This is expressed as:

$$\frac{\mathrm{D}\widehat{e}_0}{\mathrm{d}\tau} = 0, \tag{5.201}$$

indicating that the tetrad field along the geodesic defines a local inertial frame. The evolution of $\widehat{e}_a(\tau) = (\widehat{e}_a)^\mu(\tau)\partial_\mu$ along an arbitrary worldline is governed by:

$$\frac{\mathrm{D}(\widehat{e}_a)^\mu}{\mathrm{d}\tau} = \frac{\mathrm{d}(\widehat{e}_a)^\mu}{\mathrm{d}\tau} + \Gamma^\mu_{\nu\sigma}(\widehat{e}_a)^\nu u^\sigma = 0. \tag{5.202}$$

This equation plays a crucial role in determining the observations made by a freely falling observer at a given spacetime event. Once the initial tetrad field is specified, its evolution along the geodesic is fully determined, allowing for a consistent description of the observer's local reference frame at all subsequent times.

For an accelerated observer O relative to the local inertial frame S, the following quantities are defined:

$$\begin{cases} \widehat{u}(\tau), & \tau \text{ is the proper time} \\ \widehat{a}(\tau) = \dfrac{\mathrm{d}\widehat{u}}{\mathrm{d}\tau}, & \text{satisfying } \widehat{a}\cdot\widehat{u} = \dfrac{\mathrm{d}}{\mathrm{d}\tau}\left(\dfrac{1}{2}\widehat{u}\cdot\widehat{u}\right) = 0. \end{cases} \tag{5.203}$$

Unlike inertial observers, O cannot be at rest in a global inertial frame. However, at any instant, there exists an instantaneous rest frame S', where O is momentarily at rest. In this frame, the timelike basis vector $\widehat{e}'_0$ aligns with the 4-velocity $\widehat{u}$, and the remaining spatial basis vectors $\widehat{e}'_i$ are orthogonal to $\widehat{e}'_0$ and to each other, forming a local orthonormal tetrad. Thus, the observation made by O at an event P correspond to measurements in the instantaneous rest frame S'. The observer's local laboratory can be idealized as a tetrad field $\widehat{e}'_a(\tau)$, carried along the worldline and satisfying the orthogonality $\widehat{e}'_a \cdot \widehat{e}'_b = \eta_{ab}$ and timelike alignment $\widehat{e}'_0(\tau) = \widehat{u}(\tau)$. The tetrad evolves along the worldline, and physical quantities measured by O at P are obtained by projecting them onto these basis vectors.

Example 5.10. At the event point P, the worldline of particle Q intersects with that of observer O. Let $\widehat{p}$ denote the 4-momentum of particle Q. The energy E' and momentum measured by O at P are determined by projecting $\widehat{p}$ onto the observer's local tetrad frame $\widehat{e}'_a(\tau)$:

$$\frac{E'}{c} = -\widehat{p} \cdot \widehat{e}'_0(\tau) \quad \Rightarrow \quad E' = -\widehat{p} \cdot \widehat{u}(\tau), \quad p'_i = \widehat{p} \cdot \widehat{e}'_i(\tau). \tag{5.204}$$

Below, we examine two physical effects.

(1) Cosmological redshift.

An expanding universe is described by the metric

$$\mathrm{d}s^2 = -\mathrm{d}t^2 + a^2(t)(\mathrm{d}x^2 + \mathrm{d}y^2 + \mathrm{d}z^2), \tag{5.205}$$

with a comoving observer whose 4-velocity is given by $u^\mu = (1, 0, 0, 0)$. What is the photon energy observed by this observer? The timelike component of u^μ is normalized such that $g_{\mu\nu}u^\mu u^\nu = -1$, ensuring proper time parametrization. For a photon, its 4-momentum is given by $p^\mu = \dfrac{\mathrm{d}x^\mu}{\mathrm{d}\lambda}$, where λ is an affine parameter. In this spacetime, the null geodesic is $\dfrac{\mathrm{d}t}{\mathrm{d}\lambda} = \dfrac{\omega_0}{a}$, where ω_0 is the photon's intrinsic frequency. The observed photon energy by the comoving observer is then

$$E = -p_\mu u^\mu = -g_{00}\frac{\mathrm{d}x^0}{\mathrm{d}\lambda}u^0 = \frac{\omega_0}{a} \text{ (natural units)}. \tag{5.206}$$

When $a = 1$, the metric reduces to Minkowski spacetime, and the observed photon energy coincides with its intrinsic frequency ω_0. Consider a photon emitted at scale factor a_1 with energy E_1 and observed at scale factor a_2 with energy E_2. The ratio of observed to emitted energy is

$$\frac{E_2}{E_1} = \frac{a_1}{a_2}. \tag{5.207}$$

This implies that the observed frequency differs from the emitter's rest-frame frequency due to the expansion of the universe. Since $E \propto \frac{1}{\lambda}$, we have $\lambda_i \propto a_i$. By measuring the wavelength shift in spectral lines of distant galaxies, we can infer the change in the scale factor between emission and observation. In an expanding universe, a larger scale factor $a(t)$ corresponds to greater cosmological distances. Thus, the redshift z of a galaxy, defined as

$$z = \frac{\omega_1 - \omega_2}{\omega_2} = \frac{a_2}{a_1} - 1, \tag{5.208}$$

provides a measure of its distance—greater redshift indicates a more distant galaxy.

It should be noted that cosmological redshift differs fundamentally from the conventional Doppler redshift. The light observed from distant galaxies exhibits a redshift compared to that emitted by nearby sources. However, the relative velocity of these distant galaxies with respect to us cannot be rigorously defined, as velocity is inherently a local quantity. In the absence of a well-defined velocity, the Doppler effect—which arises due to relative motion—does not apply. Nevertheless, the notion of galactic "velocity" remains a useful and convenient conceptual tool in cosmological discussions.

To distinguish cosmological redshift from Doppler shift, consider a flat spacetime where two galaxies, A and B, are initially at rest in an inertial frame. If the proper distance between A and B doubles during the photon's propagation, the redshift observed at B is purely cosmological—there is no Doppler effect, as the galaxies have no relative motion. If the galaxies remain at fixed comoving coordinates while the scale factor $a(t)$ increases, (e.g. $a(t_0) \to 2a(t_0)$), there is again no relative motion between them. Thus, no Doppler shift occurs, but the photon still experiences a cosmological redshift due to the expansion of spacetime itself.

(2) Thomas precession.

In electrodynamics, an electron orbiting an atomic nucleus exhibits an additional precession of its spin vector, known as Thomas precession. This phenomenon arises from purely kinematic effects, originating in the relativistic transformation between reference frames as the electron undergoes circular motion. To analyze this effect, we employ the framework of Fermi–Walker transport, treating the electron's spin classically—akin to the spin vector of a gyroscope. This spin vector is intrinsic and can be promoted to a 4-vector $\widehat{S}$. Consider an electron orbiting the nucleus in the x-y plane with an angular velocity ω. First, the spin 4-vector evolves according to

Fermi–Walker transport,

$$\frac{\mathrm{D}S^{\mu}}{\mathrm{d}\tau} = u^{\nu}\nabla_{\nu}S^{\mu} = (u^{\mu}a^{\nu} - u^{\nu}a^{\mu})S_{\nu}. \tag{5.209}$$

Second, the spin 4-vector satisfies the constraints

$$\widehat{S}\cdot\widehat{u} = 0, \quad \widehat{S}\cdot\widehat{S} = \frac{3}{4}\hbar^2. \tag{5.210}$$

The first condition in (5.210) enforces orthogonality between the spin and the velocity 4-vector, while the second specifies the spin magnitude for a spin-1/2 particle. In the laboratory reference frame at $t = 0$, with initial conditions $x = r$, $y = 0$, $z = 0$, the spin 4-vector components are:

$$S^0 = 0, \quad S^x = \frac{1}{\sqrt{2}}\hbar, \quad S^y = 0, \quad S^z = \frac{1}{2}\hbar. \tag{5.211}$$

Solving the Fermi–Walker transport equation yields the time evolution of the spin 4-vector:

$$\begin{aligned} S^0 &= -\frac{1}{\sqrt{2}}\hbar v\gamma \sin\omega\gamma t, \\ S^x &= \frac{1}{\sqrt{2}}\hbar(\cos\omega t\cos\omega\gamma t + \gamma\sin\omega t\sin\omega\gamma t), \\ S^y &= \frac{1}{\sqrt{2}}\hbar(\sin\omega t\cos\omega\gamma t - \gamma\cos\omega t\sin\omega\gamma t), \\ S^z &= \frac{1}{2}\hbar, \end{aligned} \tag{5.212}$$

where $v = \omega r$ and $\gamma = (1 - v^2/c^2)^{-1/2}$. Rewriting the transverse components in complex form:

$$S^x + \mathrm{i}S^y = \frac{1}{\sqrt{2}}\hbar(\mathrm{e}^{-1(\gamma-1)\omega t} - \mathrm{i}(\gamma - 1)\sin\omega\gamma t\mathrm{e}^{\mathrm{i}\omega t}). \tag{5.213}$$

The first term describes a precession with angular frequency

$$\omega_{\text{Thomas}} = (\gamma - 1)\omega \approx \frac{v^2}{2c^2}\omega, \tag{5.214}$$

which is the Thomas precession frequency. The second term oscillates with handedness that alternates between right- and left-handed over each orbit: it is right-handed for $0 < \omega\gamma t < \pi$ and left-handed for $\pi < \omega\gamma t < 2\pi$. While its time average vanishes, this term ensures the preservation of the constraints in (5.210).

5.6 *Uniformly Accelerated Observer's Reference Frame

Indeed, Fermi–Walker transport offers a rigorous framework for constructing the coordinate system of a non-uniformly moving observer. For such an observer, their frame field is propagated along their worldline via Fermi–Walker transport. Consequently, it is natural for them to adopt proper time as the temporal coordinate and to define the three spatial coordinates along directions orthogonal to the tangent vector of the worldline. By fixing the origin of proper time and the initial spatial orientations, their frame at each subsequent proper time is uniquely determined through Fermi–Walker transport. At every instant of proper time, the spacelike frame vectors span a spacelike hypersurface, thereby constructing the observer's coordinate grid. For any given event P, its position must reside on one such hypersurface, and its coordinates relative to the observer are thus well-defined. It is important to note, however, that the observer's coordinate system is generally local and cannot encompass the entire spacetime manifold, being subject to certain limitations in scope and applicability.

Let us re-examine the case of uniform acceleration in flat spacetime, with a focus on observers within a uniformly accelerating spaceship. For such observers, the spacetime they experience cannot be adequately described by the standard Minkowski metric. Consider an observer undergoing uniform acceleration in the $+x$ direction with magnitude g (Earth's gravitational acceleration). In a static coordinate system, their worldline is given by:

$$x = g^{-1}\cosh(g\tau), \quad t = g^{-1}\sinh(g\tau), \tag{5.215}$$

where τ is the observer's proper time. We fix the origin of proper time such that at $\tau = 0$, the observer's position is $x(0) = g^{-1}$ and $t(0) = 0$. For this observer, the coordinate time T in their local frame coincides with their proper time τ, as measured by an ideal clock (e.g., an atomic clock):

$$T = \tau. \tag{5.216}$$

As shown in Figure 5.12, the spacelike hypersurface at point P_0 contains the event P. In the orthogonal frame basis $\widehat{\Sigma}$ at P_0, the position vector of P is given by

$$\widehat{X} = (0, X, Y, Z), \tag{5.217}$$

meaning P has spatial coordinates (X, Y, Z) relative to P_0 and a time coordinate of zero, since P is simultaneous with P_0 in this frame. To relate the coordinates in the P_0-based frame to the original static reference system, we first note that P_0 itself has coordinates $\widehat{x}_0$ in the static frame:

$$x_0 = g^{-1}\cosh(g\tau), \quad t_0 = -g^{-1}\sinh(g\tau), \quad y_0 = z_0 = 0. \tag{5.218}$$

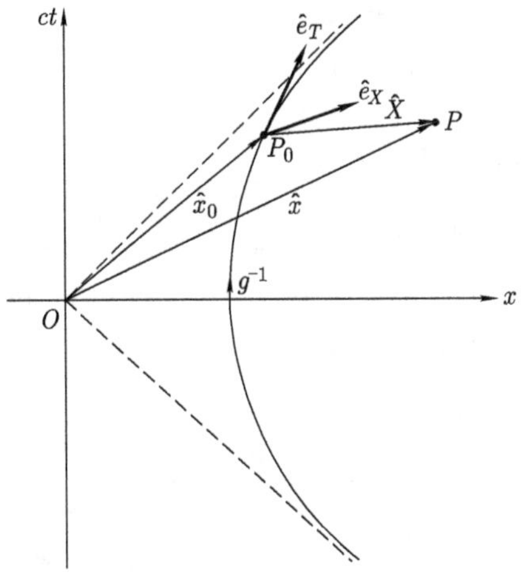

Fig. 5.12 The coordinate system of the uniformly accelerating observer P_0 can be determined through the frame field, and the coordinates of event P can be defined.

The event P is then described in the static frame by the vector $\widehat{x}$, which is the sum of $\widehat{x}_0$ and the displacement $\widehat{X}$

$$\widehat{x} = \widehat{x}_0 + \widehat{X}. \tag{5.219}$$

Next, we establish the relationship between the basis vectors of the comoving frame at point P_0 and those of the original static coordinate system. This transformation is given by a Lorentz boost

$$\begin{aligned} \widehat{e}_M &= \widehat{e}_\mu \frac{\partial x^\mu}{\partial X^M} \\ &= (\widehat{e}_t, \widehat{e}_x, \widehat{e}_y, \widehat{e}_z) \begin{pmatrix} \cosh\theta_0 & \sinh\theta_0 & 0 & 0 \\ \sinh\theta_0 & \cosh\theta_0 & 0 & 0 \\ 0 & 0 & 1 & 0 \\ 0 & 0 & 0 & 1 \end{pmatrix}, \end{aligned} \tag{5.220}$$

where θ_0 is the rapidity of the particle at P_0 relative to the static frame. The rapidity is determined by:

$$\tanh\theta_0 = \frac{v_0}{c} = \frac{1}{c}\frac{\mathrm{d}x_0}{\mathrm{d}t_0} = \tanh(gT/c), \tag{5.221}$$

where v_0 is the instantaneous velocity of the accelerating observer at proper time T. The basis vectors of the comoving frame at proper time T are related to the static frame through:

$$\begin{aligned}\widehat{e}_T &= \widehat{e}_t \cosh(gT/c) + \widehat{e}_x \sinh(gT/c),\\ \widehat{e}_X &= \widehat{e}_t \sinh(gT/c) + \widehat{e}_x \cosh(gT/c),\\ \widehat{e}_Y &= \widehat{e}_y,\\ \widehat{e}_Z &= \widehat{e}_z.\end{aligned} \tag{5.222}$$

From Eq. (5.219), we extract (let $c = 1$) the term proportional to $\widehat{e}_t$

$$t\widehat{e}_t = t_0\widehat{e}_t + X \sinh(gT)\widehat{e}_t, \tag{5.223}$$

yielding the explicit transformation

$$t = (g^{-1} + X)\sinh(gT). \tag{5.224}$$

The complete coordinate transformation is then

$$x = (g^{-1} + X)\cosh(gT), \quad y = Y, \quad z = Z. \tag{5.225}$$

It is evident that

$$\frac{t}{x} = \tanh(gT), \tag{5.226}$$

which implies that all events on the observer's simultaneity surfaces lie on radial lines through the origin. For P_0, we also have $\dfrac{t_0}{x_0} = \tanh(gT)$, confirming consistency. The invariant

$$x^2 - t^2 = (g^{-1} + X)^2, \tag{5.227}$$

implies that each constant X defines a hyperbola with asymptotes

$$x \pm t = 0. \tag{5.228}$$

Each hyperbola represents the worldline of a particle undergoing uniform acceleration in a straight line, with proper acceleration:

$$a = \frac{1}{g^{-1} + X}. \tag{5.229}$$

In the reference frame of the uniformly accelerating observer, these particle worldlines correspond to $X =$ constant. On the other hand, $T =$ constant corresponds to simultaneity surfaces of these particles, which are radial lines through the origin in the original static coordinate system.

By applying coordinate transformations, we derive the spacetime metric in the coordinate system of a uniformly accelerating observer:

$$\begin{aligned}ds^2 &= -dt^2 + dx^2 + dy^2 + dz^2\\ &= -(1 + gX)^2 dT^2 + dX^2 + dY^2 + dZ^2.\end{aligned} \tag{5.230}$$

At first glance, this metric suggests spacetime curvature, but in reality, the spacetime remains flat. As we will demonstrate later, the apparent curvature arises from the observer's accelerated frame of reference. For such an observer, the proper time is given by

$$\mathrm{d}\tau = \left(1 + \frac{gX}{c^2}\right)\mathrm{d}T = \left(1 + \frac{gX}{c^2}\right)\mathrm{d}\tau_0, \tag{5.231}$$

where τ_0 denotes the proper time at the reference position $X = 0$. In the Newtonian limit, this metric can be interpreted as a perturbation of the flat metric, where the (00)-component of the metric perturbation corresponds to the Newtonian gravitational potential. Specifically, we have

$$\mathrm{e}^{2\Phi} = \left(1 + \frac{gX}{c^2}\right)^2, \tag{5.232}$$

so the Newtonian gravitational potential is $\Phi = \ln(1 + gX)$, and the Newtonian gravitational acceleration is $g = |\nabla\Phi| = 1/(g^{-1} + X)$. This confirms that the derived acceleration indeed matches the Newtonian gravitational acceleration. Thus, within the reference frame of a uniformly accelerating observer, it is impossible to distinguish between uniform acceleration and the presence of a gravitational field—an early indication of the equivalence principle at work.

Indeed, our analysis extends beyond merely describing the coordinate system of a uniformly accelerating observer. Using the coordinate transformations introduced earlier, we can rigorously define Rindler spacetime. As previously discussed, a uniformly accelerating observer perceives a horizon—a boundary beyond which events become causally inaccessible. More precisely, the observable region is confined between the two asymptotes of the hyperbolic motion, corresponding to the spacetime domain:

$$0 < x < \infty, \quad -x < t < x. \tag{5.233}$$

Remarkably, this region covers only one-fourth of the full Minkowski spacetime, highlighting the profound limitations imposed by uniform acceleration on an observer's causal patch. We now introduce the Rindler coordinates, defined by the transformation

$$T = g^{-1}\mathrm{arctanh}(t/x), \quad X = \sqrt{x^2 - t^2}, \quad Y = y, \quad Z = z. \tag{5.234}$$

Observers following these worldlines are momentarily at rest relative to the static reference frame at $t = 0$, but they experience varying proper accelerations given by $a = c^2/X$. The inverse transformation yields

$$t = X\sinh(gT), \quad x = X\cosh(gT), \tag{5.235}$$

which, upon comparison with Eqs. (5.224) and (5.225), reveals only a constant shift in X. In these coordinates, the spacetime metric takes the form

$$\mathrm{d}s^2 = -(gX)^2\mathrm{d}T^2 + \mathrm{d}X^2 + \mathrm{d}Y^2 + \mathrm{d}Z^2, \tag{5.236}$$

where g is merely a scaling parameter, which we can set to unit by suitable coordinate redefinition. For Rindler observers, a natural choice of orthonormal frame field is

$$\widehat{e}_0 = \frac{1}{X}\partial_T, \quad \widehat{e}_i = \partial_i, \quad i = 1, 2, 3, \tag{5.237}$$

where $\widehat{e}_0$ represents the timelike direction of proper time flow, and $\widehat{e}_i$ spans the spatial axes comoving with the accelerated observers.

The timelike vector field $\widehat{e}_0$ can be integrated to yield a family of timelike curves, representing the worldlines of observers with varying proper accelerations. By studying the evolution of this congruence of curves,[4] we observe that the three-dimensional hypersurfaces formed by these curves exhibit neither expansion nor twist. Mathematically, this means both the expansion scalar and vorticity tensor vanish identically. Consequently, the spatial separation between these uniformly accelerated observers remains constant over time. However, despite this rigid motion, each observer experiences a nonzero proper acceleration. Computing the acceleration vector yields

$$\nabla_{\widehat{e}_0}\widehat{e}_0 = \frac{1}{X}\widehat{e}_1, \tag{5.238}$$

indicating that every observer undergoes uniform acceleration along the x-direction, with magnitude inversely proportional to their X-coordinate. The vanishing vorticity has an important geometric implication: the timelike integral curves of $\widehat{e}_0$ must be hypersurface orthogonal. Indeed, these curves are everywhere orthogonal to the spacelike hypersurfaces of constant Rindler time $T = T_0$, which foliate the spacetime. This reflects the static nature of the Rindler frame—while the observers accelerate, their mutual distances remain fixed, and their simultaneity surfaces are globally well-defined.

This scenario can be vividly illustrated using the example of a rigid spaceship with finite length. As depicted in Figure 5.13, the bold black line segment represents the spaceship at rest at $t = 0$, where its instantaneous velocity vanishes. However, due to relativistic effects, the required

[4] The evolution of a congruence of geodesics with respect to proper time is governed by the Raychaudhuri equation. For details, see [64].

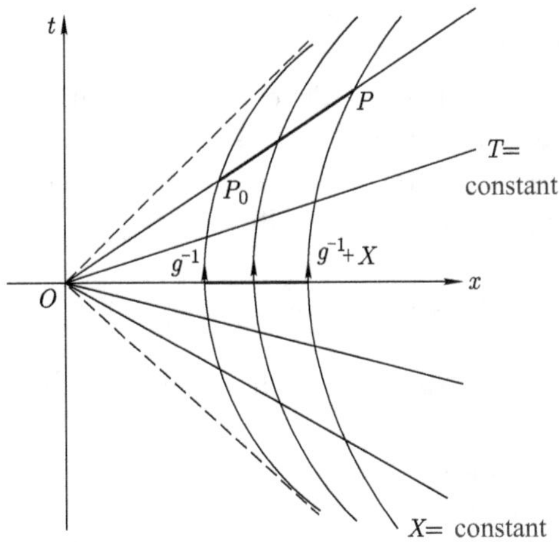

Fig. 5.13 Using Rindler coordinates, we can cover 1/4 of the flat spacetime. The lines where X is constant correspond to the worldlines of particles with different constant proper accelerations, while T being constant corresponds to the different simultaneity surfaces of uniformly accelerated observers. Using T, X, Y, Z, we can transform the Minkowski metric of flat spacetime into the Rindler spacetime metric. The thick black line in the figure represents a spaceship of finite length, with the acceleration at the rear end of the spaceship being greater than that at the front end.

proper acceleration differs along its length: the rear (left) end must accelerate more strongly than the front (right) end to maintain rigidity. This is a direct consequence of Lorentz contraction—the rear end's greater acceleration compensates for length contraction, preserving the ship's proper length during motion. The physical origin of this effect becomes clear when examining the Rindler coordinates: the rear end corresponds to smaller$X$ value, which by $a = c^2/X$ implies larger proper acceleration. Remarkably, the spaceship's worldlines in spacetime all lie along surfaces of constant T. Calculating the coordinate velocity:

$$\frac{\mathrm{d}x}{\mathrm{d}t} = \tanh(gT), \tag{5.239}$$

we find it depends only on T, not X—all points share the same instantaneous velocity at any given T, despite their different accelerations. While Figure 5.13 might suggest the spaceship appears to lengthen with increasing T, its proper length ΔX (the distance from P_0 to P) remains constant—precisely equal to its rest length at $t = T = 0$. This demonstrates the remarkable property of Born rigidity: although each point accelerates differently, their spatial separation remains unchanged in the comoving frame.

This scenario provides an excellent analogy for a skyscraper in a uniform gravitational field. The varying acceleration $g(X) = c^2/X$ along the spaceship corresponds to the position-dependent gravitational acceleration at different heights in the building. Just as the spaceship's rear end experiences stronger acceleration, the base of the skyscraper feels greater effective gravity than its upper floors—a beautiful manifestation of the equivalence principle in action.

Next, let us discuss the motion of particles in a uniformly accelerating reference frame.

(1) Free motion of massive particles in a uniformly accelerating reference frame.

Consider a spaceship undergoing uniform acceleration with a constant proper acceleration g. Inside the spaceship, a particle is initially located at $T = 0$, $X = c^2 g^{-1}$ and moves in the $+x$ direction with an initial 4-velocity $\widehat{u}_0 = \gamma(c, v, 0, 0)$, where $\gamma = 1/\sqrt{1 - v^2/c^2}$ is the Lorentz factor. The dynamics of the particle can be described by the Lagrangian

$$K = \frac{1}{2} g_{\mu\nu} \dot{X}^\mu \dot{X}^\nu. \tag{5.240}$$

where $g_{\mu\nu}$ is the metric tensor. Additionally, the normalization condition for the 4-velocity, $\widehat{u} \cdot \widehat{u} = -c^2$, yields

$$\dot{X}^2 = (gX/c)^2 \dot{T}^2 - c^2. \tag{5.241}$$

Since K does not explicitly depend on T, there exists a conserved quantity associated with time translation symmetry:

$$p_T = c\frac{\partial K}{\partial \dot{T}} = -(gX/c)^2 c\dot{T} = u_0^0. \tag{5.242}$$

Solving for $\dot{X}$, we obtain

$$\dot{X} = c\sqrt{\left(\frac{u_0^0 c}{gX}\right)^2 - 1}. \tag{5.243}$$

The particle reaches its maximum distance when its velocity vanishes ($\dot{X} = 0$), giving $X_{\max} = \dfrac{cu_0^0}{g} = \dfrac{c^2}{g}\gamma$. The distance the particle travels is $\Delta X = \dfrac{c^2}{g}(\gamma - 1)$. In the non-relativistic limit, $v \ll c$, this reduces to $\Delta X = \dfrac{v^2}{2g}$, which coincides with the classical result for the maximum height attained by a particle projected upward in a uniform gravitational field g.

This consistency further demonstrates how a uniformly accelerated frame can effectively simulate the effects of a gravitational field.

Let us now examine the motion of a freely falling particle in Rindler spacetime. Suppose the particle is initially at rest at $X = c^2g^{-1}$, with an initial 4-velocity $\widehat{u}_0 = (c, 0, 0, 0)$. From the equations of motion, the particle's spatial velocity in the X-direction is given by

$$\dot{X} = c\sqrt{\left(\frac{c^2}{gX}\right)^2 - 1}, \tag{5.244}$$

where the dot denotes differentiation with respect to the particle's proper time τ. The particle will accelerate toward the Rindler horizon at $X = 0$. Remarkably, the proper time required for the particle to reach the horizon is finite and can be computed as

$$\Delta\tau = \frac{c}{g} \tag{5.245}$$

which indicates that, from the particle's perspective, it reaches the horizon in a finite proper time. However, for an observer on the uniformly accelerating spaceship, the situation appears drastically different. The Rindler coordinate time T (measured by the spaceship's observer) and the particle's proper time τ are related via $\mathrm{d}T = \dot{T}\mathrm{d}\tau$. Integrating this relation, we find that the time taken for the particle to reach a final position X is

$$\Delta T = \frac{c}{g}\ln\frac{1 + \sqrt{1 - (gX/c^2)^2}}{gX/c^2}. \tag{5.246}$$

As the particle approaches the horizon, $X \to 0$, the Rindler time diverges: $\Delta T \to \infty$. This means that, from the perspective of the accelerating observer, the particle never actually reaches the horizon but instead appears to asymptotically approach it.

(2) Motion of photons in a uniformly accelerating reference frame.

Consider a photon emitted at time $T = 0$ from the position $X = g^{-1}$ in a reference frame undergoing uniform proper acceleration g. The photon is initially directed along the Y direction. For massless particles, we define the action as

$$K = -\frac{1}{2}\left(\frac{gX}{c}\right)^2\dot{T}^2 + \frac{1}{2}\dot{X}^2 + \frac{1}{2}\dot{Y}^2, \tag{5.247}$$

where dots denote derivatives with respect to an affine parameter λ along the photon's worldline. The null path condition yields

$$\dot{X}^2 = \left(\frac{gX}{c}\right)^2\dot{T}^2 - \dot{Y}^2. \tag{5.248}$$

Due to the spacetime symmetries, we identify two conserved quantities:

$$p_T = -\left(\frac{gX}{c^2}\right)^2 c\dot{T} = -c\dot{T}(0), \quad p_Y = \dot{Y}. \tag{5.249}$$

Applying initial conditions where the photon is emitted purely in the Y-direction at $X = g^{-1}$, we have

$$\dot{T}(0) = -\left(\frac{gX(0)}{c^2}\right)^2, \quad \dot{Y}(0) = c. \tag{5.250}$$

Combining these results, we derive the X-velocity

$$\dot{X} = c\frac{\sqrt{1-(gX/c^2)^2}}{gX/c^2}. \tag{5.251}$$

The photon's path in the X-Y plane is then described by

$$\frac{\mathrm{d}Y}{\mathrm{d}X} = \frac{\dot{Y}}{\dot{X}} = \frac{gX/c^2}{\sqrt{1-(gX/c^2)^2}}. \tag{5.252}$$

Integrating with initial condition $Y(0) = 0$, $X(0) = g^{-1}$, we obtain the trajectory of the photon as

$$X^2 + Y^2 = \frac{c^4}{g^2}, \tag{5.253}$$

revealing that the photon follows a circular path of radius c^2/g in the X-Y plane. Considering the fact that for an observer on the spaceship, the region $X < 0$ is behind the horizon, the photon's trajectory is the part of the circular orbit where $X > 0$ and $Y > 0$.

5.7 *Rotating Reference Frame

In the previous section, we examined the reference frame of a uniformly accelerating observer and uncovered several intriguing physical effects. Here, we turn our attention to the physical effects experienced by a rotating observer in flat spacetime. To analyze rotation in the x-y plane, we use polar coordinates (r, θ). In these coordinates, the flat spacetime metric takes the form

$$\mathrm{d}s^2 = -c^2\mathrm{d}t^2 + \mathrm{d}r^2 + r^2\mathrm{d}\theta^2 + \mathrm{d}z^2. \tag{5.254}$$

Now, consider a uniformly rotating disk with angular velocity ω. Introducing the comoving coordinates

$$T = t, \quad R = r, \quad \phi = \theta - \omega t, \quad Z = z, \tag{5.255}$$

we derive the metric relative to the rotating observer

$$\mathrm{d}s^2 = -\left(1 - \frac{R^2\omega^2}{c^2}\right) c^2\mathrm{d}T^2 + \mathrm{d}R^2 + 2R^2\omega\mathrm{d}T\mathrm{d}\phi + R^2\mathrm{d}\phi^2 + \mathrm{d}Z^2, \quad (5.256)$$

where

$$g_{TT} = -\gamma^{-2}, \quad \gamma = \left(1 - \frac{R^2\omega^2}{c^2}\right)^{-1/2}. \quad (5.257)$$

By applying a coordinate transformation, we express the basis vectors in the rotating frame as

$$\widehat{e}_T = \widehat{e}_t + \omega\widehat{e}_\theta, \quad \widehat{e}_R = \widehat{e}_r, \quad \widehat{e}_\phi = \widehat{e}_\theta, \quad \widehat{e}_Z = \widehat{e}_z. \quad (5.258)$$

It is evident that due to the rotation, even though $T = t$, the temporal basis vectors differ, $\widehat{e}_T \neq \widehat{e}_t$. Finally, we construct the orthogonal basis vector field in the rotating coordinate system

$$\widehat{e}_0 = \gamma\widehat{e}_T, \quad \widehat{e}_1 = \widehat{e}_R, \quad \widehat{e}_2 = \gamma^{-1}R^{-1}\widehat{e}_\phi + \gamma R\omega\widehat{e}_T, \quad \widehat{e}_3 = \widehat{e}_Z. \quad (5.259)$$

To examine the spatial distance between two points at constant time in this reference system, we consider the spatial line element derived from the metric. Setting $T =$ constant, the spatial part of the metric reduces to

$$\mathrm{d}l^2 = \mathrm{d}R^2 + \gamma^2R^2\mathrm{d}\phi^2 + \mathrm{d}Z^2. \quad (5.260)$$

This line element arises naturally from the orthogonal spatial frame field

$$\widehat{e}_1 = \widehat{e}_R, \quad \widehat{e}_2 = \gamma^{-1}R^{-1}\widehat{e}_\phi, \quad \widehat{e}_3 = \widehat{e}_Z, \quad (5.261)$$

which defines a local three-dimensional Euclidean space. However, the metric in (5.260) is necessary to maintain the orthonormality of these frame fields; any other choice would disrupt their orthogonality and normalization. For two points at the same radius, the arc length line element along the azimuthal direction is

$$\mathrm{d}l_\phi = \gamma R\mathrm{d}\phi. \quad (5.262)$$

Integrating this yields the circumference of a circle with radius R:

$$C = 2\pi\gamma R. \quad (5.263)$$

The radial line element remains unchanged:

$$\mathrm{d}l_R = \mathrm{d}R, \quad (5.264)$$

so the radius is simply R. The ratio of circumference to radius,

$$\frac{C}{R} = 2\pi\gamma > 2\pi, \quad (5.265)$$

deviates from the Euclidean result, indicating that the rotating observer perceives spatial curvature. This is a purely kinematic effect arising from the observer's non-inertial motion.

Furthermore, we can examine how proper time evolves for an observer at rest in the rotating frame. Consider an observer fixed at coordinates R, ϕ, Z. Their proper time is determined by the temporal component of the metric:

$$\mathrm{d}\tau^2 = -g_{TT}\mathrm{d}T^2, \tag{5.266}$$

where $g_{TT} = -\gamma^{-2}$. Substituting this yields

$$\mathrm{d}\tau = \gamma^{-1}\mathrm{d}T < \mathrm{d}T. \tag{5.267}$$

This indicates that clocks carried by rotating observers tick slower than those in the non-rotating (inertial) frame. Since $\gamma = (1 - R^2\omega^2/c^2)^{-1/2}$ increases with R, the time dilation effect becomes more pronounced at larger radii. A clock at the edge of the rotating disk runs slower than one at the center of the disk.

The proper time at different radial positions in the rotating reference frame can be understood from different perspectives. For an observer in the static (non-rotating) reference frame, the proper time τ at a radial distance R from the axis of rotation is given by

$$\tau = \sqrt{1 - \frac{R^2\omega^2}{c^2}}\tau_0, \tag{5.268}$$

where τ_0 denotes the proper time on the rotation axis. This expression reflects the time dilation effect in special relativity, with $R\omega$ representing the linear velocity of the rotating frame. Conversely, from the perspective of an observer within the rotating reference frame, the proper time interval is described by

$$\mathrm{d}\tau = \sqrt{1 - \frac{R^2\omega^2}{c^2}}\mathrm{d}T, \tag{5.269}$$

which arises from the modification of the spacetime metric due to rotation. This demonstrates that the interpretation of the same physical phenomenon depends on the reference frame. In the rotating frame, the observer experiences a centrifugal force field. By the equivalence principle, this centrifugal acceleration can be treated as an effective gravitational acceleration, with an associated gravitational potential

$$\Phi = -\int_0^R R'\omega^2\mathrm{d}R' = -\frac{1}{2}R^2\omega^2, \tag{5.270}$$

where we set the gravitational potential $\Phi|_{R=0} = 0$. Consequently, the proper time interval can also be expressed as

$$\mathrm{d}\tau = \sqrt{1 + \frac{2\Phi}{c^2}}\mathrm{d}\tau_0. \tag{5.271}$$

This is precisely the effect of time dilation in a gravitational potential, also known as the gravitational redshift effect.

Another interesting phenomenon in the rotating reference frame is the issue of clock synchronization, particularly the definition of simultaneity. In standard relativity, the simultaneity between two clocks at points A and B can be established using light signals, as illustrated in Figure 5.14. The procedure is as follows: a light signal is emitted from A to B, and then from B back to A; If the time coordinate at B is x^0, the coordinate time when emitted from A is $x^0 + \mathrm{d}x^{0(2)}$, and the coordinate time when received at A is $x^0 + \mathrm{d}x^{0(1)}$, then the coordinate time on the worldline of A that is simultaneous with x^0 on the worldline of B is then given by the midpoint:

$$x^0 + \frac{\mathrm{d}x^{0(1)} + \mathrm{d}x^{0(2)}}{2}. \tag{5.272}$$

For a general rotating reference frame, its spacetime line element is

$$\mathrm{d}s^2 = g_{00}(\mathrm{d}x^0)^2 + 2g_{0i}\mathrm{d}x^0\mathrm{d}x^i + g_{ij}\mathrm{d}x^i\mathrm{d}x^j. \tag{5.273}$$

Photons travel along null paths, so from $\mathrm{d}s^2 = 0$, two solutions can be obtained, one with $\mathrm{d}x^{0(2)} < 0$ and the other with $\mathrm{d}x^{0(1)} > 0$. The point

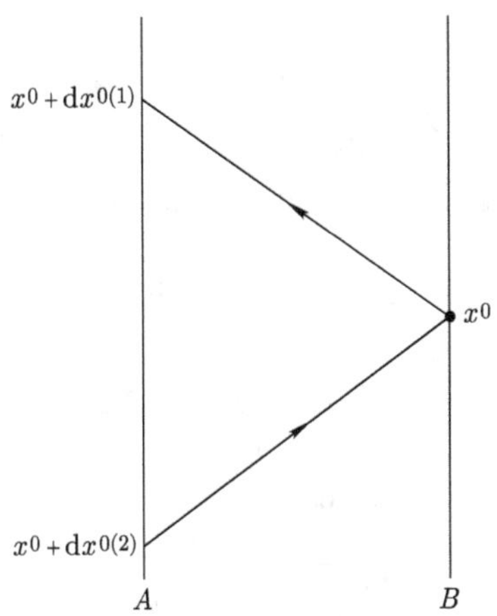

Fig. 5.14 The simultaneity of two adjacent points A and B in spacetime can be achieved by emitting a photon from A to B and then having the photon return to A.

on the worldline of A simultaneous with the coordinate time x^0 on the worldline of B is then given by

$$x^0 + \Delta x^0 = x^0 - \frac{g_{0i}\mathrm{d}x^i}{g_{00}}. \tag{5.274}$$

Crucially, in a rotating reference frame $g_{0i} \neq 0$, leading to $\Delta x^0 \neq 0$. While this synchronization method works for open line segments, it fails for closed curves. If the spatial path x^i is closed (e.g., a circular loop), the photon emission and reception occur at the same point, resulting in a non-zero discrepancy

$$\triangle x^0 = -\oint \frac{g_{0i}}{g_{00}}\mathrm{d}x^i \neq 0. \tag{5.275}$$

This implies that the coordinate time on A simultaneous with x^0 on B is not single-valued, making global clock synchronization impossible in a rotating frame.

In general, for the spacetime observed by an observer described by Eq. (5.273), we can analyze the spatial geometry from the observer's perspective. In the observer's coordinate system, the temporal direction is defined by the observer's 4-velocity

$$\widehat{e}_0 = \widehat{u}. \tag{5.276}$$

The spacelike basis vectors $\widehat{e}_i$ are not necessarily orthogonal to $\widehat{e}_0$. To construct an orthogonal spatial frame, we require

$$\widehat{e}_{i\perp} \cdot \widehat{e}_0 = 0. \tag{5.277}$$

The spatial components of the metric tensor are then given by

$$\gamma_{ij} = \widehat{e}_{i\perp} \cdot \widehat{e}_{j\perp}, \quad \gamma_{i0} = 0, \quad \gamma_{00} = 0. \tag{5.278}$$

Decomposing the spacelike vectors into orthogonal and parallel components

$$\widehat{e}_{i\perp} = \widehat{e}_i - \widehat{e}_{i\|}, \tag{5.279}$$

where

$$\widehat{e}_{i\|} = \frac{\widehat{e}_i \cdot \widehat{e}_0}{\widehat{e}_0 \cdot \widehat{e}_0}\widehat{e}_0 = \frac{g_{i0}}{g_{00}}\widehat{e}_0, \tag{5.280}$$

we derive the spatial metric components

$$\gamma_{ij} = g_{ij} - \frac{g_{i0}g_{j0}}{g_{00}}. \tag{5.281}$$

Consequently, the spatial line element takes the form

$$\mathrm{d}l^2 = \gamma_{ij}\mathrm{d}x^i\mathrm{d}x^j. \tag{5.282}$$

The impossibility of defining simultaneity in a rotating reference frame is experimentally confirmed by the Sagnac effect. Consider an experimental setup on a uniformly rotating disk: a photon emitter located at some point (R, ϕ, Z) simultaneously emits two photons, one propagating in the $+\phi$ direction and the other in the $-\phi$ direction. Both photons traverse a circular path of radius R and eventually return to the emitter, where they interfere with each other. The resulting interference fringes depend on both the radius of the circle and the angular velocity of the disk's rotation. The time difference ΔTbetween the arrivals of the two photons at the emitter is given by

$$\Delta T = \frac{4\pi\gamma^2 R^2\omega}{c^2}, \tag{5.283}$$

where γ is the Lorentz factor. This time difference can be derived using two distinct approaches:

(1) First, we consider the experiment from the perspective of the inertial (static) reference frame. As the photons propagate along the circular path of radius R, the emitter moves with angular velocity ω. During the photons' travel time T, the emitter rotates by an angle ωT, covering an arc length $R\omega T$. The two photons traverse different effective path lengths due to the disk's rotation: The co-rotating photon (moving in the $+\phi$ direction) must travel slightly farther than $2\pi R$ to complete a full loop, as the emitter moves ahead during its transit; The counter-rotating photon (moving in the $-\phi$ direction) must travel slightly shorter distance, as the emitter moves toward it. Assuming slow rotation ($R\omega \ll c$), the travel times T_1 (co-rotating) and T_2 (counter-rotating) satisfy

$$\begin{aligned} 2\pi R + R\omega T_1 &= cT_1, \\ 2\pi R - R\omega T_2 &= cT_2. \end{aligned} \tag{5.284}$$

The time difference is

$$\Delta T = T_1 - T_2 = \frac{4\gamma^2 A\omega}{c^2}, \tag{5.285}$$

where $A = \pi R^2$.

(2) We can also analyze this problem from the perspective of the rotating reference system. When $\mathrm{d}s = \mathrm{d}R = \mathrm{d}Z = 0$, we obtain

$$R^2\mathrm{d}\phi^2 + 2R^2\omega\mathrm{d}T\mathrm{d}\phi - (c^2 - R^2\omega^2)\mathrm{d}T^2 = 0. \tag{5.286}$$

Solving this yields the photon velocities:

$$v_\pm = R\frac{\mathrm{d}\phi}{\mathrm{d}T} = -R\omega \pm c, \tag{5.287}$$

and the time difference

$$\Delta T = \frac{2\pi R}{|v_+|} - \frac{2\pi R}{|v_-|} = \frac{4\gamma^2 A\omega}{c^2}. \tag{5.288}$$

This temporal difference results in an optical path difference between the two beams. For slow rotation ($\gamma \approx 1$), the path lengths become:

$$\begin{aligned} l + \Delta l &= 2\pi R + \frac{2A\omega}{c}, \\ l - \Delta l &= 2\pi R - \frac{2A\omega}{c}, \end{aligned} \tag{5.289}$$

yielding a net path difference of $4A\omega/c$. Consequently, this produces an interference fringe shift

$$\Delta N = \frac{4A\omega}{c\lambda}, \tag{5.290}$$

where λ represents the photon wavelength. In Sagnac's experiment in 1931, $\omega = 14\ \text{rad} \cdot \text{s}^{-1}$, $A = 0.0863\ \text{m}^2$, and $\lambda = 0.436 \times 10^{-6}$ m, the predicted fringe shift was $\Delta N = 0.036$. This theoretical prediction was successfully confirmed by experimental observations.

The Sagnac effect enables the optical measurement of angular velocity, forming the theoretical foundation for optical gyroscopes widely employed in inertial navigation systems. This effect reveals a fundamental distinction in special relativity: while linear velocity is relative, angular velocity has an absolute meaning. Any non-accelerating observer can always consider their reference frame to be at rest with respect to linear motion—velocity can be nullified by an appropriate choice of reference frame. However, this is not possible for angular velocity, as it produces locally observable effects independent of the reference frame. This non-relativistic nature of rotation manifests not only in optical measurements (such as the Sagnac effect) but also in mechanical systems, as demonstrated by the Foucault pendulum.

Next, we examine the motion of a particle relative to the rotating reference frame described by the metric (5.256) and compare it with its motion in the static frame. First, we compute the Christoffel symbols associated with the metric (5.256) (see Exercise 9). Using these, we analyze the spatial acceleration of the particle, given by the time derivative of its velocity:

$$\ddot{\boldsymbol{r}} = \dot{\boldsymbol{v}} = (\dot{v}^i + \Gamma^i_{jk} v^j v^k)\boldsymbol{e}_i. \tag{5.291}$$

In polar coordinates, the velocity components are $v^r = \dfrac{\mathrm{d}r}{\mathrm{d}t}$, $v^\phi = \dfrac{\mathrm{d}\theta}{\mathrm{d}t}$. The particle's acceleration in the static (inertial) frame is

$$\ddot{\boldsymbol{r}}_{\text{iner}} = (\ddot{r} - r\dot{\theta}^2)\widehat{e}_1 + (r\ddot{\theta} + 2\dot{r}\dot{\theta})\widehat{e}_2, \tag{5.292}$$

while in the rotating frame, it becomes

$$\begin{aligned}\ddot{\boldsymbol{r}}_{\text{rot}} &= (\ddot{r} - r\dot{\theta}^2 - r\omega^2 - 2r\omega\dot{\theta}\widehat{e}_1 + (r\ddot{\theta} + 2\dot{r}\dot{\theta} + 2\dot{r}\omega)\widehat{e}_2 \\ &= \ddot{\boldsymbol{r}}_{\text{iner}} - (r\omega^2 + 2r\omega\dot{\theta})\widehat{e}_1 + 2\dot{r}\omega\widehat{e}_2.\end{aligned} \tag{5.293}$$

Introducing the angular velocity vector, the position vector, and the velocity

$$\boldsymbol{\omega} = \omega\widehat{e}_z, \quad \boldsymbol{r} = r\widehat{e}_1, \quad \dot{\boldsymbol{r}} = \dot{r}\widehat{e}_1 + r\dot{\theta}\widehat{e}_2, \tag{5.294}$$

we can express the rotating-frame acceleration as

$$\ddot{\boldsymbol{r}}_{\text{rot}} = \ddot{\boldsymbol{r}}_{\text{iner}} + \boldsymbol{\omega} \times (\boldsymbol{\omega} \times \boldsymbol{r}) + 2\boldsymbol{\omega} \times \dot{\boldsymbol{r}}, \tag{5.295}$$

where the second term represents the centrifugal acceleration and the third term corresponds to the Coriolis acceleration.

Exercises

1. For a general second-order tensor $T^{\mu\nu}$, prove that its covariant divergence is

$$\nabla_\mu T^{\mu\nu} = \frac{1}{\sqrt{|g|}}\partial_\mu(\sqrt{|g|}T^{\mu\nu}) + \Gamma^\nu_{\sigma\mu}T^{\mu\sigma}. \tag{5.296}$$

Furthermore, if this tensor is an antisymmetric tensor $A^{\mu\nu} = -A^{\nu\mu}$, then

$$\nabla_\mu A^{\mu\nu} = \frac{1}{\sqrt{|g|}}\partial_\mu(\sqrt{|g|}A^{\mu\nu}). \tag{5.297}$$

2. Prove that for a diagonal metric $g_{\mu\nu}$, the Christoffel symbols are

$$\Gamma^\mu_{\sigma\rho} = 0, \quad \Gamma^\sigma_{\mu\mu} = -\frac{1}{2g_{\sigma\sigma}}\partial_\sigma g_{\mu\mu}, \tag{5.298}$$

$$\Gamma^\mu_{\sigma\mu} = \Gamma^\mu_{\mu\sigma} = \partial_\mu\left(\ln\sqrt{|g_{\mu\mu}|}\right), \quad \Gamma^\mu_{\mu\mu} = \partial_\mu\left(\ln\sqrt{|g_{\mu\mu}|}\right). \tag{5.299}$$

Here $\mu \neq \sigma \neq \rho$, and repeated indices are not summed.

3. In an n-dimensional manifold, a region Σ is bounded by a hypersurface S. Prove that if a vector field $\widehat{V}$ is zero on S, then

$$\int_\Sigma (\nabla_\mu V^\mu)\sqrt{-g}\mathrm{d}^n x = 0. \tag{5.300}$$

4. The metric of a two-dimensional sphere with radius a is

$$\mathrm{d}s^2 = a^2(\mathrm{d}\theta^2 + \sin^2\theta\mathrm{d}\phi^2). \tag{5.301}$$

(1) Calculate the Christoffel symbols of this metric;
(2) Discuss the geodesics on the sphere.

5. For the following metric, calculate its Christoffel symbols:

$$\mathrm{d}s^2 = -N^2\mathrm{d}t^2 + h_{ij}(\mathrm{d}x^i + N^i\mathrm{d}t)(\mathrm{d}x^j + N^j\mathrm{d}t), \tag{5.302}$$

where N, N^i, and h_{ij} are functions of (t, x^i).

6. Consider the parallel transport of a vector on the two-dimensional sphere in Exercise 4. If there is a vector $\widehat{V}$ at the North Pole tangent to the curve $\phi = 0$, parallel transport this vector along the following path:

   (1) First, move along the curve $\phi = 0$ to the point $\theta = \theta_0$;
   (2) Then, move along the curve $\theta = \theta_0$ from $\phi = 0$ to $\phi = \phi_0$;
   (3) Finally, move along the curve $\phi = \phi_0$ from $\theta = \theta_0$ back to the North Pole.

   Calculate the angle between the final vector and the original vector.

7. Consider a two-dimensional spacetime with the metric

$$\mathrm{d}s^2 = -x^2\mathrm{d}t^2 + \mathrm{d}x^2. \tag{5.303}$$

   Find all timelike geodesics in this spacetime of the form $x(t)$.

8. If two metrics $\mathrm{d}s^2 = g_{\mu\nu}\mathrm{d}x^\mu\mathrm{d}x^\nu$ and $\mathrm{d}\widetilde{s}^2 = \widetilde{g}_{\mu\nu}\mathrm{d}x^\mu\mathrm{d}x^\nu$ are related on the same coordinate chart by $\widetilde{g}_{\mu\nu} = \mathrm{e}^{\Omega(x)}g_{\mu\nu}$, where Ω is a non-zero function, then these two metrics are said to be conformally related. Prove that if a curve $x^\mu(\lambda)$ is a null geodesic of the metric $\mathrm{d}s^2$, then it is also a null geodesic of the metric $\mathrm{d}\widetilde{s}^2$, that is, under conformal transformation of the metric, null geodesics do not change.

9. Calculate the Christoffel symbols of the metric (5.256) and discuss the geodesic equations, proving that

$$m\frac{\mathrm{d}^2\boldsymbol{r}}{\mathrm{d}t^2} = -m\boldsymbol{\omega} \times (\boldsymbol{\omega} \times \boldsymbol{r}) - 2m\boldsymbol{\omega} \times \frac{\mathrm{d}\boldsymbol{r}}{\mathrm{d}t}. \tag{5.304}$$

10. Consider the metric

$$\mathrm{d}s^2 = \mathrm{d}r^2 + \frac{r^2}{1 - \omega^2 r^2/c^2}\mathrm{d}\theta^2 + \mathrm{d}z^2. \tag{5.305}$$

    Find the geodesics in the plane $z =$ constant.

11. A two-dimensional space with the line element

$$\mathrm{d}s^2 = \frac{\mathrm{d}r^2 + r^2\mathrm{d}\theta^2}{r^2 - a^2} - \frac{r^2\mathrm{d}r^2}{(r^2 - a^2)^2}, \quad r > a. \tag{5.306}$$

    Prove that the differential equation of the geodesics can be written as

$$a^2\left(\frac{\mathrm{d}r}{\mathrm{d}\theta}\right)^2 + a^2r^2 = Kr^4, \tag{5.307}$$

where K is a constant, and when $K = 1$ the geodesics are null. By letting $r\mathrm{d}\theta/\mathrm{d}r = \tan\phi$, prove that this two-dimensional space is mapped to a two-dimensional Euclidean plane, where (r, ϕ) are polar coordinates, and the above geodesics are mapped to straight lines.

12. We will see that outside a spherically symmetric star, the spacetime geometry can be described by the Schwarzschild metric:

$$\mathrm{d}s^2 = -\left(1 - \frac{2GM}{r}\right)\mathrm{d}t^2 + \left(1 - \frac{2GM}{r}\right)^{-1}\mathrm{d}r^2 + r^2\mathrm{d}\Omega^2, \qquad (5.308)$$

where M is the mass of the star.

(1) Prove that through the coordinate transformation

$$r = \rho\left(1 + \frac{GM}{2\rho}\right)^2, \quad 2GM < r < \infty, \qquad (5.309)$$

the spatial part of the Schwarzschild metric can be transformed into a conformally flat form. The coordinates (t, ρ, θ, ϕ) are called isotropic coordinates.

(2) Prove that in the new coordinates, the metric has an effective refractive index

$$n(x) = \frac{(1 + (GM/2|x|))^3}{1 - (GM/2|x|)}. \qquad (5.310)$$

The deceleration of light due to the refractive index has been verified by satellite experiments. When the Cassini spacecraft flew to the side of the solar system opposite to the Earth, a signal was sent from the Earth to the spacecraft, and then the reflected signal was received. Due to the presence of the gravitational field, the reflected signal was delayed. Assuming the Earth is located at r_1 and the spacecraft is located at r_2, calculate the line integral in isotropic coordinates $\int 2n\mathrm{d}l$, and prove that the time delay is

$$\Delta t = \frac{4GM}{c^3}\ln\left(\frac{4r_1 r_2}{b^2}\right), \qquad (5.311)$$

where the straight line is located from a point at radius r_1 to a relative point at radius r_2, and b is the impact parameter of the light.

(3) In the Schwarzschild spacetime, the circumference of a circle at coordinate radius ρ is $2\pi\rho n(\rho)$. Prove that this circumference has a minimum value at $r = 3GM$. What does this imply for zero geodesics in circular motion?

13. In general relativity, a particle falling from a certain height towards the surface of the Earth is actually acceleration-free and follows a geodesic. On the other hand, a particle at rest on the surface of the Earth is not moving along a geodesic and has acceleration.
    (1) Prove that a particle at rest on the surface of the Earth with x^i = constant has a 4-velocity $u^\mu = N^{-1}\delta_0^\mu$. Assume that the Earth's gravitational field is static with $g_{00} = -N^2$.
    (2) Prove that its 4-acceleration is $a^\mu = (0, a^i)$, where $a_i = -\partial_i N$. Calculate the magnitude of the acceleration.
14. The Poincaré upper half-plane can be defined by the metric

$$ds^2 = \frac{dx^2 + dy^2}{y^2}, \tag{5.312}$$

    where $x \in R$, $y \in R^+$. Provide the geodesics in this space.
15. Prove that for any three vector fields $\widehat{u}$, $\widehat{v}$, and $\widehat{w}$, they satisfy the Jacobi identity

$$[\widehat{u}, [\widehat{v}, \widehat{w}]] + [\widehat{v}, [\widehat{w}, \widehat{u}]] + [\widehat{w}, [\widehat{u}, \widehat{v}]] = 0. \tag{5.313}$$

16. Prove that for any two vector fields $\widehat{u}$ and $\widehat{v}$, we have

$$\pounds_{\widehat{u}}\pounds_{\widehat{v}} - \pounds_{\widehat{v}}\pounds_{\widehat{u}} = \pounds_{[\widehat{u},\widehat{v}]}. \tag{5.314}$$

Chapter 6

Curvature

In flat Euclidean space and Minkowski spacetime, flatness manifests through several key properties. First, in a flat space, a vector parallel transported along any closed loop returns to its starting point unchanged. Equivalently, parallel transport between two points in flat spacetime is independent of the path taken. Second, the second derivatives of tensor fields commute in flat space. Lastly, geodesics that are initially parallel remain parallel in flat spacetime.

Curvature quantifies the deviation of a spacetime manifold from flatness, measuring the extent to which the manifold is "bent." Consequently, each of the above properties of flatness must be generalized in curved spacetime, with the modifications directly linked to the curvature tensor:. First, in a curved space, the parallel transport of vectors or tensors becomes path-dependent. The Riemann curvature tensor arises naturally when examining parallel transport around an infinitesimal closed loop. Second, in a curved space, the second covariant derivatives of tensors do not commute. Their commutator is proportional to the curvature tensor. Lastly, geodesics that are initially parallel in a curved space do not remain parallel. This relative acceleration is described by the geodesic deviation equation, which depends explicitly on the Riemann curvature tensor.

6.1 Curvature Tensor

6.1.1 *Definitions of curvature and torsion tensors*

We define curvature by examining the parallel transport of a vector around a closed loop in a curved space. In a curved space, after such transport, the vector undergoes a transformation dependent on the total curvature enclosed by the loop. This behavior is captured locally by the Riemann curvature tensor. To illustrate, consider parallel transporting a vector $\widehat{V}$ along a closed loop formed by two vectors $\widehat{A}$ and $\widehat{B}$, (see Figure 6.1). The net change in $\widehat{V}$ after completing the loop encodes geometric information about the curvature within the enclosed region.

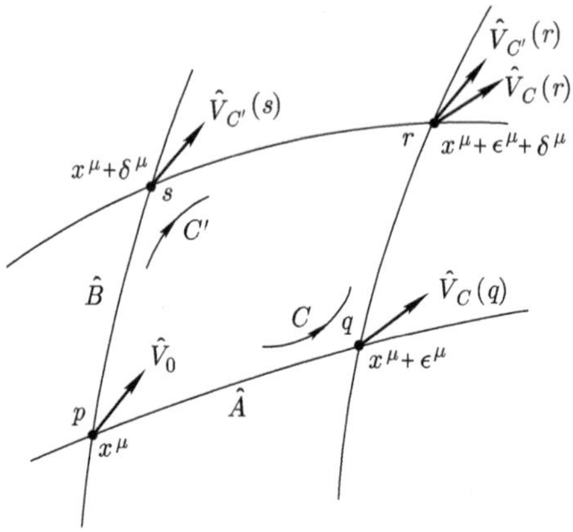

Fig. 6.1 Vector $\widehat{V}$ is parallel transported on an infinitesimal parallelogram formed by two vectors $\widehat{A}$ and $\widehat{B}$. The vectors obtained along different paths are generically different.

Note that the loop under consideration is infinitesimally small, with the vectors $\widehat{A}$ and $\widehat{B}$ primarily defining its orientation rather than its scale. When the vector $\widehat{V}$ completes a parallel transport around this loop, its transformation is characterized as follows. First, the change in $\widehat{V}$ is a linear transformation of the original vector, implying that the resulting expression carries one upper index and one lower index. Second, the transformation depends on the path defined by $\widehat{A}$ and $\widehat{B}$, necessitating the inclusion of two additional indices (μ and ν) contracted with A^ν and B^μ. Moreover, the associated tensor must be antisymmetric in the indices μ and ν, since swapping $\widehat{A}$ and $\widehat{B}$ reverses the loop's orientation, yielding an opposite change in the transported vector. Ultimately, we arrive at the expression

$$\delta V^\rho = (\delta a)(\delta b) A^\nu B^\mu R^\rho{}_{\sigma\mu\nu} V^\sigma, \tag{6.1}$$

where $R^\rho_{\sigma\mu\nu}$ is a (1, 3) type tensor referred to as the Riemann curvature tensor or simply the curvature tensor. Here, δa and δb denote the magnitudes of the two vectors forming infinitesimal the loop. A key property of this tensor is its antisymmetry under the exchange of the last two indices

$$R^\rho{}_{\sigma\mu\nu} = -R^\rho{}_{\sigma\nu\mu}. \tag{6.2}$$

The explicit form of the Riemann tensor can be derived directly from the above definition. For now, we disregard the question of whether parallel transporting two vectors along each other's directions necessarily forms a closed loop. Instead, we assume that the vectors $\widehat{A}$ and $\widehat{B}$ define an infinitesimal parallelogram, as previously discussed. Consider four points in the same coordinate system:

$$p : x^\mu, \quad q : x^\mu + \epsilon^\mu, \quad s : x^\mu + \delta^\mu, \quad r : x^\mu + \epsilon^\mu + \delta^\mu. \tag{6.3}$$

A vector V^μ is parallel transported if its covariant derivative vanishes, $\nabla_\nu V^\mu = \partial_\nu V^\mu + \Gamma^\mu_{\nu\sigma} V^\sigma = 0$, which can be rearranged to

$$\partial_\nu V^\mu = -\Gamma^\mu_{\nu\sigma} V^\sigma. \tag{6.4}$$

Suppose the vector $V^\mu(p) = V^\mu_0$ is parallel transported along $\widehat{A}$ (parametrized by ϵ^ν) to point q, then

$$\frac{V^\mu(q) - V^\mu_0}{\epsilon^\nu} = -\Gamma^\mu_{\nu\sigma} V^\sigma. \tag{6.5}$$

Using a first-order Taylor expansion and (6.4), we obtain

$$V^\mu(q) = V^\mu_0 - \Gamma^\mu_{\nu\sigma}(p) V^\sigma_0 \epsilon^\nu. \tag{6.6}$$

Further, we parallel transport the vector from q to r, then we have

$$\begin{aligned} V^\mu_C(r) &= V^\mu(q) - \Gamma^\mu_{\nu\kappa}(q) V^\kappa(q)\delta^\nu \\ &= V^\mu_0 - \Gamma^\mu_{\nu\kappa}(p) V^\kappa_0 \epsilon^\nu - (V^\kappa_0 - \Gamma^\kappa_{\sigma\rho}(p) V^\rho_0 \epsilon^\sigma)(\Gamma^\mu_{\nu\kappa}(p) + \partial_\lambda \Gamma^\mu_{\nu\kappa}(p)\epsilon^\lambda)\delta^\nu \\ &\approx V^\mu_0 - \Gamma^\mu_{\nu\kappa}(p) V^\kappa_0 \epsilon^\nu - V^\kappa_0 \Gamma^\mu_{\nu\kappa}(p)\delta^\nu \\ &\quad - V^\kappa_0 (\partial_\lambda \Gamma^\mu_{\nu\kappa}(p) - \Gamma^\rho_{\lambda\kappa}(p)\Gamma^\mu_{\nu\kappa}(p))\epsilon^\lambda \delta^\nu, \end{aligned} \tag{6.7}$$

where the subscript C represents that the path of vector transport is from p to q, and then to r, and we only retain terms up to second order in the infinitesimal displacements ϵ^μ and δ^μ. On the other hand, we may choose another path C' from p to s and then to r, then we have

$$\begin{aligned} V^\mu_{C'}(r) \approx V^\mu_0 &- \Gamma^\mu_{\nu\kappa}(p) V^\kappa_0 \epsilon^\nu - V^\kappa_0 \Gamma^\mu_{\nu\kappa}(p)\delta^\nu \\ &- V^\kappa_0 \left(\partial_\nu \Gamma^\mu_{\lambda\kappa}(p) - \Gamma^\rho_{\nu\kappa}(p)\Gamma^\mu_{\lambda\rho}(p)\right)\epsilon^\lambda \delta^\nu. \end{aligned} \tag{6.8}$$

Thus, the difference between the two transported vectors at r is

$$V^\mu_{C'}(r) - V^\mu_C(r) = V^\kappa_0 R^\mu{}_{\kappa\lambda\nu}(p)\epsilon^\lambda \delta^\nu, \tag{6.9}$$

where

$$R^\mu{}_{\kappa\lambda\nu}(p) = \partial_\lambda \Gamma^\mu_{\nu\kappa}(p) - \partial_\nu \Gamma^\mu_{\lambda\kappa}(p) + \Gamma^\rho_{\nu\kappa}(p)\Gamma^\mu_{\lambda\rho}(p) - \Gamma^\rho_{\lambda\kappa}(p)\Gamma^\mu_{\nu\rho}(p), \tag{6.10}$$

is the Riemann curvature tensor, which encodes the path dependence of parallel transport. As mentioned earlier, in this derivation, we actually assume that parallel transporting two vectors results in a closed parallelogram. This is equivalent to assuming the torsion-free condition and choosing the coordinate basis vectors, so that $[\widehat{A}, \widehat{B}] = 0$.

More fundamentally, we demonstrate that the commutator of two vector fields must be taken into account for them to form a closed polygonal path. As shown in Figure 6.2, this consideration results in a pentagonal configuration in the present case. The key quantity to examine is the change experienced by a vector (or tensor) when parallel-transported along this closed pentagonal loop back to its starting point, as shown in Figure 6.3. This leads us to the following formal definition of the Riemann curvature tensor.

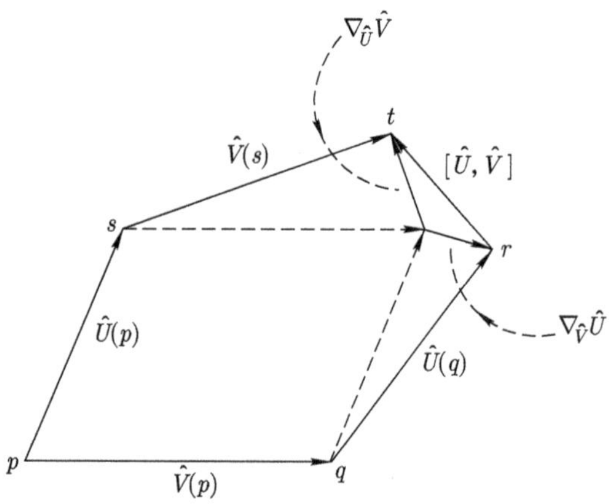

Fig. 6.2 Illustrates the commutator of two vector fields with zero torsion.

Considering the parallel transport of a vector along the pentagon loop depicted in Figure 6.3, we recall that the covariant derivative defines the parallel-transported vector $\widehat{W}pq$ from point p to q along the infinitesimal displacement pq as

$$\widehat{W}_{pq} = (1 - \epsilon\nabla_{\widehat{V}})\widehat{W}(q)), \tag{6.11}$$

where $\widehat{W}_{pq}$ represents the transported vector, differing from the original vector field $\widehat{W}$ at q. More generally, this parallel transport can be expressed via the exponential map:

$$\widehat{W}_{pq} = \exp(-\epsilon\nabla_{\widehat{V}})\widehat{W}|_q. \tag{6.12}$$

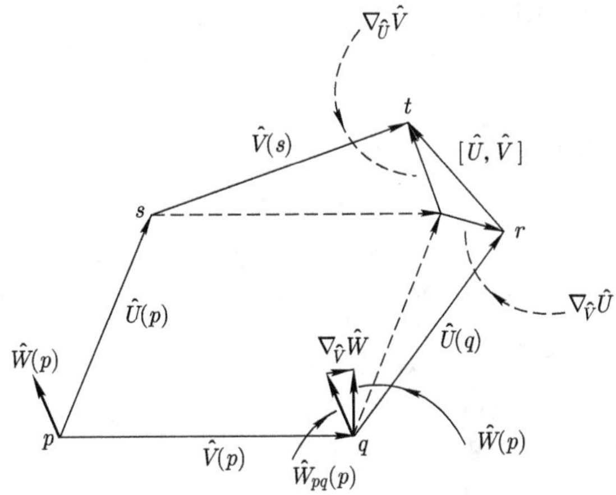

Fig. 6.3 Shows how the Riemann tensor can be obtained by considering the parallel transport of a vector along the closed pentagon.

Expanding this to second order in ϵ, we obtain

$$\widehat{W}_{pq} = \left(1 - \epsilon\nabla_{\widehat{V}} + \frac{1}{2}\epsilon^2\nabla_{\widehat{V}}\nabla_{\widehat{V}}\right)\widehat{W}(q). \tag{6.13}$$

Proceeding further, parallel transporting $\widehat{W}pq$ along qr to point r yields

$$\widehat{W}_{pqr} = \exp(-\delta\nabla_{\widehat{U}})\exp(-\epsilon\nabla_{\widehat{V}})\widehat{W}|_r. \tag{6.14}$$

Continuing this process around the pentagon and returning to the initial point p, we find

$$\begin{aligned}\widehat{W}_{pqrstp} = {} & \exp(\delta\nabla_{\widehat{U}})\exp(\epsilon\nabla_{\widehat{V}})\exp(-\epsilon\delta\nabla_{[\widehat{U},\widehat{V}]})\exp(-\delta\nabla_{\widehat{U}}) \\ & \exp(-\epsilon\nabla_{\widehat{V}})\widehat{W}|_p.\end{aligned} \tag{6.15}$$

Expanding this expression to second order in ϵ and δ, we obtain

$$\begin{aligned}\widehat{W}_{pqrstp} = {} & \left(1 + \delta\nabla_{\widehat{U}} + \frac{1}{2}\delta^2\nabla_{\widehat{U}}\nabla_{\widehat{U}}\right)\cdot\left(1 + \epsilon\nabla_{\widehat{V}} + \frac{1}{2}\epsilon^2\nabla_{\widehat{V}}\nabla_{\widehat{V}}\right) \\ & \times (1 - \epsilon\delta\nabla_{[\widehat{U},\widehat{V}]}) \\ & \times\left(1 - \delta\nabla_{\widehat{U}} + \frac{1}{2}\delta^2\nabla_{\widehat{U}}\nabla_{\widehat{U}}\right)\cdot\left(1 - \epsilon\nabla_{\widehat{V}} + \frac{1}{2}\epsilon^2\nabla_{\widehat{V}}\nabla_{\widehat{V}}\right)\widehat{W}(p).\end{aligned} \tag{6.16}$$

From this expansion, the net change in $\widehat{W}$ after completing the loop is

$$\delta\widehat{W} = \widehat{W}_{pqrstp} - \widehat{W}(p) = ([\nabla_{\widehat{U}}, \nabla_{\widehat{V}}] - \nabla_{[\widehat{U},\widehat{V}]})\delta\epsilon\widehat{W}(p). \tag{6.17}$$

This result motivates the formal definition of the Riemann curvature tensor as the operator measuring the holonomy due to parallel transport around an infinitesimal loop.

Definition The Riemann curvature tensor is a (1,3)-type tensor $\widehat{R}$: $\chi(M) \otimes \chi(M) \otimes \chi(M) \to \chi(M)$, defined for vector fields $\widehat{X}, \widehat{Y}, \widehat{Z} \in \chi(M)$ by

$$\begin{aligned} \widehat{R}(\widehat{X}, \widehat{Y}, \widehat{Z}) &= \widehat{R}(\widehat{X}, \widehat{Y})\widehat{Z}, \\ &= \nabla_{\widehat{X}} \nabla_{\widehat{Y}} \widehat{Z} - \nabla_{\widehat{Y}} \nabla_{\widehat{X}} \widehat{Z} - \nabla_{[\widehat{X}, \widehat{Y}]} \widehat{Z}. \end{aligned} \tag{6.18}$$

While two vectors may define an infinitesimal parallelogram in flat space, their parallel transports on a general manifold fail to close the quadrilateral. This non-closure is precisely characterized by the torsion tensor $\widehat{T}$. As visualized in Figure 6.4, we can directly read out the definition of the torsion tensor, measures the displacement between the tips of the parallel-transported vectors, quantifying the "gap" in the would-be parallelogram.

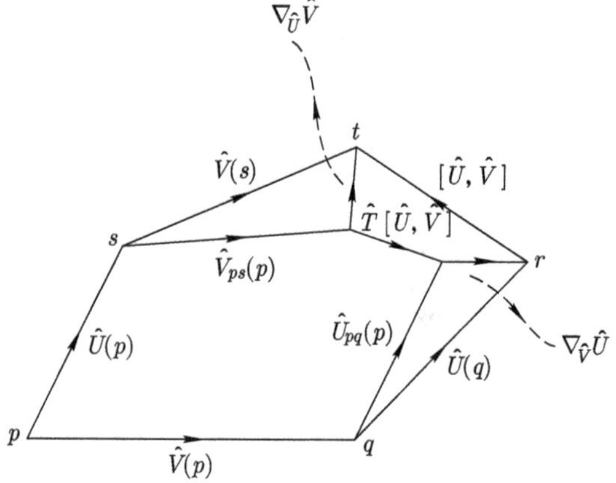

Fig. 6.4 Schematic diagram of torsion.

Definition The torsion tensor is a (1,2)-type tensor $\widehat{T} : \chi(M) \otimes \chi(M) \to \chi(M)$, defined for vector fields $\widehat{X}, \widehat{Y} \in \chi(M)$ by

$$\widehat{T}(\widehat{X}, \widehat{Y}) = \nabla_{\widehat{X}} \widehat{Y} - \nabla_{\widehat{Y}} \widehat{X} - [\widehat{X}, \widehat{Y}]. \tag{6.19}$$

From the definitions, it can be seen that

$$\begin{aligned} \widehat{R}(\widehat{X}, \widehat{Y})\widehat{Z} &= -\widehat{R}(\widehat{Y}, \widehat{X})\widehat{Z}, \\ \widehat{T}(\widehat{X}, \widehat{Y}) &= -\widehat{T}(\widehat{Y}, \widehat{X}). \end{aligned} \tag{6.20}$$

A fundamental property of the Riemann tensor $\widehat{R}$ and the torsion tensor $\widehat{T}$ is their multilinearity over the space of smooth functions $\mathcal{F}(M)$. For any

f, g, $h \in \mathcal{F}(M)$ and vector fields $\widehat{X}, \widehat{Y}, \widehat{Z} \in \chi(M)$, these tensors satisfy:

$$\widehat{R}(f\widehat{X}, g\widehat{Y})h\widehat{Z} = fgh\widehat{R}(\widehat{X}, \widehat{Y})\widehat{Z}, \tag{6.21}$$

$$\widehat{T}(f\widehat{X}, g\widehat{Y}) = fg\widehat{T}(\widehat{X}, \widehat{Y}). \tag{6.22}$$

The torsion tensor explicitly demonstrates this property:

$$\begin{aligned}\widehat{T}(f\widehat{X}, g\widehat{Y}) &= \nabla_{f\widehat{X}} g\widehat{Y} - \nabla_{g\widehat{Y}} f\widehat{X} - [f\widehat{X}, g\widehat{Y}] \\ &= f(\widehat{X}(g) + g\nabla_{\widehat{X}}\widehat{Y}) - g(\widehat{Y}(f) + f\nabla_{\widehat{Y}}\widehat{X}) \\ &\quad - f(\widehat{X}(g) + g\widehat{X}\widehat{Y}) + g(\widehat{Y}(f) + f\widehat{Y}\widehat{X}) \\ &= fg\widehat{T}(\widehat{X}, \widehat{Y}).\end{aligned} \tag{6.23}$$

The significance of multilinearity is that it ensures that the Riemann and torsion tensors are well-defined tensor fields, meaning their components transform appropriately under coordinate changes. First, this allows us to express them in terms of basis vector fields:

$$\widehat{R}(\widehat{X}, \widehat{Y})\widehat{Z} = X^\lambda Y^\mu Z^\nu \widehat{R}(\widehat{e}_\lambda, \widehat{e}_\mu)\widehat{e}_\nu, \quad \widehat{T}(\widehat{X}, \widehat{Y}) = X^\mu Y^\nu \widehat{T}(\widehat{e}_\mu, \widehat{e}_\nu). \tag{6.24}$$

Thus, it suffices to compute these tensors for basis vectors. If we choose a coordinate basis satisfying $[\widehat{e}_\mu, \widehat{e}_\nu] = 0$, then the expression for the torsion tensor is

$$\begin{aligned}T^\lambda_{\mu\nu} &= \langle \mathrm{d}x^\lambda, \widehat{T}(\widehat{e}_\mu, \widehat{e}_\nu)\rangle \\ &= \langle \mathrm{d}x^\lambda, \nabla_{\widehat{e}_\mu}\widehat{e}_\nu - \nabla_{\widehat{e}_\nu}\widehat{e}_\mu\rangle \\ &= \langle \mathrm{d}x^\lambda, (\Gamma^\eta_{\mu\nu} - \Gamma^\eta_{\nu\mu})\widehat{e}_\eta\rangle \\ &= \Gamma^\lambda_{\mu\nu} - \Gamma^\lambda_{\nu\mu},\end{aligned} \tag{6.25}$$

and the expression for the Riemann tensor is

$$\begin{aligned}R^\kappa{}_{\lambda\mu\nu} &= \langle \mathrm{d}x^\kappa, \widehat{R}(\widehat{e}_\mu, \widehat{e}_\nu)\widehat{e}_\lambda\rangle = \langle \mathrm{d}x^\kappa, \nabla_\mu\nabla_\nu\widehat{e}_\lambda - \nabla_\nu\nabla_\mu\widehat{e}_\lambda\rangle \\ &= \langle \mathrm{d}x^\kappa, \nabla_\mu(\Gamma^\sigma_{\nu\lambda}\widehat{e}_\sigma) - \nabla_\nu(\Gamma^\sigma_{\mu\lambda}\widehat{e}_\sigma)\rangle \\ &= \langle \mathrm{d}x^\kappa, \partial_\mu(\Gamma^\sigma_{\nu\lambda})\widehat{e}_\sigma + \Gamma^\sigma_{\nu\lambda}\Gamma^\rho_{\mu\sigma}\widehat{e}_\rho - (\mu \leftrightarrow \nu)\rangle \\ &= \partial_\mu(\Gamma^\kappa_{\nu\lambda}) + \Gamma^\sigma_{\nu\lambda}\Gamma^\kappa_{\mu\sigma} - (\mu \leftrightarrow \nu).\end{aligned} \tag{6.26}$$

In the above discussion, if we consider the transformation of the basis vectors under a coordinate transformation

$$\widehat{e}_\mu = \frac{\partial x^{\alpha'}}{\partial x^\mu}\widehat{e}_{\alpha'}, \tag{6.27}$$

since the transformation matrix can be regarded as a function, by multilinearity, we can directly factor out these transformation matrices, thereby ensuring the tensorial nature of curvature and torsion.

The above discussion is not only applicable to the parallel transport of vectors but also to the parallel transport of any tensor $\widehat{T}$, with the Riemann tensor given by

$$\begin{aligned}\widehat{R}(\widehat{X},\widehat{Y},\widehat{T}) &= \widehat{R}(\widehat{X},\widehat{Y})\widehat{T},\\ &= \nabla_{\widehat{X}}\nabla_{\widehat{Y}}\widehat{T} - \nabla_{\widehat{Y}}\nabla_{\widehat{X}}\widehat{T} - \nabla_{[\widehat{X},\widehat{Y}]}\widehat{T}.\end{aligned} \tag{6.28}$$

For a coordinate basis, we have

$$\begin{aligned}[\nabla_\rho,\nabla_\sigma]T^{\mu_1\cdots\mu_k}{}_{\nu_1\cdots\nu_l} &= R^{\mu_1}{}_{\lambda\rho\sigma}T^{\lambda\mu_2\cdots\mu_k}{}_{\nu_1\cdots\nu_l} + R^{\mu_2}{}_{\lambda\rho\sigma}T^{\mu_1\lambda\cdots\mu_k}{}_{\nu_1\cdots\nu_l} + \cdots\\ &\quad - R^{\lambda}{}_{\nu_1\rho\sigma}T^{\mu_1\cdots\mu_k}{}_{\lambda\nu_2\cdots\nu_l} - R^{\lambda}{}_{\nu_2\rho\sigma}T^{\mu_1\cdots\mu_k}{}_{\nu_1\lambda\cdots\nu_l} - \cdots .\end{aligned}$$

In the above expression, we use the coordinate basis $\widehat{e}_\mu = \partial_\mu$, $\widehat{\theta}^\mu = \mathrm{d}x^\mu$. Consequently the metric tensor takes the form $\widehat{g} = g_{\mu\nu}\mathrm{d}x^\mu\mathrm{d}x^\nu$, where $g_{\mu\nu}$ is a function of x. Notably, the computation of the Riemann tensor is not restricted to a coordinate basis or an orthonormal frame—it can be evaluated in an arbitrary basis. In a general basis, the Riemann tensor is given by

$$R^\kappa{}_{\lambda\mu\nu} = \partial_\mu(\Gamma^\kappa_{\lambda\nu}) + \Gamma^\kappa_{\sigma\mu}\Gamma^\sigma_{\lambda\nu} - (\mu \leftrightarrow \nu) - C^\rho_{\mu\nu}\Gamma^\kappa_{\rho\lambda}, \tag{6.29}$$

where $C^\rho_{\mu\nu}$ denote the structure constants of the chosen basis. Remarkably, the definition of the Riemann tensor remains independent of torsion, meaning that this expression holds even when the connection is not torsion-free. For completeness, the torsion tensor in a general basis is expressed as

$$T^\lambda_{\mu\nu} = \Gamma^\lambda_{\mu\nu} - \Gamma^\lambda_{\nu\mu} - C^\lambda_{\mu\nu}. \tag{6.30}$$

In many situations, it is advantageous to work with a frame field (or non-coordinate basis) $\widehat{e}_a$, where the basis vectors are orthonormal with respect to the metric, satisfying $\widehat{e}_a \cdot \widehat{e}_b = \eta_{ab}$. These frame fields generally cannot be expressed as partial derivatives of any coordinate system. Consequently, their commutator is non-vanishing: $[\widehat{e}_a, \widehat{e}_b] = C^c_{ab}\widehat{e}_c$.

Having explored the Riemann tensor's role in curvature, we now turn to the torsion tensor and its geometric significance. Torsion measures the failure of infinitesimal parallelograms to close under parallel transport, reflecting a twist in the connection's structure. Geometrically, we can use Figure 6.5 to calculate the torsion. Consider a point $p \in M$ and two infinitesimal vectors $\widehat{U} = \epsilon^\mu\widehat{e}_\mu$, $\widehat{V} = \delta^\nu\widehat{e}_\nu$ in the tangent space $T_p(M)$. In a local coordinate chart, the coordinates of points p, q, and s are

$$p : x^\mu, \quad q : x^\mu + \epsilon^\mu, \quad s : x^\mu + \delta^\mu, \tag{6.31}$$

so that $\widehat{U} = \overrightarrow{pq}$, $\widehat{V} = \overrightarrow{ps}$. If we parallel transport $\widehat{U}$ along $\overrightarrow{ps}$, we obtain the vector $\overrightarrow{sr_1} = \epsilon^\mu - \epsilon^\lambda\Gamma^\mu_{\nu\lambda}\delta^\nu$, resulting in $\overrightarrow{pr_1} = \delta^\mu + \epsilon^\mu - \epsilon^\lambda\Gamma^\mu_{\nu\lambda}\delta^\nu$.

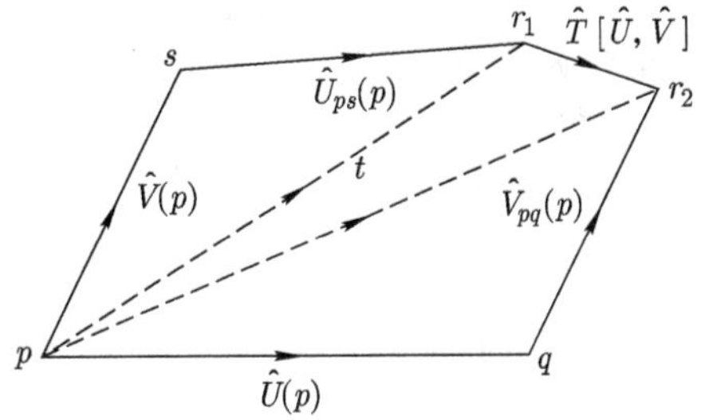

Fig. 6.5 Illustration of torsion computation.

On the other hand, if we parallel transport $\widehat{V}$ along $\overrightarrow{pq}$, we get $\overrightarrow{pr_2} = \epsilon^\mu + \delta^\mu - \epsilon^\lambda \Gamma^\mu_{\lambda\nu}\delta^\nu$. Generally, $r_1 \neq r_2$, and their difference is $\overrightarrow{r_1 r_2} = (\Gamma^\mu_{\nu\lambda} - \Gamma^\mu_{\lambda\nu})\epsilon^\lambda\delta^\nu = T^\mu_{\nu\lambda}\epsilon^\lambda\delta^\nu$, which is proportional to the torsion tensor.

The Riemann curvature tensor $R^\mu{}_{\nu\rho\sigma} \equiv 0$ if and only if the spacetime is globally flat, meaning the metric can be reduced to a constant (typically the Minkowski metric $\eta_{\mu\nu}$ in some coordinate system. On one hand, if the metric is constant, i.e., $\partial_\sigma g_{\mu\nu} = 0$, the Christoffel symbols vanish, $\Gamma^\rho_{\mu\nu} = 0$ and $\partial_\sigma \Gamma^\rho_{\mu\nu} = 0$, and consequently $R^\mu{}_{\nu\rho\sigma} = 0$. Conversely, if $R^\mu{}_{\nu\rho\sigma} \equiv 0$, the spacetime must admit a globally flat coordinate system. At any point p, we can choose coordinates where $g_{\mu\nu}|_p = \eta_{\mu\nu}$ and $\partial_\sigma g_{\mu\nu}(p) = 0$. The basis vectors satisfy

$$g_{\sigma\rho}(\widehat{e}_\mu)^\sigma(\widehat{e}_\nu)^\rho(p) = \eta_{\mu\nu}. \tag{6.32}$$

The vanishing Riemann tensor ensures that parallel transport is path-independent. Transporting the orthonormal basis from p to any other point q preserves their inner products

$$g_{\sigma\rho}(\widehat{e}_\mu)^\sigma(\widehat{e}_\nu)^\rho(q) = \eta_{\mu\nu}. \tag{6.33}$$

This yields a globally defined set of vector fields in which the metric is everywhere constant. For these parallel-transported vector fields, the commutator vanishes:

$$[\widehat{e}_\mu, \widehat{e}_\nu] = 0. \tag{6.34}$$

This follows because (in a torsion-free setting)

$$[\widehat{e}_\mu, \widehat{e}_\nu] = \nabla_{\widehat{e}_\mu}\widehat{e}_\nu - \nabla_{\widehat{e}_\nu}\widehat{e}_\mu - T(\widehat{e}_\mu, \widehat{e}_\nu), \tag{6.35}$$

where $\nabla\widehat{e}_\nu = 0$ (parallel transport) and $\widehat{T} = 0$ (no torsion). By Frobenius' theorem, the vanishing commutator implies $\{\widehat{e}_\mu\}$ is a coordinate basis, and the metric in these coordinates reduces globally to $\eta_{\mu\nu}$.

In Einstein's general relativity, the connection is assumed torsion-free, whether for the Levi–Civita or spin connection. However, theories like Einstein-Cartan gravity allow for torsion while maintaining zero curvature. Such frameworks—though physically interesting—lie beyond the scope of our discussion.

6.1.2 *Symmetries and identities of the Riemann tensor*

The Riemann tensor exhibits several symmetries among its components. Below, we analyze these symmetries to determine the number of independent components it possesses. Consider an n-dimensional space. Since the Riemann tensor $R^{\mu}{}_{\nu\rho\sigma}$ carries four indices, one might naively expect it to have n^4 independent components. However, this is not the case due to its inherent symmetries.

First, the antisymmetry in its last two indices, $R^{\mu}{}_{\nu\rho\sigma} = -R^{\mu}{}_{\nu\sigma\rho}$, reduces the number of possible components to at most $n^2 \times \dfrac{n(n-1)}{2}$ components. Secondly, the Riemann tensor satisfies further symmetries. Introducing the fully covariant (0,4)-type tensor $R_{\mu\nu\sigma\rho} = g_{\mu\lambda}R^{\lambda}{}_{\nu\sigma\rho}$, we find that it obeys the following relations:

(1) The last two indices are antisymmetric, $R_{\mu\nu\sigma\rho} = -R_{\mu\nu\rho\sigma}$;
(2) The first two indices are antisymmetric, $R_{\mu\nu\sigma\rho} = -R_{\nu\mu\sigma\rho}$;
(3) The exchange of the two pairs of indices is symmetric, $R_{\mu\nu\sigma\rho} = R_{\sigma\rho\mu\nu}$;
(4) The first Bianchi identity: $R_{\mu\nu\sigma\rho} + R_{\mu\sigma\rho\nu} + R_{\mu\rho\nu\sigma} = 0$, or written as $R_{\mu[\nu\sigma\rho]} = 0$.

These symmetry relations can be derived by working in Riemann normal coordinates, where the Christoffel symbols vanish at a point. However, since these relations are tensor equations, they must hold universally and can also be verified algebraically without relying on a specific coordinate choice. In Riemann normal coordinates, the components of the Riemann tensor take the form:

$$R_{\mu\nu\sigma\rho}|_{\text{RNC}} = \frac{1}{2}(\partial_\mu\partial_\sigma g_{\nu\rho} - \partial_\nu\partial_\sigma g_{\mu\rho} + \partial_\nu\partial_\rho g_{\mu\sigma} - \partial_\mu\partial_\rho g_{\nu\sigma}). \tag{6.36}$$

From this explicit expression, it is easy to verify the above symmetry properties.

Using the established symmetries, we now systematically determine the number of independent components of the Riemann tensor in an n-dimensional space. First, we consider the antisymmetry properties. These imply that the Riemann tensor can be treated as a symmetric bilinear form acting on the space of antisymmetric 2-tensors. We can denote $R_{[\mu\nu][\sigma\rho]} \to Y_{(M,N)}$, where $M = [\mu\nu]$ and $N = [\sigma\rho]$ represent pairs of antisymmetric indices, each with $\frac{n(n-1)}{2}$ independent components. Considering the third property, M and N are symmetric, which tells us that the number of independent components might be $\frac{1}{8}n(n-1)(n(n-1)+2)$. Finally, by the first Bianchi identity $R_{\mu[\nu\sigma\rho]} = 0$ and other relations, we find $R_{[\mu\nu\sigma\rho]} = 0$. This fully antisymmetric condition gives $n(n-1)(n-2)(n-3)/4!$ constraints. Ultimately, we find that the number of independent components of the Riemann tensor is

$$\frac{1}{8}(n^4 - 2n^3 + 3n^2 - 2n) - \frac{1}{24}n(n-1)(n-2)(n-3) = \frac{1}{12}n^2(n^2-1).$$

In Riemann normal coordinates, the non-vanishing second derivatives of the metric encode the Riemann tensor's degrees of freedom. The inability to eliminate all second derivatives reflects the intrinsic curvature described by the Riemann tensor. In Riemann normal coordinates, the spacetime metric near a point can be expanded as

$$\mathrm{d}s^2 = \eta_{\mu\nu}\mathrm{d}x^\mu \mathrm{d}x^\nu + \frac{1}{12}R_{\mu\nu\sigma\rho}(x^\mu \mathrm{d}x^\nu - x^\nu \mathrm{d}x^\mu)(x^\sigma \mathrm{d}x^\rho - x^\rho \mathrm{d}x^\sigma) + \cdots. \quad (6.37)$$

This expansion confirms that the Riemann tensor precisely captures the irreducible second-order deviations from flat spacetime.

Furthermore, the Riemann tensor satisfies a fundamental differential identity involving its covariant derivative. To derive this, we work in Riemann normal coordinates (RNC), where the covariant derivative of the Riemann tensor reduces to a partial derivative:

$$\begin{aligned}\nabla_\lambda R_{\rho\sigma\mu\nu} &= \partial_\lambda R_{\rho\sigma\mu\nu} \\ &= \frac{1}{2}\partial_\lambda(\partial_\mu\partial_\sigma g_{\rho\nu} - \partial_\mu\partial_\rho g_{\nu\sigma} - \partial_\nu\partial_\sigma g_{\rho\mu} + \partial_\nu\partial_\rho g_{\mu\sigma}).\end{aligned}$$

Summing over the permutations of the first three indices, we obtain

$$\begin{aligned}&\nabla_\lambda R_{\rho\sigma\mu\nu} + \nabla_\rho R_{\sigma\lambda\mu\nu} + \nabla_\sigma R_{\lambda\rho\mu\nu} \\ &= \frac{1}{2}(\partial_\lambda\partial_\mu\partial_\sigma g_{\rho\nu} - \partial_\lambda\partial_\mu\partial_\rho g_{\nu\sigma} - \partial_\lambda\partial_\nu\partial_\sigma g_{\rho\mu} + \partial_\lambda\partial_\nu\partial_\rho g_{\mu\sigma} \\ &\quad + \partial_\rho\partial_\mu\partial_\lambda g_{\sigma\nu} - \partial_\rho\partial_\mu\partial_\sigma g_{\nu\lambda} - \partial_\rho\partial_\nu\partial_\lambda g_{\sigma\mu} + \partial_\rho\partial_\nu\partial_\sigma g_{\mu\lambda} \\ &\quad + \partial_\sigma\partial_\mu\partial_\rho g_{\lambda\nu} - \partial_\sigma\partial_\mu\partial_\lambda g_{\nu\rho} - \partial_\sigma\partial_\nu\partial_\rho g_{\lambda\mu} + \partial_\sigma\partial_\nu\partial_\lambda g_{\mu\rho}) \\ &= 0.\end{aligned} \quad (6.38)$$

Equation (6.38) can be written as

$$\nabla_{[\lambda}R_{\rho\sigma]\mu\nu} = 0. \quad (6.39)$$

It is known as the second Bianchi identity or simply the Bianchi identity. It is closely related to the Jacobi identity for covariant derivatives:

$$[[\nabla_\lambda, \nabla_\rho], \nabla_\sigma] + [[\nabla_\rho, \nabla_\sigma], \nabla_\lambda] + [[\nabla_\sigma, \nabla_\lambda], \nabla_\rho] = 0. \tag{6.40}$$

The identities discussed earlier can be written in a coordinate-independent form. First, define the (0, 4) tensor $\widehat{R}(\widehat{X}, \widehat{Y}, \widehat{Z}, \widehat{W}) = \widehat{g}(\widehat{R}(\widehat{Z}, \widehat{W})\widehat{X}, \widehat{Y})$, then the above identities are actually

$$\begin{aligned}
\widehat{R}(\widehat{X}, \widehat{Y}, \widehat{Z}, \widehat{W}) &= -\widehat{R}(\widehat{Y}, \widehat{X}, \widehat{Z}, \widehat{W}), \\
\widehat{R}(\widehat{X}, \widehat{Y}, \widehat{Z}, \widehat{W}) &= -\widehat{R}(\widehat{X}, \widehat{Y}, \widehat{W}, \widehat{Z}), \\
\widehat{R}(\widehat{X}, \widehat{Y}, \widehat{Z}, \widehat{W}) &= \widehat{R}(\widehat{Z}, \widehat{W}, \widehat{X}, \widehat{Y}), \\
\widehat{R}(\widehat{X}, \widehat{Y}, \widehat{Z}, \widehat{W}) + \widehat{R}(\widehat{Z}, \widehat{Y}, \widehat{W}, \widehat{X}) &+ \widehat{R}(\widehat{W}, \widehat{Y}, \widehat{X}, \widehat{Z}) = 0, \\
\nabla_{\widehat{X}} \widehat{R}(\widehat{Y}, \widehat{Z})\widehat{V} + \nabla_{\widehat{Z}} \widehat{R}(\widehat{X}, \widehat{Y})\widehat{V} &+ \nabla_{\widehat{Y}} \widehat{R}(\widehat{Z}, \widehat{X})\widehat{V} = 0.
\end{aligned} \tag{6.41}$$

6.1.3 *Ricci tensor and Ricci scalar*

Using the Riemann tensor, we can further construct other curvature tensors via contraction. The most prominent among these is the Ricci tensor.

Definition The Ricci tensor is obtained by contracting the first and third indices of the Riemann curvature tensor:

$$R_{\mu\nu} = R^{\lambda}{}_{\mu\lambda\nu} = g^{\lambda\rho} R_{\mu\lambda\nu\rho}. \tag{6.42}$$

For general non-Christoffel connections, other independent contractions may exist. However, in the case of Christoffel connections, the Ricci tensor is the unique nontrivial contraction of the Riemann tensor. An important property of the Ricci tensor is that it is symmetric:

$$R_{\mu\nu} = R_{\nu\mu}. \tag{6.43}$$

Definition A further contraction of the Ricci tensor yields the Ricci scalar (or scalar curvature):

$$R = R^{\mu}_{\mu} = g^{\mu\nu} R_{\mu\nu}. \tag{6.44}$$

The Ricci scalar, also known as the scalar curvature, is a coordinate-invariant quantity that characterizes the intrinsic curvature of the manifold. Notably, its definition remains valid even in a non-coordinate basis, ensuring its invariance under changes of frame.

Using the Ricci tensor and Ricci scalar, we can define the Einstein tensor, which plays a fundamental role in general relativity.

Definition The Einstein tensor is constructed from the Ricci tensor and the Ricci scalar as follows:

$$G_{\mu\nu} = R_{\mu\nu} - \frac{1}{2} R g_{\mu\nu}. \tag{6.45}$$

A key property of the Einstein tensor arises from the contracted Bianchi identity. Starting from

$$\begin{aligned} 0 &= g^{\nu\sigma} g^{\mu\lambda} (\nabla_\lambda R_{\rho\sigma\mu\nu} + \nabla_\rho R_{\sigma\lambda\mu\nu} + \nabla_\sigma R_{\lambda\rho\mu\nu}) \\ &= \nabla^\mu R_{\rho\mu} - \nabla_\rho R + \nabla^\nu R_{\rho\nu}, \end{aligned} \tag{6.46}$$

we obtain, after contraction and simplification,

$$\nabla^\mu R_{\rho\mu} = \frac{1}{2} \nabla_\rho R.$$

This implies that the Einstein tensor is divergence-free:

$$\nabla^\mu G_{\mu\nu} = 0. \tag{6.47}$$

The Einstein tensor is a symmetric tensor due to the symmetries of $R_{\mu\nu}$ and $g_{\mu\nu}$.

The Riemann curvature tensor can be decomposed into its trace (contraction) and traceless parts. By removing all possible trace components, we obtain the Weyl tensor, which represents the conformally invariant part of the curvature.

Definition The Weyl tensor in an n-dimensional manifold is defined as the completely traceless part of the Riemann curvature tensor:

$$C_{\rho\sigma\mu\nu} = R_{\rho\sigma\mu\nu} - \frac{2}{n-2}(g_{\rho[\mu} R_{\nu]\sigma} - g_{\sigma[\mu} R_{\nu]\rho}) + \frac{2}{(n-1)(n-2)} R g_{\rho[\mu} g_{\nu]\sigma}. \tag{6.48}$$

By definition, all possible contractions of $C_{\rho\sigma\mu\nu}$ are zero, but it retains all the symmetries of the Riemann tensor:

$$\begin{aligned} C_{\rho\sigma\mu\nu} &= C_{[\rho\sigma][\mu\nu]}, \\ C_{\rho\sigma\mu\nu} &= C_{\mu\nu\rho\sigma}, \\ C_{\rho[\sigma\mu\nu]} &= 0. \end{aligned} \tag{6.49}$$

It is important to note that the Weyl tensor can only be properly defined in spaces of three or more dimensions. In the special case of three-dimensional spaces, the Weyl tensor vanishes identically. However, for dimensions four and higher ($n \geq 4$), it satisfies a particular form of the Bianchi identity:

$$\nabla^\rho C_{\rho\sigma\mu\nu} = -2\frac{(n-3)}{(n-2)} \left(\nabla_{[\mu} R_{\nu]\sigma} + \frac{1}{2(n-1)} g_{\sigma[\nu} \nabla_{\mu]} R \right). \tag{6.50}$$

One of the most significant properties of the Weyl tensor is its conformal invariance. Under conformal transformations of the metric:

$$g_{\mu\nu} \to \Omega^2(x) g_{\mu\nu}, \tag{6.51}$$

the Weyl tensor remains unchanged.

A metric is said to be conformally flat if it differs from the flat spacetime metric by a conformal factor, $g_{\mu\nu} = \Omega^2 \eta_{\mu\nu}$, where $\eta_{\mu\nu}$ is the flat space metric.

Theorem A metric is conformally flat if and only if its corresponding Weyl tensor vanishes identically.

For example, all two-dimensional Riemannian manifolds are conformally flat.

In three dimensions, conformal flatness is governed by the Cotton tensor instead of the Weyl tensor, which vanishes identically. In three dimensions, the Cotton tensor is defined as:

$$C^{ij} = \varepsilon^{ikl} \nabla_k \left(R^j_l - \frac{1}{4} R \delta^j_l \right), \tag{6.52}$$

which satisfies the following key properties:

(1) Symmetric and traceless;
(2) Transverse (or covariantly conserved): $\nabla_i C^{ij} = 0$;
(3) Under conformal transformations, $C^{ij} \to \Omega^{-5} C^{ij}$, meaning that it has conformal weight $-5/2$.

6.1.4 *Intrinsic curvature*

In 1, 2, 3, and 4 dimensions, the Riemann tensor has 0, 1, 6, and 20 independent components, respectively. This implies that one-dimensional manifolds (such as $\mathbf{S}^1$) are intrinsically flat and cannot exhibit curvature in the Riemannian sense. At first glance, it may seem counterintuitive to claim that a circle $\mathbf{S}^1$ is not curved. However, from a purely intrinsic geometric perspective, it is indeed flat. Our everyday intuition about the curvature of a circle stems from its embedding in a higher-dimensional space, where it appears bent. This perceived curvature is extrinsic, arising solely from the way the circle is immersed in its ambient space.

In two dimensions, the Riemann curvature tensor has only one independent component, R_{1212}. Since all curvature information is encoded in the

Ricci scalar R, the Riemann tensor must be expressible entirely in terms of R and $g_{\mu\nu}$. This leads to the key relationship:

$$R_{\mu\nu\sigma\rho} = \frac{1}{2}R(g_{\mu\sigma}g_{\nu\rho} - g_{\nu\sigma}g_{\mu\rho}). \tag{6.53}$$

To understand how this formula arises, we consider essential properties of the Riemann tensor: First, we must consider the symmetries satisfied by the Riemann tensor, and second, the contraction of the Riemann tensor gives the Ricci scalar. Given the symmetries and the fact that contraction must reproduce R, Eq. (6.53) is the unique tensor form that satisfies all constraints while linearly relating the Riemann tensor to the Ricci scalar.

Example 6.1. Cylinder.

Consider a cylinder $\mathbf{I} \times \mathbf{S}^1$, where $\mathbf{I}$ is a closed interval, as shown in Figure 6.6. This surface can be constructed by identifying (gluing) a pair of opposite edges of a flat rectangular strip. The induced metric on it is constant, meaning the cylinder is intrinsically flat. This aligns with our earlier observation: In two dimensions, intrinsic curvature is fully determined by the Ricci scalar, which here is zero. This presents another example showing the difference between extrinsic and intrinsic curvature. While the cylinder appears "curved" when viewed in $\mathbf{R}^3$, this bending is purely an extrinsic property—it arises from how the cylinder is embedded in higher-dimensional space, not from its intrinsic geometry. As discussed in Section 4.4.1, a bug confined to the surface of a cylinder would not detect any curvature.

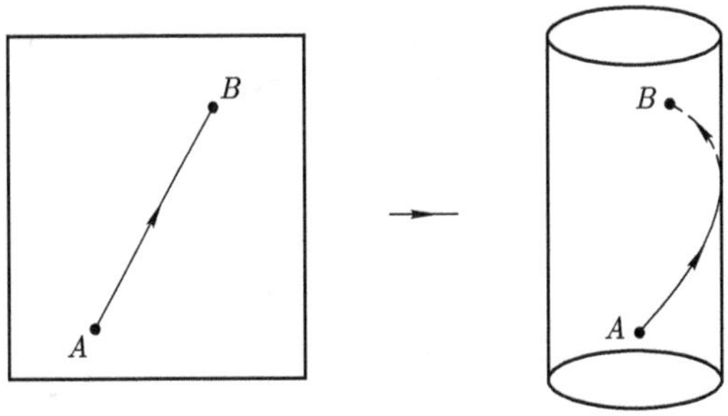

Fig. 6.6 Gluing one pair of opposite edges of a rectangle will result in a cylindrical surface, which is flat.

Example 6.2. Torus of genus 1.

A torus can be further constructed by identifying both pairs of opposite edges of a flat rectangle (i.e., gluing the top and bottom circles of the cylinder), meaning it is $\mathbf{S}^1 \times \mathbf{S}^1$, as shown in Figure 6.7. The torus is also flat.

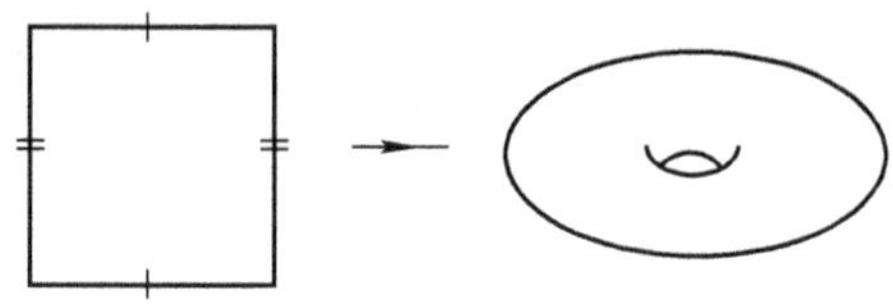

Fig. 6.7 A torus can be obtained by gluing both pairs of opposite edges of a rectangle, and it is also intrinsically flat.

Example 6.3. Cone.

As shown in Figure 6.8(a), a cone constitutes a two-dimensional manifold exhibiting non-zero curvature solely at its apex. Geometrically, the cone can be constructed by excising a wedge of a specific angle from a flat plane and identifying (gluing) the resulting edges. More precisely, it represents a manifold with a singular point at its vertex. As shown in Figure 6.8(b), the coordinates on the cone can be characterized by ρ and θ, with the metric given by

$$\mathrm{d}s^2 = \mathrm{d}\rho^2 + r^2\mathrm{d}\theta^2 = \mathrm{d}\rho^2 + \left(\frac{R}{\rho_0}\right)^2 \rho^2\mathrm{d}\theta^2, \tag{6.54}$$

where $0 < \rho < \rho_0$. Notably, despite its conical geometry, the scalar curvature of this manifold vanishes everywhere except at the singularity.

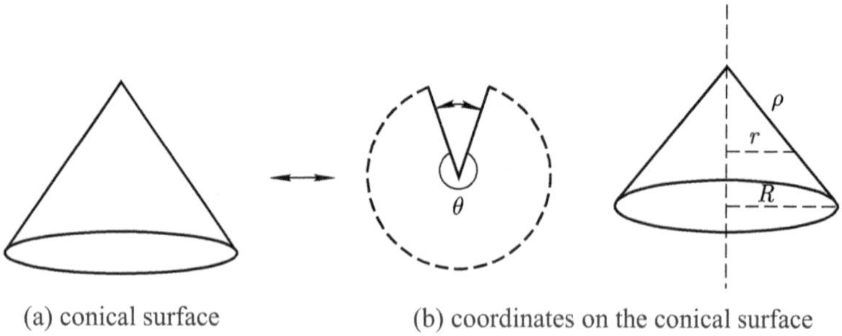

Fig. 6.8 A conical surface can be obtained by gluing a circular disk with a deficit angle.

Example 6.4. Two-dimensional sphere.

For a two-dimensional sphere of radius a, the metric is

$$\mathrm{d}s^2 = a^2(\mathrm{d}\theta^2 + \sin^2\theta \mathrm{d}\phi^2), \tag{6.55}$$

with the non-zero Christoffel symbols being

$$\begin{aligned} \Gamma^\theta_{\phi\phi} &= -\sin\theta\cos\theta, \\ \Gamma^\phi_{\theta\phi} &= \Gamma^\phi_{\phi\theta} = \cot\theta. \end{aligned} \tag{6.56}$$

The only possibly non-zero Riemann tensor components are

$$\begin{aligned} R^\theta{}_{\phi\theta\phi} &= \partial_\theta \Gamma^\theta_{\phi\phi} - \partial_\phi \Gamma^\theta_{\theta\phi} + \Gamma^\theta_{\theta\lambda}\Gamma^\lambda_{\phi\phi} - \Gamma^\theta_{\phi\lambda}\Gamma^\lambda_{\theta\phi} \\ &= (\sin^2\theta - \cos^2\theta) - (0) + (0) - (-\sin\theta\cos\theta)(\cot\theta) \\ &= \sin^2\theta. \end{aligned} \tag{6.57}$$

Thus,

$$\begin{aligned} R_{\theta\phi\theta\phi} &= g_{\theta\lambda} R^\lambda_{\phi\theta\phi} \\ &= g_{\theta\theta} R^\theta_{\phi\theta\phi} \\ &= a^2 \sin^2\theta. \end{aligned} \tag{6.58}$$

The Ricci tensor is

$$\begin{aligned} R_{\theta\theta} &= g^{\phi\phi} R_{\phi\theta\phi\theta} = 1, \\ R_{\theta\phi} &= R_{\phi\theta} = 0, \\ R_{\phi\phi} &= g^{\theta\theta} R_{\theta\phi\theta\phi} = \sin^2\theta. \end{aligned} \tag{6.59}$$

The Ricci scalar is

$$R = g^{\theta\theta} R_{\theta\theta} + g^{\phi\phi} R_{\phi\phi} = \frac{2}{a^2}. \tag{6.60}$$

Thus, the two-dimensional sphere is a space of constant curvature and is also a maximally symmetric space.[1] It has positive curvature, inversely proportional to the square of the radius. Are there spaces with negative curvature?

Example 6.5. Poincaré upper half-plane $\mathbf{H}^2$.

The Poincaré upper half-plane is a two-dimensional Riemannian manifold equipped with the metric

$$\mathrm{d}s^2 = \frac{a^2}{y^2}(\mathrm{d}x^2 + \mathrm{d}y^2), \quad y > 0. \tag{6.61}$$

[1]The definition of maximally symmetric spaces will be given in Section 6.3.

For a fixed x, the distance between two points (x, y_1) and (x, y_2) is given by

$$\delta s = \int_{y_1}^{y_2} \sqrt{g_{\mu\nu} \frac{\mathrm{d}x^\mu}{\mathrm{d}y} \frac{\mathrm{d}x^\nu}{\mathrm{d}y}} \mathrm{d}y = a \int_{y_1}^{y_2} \frac{\mathrm{d}y}{y} = a \ln\left(\frac{y_2}{y_1}\right). \tag{6.62}$$

The non-zero Christoffel symbols for this metric are

$$\Gamma^x_{xy} = \Gamma^x_{yx} = -y^{-1}, \quad \Gamma^y_{xx} = y^{-1}, \quad \Gamma^y_{yy} = y^{-1}. \tag{6.63}$$

The geodesics in the Poincaré upper half-plane fall into two classes: one with x constant and $y = e^{\lambda/a}$; the other being semi-circles, $(x - x_0)^2 + y^2 = l^2$, where l is a constant. For this two-dimensional manifold, the only non-zero Riemann tensor component is $R^x{}_{yxy} = -y^{-2}$, and the Ricci tensor is

$$R_{xx} = -y^{-2}, \quad R_{xy} = 0, \quad R_{yy} = -y^{-2}, \tag{6.64}$$

with the Ricci scalar being

$$R = -\frac{2}{a^2}. \tag{6.65}$$

This space, known as the Poincaré upper half-plane or Lobachevsky space, is a prototypical example of a negatively curved maximally symmetric space. It has the same topology as two-dimensional Euclidean space but cannot be isometrically embedded in three-dimensional Euclidean space R^3. Since it has constant curvature, it is also a maximally symmetric space. In higher dimensions, negatively curved maximally symmetric spaces are called hyperbolic spaces, denoted as H^n.

In two dimensions, the Einstein tensor vanishes identically,

$$G_{\mu\nu} = 0, \tag{6.66}$$

which can be easily proven by using the relationship between the Riemann tensor and the Ricci scalar discussed previously. The vanishing of Einstein tensor $G_{\mu\nu}$ implies that two-dimensional gravity is trivial.

In three dimensions, the Riemann tensor has six independent components. Since the Ricci tensor is symmetric, it also possesses six independent components. This equality in the number of components implies that the Riemann tensor can be fully expressed in terms of the Ricci tensor and the Ricci scalar R. Unlike high-dimensional general gravity and two-dimensional gravity, three-dimensional gravity exhibits no local dynamical degrees of freedom (e.g., gravitational waves). However, it admits global (topological) degrees of freedom, which lead to non-trivial physical phenomena. In particular, it has been found that three-dimensional gravity in the presence of a negative cosmological constant has global boundary degrees of freedom, hence non-trivial black hole solutions exist.

6.2 Killing Vectors

The conserved quantities in particle motion are fundamentally linked to the symmetries of spacetime. Specifically, time-translation symmetry gives rise to energy conservation, while space-translation symmetry ensures momentum conservation for a point particle. Similarly, rotational symmetry underlies the conservation of angular momentum. Consequently, analyzing the symmetries of spacetime carries profound physical implications. In curved spacetime, however, global coordinates that cover the entire manifold are generally absent. Thus, it becomes essential to define coordinate-independent vectors to characterize the symmetries of the spacetime manifold.

Under a general coordinate transformation $x \to x'$, the metric tensor transforms as

$$g'_{\alpha\beta}(x'(x)) = \frac{\partial x^\mu}{\partial x'^\alpha}\frac{\partial x^\nu}{\partial x'^\beta}g_{\mu\nu}(x), \tag{6.67}$$

or equivalently

$$g_{\mu\nu}(x) = \frac{\partial x'^\alpha}{\partial x^\mu}\frac{\partial x'^\beta}{\partial x^\nu}g'_{\alpha\beta}(x'(x)). \tag{6.68}$$

Here x' and x describe the same physical event in different coordinate charts. An isometry is a transformation under which the functional form of the metric remains unchanged, i.e., $g'_{\alpha\beta}(x'(x)) = g_{\alpha\beta}(x')$. Such a transformation preserves distances and thus defines a symmetry of the spacetime. The set of all isometries forms a group, known as the isometry group. The generators of this group can be obtained by considering an infinitesimal coordinate transformation. Consider an infinitesimal coordinate transformation

$$x'^\alpha = x^\alpha + \epsilon\xi^\alpha(x), \tag{6.69}$$

where ϵ is an infinitesimal parameter and $\xi^\alpha(x)$ is the generator of the transformation. Substituting this into the metric transformation law and expanding to first order in ϵ, we obtain

$$\begin{aligned} g_{\mu\nu}(x) &= (\delta^\alpha_\mu + \epsilon\partial_\mu\xi^\alpha)(\delta^\beta_\nu + \epsilon\partial_\nu\xi^\beta)(g_{\alpha\beta} + \epsilon\xi^\gamma\partial_\gamma g_{\alpha\beta}) \\ &= g_{\mu\nu} + \epsilon\delta^\beta_\nu\partial_\mu\xi^\alpha g_{\alpha\beta} + \epsilon\delta^\alpha_\mu\partial_\nu\xi^\beta g_{\alpha\beta} + \epsilon\delta^\beta_\nu\partial^\alpha_\mu\xi^\gamma\partial_\gamma g_{\alpha\beta}. \end{aligned}$$

For the metric to remain invariant, the first-order term must vanish, which leads to

$$\begin{aligned} &0 = \epsilon(g_{\alpha\nu}\partial_\mu\xi^\alpha + g_{\beta\mu}\partial_\nu\xi^\beta + \partial_\gamma g_{\mu\nu}\xi^\gamma) \\ &\Rightarrow \partial_\mu\xi_\nu + \partial_\nu\xi_\mu - \xi^\alpha\partial_\mu g_{\alpha\nu} - \xi^\beta\partial_\nu g_{\beta\mu} + \xi^\gamma\partial_\gamma g_{\mu\nu} = 0 \\ &\Rightarrow 2\partial_{\{\mu}\xi_{\nu\}} - 2\xi_\gamma\Gamma^\gamma_{\mu\nu} = 0 \\ &\Rightarrow \xi_{\nu;\mu} + \xi_{\mu;\nu} = 0. \end{aligned} \tag{6.70}$$

Thus, the generator of the infinitesimal transformation for isometry satisfies

$$\nabla_\mu \xi_\nu + \nabla_\nu \xi_\mu = 0. \tag{6.71}$$

This equation is known as the Killing equation.

Definition A Killing vector is the infinitesimal generator of the isometry group, satisfying the Killing equation

$$\nabla_\mu \xi_\nu + \nabla_\nu \xi_\mu = 0. \tag{6.72}$$

Using the Lie derivative, the Killing equation can be expressed compactly as

$$(\pounds_{\widehat{\xi}}\, g)_{\mu\nu} = 0, \tag{6.73}$$

where $\pounds_{\widehat{\xi}}$ denotes the Lie derivative along the vector field $\widehat{\xi}$. Let $\phi_t : M \to M$ be the one-parameter diffeomorphisms (or flow) generated by the Killing vector $\widehat{\xi}$. Equation (6.73) implies that the metric is invariant under the action of ϕ_t, meaning the local geometry remains unchanged as we propagate along the integral curves of $\widehat{\xi}$. Geometrically, the Killing vector field represents a direction along which the manifold's metric structure remains invariant.

Killing vectors care associated with specific symmetries of the spacetime manifold. To analyze them effectively, we can adopt a suitably chosen coordinate system that simplifies their description. Specifically, for any given Killing vector $\widehat{\xi}$, there always exists a coordinate system $\{\widetilde{x}\}$ in which the components of the Killing vector take the particularly simple form

$$\widetilde{\xi}^1 = [\text{constant}] = b, \quad \widetilde{\xi}^\mu = 0 \ (\mu \neq 1). \tag{6.74}$$

This can be proven as follows: Suppose $\{\xi^\mu\}$ are the components of the Killing vector in an arbitrary coordinate system $\{x^\mu\}$. Under a coordinate transformation, the vector components transform as:

$$\xi^\mu = \frac{\partial \widetilde{x}^\mu}{\partial x^\alpha}\xi^\alpha \Rightarrow \begin{cases} \dfrac{\partial \widetilde{x}^1}{\partial x^\alpha}, & \xi^\alpha = b, \\ \dfrac{\partial \widetilde{x}^\mu}{\partial x^\alpha}, & \xi^\alpha = 0, \quad \mu \neq 1. \end{cases} \tag{6.75}$$

This constitutes a system of linear partial differential equations with coefficients ξ^α, which is always solvable. Therefore, we can always find a coordinate system $\{\widetilde{x}\}$ in which the components of the Killing vector take the form (6.74). Furthermore, in this coordinate system, the first equation of the Killing Eq. (6.70) gives

$$\frac{\partial \widetilde{g}_{\mu\nu}}{\partial \widetilde{x}^1} = 0, \tag{6.76}$$

indicating that the metric components $\widetilde{g}_{\mu\nu}$ are independent of $\widetilde{x}^1$. Consequently, finite transformation along $\widetilde{x}^1$ correspond to isometries of the metric. This leads to a practical criterion for identifying Killing vectors: If the metric $g_{\mu\nu}$ exhibits no dependence on a particular coordinate, then the vector field generating translations along that coordinate is a Killing vector. While such a "symmetry-adapted" coordinate system always exists in principle, it may not always be the most convenient choice for computational purposes. Nevertheless, its existence underscores the deep connection between Killing vectors and spacetime symmetries.

Example 6.6. Killing vectors in three-dimensional Euclidean space $\mathbf{R}^3$.

The metric of three-dimensional Euclidean space R^3 is $\mathrm{d}s^2 = \mathrm{d}x^2 + \mathrm{d}y^2 + \mathrm{d}z^2$. Clearly, we have the Killing vectors

$$X^\mu = (1,0,0), \quad Y^\mu = (0,1,0), \quad Z^\mu = (0,0,1). \tag{6.77}$$

Euclidean space should also have rotational invariance. To see this, we switch to spherical coordinates, where the metric is

$$\mathrm{d}s^2 = \mathrm{d}r^2 + r^2\mathrm{d}\theta^2 + r^2\sin^2\theta\mathrm{d}\phi^2. \tag{6.78}$$

It is easy to see that $\widehat{R} = \partial_\phi = -y\partial_x + x\partial_y$ is a Killing vector, representing rotation around the z-axis. Its components in Cartesian coordinates are

$$R^\mu = (-y, x, 0). \tag{6.79}$$

Similarly, we can obtain the Killing vectors corresponding to rotations around the x-axis and y-axis:

$$S^\mu = (z, 0, -x), \quad T^\mu = (0, -z, y). \tag{6.80}$$

Example 6.7. Four-dimensional flat spacetime.

In Cartesian coordinates, the Minkowski metric is $\eta_{\mu\nu}$, whose components are independent of all coordinates. Consequently, the four translation generators along each coordinate direction are Killing vectors, corresponding to spacetime translations. In spherical coordinates, additional symmetries become apparent—namely, rotational invariance (including both spatial rotations and Lorentz boosts). These transformations generate the Lorentz group. Combining these results, we conclude that the full isometry group of four-dimensional Minkowski spacetime is the Poincaré group.

The existence of Killing vectors on a spacetime manifold allows us to define conserved quantities associated with them. To illustrate this, consider the motion of a massive particle with 4-momentum $p^\mu = mu^\mu$. The particle's geodesic is governed by the equation $p^\lambda \nabla_\lambda p_\mu = 0$, which expands to

$$p^\lambda \partial_\lambda p_\mu - \Gamma^\alpha_{\lambda\mu} p^\lambda p_\alpha = 0. \tag{6.81}$$

The first term in (6.81), $p^\lambda \partial_\lambda p_\mu = m \frac{\mathrm{d}x^\lambda}{\mathrm{d}\tau} \partial_\lambda p_\mu = m \frac{\mathrm{d}p_\mu}{\mathrm{d}\tau}$, describes the change in momentum components along the geodesic, while the second term can be rewritten using the Christoffel symbols:

$$\begin{aligned}
\Gamma^\alpha_{\lambda\mu} p^\lambda p_\alpha &= \frac{1}{2} g^{\sigma\alpha} (\partial_\lambda g_{\sigma\mu} + \partial_\mu g_{\lambda\sigma} - \partial_\sigma g_{\lambda\mu}) p^\lambda p_\alpha \\
&= \frac{1}{2} (\partial_\lambda g_{\sigma\mu} + \partial_\mu g_{\lambda\sigma} - \partial_\sigma g_{\lambda\mu}) p^\lambda p^\sigma \\
&= \frac{1}{2} \partial_\mu g_{\lambda\sigma} p^\lambda p^\sigma.
\end{aligned} \tag{6.82}$$

Thus, the geodesic equation simplifies to

$$m \frac{\mathrm{d}p_\mu}{\mathrm{d}\tau} = \frac{1}{2} \partial_\mu g_{\lambda\sigma} p^\lambda p^\sigma. \tag{6.83}$$

If the metric $g_{\lambda\sigma}$ is independent of a certain coordinate x^μ, then the corresponding momentum p_μ remains conserved along the geodesic,

$$\partial_\mu g_{\lambda\sigma} = 0 \quad \Rightarrow \quad \frac{\mathrm{d}p_\mu}{\mathrm{d}\tau} = 0. \tag{6.84}$$

This demonstrates that the Killing vectors of the spacetime manifold yield conserved quantities for particle motion. More generally, we can examine the evolution of the inner product between a Killing vector $\widehat{\xi}$ and the particle's 4-momentum along the geodesic:

$$\begin{aligned}
p^\mu \nabla_\mu (\xi_\nu p^\nu) &= p^\mu \xi_\nu \nabla_\mu p^\nu + p^\mu p^\nu \nabla_\mu \xi_\nu \\
&= p^\mu p^\nu \nabla_\mu \xi_\nu \\
&= p^\mu p^\nu \nabla_{(\mu} \xi_{\nu)}.
\end{aligned} \tag{6.85}$$

From the Killing equation, it follows that

$$\nabla_{(\mu} \xi_{\nu)} = 0 \Rightarrow p^\mu \nabla_\mu (\widehat{\xi} \cdot \widehat{p}) = 0, \tag{6.86}$$

implying that the inner product $\widehat{\xi} \cdot \widehat{p}$ is a conserved quantity.

Theorem If a spacetime manifold has a Killing vector field $\widehat{\xi}$, and a particle moving along a geodesic (whether massive or massless) has 4-momentum $\widehat{p}$, then $\widehat{\xi} \cdot \widehat{p}$ is a conserved quantity along the geodesic,

$$\frac{\mathrm{d}}{\mathrm{d}\lambda} (\widehat{\xi} \cdot \widehat{p}) = 0. \tag{6.87}$$

Each Killing vector field on a spacetime manifold corresponds to a conserved quantity along geodesic motion. This property is profoundly useful when analyzing particle dynamics in curved spacetime. By definition, the metric remains invariant along the direction of a Killing vector ξ^μ. Physically, this means that a freely moving particle experiences no force in the ξ^μ-direction, and consequently, the corresponding momentum component is conserved. This generalizes familiar conservation laws from flat spacetime: Time-translation invariance implies energy conservation; Spatial-translation invariance implies linear momentum conservation; Rotational invariance implies angular momentum conservation. In each case, the conserved quantity arises precisely because the metric is invariant along the symmetry direction—exactly as described by the Killing vector formalism.

Furthermore, we can define Killing tensors, which satisfy

$$\nabla_{(\mu} K_{\nu_1 \nu_2 \cdots \nu_l)} = 0. \tag{6.88}$$

Analogous to Killing vectors, these tensors yield conserved quantities along geodesics:

$$p^\mu \nabla_\mu (K_{\nu_1 \nu_2 \cdots \nu_l} p^{\nu_1} \cdots p^{\nu_l}) = 0. \tag{6.89}$$

Thus, for a particle with 4-momentum p^μ, the scalar $K_{\nu_1 \nu_2 \cdots \nu_l} p^{\nu_1} \cdots p^{\nu_l}$ remains constant during free-fall motion. A canonical example of a Killing tensor arises from the symmetrized tensor product of Killing vectors.

Since Killing vectors represent the symmetries of spacetime manifolds, they can be used not only to define conserved quantities for point particles but also to establish conserved quantities for the entire spacetime. Given a timelike Killing vector $\widehat{\xi}$, we may define the associated current

$$J_T^\mu = \xi_\nu T^{\mu\nu} \Rightarrow \nabla_\mu J_T^\mu = 0. \tag{6.90}$$

This current leads to a conserved energy

$$E_T = \int_\Sigma J_T^\nu n_\mu \sqrt{\gamma} \mathrm{d}^3 x, \tag{6.91}$$

which remains invariant for any spacelike hypersurface Σ. Similarly, the presence of spacelike Killing vectors allows for the definition of conserved momentum or angular momentum components.

Before closing this section, we present two key identities involving Killing vectors:

$$\begin{aligned} \nabla_\mu \nabla_\sigma \xi_\rho &= R^\rho{}_{\sigma\mu\nu} \xi^\nu, \\ \xi^\lambda \nabla_\lambda R &= 0. \end{aligned} \tag{6.92}$$

These identities play an important role in subsequent discussions, and their derivations are left as instructive exercises for the reader.

6.3 Maximally Symmetric Spaces

Killing vectors serve as the infinitesimal generators of spacetime isometries, forming a closed Lie algebra under the commutator operation. While these are properly understood as vector fields rather than constant vectors, they exhibit rich algebraic structure. For an $n \geq 2$ dimensional manifold, the number of linearly independent Killing vector fields may exceed n. At any given point, these vectors may become linearly dependent. For $\sum_n c_n \widehat{V}^{(n)}$, if $c_n \in R$ and $\widehat{V}^{(n)}$ are Killing vectors, then the combination remains a Killing vector. However, this property fails for function-valued coefficients $c_n \in \mathcal{F}(M)$. Nevertheless, if $\widehat{X}$ and $\widehat{Y}$ are Killing vectors, their commutator $[\widehat{X}, \widehat{Y}]$ is still a Killing vector. This demonstrates that the collection of all Killing vector fields constitutes a Lie algebra, specifically the Lie algebra of the isometry group of the spacetime manifold. Below we provide a proof that the commutator preserves the Killing property.

Theorem The commutator of two Killing vector fields is still a Killing vector field.

Proof Let $\widehat{X}$ and $\widehat{Y}$ be Killing vector fields, and their commutator is

$$[\widehat{X}, \widehat{Y}]^\mu = [X^\sigma \partial_\sigma, Y^\rho \partial_\rho]^\mu = X^\sigma \partial_\sigma Y^\mu - Y^\sigma \partial_\sigma X^\mu. \tag{6.93}$$

Then we consider

$$\begin{aligned}
\nabla^{(\nu}[\widehat{X}, \widehat{Y}]^{\mu)} &= \nabla^\nu (X^\sigma \partial_\sigma Y^\mu - Y^\sigma \partial_\sigma X^\mu) + (\nu \leftrightarrow \mu) \\
&= (\nabla^\nu X^\sigma)(\nabla_\sigma Y^\mu) + X^\sigma \nabla^\nu \nabla_\sigma Y^\mu \\
&\quad - (\nabla^\nu Y^\sigma)(\nabla_\sigma X^\mu) - Y^\sigma \nabla^\nu \nabla_\sigma X^\mu + (\nu \leftrightarrow \mu) \\
&= -(\nabla^\sigma X^\nu)(\nabla_\sigma Y^\mu) - X^\sigma \nabla^\nu \nabla^\mu Y_\sigma \\
&\quad + (\nabla^\sigma Y^\nu)(\nabla_\sigma X^\mu) + Y^\sigma \nabla^\nu \nabla^\mu X_\sigma + (\nu \leftrightarrow \mu) \\
&= -X^\sigma R_\sigma{}^{\nu\mu}{}_\rho Y^\rho + Y^\sigma R_\sigma{}^{\nu\mu}{}_\rho X^\rho + (\nu \leftrightarrow \mu) \\
&= 0.
\end{aligned} \tag{6.94}$$

Therefore, $[\widehat{X}, \widehat{Y}]$ is indeed a Killing vector field. The set of Killing vector fields $\{\widehat{X}\}$ forms the Lie algebra of the isometry group.

The n-dimensional flat space $\mathbf{R}^n$ possesses a complete set of isometries generated by: n translations and $\dfrac{n(n-1)}{2}$ independent rotations. Fixing a point P, there are at most n translational generators and $\frac{n(n-1)}{2}$ possible rotations, each corresponding to the rotation in a plane formed by pairs of coordinate axes. This yields the maximal number of Killing vector fields for flat space

$$n + \frac{n(n-1)}{2} = \frac{n(n+1)}{2}. \tag{6.95}$$

While the local geometry of an n-dimensional curved spacetime approximates flat space, global symmetries are generally reduced. The flat spacetime case represents the maximal possible symmetry, leading to the definition: A spacetime is maximally symmetric if it admits exactly $\frac{n(n+1)}{2}$ independent Killing vector fields, which is the maximum number possible for an n-dimensional manifold.

Theorem An n-dimensional spacetime admits at most $n(n+1)/2$ linearly independent Killing vector fields.

Proof A Killing vector field $\widehat{\xi}$ satisfies $\nabla_{(\mu}\xi_{\nu)} = 0$. The second covariant derivative is given by the curvature relation:

$$\nabla_\sigma \nabla_\rho \xi_\mu = -R^\lambda{}_{\sigma\rho\mu}\xi_\lambda, \tag{6.96}$$

which implies that all higher derivatives of the Killing vector field $\widehat{\xi}$ at point X are completely determined by the linear combinations of ξ_σ and $\xi_{\sigma;\rho}$. Consequently, the Taylor expansion of $\xi_\mu(x)$ about X takes the form

$$\xi_\rho(x) = A_\rho^\lambda(x; X)\xi_\lambda(X) + B_\rho^{\lambda\nu}(x; X)\xi_{\lambda;\nu}(X), \tag{6.97}$$

which only includes ξ_σ and $\xi_{\sigma;\rho}$, where the coefficients A and B depend only on the metric $g_{\mu\nu}$ and the base point X, and are therefore universal for all Killing vector fields. The maximal number of linearly independent Killing vectors is determined by the number of independent parameters in the initial data at X:

(1) The values $\xi_\lambda(X)$ provide n independent components;
(2) The antisymmetric combination $\xi_{[\lambda;\nu]}(X)$ (equivalent to $\xi_{\lambda;\nu}(X)$ due to the Killing equation) yields $\frac{n(n-1)}{2}$ independent components.

The total dimension of the solution space is thus bounded by $\dfrac{n(n+1)}{2}$.

A maximally symmetric space is both homogeneous and isotropic at every point. Homogeneity means that the space looks the same at every point—no point is geometrically distinct from any other. Mathematically, this implies that there exist Killing vector fields that can map any given point to any other point in its neighborhood. Isotropy means that the space looks the same in all directions at any given point. Formally, while keeping a point X fixed, $\xi^\lambda(X) = 0$, the derivative $\xi_{\lambda;\nu}(X)$ can take all possible values, generating rotational symmetry.

Proposition A maximally symmetric space is necessarily a space of constant curvature, meaning its Riemann curvature tensor must take the form:

$$R_{\lambda\rho\sigma\nu} = K(g_{\lambda\sigma}g_{\rho\nu} - g_{\lambda\nu}g_{\rho\sigma}), \tag{6.98}$$

where K is a constant determining the curvature scale,

$$K = \frac{R}{n(n-1)}. \tag{6.99}$$

Proof Suppose $\hat{\xi}$ is a Killing vector field, whose covariant derivative is also a tensor. For the tensor $\nabla_\sigma \xi_\rho$, we have

$$(\nabla_\mu \nabla_\nu - \nabla_\nu \nabla_\mu)\nabla_\sigma \xi_\rho = R^\lambda{}_{\rho\mu\nu}\nabla_\sigma \xi_\lambda + R^\lambda{}_{\sigma\mu\nu}\nabla_\lambda \xi_\rho. \tag{6.100}$$

Using the Killing equation $\nabla_\sigma \xi_\rho = -\nabla_\rho \xi_\sigma$, and the previously mentioned equation $\nabla_\mu \nabla_\nu \xi_\rho = R^\lambda{}_{\mu\nu\rho}\xi_\lambda$, we can obtain from (6.100):

$$\begin{aligned}
0 &= \nabla_\mu(R^\lambda{}_{\nu\sigma\rho}\xi_\lambda) - \nabla_\nu(R^\lambda{}_{\mu\sigma\rho}\xi_\lambda) - R^\lambda{}_{\rho\mu\nu}\nabla_\sigma \xi_\lambda - R^\lambda{}_{\sigma\mu\nu}\nabla_\lambda \xi_\rho \\
&= (\nabla_\mu R^\lambda{}_{\nu\sigma\rho} - \nabla_\nu R^\lambda{}_{\mu\sigma\rho})\xi_\lambda \\
&\quad + R^\lambda{}_{\nu\sigma\rho}\nabla_\mu \xi_\lambda - R^\lambda{}_{\mu\sigma\rho}\nabla_\nu \xi_\lambda - R^\lambda{}_{\rho\mu\nu}\nabla_\sigma \xi_\lambda + R^\lambda{}_{\sigma\mu\nu}\nabla_\rho \xi_\lambda \\
&= (\nabla_\mu R^\lambda{}_{\nu\sigma\rho} - \nabla_\nu R^\lambda{}_{\mu\sigma\rho})\xi_\lambda \\
&\quad + (R^\lambda{}_{\nu\sigma\rho}\delta^\gamma_\mu - R^\lambda{}_{\mu\sigma\rho}\delta^\gamma_\nu - R^\lambda{}_{\rho\mu\nu}\delta^\gamma_\sigma + R^\lambda{}_{\sigma\mu\nu}\delta^\gamma_\rho)\nabla_\gamma \xi_\lambda.
\end{aligned} \tag{6.101}$$

The above discussion applies to any spacetime. The final relation includes terms proportional to the Killing vector and terms related to the first covariant derivative of the Killing vector. If we consider a maximally symmetric space, the previous discussion tells us that the number of Killing vectors plus the number of independent covariant derivatives of the Killing vectors equals the number of Killing vectors in the maximally symmetric space. Therefore, (6.101) should not link the Killing vector and its covariant derivative, otherwise these quantities would not be linearly independent. This means that the condition for (6.101) to hold in a maximally symmetric space is that the coefficients in front of the Killing vector and its covariant derivative must both be zero. Using the definition of the Killing vector, we know that its covariant derivative symmetrized does not give a restriction, so we only need to consider the coefficients in front of its antisymmetric part. This gives us

$$\begin{gathered}
\nabla_\mu R^\lambda{}_{\nu\sigma\rho} = \nabla_\nu R^\lambda{}_{\mu\sigma\rho}, \\
R^\lambda{}_{\nu\sigma\rho}\delta^\gamma_\mu - R^\lambda{}_{\mu\sigma\rho}\delta^\gamma_\nu - R^\lambda{}_{\rho\mu\nu}\delta^\gamma_\sigma + R^\lambda{}_{\sigma\mu\nu}\delta^\gamma_\rho - (\lambda \leftrightarrow \gamma) = 0.
\end{gathered} \tag{6.102}$$

By appropriately contracting the second equation of (6.102), we find

$$\begin{aligned}(n-1)R^{\lambda}{}_{\nu\sigma\rho} &= R_{\nu\rho}\delta^{\lambda}_{\sigma} - R_{\nu\sigma}\delta^{\lambda}_{\rho},\\ nR^{\mu}{}_{\rho} &= R\delta^{\mu}_{\rho}, \quad R = \text{constant}.\end{aligned} \tag{6.103}$$

This is actually equivalent to the relationship between the Riemann tensor and the Ricci scalar obtained above.

A space of constant curvature is uniquely determined by its curvature scalar K. In a fixed dimension, aside from flat spacetime, there exist other maximally symmetric spacetimes. For Riemannian manifolds, these can be classified as follows:

(1) $K = 0$, Euclidean flat space.
(2) $K > 0$, sphere ($R > 0$, positive constant curvature). In spherical coordinates, its scalar curvature scales as $R \propto 1/a^2$, where a is the radius of the sphere. This dependence can be deduced from dimensional analysis: The metric field is dimensionless, the Riemann tensor has the dimension of [length]$^{-2}$, and since the only intrinsic scale of the sphere is its radius, the scalar curvature must be inversely proportional to the square of the radius.
(3) $K < 0$, hyperbolic space. This space has constant negative curvature. A common parametrization is the Poincaré metric

$$\mathrm{d}s^2 = \frac{L^2}{r^2}(\mathrm{d}r^2 + \mathrm{d}x^2), \tag{6.104}$$

where L sets the curvature scale and is known as the radius of the hyperbolic space. Of course, since the hyperbolic space is non-compact, L is just a scale of this space and does not represent a physical radius. The scalar curvature in this case is $R \propto -1/L^2$. A typical example of hyperbolic space is $\mathbf{H}^2$: the Poincaré upper half-plane or the Lobachevsky space.

In cosmology, assuming the cosmological principle (i.e., homogeneity and isotropy on large scales), the spatial geometry of the universe must fall into one of the three constant-curvature classes discussed above. These correspond to three distinct cosmological models: flat universe ($K = 0$), closed universe ($K > 0$), and open universe ($K < 0$), respectively.

For pseudo-Riemannian manifolds (with Lorentzian signature), the classification of constant-curvature spacetimes is as follows:

(1) If $R = 0$, it corresponds to Minkowski spacetime.

(2) If $R > 0$, it corresponds to de Sitter spacetime, which has important applications in cosmology. In the standard model of cosmology, the very early universe underwent a period of inflation, driven by a dominant effective cosmological constant, resulting in a near-exponential expansion. This phase can be approximated as an approximately de Sitter epoch. In the current era,observational evidence indicates that the universe is undergoing accelerated expansion, primarily driven by dark energy. In the far future, the universe is expected to evolve into an asymptotically de Sitter spacetime. However, de Sitter spacetime exhibits several peculiar properties: It possesses a cosmological horizon, which allows for the definition of temperature and entropy. Due to the presence of this horizon, the physics within de Sitter spacetime is highly unconventional, leading to only a finite number of observable degrees of freedom for any given observer.

(3) If $R < 0$, it corresponds to anti-de Sitter spacetime. Anti-de Sitter spacetime also possesses profound physical significance, particularly in the context of quantum gravity. A major breakthrough in recent years has been the discovery of the AdS/CFT correspondence, which establishes a remarkable duality: Quantum gravity in anti-de Sitter spacetime is holographically equivalent to a conformal field theory defined on its boundary. This correspondence provides a concrete realization of the holographic principle in quantum gravity and offers a fundamentally new non-perturbative definition of quantum gravity itself.

Exercises

1. Prove the multilinearity of the formally defined Riemann tensor.
2. Prove the conformal invariance of the Weyl tensor under the conformal transformation

$$\widetilde{g}_{\mu\nu} = \mathrm{e}^{2\omega(x)} g_{\mu\nu}. \tag{6.105}$$

   Further, verify that

$$\begin{aligned} \widetilde{R}_{\sigma\nu} = R_{\sigma\nu} &- \omega^{-1}\nabla_\alpha\nabla_\beta((n-2)\delta^\alpha_\nu\delta^\beta_\sigma + g_{\sigma\nu}g^{\alpha\beta}) \\ &+ \omega^{-2}\nabla_\alpha\omega\nabla_\beta\omega(2(n-2)\delta^\alpha_\nu\delta^\beta_\sigma - (n-3)g_{\sigma\nu}g^{\alpha\beta}), \end{aligned} \tag{6.106}$$

   and thus prove that the Einstein tensor is also invariant under conformal transformations.
3. Prove that the Cotton tensor has conformal weight $-5/2$ under conformal transformations.

4. Provide the relationship between the Riemann tensor and the Ricci tensor, as well as the scalar in three dimensions.
5. If an N-dimensional spacetime has coordinates $x^A = (x^\mu, x^m)$, where Greek indices take values $\mu = 1, \ldots, n$ and Latin indices take values $m = n+1, \ldots, N$, and the metric is block-diagonal, with $g_{\mu\nu}$ depending only on x^μ and g_{mn} depending only on x^n, prove that the Riemann tensor also has a similar decomposition.
6. Using coordinate transformations, we can locally choose $g_{00} = -1$ and $g_{0i} = 0$, so that the line element takes the form

$$\mathrm{d}s^2 = -\mathrm{d}t^2 + h_{ij}\mathrm{d}x^i\mathrm{d}x^j. \tag{6.107}$$

This coordinate system is called synchronous coordinates. Prove that the Ricci tensor is

$$\begin{aligned} R^0_0 &= -\frac{1}{2}\partial_t\Omega^i_i - \frac{1}{4}\Omega^i_j\Omega^j_i, \\ R^0_i &= \frac{1}{2}(\nabla_j\Omega^j_i - \nabla_i\Omega^j_j), \\ R^j_i &= -{}^{(3)}R^j_i - \frac{1}{2\sqrt{h}}\partial_t(\sqrt{h}\Omega^j_i), \end{aligned} \tag{6.108}$$

where $\Omega_{ij} = \partial_t h_{ij}$, and ${}^{(3)}R^j_i$ is the three-dimensional Ricci tensor constructed from h_{ij}. All index raising and lowering and covariant derivatives are with respect to the three-dimensional metric h_{ij}.
7. Find a coordinate transformation in flat spacetime that makes the Lorentz rotation symmetry explicit.
8. If $\widehat{\xi}$ is a timelike Killing vector, then we can define a 4-velocity along the Killing vector $u^\mu \equiv \xi^\mu/\sqrt{-\xi^\nu\xi_\nu} = \xi^\mu/N$, where N is the norm of the Killing vector. Prove that the 4-acceleration of this 4-velocity is

$$\widehat{a} = \nabla_{\widehat{u}}\widehat{u} = \frac{1}{2}\nabla \ln N. \tag{6.109}$$

9. Given a spacetime metric

$$\mathrm{d}s^2 = -2\mathrm{d}u\mathrm{d}v + a^2(u)\mathrm{d}x^2 + b^2(u)\mathrm{d}y^2, \tag{6.110}$$

where $u = t - z$ and $v = t + z$. This metric represents a gravitational wave propagating in the z-direction. Find all Killing vectors of this metric.
10. Prove the following two equations related to the Killing vector $\widehat{\xi}$:

$$\begin{aligned} \nabla_\mu\nabla_\sigma\xi^\rho &= R^\rho{}_{\sigma\mu\nu}\xi^\nu, \\ \xi^\lambda\nabla_\lambda R &= 0. \end{aligned} \tag{6.111}$$

11. Prove that under an infinitesimal coordinate transformation $x^\mu \to x^\mu + \xi^\mu(x)$, the change in the Christoffel symbols is

$$\delta\Gamma^\sigma_{\mu\nu} = -\frac{1}{2}\nabla_{(\mu}\nabla_{\nu)}\xi^\sigma + \frac{1}{2}R^\sigma{}_{(\mu\nu)\rho}\xi^\rho, \tag{6.112}$$

and thus, if ξ^μ is a Killing vector, then $\delta\Gamma^\sigma_{\mu\nu} = 0$.

12. For a two-dimensional sphere with the metric

$$\mathrm{d}s^2 = \mathrm{d}\theta^2 + \sin^2\theta \mathrm{d}\phi^2, \tag{6.113}$$

find all Killing vectors of this spacetime.

13. Calculate the scalar curvature of hyperbolic space from its metric in Poincaré coordinates.

Chapter 7

Einstein's Equations

In the preceding chapters, we have laid the mathematical foundation for general relativity by introducing the necessary concepts in differential geometry. In this chapter, we will delve into the derivation of Einstein's equations. To establish a physical intuition for these equations, we begin with an analysis of Newtonian tidal forces. This discussion serves as a motivational bridge, demonstrating how the geodesic deviation equation in general relativity naturally generalizes the notion of tidal forces—ultimately guiding us to the vacuum Einstein equations.

For the general case, we adopt a variational approach based on the principle of least action, which provides an elegant and systematic route to the full field equations. Additionally, this chapter explores the dynamics of matter fields in curved spacetime and examines the physical constraints imposed on the energy-momentum tensor.

7.1 Geodesic Deviation and Tidal Forces

the fifth postulate (the parallel postulate) asserts that, given a straight line and a point not on that line, there exists exactly one line through the point that is parallel to the original line. However, this postulate fails in non-Euclidean geometries. For example, on a two-dimensional sphere, all geodesics eventually intersect. To analyze how initially parallel geodesics diverge or converge, we examine the behavior of deviation vectors along the geodesic curves. For this purpose, we construct a one-parameter family of geodesics $\gamma_s(t)$, where for each real number $s \in \mathbf{R}$, γ_s is a geodesic parameterized by an affine parameter t. The collection of these curves forms a two-dimensional smooth surface, which can be embedded in a manifold M of arbitrary dimension. This surface, called a geodesic congruence, consists of the set of points $x^\mu(s,t) \in M$. The tangent vector to a geodesic in the congruence is given by

$$T^\mu = \frac{\partial x^\mu}{\partial t}, \tag{7.1}$$

while the deviation vector between two infinitesimally separated geodesics is defined as

$$S^\mu = \frac{\partial x^\mu}{\partial s}. \tag{7.2}$$

Here S^μ points from one geodesic to the adjacent one, as shown in Figure 7.1.

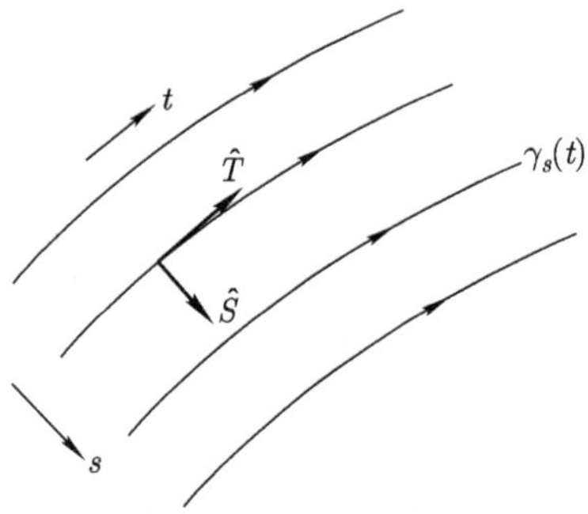

Fig. 7.1 Geodesic congruence composed of a single parameter family, where parameter s describes different geodesics, and t is the parameter describing geodesics. Thus, the directional derivative along the parameter t forms the tangent vector of the geodesic, while the adjacent geodesic can be characterized by the geodesic deviation vector.

To study how the deviation vector S^μ changes along the geodesic, we introduce the geodesic relative velocity and acceleration vectors: The covariant derivative of S^μ along T^μ defines the relative velocity

$$V^\mu = (\nabla_{\widehat{T}}\widehat{S})^\mu = T^\rho \nabla_\rho S^\mu, \tag{7.3}$$

and the covariant derivative of V^μ along T^μ defines the relative acceleration

$$a^\mu = (\nabla_{\widehat{T}}\widehat{V})^\mu = T^\rho \nabla_\rho V^\mu. \tag{7.4}$$

Since $\widehat{S}$ and $\widehat{T}$ are coordinate basis vectors, their commutator satisfy

$$[\widehat{S}, \widehat{T}] = 0.$$

Assuming a torsion-free connection, the covariant derivatives satisfy $\nabla_{\widehat{X}}\widehat{Y} - \nabla_{\widehat{Y}}\widehat{X} - [\widehat{X}, \widehat{Y}] = 0$, which simplifies to $\nabla_{\widehat{S}}\widehat{T} = \nabla_{\widehat{T}}\widehat{S}$, meaning

$$S^\rho \nabla_\rho T^\mu = T^\rho \nabla_\rho S^\mu. \tag{7.5}$$

The relative 4-acceleration a^μ can be expanded as

$$\begin{aligned}
a^\mu &= T^\rho\nabla_\rho(T^\sigma\nabla_\sigma S^\mu) \\
&= T^\rho\nabla_\rho(S^\sigma\nabla_\sigma T^\mu) \\
&= (T^\rho\nabla_\rho S^\sigma)(\nabla_\sigma T^\mu) + T^\rho S^\sigma\nabla_\rho\nabla_\sigma T^\mu \\
&= (S^\rho\nabla_\rho T^\sigma)(\nabla_\sigma T^\mu) + T^\rho S^\sigma(\nabla_\sigma\nabla_\rho T^\mu + R^\mu{}_{\nu\rho\sigma}T^\nu) \\
&= (S^\rho\nabla_\rho T^\sigma)(\nabla_\sigma T^\mu) + S^\rho\nabla_\sigma(T^\rho\nabla_\rho T^\mu) \\
&\quad - (S^\sigma\nabla_\sigma T^\rho)\nabla_\rho T^\mu + R^\mu{}_{\nu\rho\sigma}T^\nu T^\rho S^\sigma \\
&= R^\mu{}_{\nu\rho\sigma}T^\nu T^\rho S^\sigma.
\end{aligned} \tag{7.6}$$

Thus, we obtain the geodesic deviation equation.

Geodesic Deviation Equation

$$a^\mu = \frac{\mathrm{D}^2}{\mathrm{d}t^2}S^\mu = R^\mu{}_{\nu\rho\sigma}T^\nu T^\rho S^\sigma. \tag{7.7}$$

This equation is also known as the Jacobi equation. It reveals that the relative acceleration between two infinitesimally separated geodesics is governed by the Riemann curvature tensor.

In the above discussion, there are several remarkable points. First, the vanishing commutator of $\widehat{S}$ and $\widehat{T}$ implies that the vector fields are surface-filling. This arises because the two vectors can be expressed as derivatives of coordinate bases. Second, while $\widehat{S}$ and $\widehat{T}$ need not be orthogonal, we can always construct an orthogonal vector via projection:

$$\widehat{\eta} = \widehat{S} + \widehat{T}(\widehat{T}\cdot\widehat{S}), \tag{7.8}$$

which satisfies $\widehat{\eta}\cdot\widehat{T} = 0$. For $\widehat{\eta}$, the geodesic deviation equation retains the same form

$$\frac{\mathrm{D}^2}{\mathrm{d}t^2}\eta^\mu - R^\mu{}_{\nu\rho\sigma}T^\nu T^\rho\eta^\sigma = 0. \tag{7.9}$$

Consider now a freely falling observer's inertial frame, moving along the geodesic with 4-velocity $\widehat{T}$. In this frame, we may choose a tetrad $e^\mu_{\widehat{a}}$ such that $e^\mu_{\widehat{0}} = T^\mu$, $\{\widehat{e}_1, \widehat{e}_2, \widehat{e}_3\}$ are spacelike basis vectors orthogonal to $\widehat{T}$, and all the frame fields are parallel transported along the geodesic: $\dfrac{\mathrm{D}}{\mathrm{D}\tau}\widehat{e}_m = 0$. Within this frame, the deviation vector can be expressed as

$$\eta^m = e^m_\mu\eta^\mu, \tag{7.10}$$

where the time component vanishes due to orthogonality

$$\eta^0 = e^0_\mu\eta^\mu = T_\mu\eta^\mu = 0. \tag{7.11}$$

The remaining three spacelike components satisfy the geodesic deviation equation

$$\frac{\mathrm{D}^2}{\mathrm{d}\tau^2}\eta^i - R^\mu{}_{\nu\rho\sigma} e^i_\mu T^\nu T^\rho \eta^j e^\sigma_j = 0, \tag{7.12}$$

which can be compactly rewritten as

$$\frac{\mathrm{D}^2}{\mathrm{d}\tau^2}\eta^i + K^i{}_j \eta^j = 0, \tag{7.13}$$

with the tidal tensor defined by

$$K^i{}_j = -R^\mu{}_{\nu\rho\sigma} e^i_\mu T^\nu T^\rho e^\sigma_j. \tag{7.14}$$

This formulation bears a direct analogy to the Newtonian description of tidal forces.

In Newtonian mechanics, consider two particles moving along neighboring trajectories C_1 and C_2, as shown in Figure 7.2. Their positions at time t are given by

$$x^i(t), \quad x'^i(t) = x^i(t) + \eta^i(t), \tag{7.15}$$

where $\eta^i(t)$ is the deviation vector connecting the two curves. For the first particle, the equation of motion is governed by the gravitational potential $\Phi(\boldsymbol{x})$

$$\frac{\mathrm{d}^2 x^i}{\mathrm{d}t^2} = -\delta^{ij}\frac{\partial \Phi(\boldsymbol{x})}{\partial x^j}, \tag{7.16}$$

and for the second particle, expanding the potential to first order in η^i yields

$$\begin{aligned}\frac{\mathrm{d}^2 x'^i}{\mathrm{d}t^2} = \frac{\mathrm{d}^2(x^i + \eta^i)}{\mathrm{d}t^2} &= -\delta^{ij}\frac{\partial \Phi(\boldsymbol{x}+\boldsymbol{\eta})}{\partial x^j} \\ &= -\delta^{ij}\left(\frac{\partial \Phi(\boldsymbol{x})}{\partial x^j} + \frac{\partial}{\partial x^k}\frac{\partial \Phi(\boldsymbol{x})}{\partial x^j}\eta^k + \cdots\right).\end{aligned}$$

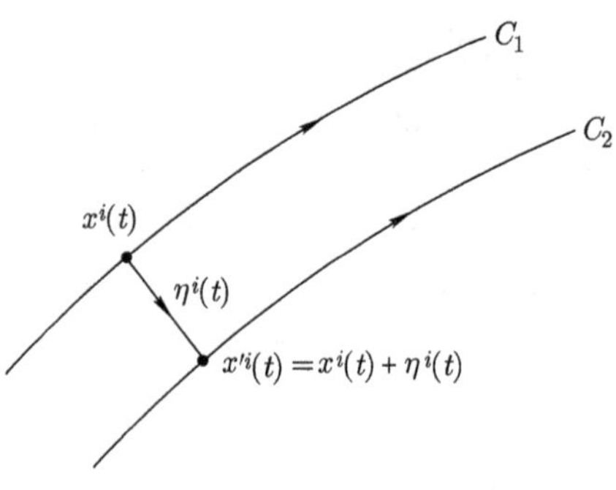

Fig. 7.2 Geodesic motion of two adjacent particles in Newtonian mechanics.

The relative acceleration of the two particles, which defines the Newtonian deviation equation, is

$$\ddot{\eta}^i = -\delta^{ij}\left(\frac{\partial^2\Phi}{\partial x^j \partial x^k}\right)_x \eta^k. \tag{7.17}$$

Introducing the tidal tensor

$$K^i{}_j \equiv \partial^i \partial_j \Phi, \tag{7.18}$$

the deviation equation takes the compact form

$$\ddot{\eta}^i + K^i{}_j \eta^j = 0. \tag{7.19}$$

The tensor K characterizes the differential acceleration, determining the force that pushes apart or pulls together adjacent particles. In gravitational physics, these forces are known as tidal forces, and K is correspondingly referred to as the tidal acceleration tensor. An important property arises from the Newtonian vacuum condition. From Laplace's equation for the gravitational potential in vacuum

$$\nabla^2 \Phi = 0, \tag{7.20}$$

we immediately deduce that the tidal tensor is traceless

$$K^i{}_i = 0. \tag{7.21}$$

Example 7.1. Tidal forces outside a spherically distributed mass.

The Newtonian gravitational potential in this case is $\Phi = -GM/r$, and the tidal acceleration is

$$a_{ij} = -\frac{\partial^2\Phi}{\partial x^i \partial x^j} = -(\delta_{ij} - 3n_i n_j)\frac{GM}{r^3}, \tag{7.22}$$

where $n^i = x^i/r$. In spherical coordinates (r, θ, ϕ),

$$a_{rr} = \frac{2GM}{r^3}, \quad a_{\theta\theta} = a_{\phi\phi} = -\frac{GM}{r^3}. \tag{7.23}$$

Clearly, the tensor K is traceless. For an observer freely falling radially, taking the above basis, we get

$$\frac{\mathrm{d}^2\eta^r}{\mathrm{d}t^2} = \frac{2GM}{r^3}\eta^r, \quad \frac{\mathrm{d}^2\eta^\theta}{\mathrm{d}t^2} = -\frac{GM}{r^3}\eta^\theta, \quad \frac{\mathrm{d}^2\eta^\phi}{\mathrm{d}t^2} = -\frac{GM}{r^3}\eta^\phi. \tag{7.24}$$

Thus, this observer will feel stretched in the radial direction and compressed in the transverse direction by the tidal forces.

Example 7.2. Tidal forces on Earth.

The tidal phenomena on Earth are mainly due to the influence of the Moon, although the Sun's influence is also significant. To simplify the problem, we only discuss the Moon's influence here. Suppose the Earth is at the origin $(0, 0, 0)$ of our chosen coordinate system, and the Moon is at $(0, 0, d)$ on the z-axis. The Newtonian gravitational potential of the Moon is

$$\Phi_m(z) = -\frac{GM_{\mathrm{m}}}{(x^2 + y^2 + (z-d)^2)^{1/2}}, \tag{7.25}$$

and the tidal acceleration is

$$\left.\left(\frac{\partial^2 \Phi}{\partial x^i \partial x^j}\right)\right|_0 = \frac{GM_{\mathrm{m}}}{d^3}\mathrm{diag}(1, 1, -2). \tag{7.26}$$

Consider a unit of ocean mass m located at $r = r(0, -\sin\theta, \cos\theta)$ in the (y, z)-plane, where $r \ll d$. The tidal force is

$$\boldsymbol{F}_{\mathrm{tidal}} = \frac{GmM_{\mathrm{m}}}{d^2}\frac{r}{d}(0, -\sin\theta, 2\cos\theta). \tag{7.27}$$

Along the z-axis, when $\theta = 0$, the force is in the $+z$ direction, and when $\theta = \pi$, the force is in the $-z$ direction. That is, the tidal force pulls the ocean apart along the z-direction. Conversely, the tidal force squeezes the ocean in the transverse direction.

Einstein's equivalence principle states that an astronaut inside a freely falling spacecraft experiences a local absence of gravity. While this suggests gravity is canceled for a point-like observer, real extended objects—such as a drop of liquid—reveal subtle gravitational effects through tidal forces. For a drop of liquid in free fall, if there are no external forces (including gravity), surface tension would maintain it in a perfect spherical shape. However, due to tidal forces, the water droplet in the spacecraft exhibits a slight but measurable deformation. Within the framework of general relativity, this deformation arises from geodesic deviation, which provides a geometric interpretation of tidal forces. A direct comparison between the geodesic deviation equations in general relativity and Newtonian mechanics yields

$$\partial^i \partial_j \Phi = K^i{}_j = -R^\mu{}_{\nu\rho\sigma} e^i_\mu T^\nu T^\rho e^\sigma_j, \tag{7.28}$$

where the Riemann tensor encodes the tidal forces experienced by an object (e.g., the water droplet) moving along a geodesic. The Riemann tensor can be decomposed into three fundamental components: the Weyl tensor, governing shape distortion due to tidal forces; the Ricci tensor, describing volume changes induced by spacetime curvature; and the Ricci scalar, contributing to the overall curvature scaling. Thus, while the Ricci tensor precisely determines how an object's volume changes under tidal forces, the Weyl tensor dictates how its shape is deformed—a manifestation of the non-local, differential nature of gravity in general relativity.

7.2 Einstein's Equations

By introducing a coordinate system comoving with the astronaut, where the basis vectors satisfy $e_i^\mu = \delta_i^\mu$ and the tangent vector is $T^\sigma = (1, 0, 0, 0)$, we derive

$$R^i{}_{00i} = 0, \tag{7.29}$$

a consequence of the traceless nature of the tensor K^i_j. Since this must hold as a tensor equation, we generalize it to

$$R^\rho{}_{00\rho} = 0, \tag{7.30}$$

which can be rewritten as

$$R^\rho{}_{00\rho} = R^\rho{}_{\mu\nu\rho} T^\mu T^\nu = -R_{\mu\nu} T^\mu T^\nu = 0. \tag{7.31}$$

This is a scalar relation and must therefore hold in all coordinate systems. Consequently, we arrive at

$$R_{\mu\nu} = 0.$$

This is known as the vacuum Einstein equations. Strictly speaking, the above argument is not a fully rigorous derivation. The underlying logic proceeds as follows: first, in vacuum, the Poisson equation $\nabla^2\Phi = 0$ imposes the traceless condition on $K^i{}_j$; This leads to the tensor equation

$$R_{\mu\nu} T^\mu T^\nu = 0, \tag{7.32}$$

which must hold for all observers, i.e., for arbitrary timelike vectors T^μ; The only consistent solution under this requirement is $R_{\mu\nu} = 0$, the vacuum Einstein equations.

Vacuum Einstein equations

$$R_{\mu\nu} = 0. \tag{7.33}$$

The vacuum Einstein equations consist of ten independent, highly nonlinear partial differential equations, making their exact solutions notoriously difficult to obtain. Despite their complexity, two key pieces of evidence strongly support their validity:

(1) The Newtonian limit. In the weak-field, low-velocity regime, general relativity must reduce to Newtonian gravity.
(2) The spherically symmetric Schwarzschild solution, which enables precise tests of general relativity, such as the perihelion precession of Mercury and the bending of light by gravity.

Here, we first discuss the Newtonian limit. To establish the Newtonian limit of general relativity, the following three conditions must be satisfied.

Newtonian limit: fundamental requirements:

(1) Weak gravity field. The spacetime metric exhibits only small deviations from flat Minkowski spacetime, expressible as: $g_{\mu\nu} = \eta_{\mu\nu} + h_{\mu\nu}$, where $h_{\mu\nu} = O(\epsilon)$;
(2) Slow temporal variation. The metric perturbations change adiabatically with time, characterized by: $\partial_0 h_{\mu\nu} = O(\epsilon^2)$;
(3) Non-relativistic motion. All particles move at velocities significantly slower than the speed of light, allowing the neglect of relativistic effects.

Since the perturbation field is a small quantity, we can obtain the inverse of the metric field

$$g^{\mu\nu} = \eta^{\mu\nu} - h^{\mu\nu}, \tag{7.34}$$

where $h^{\mu\nu} = \eta^{\mu\sigma}\eta^{\nu\rho}h_{\sigma\rho}$ is raised and lowered using the background flat spacetime metric. Thus, the Christoffel symbols are

$$\begin{aligned} \Gamma^{\mu}_{\sigma\rho} &= \frac{1}{2}g^{\mu\nu}(\partial_\sigma g_{\nu\rho} + \partial_\rho g_{\nu\sigma} - \partial_\nu g_{\sigma\rho}) \\ &= \frac{1}{2}\eta^{\mu\nu}(\partial_\sigma h_{\nu\rho} + \partial_\rho h_{\nu\sigma} - \partial_\nu h_{\sigma\rho}) + O(\epsilon^2), \end{aligned} \tag{7.35}$$

where we have explicitly written out the first-order small quantities. The geodesic equation reduces to

$$\frac{\mathrm{d}^2 x^\mu}{\mathrm{d}\tau^2} + \Gamma^{\mu}_{00}\left(\frac{\mathrm{d}t}{\mathrm{d}\tau}\right)^2 = 0. \tag{7.36}$$

Since $\Gamma^{\mu}_{00} = -\frac{1}{2}\eta^{\mu\lambda}\partial_\lambda h_{00}$, the geodesic equation becomes

$$\frac{\mathrm{d}^2 x^\mu}{\mathrm{d}\tau^2} = \frac{1}{2}\eta^{\mu\lambda}\partial_\lambda h_{00}\left(\frac{\mathrm{d}t}{\mathrm{d}\tau}\right)^2. \tag{7.37}$$

The $\mu = 0$ component gives

$$\frac{\mathrm{d}^2 t}{\mathrm{d}\tau^2} = 0. \tag{7.38}$$

That is, $\frac{dt}{d\tau}$ is a constant, and we can simply take $t = \tau$. Physically, since the particle is moving slowly, $\frac{dx^i}{d\tau} \ll \frac{dt}{d\tau}$ we can neglect the difference between coordinate time and proper time. For the $\mu = i$ component, we have

$$\frac{d^2x^i}{d\tau^2} = \frac{1}{2}\left(\frac{dt}{d\tau}\right)^2 \partial_i h_{00}, \tag{7.39}$$

which can be rewritten as

$$\frac{d^2x^i}{dt^2} = \frac{1}{2}\partial_i h_{00}. \tag{7.40}$$

Comparing this with the motion of a point particle in Newtonian gravitational potential, we have

$$\frac{d^2x^i}{dt^2} = -\partial^i\Phi, \tag{7.41}$$

we obtain the equivalence

$$h_{00} = -2\Phi, \tag{7.42}$$

or

$$g_{00} = -(1 + 2\Phi). \tag{7.43}$$

Furthermore, we can calculate the Riemann tensor and obtain its first-order non-zero component

$$R^\mu{}_{0\nu 0} = -\frac{1}{2}g^{\mu\sigma}\partial_\sigma\partial_\nu h_{00} + O(\epsilon^2). \tag{7.44}$$

The (00) component of $R_{\mu\nu} = 0$ gives

$$R^\mu{}_{0\mu 0} = -\frac{1}{2}\nabla^2 h_{00}, \tag{7.45}$$

which is actually the Laplace equation

$$\nabla^2\Phi = 0. \tag{7.46}$$

Thus, we see that if we regard the deviation of the metric field from flat spacetime (or perturbation) as the Newtonian gravitational potential, then the geodesic equation yields Newton's second law, and the vacuum Einstein equations yield the Poisson equation for the Newtonian gravitational potential in vacuum. In short, in the Newtonian limit, Einstein's theory of general relativity reverts to Newtonian gravity.

In the presence of matter distribution, the spacetime is no longer a vacuum. In Newtonian gravity, the gravitational potential Φ is governed by the Poisson equation

$$\nabla^2\Phi = 4\pi G\rho. \tag{7.47}$$

How can we generalize it to a relativistic tensor equation? Now we know that mass density is just one component of the energy-momentum tensor, and the gravitational potential is a component of the perturbation of the metric field, so we expect the following naive ansatz

$$\rho \to T_{\mu\nu}, \quad \nabla^2 \to \nabla^\mu\nabla_\mu, \quad \Phi \to g_{\mu\nu}? \tag{7.48}$$

However, this approach fails because the covariant derivative of the metric vanishes identically $\nabla_\mu g_{\sigma\rho} = 0$.

Historically, several approaches were proposed to generalize the Poisson Eq. (7.47) into a relativistic theory of gravity. One early idea was to retain the gravitational potential Φ within a scalar theory of gravity, modifying the Newtonian framework covariantly. For instance, in this scheme, the Laplacian ∇^2 was replaced by the covariant d'Alembert operator acting on Φ, while the source term was generalized to the trace of the energy-momentum tensor, i.e.,

$$\Box\Phi = 4\pi G T^\mu_\mu. \tag{7.49}$$

However, this theory did not agree with experiments, as its prediction for the perihelion precession of Mercury was smaller than the experimental results. Moreover, since the trace of the electromagnetic energy-momentum tensor $(T^{\mathrm{EM}})^\mu_\mu$ is zero, electromagnetic theory could not couple with gravity. In other words, in this theory, photons would not experience gravitational redshift, and light rays would not be deflected. Therefore, using a scalar field to describe gravity is incorrect. Another viewpoint considered gravity as a vector theory, but such a theory could only produce repulsive forces.

As a tensor theory, if the right-hand side of the field equation is replaced by the energy-momentum tensor, then the left-hand side must correspondingly be a (0, 2)-tensor involving second-order derivatives of the metric. The most natural candidate satisfying these requirements is the Ricci tensor $R_{\mu\nu}$, suggesting the provisional field equation: $R_{\mu\nu} = \kappa T_{\mu\nu}$. However, this ansatz violates the energy-momentum conservation law $\nabla^\mu T_{\mu\nu} = 0$. Fortunately, we only need to slightly modify this anstz to satisfy the energy-momentum conservation. The resolution lies in modifying the geometric side through the Einstein tensor $G_{\mu\nu}$, which inherently satisfies $\nabla^\mu G_{\mu\nu} = 0$. This leads to the conjectured form:

$$G_{\mu\nu} = R_{\mu\nu} - \frac{1}{2} g_{\mu\nu} R = \kappa T_{\mu\nu}. \tag{7.50}$$

There is an undetermined factor κ on the right-hand side of the equation, which can be determined through the Newtonian limit. First, taking the trace of both sides of Eq. (7.50), in an arbitrary D-dimensional spacetime,

$$R = \frac{2}{2-D} \kappa T, \tag{7.51}$$

where $T = T^\mu_\mu$ is the trace of the energy-momentum tensor. Substituting back gives the Ricci tensor expression

$$R_{\mu\nu} = \kappa \left(T_{\mu\nu} - \frac{1}{D-2} T g_{\mu\nu} \right). \tag{7.52}$$

For $D = 4$, this simplifies to

$$R_{\mu\nu} = \kappa \left(T_{\mu\nu} - \frac{1}{2} T g_{\mu\nu} \right). \tag{7.53}$$

On the other hand, considering dust-dominated matter where

$$T_{00} = \rho, \quad T = -\rho, \tag{7.54}$$

dominates the energy-momentum tensor, the Newtonian limit of the (00) component of the Einstein equation yields

$$R_{00} = -\frac{1}{2} \nabla^2 h_{00} = \frac{1}{2} \kappa T_{00}. \tag{7.55}$$

Matching with Poisson's equation fixes the coupling constant

$$\kappa = 8\pi G. \tag{7.56}$$

Finally, we arrive at the Einstein equation with the energy-momentum tensor:[1]

[1]The development of general relativity was a tumultuous intellectual journey. Between October–November 1915, Einstein refined his earlier framework through a series of breakthroughs, culminating in four seminal papers. These included the quantitative explanation of Mercury's perihelion advance (November 18) and the final formulation of the field equations (November 25). For detailed historical accounts, see Pais's authoritative biography [28].

Einstein equation

$$R_{\mu\nu} - \frac{1}{2}g_{\mu\nu}R = 8\pi G T_{\mu\nu}, \tag{7.57}$$

sometimes also written as

$$G_{\mu\nu} = 8\pi G T_{\mu\nu}. \tag{7.58}$$

In summary, our derivation begins with the conservation equation of the energy-momentum tensor and leads to the selection of the Einstein tensor. However, this approach contains an important subtlety: the metric-compatible connection ($\nabla_\mu g_{\sigma\rho} = 0$) we adopted allows for the inclusion of an additional term proportional to the metric tensor on the equation's left-hand side—the well-known cosmological constant term. This modification yields the field equations:

$$G_{\mu\nu} + \Lambda g_{\mu\nu} = 8\pi G T_{\mu\nu}. \tag{7.59}$$

It can be transformed into

$$R_{\mu\nu} = 8\pi G\left(T_{\mu\nu} - \frac{1}{2}T g_{\mu\nu}\right) + \Lambda g_{\mu\nu}. \tag{7.60}$$

Applying the Newtonian approximation to this framework gives

$$\nabla^2\Phi = 4\pi G_\rho - \Lambda. \tag{7.61}$$

For the gravitational acceleration outside a spherically symmetric mass distribution, we obtain:

$$\boldsymbol{g} = -\nabla\Phi = \left(-\frac{GM}{r^2} + \frac{\Lambda r}{3}\right)\boldsymbol{e}_r. \tag{7.62}$$

Therefore, if the cosmological constant is negative, it will induce a linearly increasing attractive force, and if the cosmological constant is positive, it will induce a repulsive force that increases linearly with distance.

Einstein originally introduced the cosmological constant term to achieve a static universe. However, this model proved to be unstable. The subsequent discovery of cosmic expansion by Hubble rendered the static universe obsolete, leading Einstein to famously dismiss the cosmological constant as "the greatest blunder of [his] life." Remarkably, modern cosmology has dramatically reversed this narrative. Observational evidence now overwhelmingly indicates that the universe is not only expanding but doing so at an accelerating rate. This acceleration is attributed to a mysterious component called dark energy, which constitutes about 68% of the universe's energy density. The simplest and most compelling explanation for dark energy is a positive cosmological constant—precisely the term Einstein had proposed and later rejected. Thus, what Einstein considered his greatest mistake has ironically become one of the most profound discoveries in cosmology, shaping our current understanding of the universe's evolution and composition.

7.3 Weak Field Approximation

The weak field approximation serves as a fundamental tool in general relativity, enabling the analysis of numerous physical scenarios where gravitational effects are relatively small. This approximation treats the spacetime metric as a slight perturbation about the flat Minkowski background:

$$g_{\mu\nu} = \eta_{\mu\nu} + h_{\mu\nu}, \quad |h_{\mu\nu}| \ll 1. \tag{7.63}$$

This framework is commonly called linearized gravity since it keeps only first-order terms in $h_{\mu\nu}$. The weak field approximation is a further generalization of the Newtonian approximation, relaxing the requirements that the perturbation in the Newtonian approximation is static and the motion of particles is slow, meaning that $h_{\mu\nu}$ can be time-dependent, and the motion of particles can be at the speed of light, allowing us to discuss the motion of photons.

To first-order approximation, the inverse of the metric field is $g^{\mu\nu} = \eta^{\mu\nu} - h^{\mu\nu}$, where $h^{\mu\nu} = \eta^{\mu\rho}\eta^{\nu\sigma}h_{\rho\sigma}$. In other words, the indices of the perturbation field are raised and lowered using the Minkowski metric $\eta^{\mu\nu}$ and its inverse $\eta_{\mu\nu}$. Consequently, linearized gravity can be interpreted as a theory of a symmetric tensor field $h_{\mu\nu}$ propagating in flat spacetime. This theory exhibits Lorentz invariance in the context of special relativity: while the flat spacetime metric is invariant, the perturbation field transforms as

$$h_{\mu'\nu'} = \Lambda^{\mu}_{\mu'}\Lambda^{\nu}_{\nu'}h_{\mu\nu}. \tag{7.64}$$

At linear order, the Christoffel symbols are

$$\begin{aligned}\Gamma^{\rho}_{\mu\nu} &= \frac{1}{2}g^{\rho\lambda}(\partial_\mu g_{\nu\lambda} + \partial_\nu g_{\lambda\mu} - \partial_\lambda g_{\mu\nu})\\ &= \frac{1}{2}\eta^{\rho\lambda}(\partial_\mu h_{\nu\lambda} + \partial_\nu h_{\lambda\mu} - \partial_\lambda h_{\mu\nu}).\end{aligned} \tag{7.65}$$

The Riemann tensor is

$$\begin{aligned}R_{\mu\nu\rho\sigma} &= \eta_{\mu\lambda}\partial_\rho\Gamma^{\lambda}_{\nu\sigma} - \eta_{\mu\lambda}\partial_\sigma\Gamma^{\lambda}_{\nu\rho}\\ &= \frac{1}{2}(\partial_\rho\partial_\nu h_{\mu\sigma} + \partial_\sigma\partial_\mu h_{\nu\rho} - \partial_\sigma\partial_\nu h_{\mu\rho} - \partial_\rho\partial_\mu h_{\nu\sigma}).\end{aligned}$$

The Ricci tensor is

$$R_{\mu\nu} = \frac{1}{2}(\partial_\sigma\partial_\nu h^{\sigma}_{\mu} + \partial_\sigma\partial_\mu h^{\sigma}_{\nu} - \partial_\mu\partial_\nu h - \Box h_{\mu\nu}), \tag{7.66}$$

where $h = \eta^{\mu\nu}h_{\mu\nu} = h^{\mu}_{\mu}$, and the d'Alembert operator is defined relative to the flat spacetime background, $\Box = -\partial_t^2 + \partial_x^2 + \partial_y^2 + \partial_z^2$. The Ricci scalar is

$$R = \partial_\mu\partial_\nu h^{\mu\nu} - \Box h. \tag{7.67}$$

The Einstein tensor is

$$G_{\mu\nu} = \frac{1}{2}(\partial_\sigma\partial_\nu h^\sigma_\mu + \partial_\sigma\partial_\mu h^\sigma_\nu - \partial_\mu\partial_\nu h - \Box h_{\mu\nu} - \eta_{\mu\nu}\partial_\mu\partial_\nu h^{\mu\nu} + \eta_{\mu\nu}\Box h).$$

This linearized Einstein tensor can be obtained from the following Lagrangian density (defined in Section 7.6) through variation:

$$\mathcal{L} = \frac{1}{2}\left[(\partial_\mu h^{\mu\nu})(\partial_\nu h) - (\partial_\mu h^{\rho\sigma})(\partial_\rho h^\mu_\sigma) + \frac{1}{2}\eta^{\mu\nu}(\partial_\mu h^{\rho\sigma})(\partial_\nu h_{\rho\sigma}) \right. \\ \left. -\frac{1}{2}\eta^{\mu\nu}(\partial_\mu h)(\partial_\nu h)\right].$$

The energy-momentum tensor $T_{\mu\nu}$ in Einstein's equations must itself be a first-order small quantity, as it governs the dynamics of the metric perturbation. Consequently, energy-momentum conservation reduces to the flat-space continuity equation, $\partial_\mu T^{\mu\nu} = 0$.

In full general relativity, the theory is invariant under general coordinate transformations (diffeomorphisms). However, in the weak-field approximation, these transformations are constrained to ensure that the metric retains its form as a small perturbation over the Minkowski background. Specifically, after a coordinate transformation, the metric should still decompose as $g'_{\mu\nu} = \eta_{\mu\nu} + h'_{\mu\nu}$, although the perturbation field $h'_{\mu\nu}$ may be different from the original one. Consider an infinitesimal coordinate transformation of the form

$$x'^\mu = x^\mu + \zeta^\mu(x) \quad \Rightarrow \quad \frac{\partial x'^\mu}{\partial x^\nu} = \delta^\mu_\nu + \partial_\nu\zeta^\nu, \tag{7.68}$$

it leads to

$$g'_{\mu\nu} = \frac{\partial x^\rho}{\partial x'^\mu}\frac{\partial x^\sigma}{\partial x'^\nu}g_{\rho\sigma} = \eta_{\mu\nu} + h_{\mu\nu} - \partial_\mu\zeta_\nu - \partial_\nu\zeta_\mu, \tag{7.69}$$

where $\zeta_\mu = \eta_{\mu\nu}\zeta^\nu$. Thus, the gauge transformation on the perturbation field is

$$h'_{\mu\nu} = h_{\mu\nu} - \partial_\mu\zeta_\nu - \partial_\nu\zeta_\mu. \tag{7.70}$$

It can be proven that under this transformation, the Riemann tensor is invariant, $\delta R_{\mu\nu\rho\sigma} = 0$.

Due to gauge invariance, we can choose an appropriate gauge to discuss the problem. A particularly useful choice is the harmonic gauge:[2] defined by the condition

$$\Box x^\mu = 0, \tag{7.71}$$

[2]Functions f that satisfy $\Box f = 0$ are called harmonic functions, which is why this gauge is named the harmonic gauge. In various literatures, this gauge is also often referred to as the Lorenz gauge, Einstein gauge, Hilbert gauge, de Donder gauge, or Fock gauge.

where x^μ should be treated as scalar coordinate functions. This is equivalent to requiring

$$\partial_\nu(\sqrt{-g}g^{\mu\nu}) = 0, \tag{7.72}$$

or equivalently,

$$g^{\sigma\rho}\Gamma^\mu_{\sigma\rho} = 0. \tag{7.73}$$

The harmonic gauge condition is not limited to perturbations in flat spacetime but can be naturally extended to weak-field approximations in general background geometries. In the weak field limit, the harmonic gauge condition reduces to

$$\frac{1}{2}\eta^{\mu\nu}\eta^{\lambda\rho}(\partial_\mu h_{\nu\lambda} + \partial_\nu h_{\lambda\mu} - \partial_\lambda h_{\mu\nu}) = 0, \tag{7.74}$$

which imposes the following constraint on the metric perturbation

$$\partial_\mu h^\mu_\lambda - \frac{1}{2}\partial_\lambda h = 0, \tag{7.75}$$

where $h = \eta^{\mu\nu}h_{\mu\nu}$ is the trace of the perturbation. The harmonic gauge does not completely fix all gauge degrees of freedom. A residual symmetry remains: the gauge condition is preserved under coordinate transformations of the form $x^\mu \to x'^\mu = x^\mu + \xi^\mu$, where the transformation vector ξ^μ satisfies the harmonic condition $\Box\xi^\mu = 0$. This means we retain the freedom to make harmonic coordinate adjustments while maintaining the gauge condition. When working in the harmonic gauge, the linearized Einstein field equations simplify significantly. The general form becomes:

$$\Box h_{\mu\nu} - \frac{1}{2}\eta_{\mu\nu}\Box h = -16\pi G T_{\mu\nu}. \tag{7.76}$$

For the vacuum case, these reduce to the particularly simple form:

$$\Box h_{\mu\nu} = 0. \tag{7.77}$$

This is precisely a relativistic wave propagation equation, showing that metric perturbations propagate as waves at the speed of light in vacuum.

In practical applications, the metric perturbation can be transformed to simplify the problem. We define the "trace-reversed" perturbation as

$$\bar{h}_{\mu\nu} = h_{\mu\nu} - \frac{1}{2}\eta_{\mu\nu}h, \tag{7.78}$$

whose trace satisfies $\bar{h}^\mu_\mu = -h^\mu_\mu$, justifying its name. The harmonic gauge condition now becomes

$$\partial_\mu \bar{h}^\mu{}_\lambda = 0. \tag{7.79}$$

It can be shown that suitable gauge parameters always exist to enforce this condition. Under a gauge transformation, the trace-reversed perturbation transforms as

$$\overline{h}'_{\mu\nu} = \overline{h}_{\mu\nu} - \partial_\mu \zeta_\nu - \partial_\nu \zeta_\mu + \eta_{\mu\nu}\partial_\lambda \zeta^\lambda, \tag{7.80}$$

leading to

$$\partial^\mu \overline{h}'_{\mu\nu} = \partial^\mu \overline{h}_{\mu\nu} - \Box \zeta_\nu. \tag{7.81}$$

Thus, by choosing $\Box \zeta_\nu = \partial^\mu \overline{h}_{\mu\nu}$, we ensure that the transformed perturbation satisfies the harmonic gauge condition (7.79).

Using the trace-reversed perturbation, the complete field equations simplify to

$$\Box \overline{h}_{\mu\nu} = -16\pi G T_{\mu\nu}, \tag{7.82}$$

while in vacuum, they reduce to

$$\Box \overline{h}_{\mu\nu} = 0. \tag{7.83}$$

Notably, these equations closely resemble those governing the electromagnetic gauge potential in the Lorenz gauge. This analogy allows us to leverage well-established techniques from electrodynamics—such as wave solutions, retarded potentials, and radiation theory—to analyze gravitational perturbations in the weak-field regime. An important point is that Eq. (7.82) is a linear differential equation, with the left side being the d'Alembert operator in flat spacetime. This linearity permits the use of superposition principles and Green's function methods to construct general solutions.

For a steady-state spherically symmetric source in the weak-field regime, where gravitational effects are small and the mass density ρ is not extreme, we recover the Newtonian limit. In this approximation, the time-time component of the metric perturbation satisfies

$$h_{00} = -2\Phi, \tag{7.84}$$

where Φ is the traditional Newtonian gravitational potential, given by $\Phi = -GM/r$. If the energy-momentum tensor of the source is dominated by its rest energy density $\rho = T_{00}$ with other components negligible, the Einstein field equations imply that $\overline{h}_{\mu\nu}$ is similarly dominated by its $\overline{h}_{00}$ component. Consequently,

$$h = -\overline{h} = -\eta^{\mu\nu}\overline{h}_{\mu\nu} = \overline{h}_{00}, \tag{7.85}$$

and

$$\overline{h}_{00} = 2h_{00} = -4\Phi. \tag{7.86}$$

Other components of $\bar{h}_{\mu\nu}$ can be neglected, so $\bar{h} = 4\Phi$, from which we get

$$h_{i0} = \bar{h}_{i0} - \frac{1}{2}\eta_{i0}\bar{h} = 0, \tag{7.87}$$

and

$$h_{ij} = \bar{h}_{ij} - \frac{1}{2}\eta_{ij}\bar{h} = -2\Phi\delta_{ij}. \tag{7.88}$$

Thus, in the weak field limit, the spacetime metric for a non-rotating star or planet takes the form

$$\mathrm{d}s^2 = -(1+2\Phi)\mathrm{d}t^2 + (1-2\Phi)(\mathrm{d}x^2 + \mathrm{d}y^2 + \mathrm{d}z^2). \tag{7.89}$$

If the object rotates slowly, its effect on spacetime can still be treated perturbatively within the weak-field approximation, which will be discussed in Chapter 10. In addition, the weak-field formalism also provides the foundation for analyzing gravitational radiation, which will be discussed in Chapter 11.

7.4 Review of the Equivalence Principle

Let us summarize our discussion of general relativity thus far. In general relativity, gravity is manifested as the curvature of spacetime. In curved spacetime, the motion of particles is geodesic motion. This can also be obtained from the action of a point particle in curved spacetime. On the other hand, the energy-momentum tensor causes the curvature of spacetime. This can be clearly seen from the Einstein equations: the left side of the equation is a purely geometric quantity, while the right side is proportional to the energy-momentum tensor.

The equivalence principle has been pivotal in the development of general relativity. It begins with the weak equivalence principle, rooted in the insights of Galileo and Newton, which states that an object's inertial mass and gravitational mass are equal. As a consequence, all objects respond identically to a gravitational field, accelerating uniformly according to the gradient of the gravitational potential. This universality of free-fall implies that the effects of gravity are locally indistinguishable from those of a uniformly accelerated reference frame—a foundational idea that shaped Einstein's thinking. Einstein elevated this principle further, proposing that no local experiment—whether mechanical, electromagnetic, or otherwise—can differentiate between a uniform gravitational field and constant acceleration. This stronger statement, known as Einstein's equivalence principle, asserts that in a sufficiently small spacetime region, the laws of physics reduce to those of special relativity, effectively "erasing" any detectable influence of gravity. Crucially, this principle redefines the notion of inertial frames. In general relativity, global inertial frames do not exist due to spacetime curvature; instead, we can only construct local inertial frames—those of freely falling observers over small regions. Mathematically, such frames correspond to Riemann normal coordinates, where the metric assumes the Minkowski form and the Christoffel symbols vanish, locally mimicking flat spacetime.

Einstein's equivalence principle dictates that physical laws in a local inertial frame must coincide with those of special relativity. This correspondence becomes manifest when employing Riemann normal coordinates, where physical laws adopt their flat spacetime form. In flat spacetime, the equation of motion for a free particle is given by

$$\frac{\mathrm{d}^2 x^\mu}{\mathrm{d}\lambda^2} = 0. \tag{7.90}$$

In curved spacetime, this generalizes to the geodesic equation

$$\frac{\mathrm{d}^2 x^\mu}{\mathrm{d}\lambda^2} + \Gamma^\mu_{\rho\sigma} \frac{\mathrm{d}x^\rho}{\mathrm{d}\lambda} \frac{\mathrm{d}x^\sigma}{\mathrm{d}\lambda} = 0, \tag{7.91}$$

which naturally reduces to Eq. (7.90) in Riemann normal coordinates. This example inspires us on how to discuss physics in curved spacetime. When a physical law is known in flat spacetime, we may try to express it in a covariant tensor form and consider the resulting expression as valid in curved spacetime. This prescription constitutes the principle of covariance, which can be succinctly stated as follows:

"Comma goes to semicolon rule"

$$\eta_{\mu\nu} \to g_{\mu\nu}, \quad \partial_\mu \to \nabla_\mu. \tag{7.92}$$

In other words, to transition from flat to curved spacetime, one need only replace the Minkowski metric with a general metric field $g_{\mu\nu}(x)$ of curved space and promote partial derivative to the covariant derivative. This prescription is often called the principle of minimal coupling. Minimal coupling means that the interaction between matter fields and the gravitational field is introduced in the simplest possible way, without explicit dependence on curvature tensors such as the Riemann tensor, or Ricci tensor, or Ricci scalar. Instead, the coupling arises solely through the replacement of flat-spacetime operations with their covariant counterparts.

Example 7.3. Geodesic equation.

This represents the most direct application of the "comma goes to semicolon rule":

$$\frac{\mathrm{d}^2 x^\mu}{\mathrm{d}\lambda^2} = \frac{\mathrm{d}x^\nu}{\mathrm{d}\lambda}\partial_\nu \frac{\mathrm{d}x^\mu}{\mathrm{d}\lambda} \to \frac{\mathrm{d}x^\nu}{\mathrm{d}\lambda}\nabla_\nu \frac{\mathrm{d}x^\mu}{\mathrm{d}\lambda}, \tag{7.93}$$

that is, $\frac{\mathrm{d}}{\mathrm{d}\lambda} \to \frac{\mathrm{D}}{\mathrm{d}\lambda}$. Therefore,

$$\frac{\mathrm{d}^2 x^\mu}{\mathrm{d}\lambda^2} = 0 \quad \to \quad \frac{\mathrm{D}}{\mathrm{d}\lambda}\frac{\mathrm{d}}{\mathrm{d}\lambda}x^\mu = 0. \tag{7.94}$$

This is precisely the geodesic equation.

Example 7.4. Conservation of energy-momentum.

$$\partial_\mu T^{\mu\nu} = 0 \quad \to \quad \nabla_\mu T^{\mu\nu} = 0. \tag{7.95}$$

This extension has been used in our previous discussions.

Example 7.5. Maxwell's equations.

$$\begin{aligned} &\partial_\mu F^{\nu\mu} = 4\pi J^\nu \to \nabla_\mu F^{\nu\mu} = 4\pi J^\nu, \\ &\partial_{[\mu} F_{\nu\lambda]} = 0 \to \nabla_{[\mu} F_{\nu\lambda]} = 0, \end{aligned} \tag{7.96}$$

there is actually no change. This is because Maxwell's equations can be expressed in differential form:

$$\mathrm{d}(*\widehat{F}) = 4\pi(*\widehat{J}), \quad \mathrm{d}\widehat{F} = 0, \tag{7.97}$$

which are tensor equations themselves.

It is important to note that the minimal coupling prescription can sometimes introduce ambiguities in its application. For example, consider the expression $Y^\mu \partial_\mu \partial_\nu X^\nu = 0$. When applying the covariant derivative replacement rule, there are at least two possible generalizations:

$$Y^\mu \partial_\mu \partial_\nu X^\nu = 0 \to \begin{cases} Y^\mu \nabla_\mu \nabla_\nu X^\nu, \\ Y^\mu \nabla_\nu \nabla_\mu X^\nu, \end{cases} \tag{7.98}$$

which yield different results because, unlike partial derivatives, covariant derivatives do not generally commute. This ambiguity is analogous to the operator ordering problem in quantum mechanics, where multiple quantizations of a classical expression may exist. The correct choice among possible generalizations must ultimately be determined by experimental verification.

Is Einstein's equivalence principle a fundamental principle of nature? The modern view suggests that we should not expect it to hold as an exact law of nature. To explore this, consider the following modified electromagnetic field equation:

$$\nabla_\mu[(1+\alpha R)F^{\nu\mu}] = 4\pi J^\nu, \tag{7.99}$$

where R is the Ricci scalar and α is a coupling coefficient. The equivalence principle strictly demands $\alpha = 0$, yet the equation above appears physically consistent: it preserves electric charge conservation and reduces to conventional Maxwell's equations in flat spacetime. Why, then, must α vanish? The answer lies in the energy scale of the phenomenon. Dimensional analysis reveals that α must have the dimension of length squared. Since the α-dependent term couples electromagnetism to gravity, the only natural length scale is the Planck length,

$$\alpha \sim l_{\mathrm{P}}^2, \tag{7.100}$$

where

$$l_{\mathrm{P}} = \left(\frac{G\hbar}{c^3}\right)^{1/2} \approx 1.6 \times 10^{-33}\ \mathrm{cm}. \tag{7.101}$$

Consequently, any observable effects arising from the α term would emerge only near the Planck scale. At ordinary energy scales, such corrections are greatly suppressed, rendering them undetectable with current experimental precision.

The meaning and foundational role of the equivalence principle in general relativity have long been subjects of debate. Prominent physicists such as Charles W. Misner *et al.* argue for its indispensability, while others, like John L. Synge, maintain a contrasting view. We shall not engage deeply in this dispute here, but one fact remains undeniable: the equivalence principle played a pivotal role in Einstein's formulation of general relativity. Today, however, its status as an absolute necessity appears less certain.

In addition to general relativity, there are other theories of gravity, which can be roughly divided into two categories:

(1) Metric-type theories: based on metrics, geodesics, etc.;
(2) Non-metric-type theories: based on torsion, etc.

In this book, we only discuss metric-type theories, including Einstein's general relativity, Brans–Dicke theory, and related frameworks. A key implication of Einstein's equivalence principle is that it uniquely selects metric-type theories as viable descriptions of gravity. Among these, general relativity appears to be the only theory that fully satisfies the strong equivalence principle.

7.5 General Discussion of Einstein's Equations

Let us revisit Einstein's field equations

$$G_{\mu\nu} = R_{\mu\nu} - \frac{1}{2} g_{\mu\nu} R = 8\pi G T_{\mu\nu}. \tag{7.102}$$

These equations can be equivalently expressed in the alternative form

$$R_{\mu\nu} = 8\pi G \left(T_{\mu\nu} - \frac{1}{D-2} g_{\mu\nu} T \right), \tag{7.103}$$

where D denotes the dimension of spacetime. As these constitute a set of local equations, their solutions vary across different spacetime regions. In many cases, proper junction conditions must be imposed to smoothly connect these solutions across region boundaries. In vacuum, Einstein's equations simplify to

$$R_{\mu\nu} = 0. \tag{7.104}$$

When spacetime dimensions $D < 4$, $R_{\mu\nu} = 0$ necessarily implies $R_{\mu\nu\sigma\rho} = 0$, so there are only flat spacetime solutions.[3] In higher dimensions $D \geq 4$, non-vanishing Weyl tensors permit more complex spacetime geometries. Particularly in four dimensions, the vacuum Einstein equations admit physically significant solutions, which include the Schwarzschild solution describing the static spherically symmetric spacetime and the Kerr solution describing the external spacetime of a rotating star.

The Einstein tensor in Einstein's equations consists of 10 independent components in four dimensions, suggesting 10 independent equations at first glance. However, the Bianchi identities require $\nabla^\mu G_{\mu\nu} = 0$, giving 4 constraints and effectively reducing the number of independent equations to 6. The metric field which is to be determined via Einstein's equations, also has 10 independent components. At first sight, this might suggest an underdetermined system, since there appear to be fewer equations than unknowns. However, this discrepancy is resolved by accounting for the gauge freedom inherent in general relativity. Specifically, if $g_{\mu\nu}$ is a solution to Einstein's equations in coordinates x_μ, then under an arbitrary coordinate transformation $x^\mu \to x'^\mu(x)$, the transformed metric $g'_{\mu\nu}(x')$ remains a solution. This is because Einstein's equations are tensor equations and thus manifestly covariant—their form is invariant under general coordinate

[3] An exception occurs in three dimensions with a negative cosmological constant, where non-trivial black hole solutions exist. This special case arises from the unique properties of three-dimensional AdS gravity, which possesses global degrees of freedom despite lacking local dynamical degrees of freedom.

transformations. This coordinate freedom implies that there are four non-physical degrees of freedom in $g_{\mu\nu}$, corresponding to the choice of gauge, i.e. the 4 arbitrary functions $x'_\mu(x^\nu)$ defining the coordinate system. A more physical interpretation is that not all metric degrees of freedom are dynamical; some are redundant due to diffeomorphism invariance. Consequently, Einstein's equations ultimately govern only 6 independent, gauge-invariant degrees of freedom in the metric, matching the number of truly independent field equations.

Einstein's equations, formulated as a set of non-linear partial differential equations, pose significant challenges in obtaining exact solutions. The complexity arises from the structure of the Ricci tensor and Ricci scalar, which are contractions of the Riemann tensor. These quantities involve not only derivative terms but also products of Christoffel symbols—effectively introducing non-linear dependencies on the inverse metric and its derivatives. Furthermore, the stress-energy tensor $T_{\mu\nu}$ typically couples to the metric itself, rendering Einstein's equations highly non-linear. Consequently, conventional techniques such as the linear superposition principle and Green's function method are inapplicable. Despite these challenges, the search for exact solutions to Einstein's equations remains a central pursuit in general relativity. For comprehensive treatments of this subject, we refer the reader to [35, 36]. Remarkably, the first exact solution—describing a static, spherically symmetric spacetime—was derived by the German physicist Karl Schwarzschild within a year of general relativity's publication. This solution, now foundational in relativistic astrophysics, models the exterior geometry of non-rotating stars or black holes. In contrast, the solution for a rotating star, which exhibits much richer structure, was not discovered until 1962 by the New Zealand physicist Roy Kerr.

In Newtonian mechanics, the gravitational potential satisfies Poisson equation

$$\nabla^2\Phi = 4\pi G_\rho, \tag{7.105}$$

which is a linear differential equation. This linearity implies that the total gravitational potential produced by multiple sources is simply the linear superposition of the potentials generated by each source individually. The situation in general relativity is fundamentally different. Einstein's theory requires that the gravitational field must couple to itself, a necessity that stems directly from the equivalence principle. To understand why, consider a bound gravitational system (what one might call a "gravitational atom") consisting of two particles held together by their mutual gravitational attraction. If gravity did not self-couple, the system's gravitational mass—which determines how it responds to an external gravitational field—would differ from its inertial mass due to the negative contribution from gravitational binding energy. This would violate the equivalence principle. Thus, consistency with the equivalence principle demands that the gravitational field must act as its own source. This self-interaction is mathematically manifested in the nonlinear nature of Einstein's field equations, where the gravitational field energy itself contributes to the total stress-energy tensor that generates spacetime curvature.

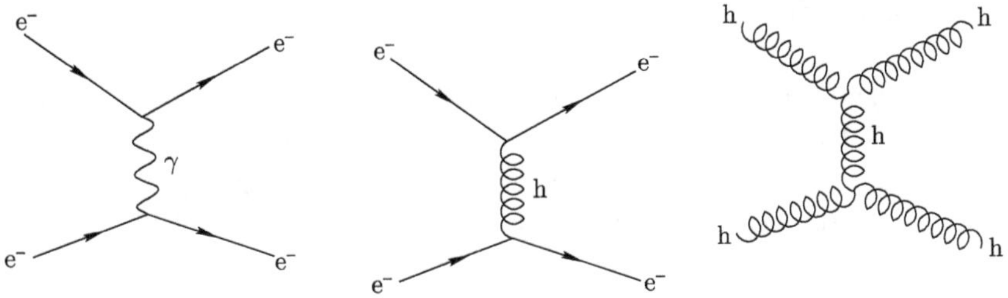

Left to right: (a) The interaction of electrons exchanging photons. (b) The interaction of electrons exchanging gravitons. (c) The interaction of gravitons.

Fig. 7.3 Shows that in electromagnetic theory, charged particles interact by exchanging virtual photons, which is a linear theory. At the tree level, electrons can also interact by exchanging gravitons, but the interaction strength is very weak. In general relativity, due to the self-interaction of gravitons, the gravitational interaction is nonlinear.

We can compare the non-linearity of gravity with other interactions. In quantum electrodynamics, the electromagnetic interaction between electrons is mediated by virtual photon exchange, as depicted in the Feynman diagram of Figure 7.3(a). The gravitational interaction, when treated as a quantum field theory in flat spacetime, involves both linear and nonlinear processes represented by the Feynman diagrams in Figures 7.3(b) and 7.3(c) respectively, with Figure 7.3(c) explicitly demonstrating gravity's nonlinear character. From a quantum field theory perspective, gravity emerges from the exchange of spin-2 gravitons, leading to nonlinear interactions qualitatively similar to those in non-Abelian Yang–Mills theories like QCD. However, unlike non-Abelian gauge theories, general relativity is non-renormalizable. Attempts to construct a renormalizable gravitational theory through higher-order derivative terms have proven unsuccessful, fundamentally because the gravitational coupling constant is dimensionful.

This suggests that a UV-complete theory of quantum gravity cannot be formulated purely in terms of local interactions.[4]

To construct a UV-complete theory of quantum gravity, we must move beyond local interactions. Non-local modifications can significantly improve the high-energy behavior of gravitational theories. Among possible candidates, string theory provides the most developed framework for a consistent quantum theory of gravity. In string theory, the fundamental constituents of nature are not point particles but extended one-dimensional objects. At high energy scales, all fundamental entities—including leptons, gauge bosons mediating interactions, and gravitons—emerge as vibrational modes of these strings. The characteristic string scale introduces an inherent non-locality in interactions, manifested as a natural "fuzziness" that regularizes the UV divergences plaguing point-particle theories of gravity. Remarkably, in the low-energy limit, string theory naturally reproduces Einstein's general relativity as an effective field theory. The massless spin-2 mode of the closed string spectrum precisely corresponds to the graviton, while the stringy corrections to Einstein's equations appear as higher-derivative terms suppressed by the string scale.

7.6 Principle of Action

In this section, we introduce how to derive Einstein's equations from the principle of action. Let us first review the action of a point particle in analytical mechanics. Consider a particle described by coordinates $q^a(t)$ and velocities $\dot{q}^a(t)$, where t is non-relativistic time. The action of the particle is given by

$$S = S[q^a(t)] = \int \mathrm{d}t L(q^a, \dot{q}^a), \tag{7.106}$$

where $L(q^a, \dot{q}^a)$ is the Lagrangian. From the Lagrangian, we define the canonical conjugate momentum as $p^a \equiv \dfrac{\partial L}{\partial \dot{q}^a}$, and the Hamiltonian can be further constructed as

$$H \equiv \sum_a p_a \dot{q}^a - L = H(q^a, p_b). \tag{7.107}$$

In analytical mechanics, both the Hamilton principle and the principle of action (Lagrangian) can be used to derive the equations of motion. Under

[4]Recent research has revealed that maximal ($N = 8$) supergravity may exhibit perturbative finiteness. While this property remains unproven, even if established it would not guarantee UV-completeness, as non-perturbative effects could still lead to divergences.

the action principle, we consider variations of coordinates and require the action attain an extremum. This variational procedure leads to the Euler–Lagrange equations

$$\frac{\partial L}{\partial q^a} - \frac{\mathrm{d}}{\mathrm{d}t}\left(\frac{\partial L}{\partial \dot{q}^a}\right) = 0. \tag{7.108}$$

Equivalently, the dynamics can be expressed in terms of Hamilton's equations:

$$\frac{\mathrm{d}p_a}{\mathrm{d}t} = \{p_a, H\}, \quad \frac{\mathrm{d}q_a}{\mathrm{d}t} = \{q_a, H\}, \tag{7.109}$$

where the Poisson bracket is defined as

$$\{f, g\}_{\mathrm{PB}} = \frac{\partial f}{\partial q^a}\frac{\partial g}{\partial p_a} - \frac{\partial f}{\partial p_a}\frac{\partial g}{\partial q^a}. \tag{7.110}$$

Unlike point particles, classical field theory possesses infinitely many degrees of freedom. When the theory is formulated on spacetime, the field depends not only on time but on all spacetime coordinates. Moreover, the coordinates describing the field space are the fields themselves, rather than the conventional spacetime coordinates. Consequently, the following replacements are necessary:

$$\frac{\mathrm{d}}{\mathrm{d}t} \to \partial_\mu, \quad q^a(t) \to \phi^a(x^\mu), \tag{7.111}$$

where ϕ^a represents the field, which may be a scalar, vector, or tensor field, and $\phi^a(x^\mu)$ is defined over the entire spacetime. In this framework, the action for the field can be expressed as:

$$S[\phi^a(x^\mu)] = \int \mathrm{d}^D x \mathcal{L} = \int \mathrm{d}t L. \tag{7.112}$$

Here $\mathcal{L}$ is the Lagrangian density, and its spatial integration gives the Lagrangian L. Applying the principle of least action, we derive the Euler–Lagrange equations for classical field theory

$$\frac{\partial \mathcal{L}}{\partial \phi^a} - \partial_\mu\left(\frac{\partial \mathcal{L}}{\partial(\partial_\mu \phi^a)}\right) = 0. \tag{7.113}$$

7.6.1 *Einstein–Hilbert action*

To formulate the action for the gravitational field, we must first ensure that it remains invariant under general coordinate transformations. The ordinary integral measure $\int \mathrm{d}^D x$ is not a true scalar but a scalar density. To define a proper invariant volume element, we introduce the determinant of the metric to define a good integral element, leading to the modified action:

$$S = \int \mathrm{d}^D x \mathcal{L} \quad \rightarrow \quad \int \mathrm{d}^D x \sqrt{-g} \mathcal{L}_{\mathrm{H}}, \tag{7.114}$$

where $\mathcal{L}_{\mathrm{H}}$ must now be a scalar to maintain invariance. How should $\mathcal{L}_{\mathrm{H}}$ be constructed? It should be constructed from tensor fields, their derivatives, and geometric invariants. The basic elements include ∇_μ, $g_{\mu\nu}$, $R_{\alpha\beta\gamma\delta}$, etc. Since $g^{\mu\nu} g_{\mu\nu} = D$, the action can include a cosmological constant term $S_\Lambda = (\text{constant}) \int \mathrm{d}^D x \sqrt{-g} \Lambda$. Moreover, the identity $\nabla^\mu g^{\mu\nu} = 0$ implies that derivatives of the metric itself cannot contribute dynamically.

In conventional field theories, whether scalar fields or gauge fields, the action typically contains only up to second-order derivatives of the field. This ensures that the resulting equations of motion have dynamics and admit a well-posed initial value problem. By analogy, we expect that the gravitational action to depend only on second-order derivatives of the metric field (up to a total divergence). The only choice is the Ricci scalar: $R_{\alpha\beta\gamma\delta} g^{\alpha\gamma} g^{\beta\delta} = R$,which is the lowest-order derivative term that can be obtained through the covariant derivative ∇_μ, the metric field $g_{\mu\nu}$, and the Riemann tensor $R_{\alpha\beta\gamma\delta}$. In this way, we obtain the famous Einstein–Hilbert action[5]

$$S_{\mathrm{EH}} = (\text{constant}) \int \mathrm{d}^D x \sqrt{-g} R. \tag{7.115}$$

In principle, one can construct additional curvature invariants using the Riemann tensor, Ricci tensor, Ricci scalar, and their covariant derivative terms, such as

$$R^n, \quad f(R), \quad R^{\mu\nu} R_{\mu\nu} = R_2, \quad R^{\alpha\beta\gamma\delta} R_{\alpha\beta\gamma\delta} = R_4, \quad \cdots. \tag{7.116}$$

However, such terms introduce higher-order derivatives of the metric, leading to modified gravitational theories collectively known as higher-derivative gravity.

The constant before the integral in the Einstein–Hilbert action can be determined via dimensional analysis. First, observe that the metric field $g_{\mu\nu}$ is dimensionless, i.e., $[g_{\mu\nu}] = 0$. Consequently $[\sqrt{-g}] = 0$, and due to the existence of partial derivatives, the Christoffel symbols

[5]Strictly speaking, this action was first discovered by Hilbert. However, as Hilbert himself remarked: "Maybe the kids in Gottingen know more differential geometry than Einstein, but general relativity remains Einstein's theory." Hence, the action is commonly attributed to both as the Einstein–Hilbert action.

$[\Gamma] \sim 1/\text{length} = 1/L$. The Ricci scalar R, constructed from second derivatives of the metric, thus has $[R] = 1/L^2$. On the other hand, the dimensions of the action and the integral are $[S] = [\hbar]$, $[\mathrm{d}^D x] = L^D$, so the dimension of the undetermined constant is

$$[\text{constant}] = L^{2-D}[\hbar] = L^{2-D}MLT^{-1}L = L^{4-D}MT^{-1}, \tag{7.117}$$

where we have used the relation $E = \hbar\omega$. In D-dimensional Newtonian gravity, the gravitational potential scales as

$$V \propto -\frac{GM}{r^{D-3}}, \tag{7.118}$$

as derived from the Poisson equation and Gauss's law. The resulting acceleration $a = -\nabla V \propto \dfrac{GM}{r^{D-2}}$, leading to

$$[G] = L^{D-1}T^{-2}M^{-1}. \tag{7.119}$$

This differs from the dimension of the constant mentioned earlier. However, by introducing the speed of light c, we find that the combination $[G^{-1}c^n] = L^{1+n-D}MT^{2-n}$ and when $n = 3$ it has

$$[\text{constant}] = [G^{-1}c^3]. \tag{7.120}$$

Remarkably, this relationship holds in any spacetime dimension. In particle physics,it is conventional to adopt natural units $c = 1$, $\hbar = 1$, simplifying the dimensional analysis. Here, Newton's constant has

$$[G] \sim L^{D-2}, \tag{7.121}$$

implying that the action's prefactor must scale as

$$\text{constant} \propto G^{-1}. \tag{7.122}$$

To fix the exact coefficient, we require consistency with the coupling to matter fields. Varying the full action (gravity + matter) yields Einstein's equations, where the energy-momentum tensor must match standard field-theoretic definitions. This uniquely determines the proportionality constant, leading to the conventional form of the gravitational action

$$S_{\mathrm{g}} = S_{\mathrm{EH}} + S_{\Lambda} = \frac{1}{16\pi G}\int \mathrm{d}^D x\sqrt{-g}(R - 2\Lambda). \tag{7.123}$$

7.6.2 *Einstein's equations*

We must also account for the coupling between matter and gravity in our formulation. The complete action, incorporating both gravitational and matter field contributions, is given by:

$$S = S_{\mathrm{g}} + S_{\mathrm{m}}. \tag{7.124}$$

Applying the principle of least action to this combined action leads to the derivation of Einstein's field equations. Our analysis begins with the gravitational sector of the action, specifically focusing on the Einstein–Hilbert action. While the conventional approach involves varying the action with respect to the metric tensor, computing $\frac{\delta S}{\delta g_{\mu\nu}}$, we find it computationally advantageous to instead consider the variation with respect to the inverse metric $\frac{\delta S}{\delta g^{\mu\nu}}$, which gives equivalent results. From the action $\frac{1}{16\pi G}\int \mathrm{d}^D x\sqrt{-g}R$ or simply from $S_0\int \mathrm{d}^D x\sqrt{-g}R$, we find

$$\delta S_0 = \int \mathrm{d}^D x\sqrt{-g}(\delta R) + R\delta(\sqrt{-g}). \tag{7.125}$$

We shall now examine each term in this variation separately. For the second term on the right-hand side of Eq. (7.125), we compute the variation of $\sqrt{-g}$ as follows,

$$\delta\sqrt{-g} = -\frac{1}{2}\sqrt{-g}g_{\mu\nu}g^{\mu\nu} = \frac{1}{2}\sqrt{-g}g^{\mu\nu}\delta g_{\mu\nu}. \tag{7.126}$$

The first equality follows from the general identity for the variation of a determinant,

$$\frac{\delta \det M}{\det M} = \mathrm{Tr}(M^{-1}\delta M), \tag{7.127}$$

applied to the metric tensor $M = g_{\mu\nu}$. The second equality arises from the condition

$$0 = \delta(g^{\mu\nu}g_{\nu\lambda}) = (\delta g^{\mu\nu})g_{\nu\lambda} + g^{\mu\nu}\delta g_{\nu\lambda}, \tag{7.128}$$

which implies

$$(\delta g^{\mu\nu})g_{\nu\lambda} = -g^{\mu\nu}\delta g_{\nu\lambda}. \tag{7.129}$$

Contracting indices $\mu = \lambda$ recovers the relation needed for the second step in (7.126). However, we note that only the first step is strictly necessary for the subsequent derivation. For the first term on the right-hand side of Eq. (7.125), we analyze the variation of the Ricci scalar:

$$\delta R = \delta(g^{\mu\nu}R_{\mu\nu}) = \delta g^{\mu\nu}R_{\mu\nu} + g^{\mu\nu}\delta R_{\mu\nu}. \tag{7.130}$$

The key challenge lies in evaluating $\delta R_{\mu\nu}$. To compute this systematically, we employ the Palatini identity (or first-order formalism), treating the connection $\Gamma^{\alpha}_{\beta\gamma}$ as an independent variable before imposing metric compatibility and torsion-free conditions. The chain rule gives:

$$\frac{\delta R_{\mu\nu}}{\delta g^{\sigma\rho}} = \frac{\delta R_{\mu\nu}}{\delta \Gamma^{\alpha}_{\beta\gamma}} \frac{\delta \Gamma^{\alpha}_{\beta\gamma}}{\delta g^{\sigma\rho}}. \tag{7.131}$$

If we only consider metric-compatible and torsion-free connections, this step-by-step variation is not problematic. The only difficulty might come from the fact that $\Gamma^{\alpha}_{\beta\gamma}$ is not a tensor. However, we know that the difference between two Christoffel symbols is a tensor, that is, $\delta\Gamma$ is a tensor, so there is no problem. The variation of the Riemann tensor follows from the Palatini identity:

$$\delta R^{\rho}{}_{\mu\lambda\nu} = \nabla_{\lambda}(\delta\Gamma^{\rho}_{\nu\mu}) - \nabla_{\nu}(\delta\Gamma^{\rho}_{\lambda\mu}), \tag{7.132}$$

where the covariant derivative acts on $\delta\Gamma$ as:

$$\nabla_{\lambda}(\delta\Gamma^{\rho}_{\nu\mu}) = \partial_{\lambda}(\delta\Gamma^{\rho}_{\nu\mu}) + \Gamma^{\rho}_{\lambda\sigma}\delta\Gamma^{\rho}_{\nu\mu} - \Gamma^{\rho}_{\lambda\nu}\delta\Gamma^{\rho}_{\sigma\mu} - \Gamma^{\sigma}_{\lambda\mu}\delta\Gamma^{\rho}_{\nu\sigma}. \tag{7.133}$$

Contracting indices and using the symmetry of $g^{\mu\nu}$, we find

$$\begin{aligned} g^{\mu\nu}\delta R_{\mu\nu} &= g^{\mu\nu}(\nabla_{\lambda}(\delta\Gamma^{\lambda}_{\nu\mu}) - \nabla_{\nu}(\delta\Gamma^{\lambda}_{\lambda\mu})) \\ &= \nabla_{\nu}(g^{\mu\lambda}\delta\Gamma^{\nu}_{\lambda\mu} - g^{\mu\nu}\delta\Gamma^{\lambda}_{\lambda\mu}). \end{aligned} \tag{7.134}$$

This term is a total divergence nd thus vanishes upon integration over the spacetime manifold (assuming appropriate boundary conditions). Consequently, only the $\delta g^{\mu\nu}R_{\mu\nu}$ term in (7.130) contributes to the field equations.

By combining both contributions, we arrive at the variation of the action:

$$\begin{aligned} \delta S_0 &= \int \mathrm{d}^D x\sqrt{-g}\left(R_{\mu\nu}\delta g^{\mu\nu} - \frac{1}{2}g_{\mu\nu}R\delta g^{\mu\nu}\right) \\ &= \int \mathrm{d}^D x\sqrt{-g}\left(R_{\mu\nu} - \frac{1}{2}g_{\mu\nu}R\right)\delta g^{\mu\nu}. \end{aligned} \tag{7.135}$$

On the other hand, the matter action is defined as

$$S_{\mathrm{m}} = \int \mathrm{d}^D x\mathcal{L}_{\mathrm{m}}, \tag{7.136}$$

where the integration measure does not include the $\sqrt{-g}$ factor. From this action, we can define the energy-momentum tensor of matter

$$T_{\mu\nu} \equiv -\frac{2}{\sqrt{-g}}\frac{\delta\mathcal{L}_{\mathrm{m}}}{\delta g^{\mu\nu}}. \tag{7.137}$$

Varying the full action, comprising both gravitational and matter contributions, with respect to the metric $g^{\mu\nu}$ yields

$$\begin{aligned}\delta(S_{\mathrm{EH}}+S_{\mathrm{m}}) &= \frac{1}{16\pi G}\int \mathrm{d}^D x\sqrt{-g}\left(R_{\mu\nu}-\frac{1}{2}g_{\mu\nu}R\right)\delta g^{\mu\nu}\\ &\quad -\frac{1}{2}\int \mathrm{d}^D x\sqrt{-g}T_{\mu\nu}\delta g^{\mu\nu}\\ &= \frac{1}{16\pi G}\int \mathrm{d}^D x\sqrt{-g}\left(R_{\mu\nu}-\frac{1}{2}g_{\mu\nu}R-8\pi G T_{\mu\nu}\right)\delta g^{\mu\nu}.\end{aligned} \tag{7.138}$$

Requiring stationarity of the action then leads to Einstein's field equations

$$R_{\mu\nu}-\frac{1}{2}g_{\mu\nu}R = 8\pi G T_{\mu\nu}. \tag{7.139}$$

7.6.3 *Energy-momentum tensor*

We now demonstrate that the energy-momentum tensor defined in (7.137) is consistent with the coupling coefficient $1/(16\pi G)$ in the Einstein–Hilbert action, thereby validating its form.

(1) Scalar field theory.

We begin by examining a scalar field theory described by the action

$$S_{\mathrm{scalar}} = \int \mathrm{d}^D x\sqrt{-g}\left(-\frac{1}{2}g^{\mu\nu}\nabla_\mu\phi\nabla_\nu\phi - V(\phi)\right). \tag{7.140}$$

Following the definition of the energy-momentum tensor, we compute

$$\begin{aligned}T_{\mu\nu} &\equiv -\frac{2}{\sqrt{-g}}\frac{\delta\mathcal{L}_{\mathrm{scalar}}}{\delta g^{\mu\nu}},\\ &= -\frac{2}{\sqrt{-g}}\left(\frac{\delta\sqrt{-g}}{\delta g^{\mu\nu}}\left(-\frac{1}{2}g^{\sigma\rho}\nabla_\sigma\phi\nabla_\rho\phi - V(\phi)\right)\right.\\ &\quad \left.+\sqrt{-g}\left(-\frac{1}{2}\right)\nabla_\mu\phi\nabla_\nu\phi\right)\\ &= -\frac{1}{2}g_{\mu\nu}(\nabla\phi)^2 - g_{\mu\nu}V(\phi)+\nabla_\mu\phi\nabla_\nu\phi.\end{aligned} \tag{7.141}$$

In the case of flat spacetime, the time-time component T_{00} of this tensor reduces to

$$\begin{aligned}T_{00} &= \dot{\phi}^2+\frac{1}{2}(-\dot{\phi}^2+(\partial_i\phi)^2)+V(\phi)\\ &= \frac{1}{2}(\dot{\phi}^2+(\partial_i\phi)^2)+V(\phi),\end{aligned} \tag{7.142}$$

which precisely matches the Hamiltonian density of the system.

(2) Electromagnetic field theory.

For the electromagnetic field, the action is given by

$$S_{\text{EM}} = \int \mathrm{d}^D x \sqrt{-g} \left(-\frac{1}{4} F_{\mu\nu} F^{\mu\nu} \right), \tag{7.143}$$

and its energy-momentum tensor is derived as follows:

$$\begin{aligned} T_{\mu\nu} &= -\frac{2}{\sqrt{-g}} \frac{\delta}{\delta g^{\mu\nu}} \left(-\frac{1}{4} \sqrt{-g} F_{\gamma\delta} F^{\gamma\delta} \right), \\ &= -\frac{2}{\sqrt{-g}} \left\{ \frac{\delta\sqrt{-g}}{\delta g^{\mu\nu}} \left(-\frac{1}{4} F^2 \right) + \sqrt{-g} \frac{\delta}{\delta g^{\mu\nu}} \left(-\frac{1}{4} F_{\alpha\beta} F_{\gamma\delta} g^{\alpha\gamma} g^{\beta\delta} \right) \right\} \\ &= -\frac{1}{4} g_{\mu\nu} F^2 - 2 \left(-\frac{1}{4} \right) \left[F_{\alpha\beta} F_{\gamma\delta} (\delta^\alpha_\mu \delta^\gamma_\nu g^{\beta\delta} + \delta^\beta_\mu \delta^\delta_\nu g^{\alpha\gamma}) \right] \\ &= -\frac{1}{4} g_{\mu\nu} F^2 + \frac{1}{2} (F_\mu{}^\delta F_{\nu\delta} + F_{\alpha\mu} F^\alpha_\nu) \\ &= -\frac{1}{4} g_{\mu\nu} F^2 - F_\mu{}^\alpha F_{\alpha\nu}. \end{aligned} \tag{7.144}$$

Similarly, its (00) component gives the Hamiltonian density $\mathcal{H} = \frac{1}{2}(\boldsymbol{E}^2 + \boldsymbol{B}^2)$. The trace of the above energy-momentum tensor is

$$T^\mu_\mu = F^{\mu\alpha} F_{\mu\alpha} - \frac{D}{4} F^2 = \left(1 - \frac{D}{4} \right) F^2. \tag{7.145}$$

Notably, in $D = 4$ spacetime dimensions, the trace vanishes. This indicates that four-dimensional classical electromagnetic theory is scale-invariant.

From the proceeding examples, we observe that the energy-momentum tensor derived by varying the action with respect to the metric, consistently reproduces the expected physical results. Remarkably, this definition automatically yields a symmetric tensor, circumventing ambiguities present in other constructions. Later, we shall demonstrate that it satisfies the conservation law $\nabla^\mu T_{\mu\nu} = 0$, reflecting the local conservation of energy and momentum in the theory. In quantum field theory, there is some kind of ambiguity in defining the energy-momentum tensor. The canonical energy-momentum tensor obtained via the Noether's theorem is not symmetric. While symmetrization procedures exist, they introduce arbitrariness in the choice of added terms. The gravitational minimal-coupling approach introduced above, provides a unique and symmetric tensor without ad hoc modifications. In fact, it provides a natural way to define the energy-momentum tensor in quantum field theory: starting from a field theory minimally coupled with gravity, define the energy-momentum tensor according to the formula above, and then let $g_{\mu\nu} \to \eta_{\mu\nu}$.

(3) Vacuum energy.

As discussed earlier, alongside the Einstein–Hilbert action, we may introduce the cosmological constant term:

$$S_\Lambda = \frac{1}{16\pi G}\int \mathrm{d}^D x\sqrt{-g}(-2\Lambda). \tag{7.146}$$

This term can be treated as part of the matter field action, yielding an energy-momentum tensor of the form:

$$T_{\mu\nu} = -\frac{\Lambda}{8\pi G}g_{\mu\nu}. \tag{7.147}$$

For an ideal fluid in curved space, the energy-momentum tensor takes the form

$$T_{\mu\nu} = (\rho + p)u_\mu u_\nu + pg_{\mu\nu}, \tag{7.148}$$

which implies that the cosmological constant term corresponds to a matter component with the equation of state:

$$p = -\rho. \tag{7.149}$$

This represents a fluid with negative pressure, commonly interpreted as vacuum energy. Including the cosmological constant, the Einstein field equations take the form

$$R_{\mu\nu} - \frac{1}{2}Rg_{\mu\nu} + \Lambda g_{\mu\nu} = 8\pi GT_{\mu\nu}, \tag{7.150}$$

where the energy-momentum tensor on the right-hand side comes from other matter fields.

The origin of the cosmological constant term presents a profound issue in theoretical physics. In non-gravitational physics, the vacuum potential can be shifted arbitrarily since only relative potential differences are physically meaningful. However, in general relativity, the absolute value of the vacuum energy density becomes significant, as gravity couples to all forms of energy. The cosmological constant is naturally interpreted as vacuum energy, arising from the quantum zero-point fluctuations of fields. For example, for a one-dimensional harmonic oscillator, the ground state energy is $E_0 = \frac{1}{2}\hbar\omega$. In quantum field theory, this generalizes to an infinite collection of oscillators, yielding a total zero-point energy

$$E_0 = \sum_\omega \frac{1}{2}\hbar\omega, \tag{7.151}$$

where $\omega = \sqrt{m^2 + k^2}$. This sum is ultraviolet-divergent, necessitating a cutoff $k_{\max}$, leading to an estimated vacuum energy density:

$$\rho_{\text{vac}} \sim \hbar k_{\max}^4. \tag{7.152}$$

A natural ultraviolet cutoff is the Planck scale $k_{\max} \sim m_{\text{P}} \sim 10^{18}$ GeV,the theoretical prediction becomes

$$\rho_{\text{vac}} \sim 10^{112}\ \text{erg/cm}^3. \tag{7.153}$$

However, if dark energy originates from vacuum energy, observational constraints (e.g., from cosmic acceleration) require

$$|\rho_\Lambda^{\text{obs}}| \leq 10^{-8}\ \text{erg/cm}^3. \tag{7.154}$$

This results in a staggering discrepancy of 120 orders of magnitude between theory and observation! Even adopting a lower cutoff (e.g., the TeV scale) still leaves a mismatch of nearly 60 orders of magnitude—a clear indication of a deep theoretical inconsistency. In fact, the cosmological constant problem was identified before dark energy was observationally confirmed (see Ref. [37]). It remains one of the most significant unsolved puzzles in modern physics, challenging our understanding of quantum field theory, gravity, and their unification.

7.6.4 *Conservation of energy-momentum and equations of motion*

For a point particle, the action is given by

$$S_{\text{particle}} = -m \int \mathrm{d}\lambda \sqrt{-g_{\mu\nu}\dot{x}^\mu \dot{x}^\nu}. \tag{7.155}$$

This expression represents a line integral and does not directly define the Lagrangian of the particle. To obtain the energy-momentum tensor for a point particle, we modify this integral by inserting the identity $1 = \int \mathrm{d}^D y \delta^D(y - x(\lambda))$, which transforms the action into

$$S = -m \int \mathrm{d}^D y \int \mathrm{d}\tau \sqrt{-g_{\mu\nu}\dot{x}^\mu \dot{x}^\nu}\, \delta^D(y - x(\tau)). \tag{7.156}$$

The corresponding energy-momentum tensor is then obtained as

$$\begin{aligned} T_{\mu\nu}(y) &= -\frac{2m}{\sqrt{-g}} \left(-\int \mathrm{d}\tau \frac{1}{2\sqrt{-\dot{x}^2}} (-\dot{x}_\mu \dot{x}_\nu) \right) \delta^D(y - x(\tau)) \\ &= -m \int \mathrm{d}\tau \frac{\delta^D(y - x(\tau))}{\sqrt{-g(y)}} \frac{\dot{x}_\mu \dot{x}_\nu}{\sqrt{-\dot{x}^2}}. \end{aligned} \tag{7.157}$$

Considering the energy-momentum conservation law,

$$\nabla_\mu T^{\mu\nu} = \frac{1}{\sqrt{-g}}\partial_\mu(\sqrt{-g}T^{\mu\nu}) + \Gamma^\nu_{\sigma\mu}T^{\mu\sigma},$$

we find that for a point particle,

$$\int \dot{x}^\mu \dot{x}^\nu \frac{\partial}{\partial y^\mu}\delta^4(y - x(\tau))\mathrm{d}\tau + \Gamma^\nu_{\sigma\mu}\int \dot{x}^\mu \dot{x}^\sigma \delta^4(y - x(\tau))\mathrm{d}\tau = 0. \tag{7.158}$$

Since $\delta^4(y-x(\tau))$ only depends on $y^\mu - x^\mu$, we may replace $\dfrac{\partial}{\partial y^\mu}$ with $-\dfrac{\partial}{\partial x^\mu}$. Using the identity $\dot{x}^\mu \dfrac{\partial}{\partial x^\mu}\delta^4(y - x(\tau)) = \dfrac{\mathrm{d}}{\mathrm{d}\tau}\delta^4(y - x(\tau))$, and integrating by parts, we arrive at

$$\int (\ddot{x}^\nu + \Gamma^\nu_{\sigma\mu}\dot{x}^\mu \dot{x}^\sigma)\delta^4(y - x(\tau)) = 0. \tag{7.159}$$

This yields the geodesic equation, demonstrating that the motion of the point particle can be derived from energy-momentum conservation. If we write down the complete Einstein equations, the energy-momentum tensor of the point particle appear on the right-hand side of the equation. At the location of the point particle, the presence of the delta function introduces a singularity in the field equations. However, outside the point particle, the field equations governs its motion, which coincides with that of a "test" particle in the same geometry. Thus, the geodesic equation, or more generally the equations of motion of matter, are equivalent to the conservation of energy-momentum in the action principle.

It is remarkable that Einstein's equations inherently determine the motion of a point particle, without the need for additional assumptions. This stands in contrast to electromagnetism, where the equations of motion for a charged particle (e.g., the Lorentz force law) must be introduced separately from Maxwell's equations. The key distinction arises from the nonlinearity of general relativity. Unlike electromagnetism, where the electromagnetic field does not self-interact, the gravitational field itself carries energy-momentum and thus acts as its own source. This self-coupling ensures that the field equations automatically enforce energy-momentum conservation, which in turn governs the particle's trajectory.

The above discussion of point particles can be generalized to continuous media. Consider a dust distribution described by the energy-momentum tensor: $T^{\mu\nu} = \rho u^\mu u^\nu$, then the conservation of energy-momentum implies

$$\nabla_\mu(\rho u^\mu u^\nu) = \nabla_\mu(\rho u^\mu)u^\nu + \rho u^\mu \nabla_\mu u^\nu = 0. \tag{7.160}$$

Contracting this relation with u_ν, we find

$$-\nabla_\mu(\rho u^\mu) + \rho u^\mu u_\nu \nabla_\mu u^\nu = 0. \tag{7.161}$$

From the normalization condition of the 4-velocity $u_\mu \nabla_\mu u^\nu = 0$, we obtain $\nabla_\mu(\rho u^\mu) = 0$, which is the relativistic continuity equation governing energy conservation. Substituting this equation into (7.160) yields $u^\mu \nabla_\mu u^\nu = 0$, which is the geodesic equation of dust particle.

More generally, it can shown that the conservation of energy-momentum tensor $\nabla_\mu T^{\mu\nu} = 0$ implies that any sufficiently small object, provided its self-gravity is negligible, must move along a geodesic. This demonstrates that Einstein's field equations inherently encode the "geodesic hypothesis": the worldline of a test particle is a geodesic. Notably, this geodesic equation can also be derived by extremizing the action of the test particle through variational principles. The consistency between Einstein's equations, the geodesic equation, and energy-momentum conservation thus reflects the underlying action-based formulation of general relativity. However, this conclusion assumes that the object's size and self-gravity are negligible. For extended bodies, internal stresses and gravitational self-interaction lead to deviations from geodesic motion. Nevertheless, their dynamics can still be determined either by extremizing the full action or, equivalently, by imposing local energy-momentum conservation, $\nabla_\mu T^{\mu\nu} = 0$.

The conservation of the energy-momentum tensor arises naturally as a consequence of diffeomorphism invariance, a fundamental principle of general relativity. This invariance reflects the fact that the theory does not depend on any preferred geometric background (background independence), meaning no coordinate system in spacetime holds special priority; they are merely different descriptions of the same underlying manifold. Mathematically, diffeomorphism invariance requires the action S to remain unchanged under infinitesimal coordinate transformations. Consider a small coordinate transformation $x^\mu \to x'^\mu = x^\mu + \zeta^\mu(x)$, where $\zeta^\mu(x)$ is an arbitrary smooth vector field. Under this transformation, tensor fields transform covariantly, and the metric variation is given by: $\delta g_{\mu\nu} = 2\nabla_{(\mu}\zeta_{\nu)}$. For the matter action S_m, invariance under this transformation $\delta S_\mathrm{m} = 0$ implies

$$0 = \int \mathrm{d}^D x \frac{\delta \mathcal{L}_\mathrm{m}}{\delta g_{\mu\nu}} \nabla_\mu \zeta_\nu = -\int \mathrm{d}^D x \sqrt{-g}\,\zeta_\nu \nabla_\mu \left(\frac{1}{\sqrt{-g}} \frac{\delta \mathcal{L}_\mathrm{m}}{\delta g_{\mu\nu}} \right),$$

where we have used Stokes' theorem and discarded boundary terms. Since ζ_ν is arbitrary, the integrand must vanish identically, we have $\nabla_\mu T^{\mu\nu} = 0$.

The previous derivation of energy-momentum conservation contains a subtlety: under a diffeomorphism $x^\mu \to x^\mu + \zeta^\mu(x)$, not only does the metric transform, but the matter fields ϕ also undergo a variation $\delta\phi$. Consequently, the full variation of the matter action S_{m} must account for this additional term: $\left(\dfrac{\delta S_{\text{m}}}{\delta\phi}\right)_{\text{g}}\delta\phi$, where the subscript g indicates that the functional derivative is taken with the metric held fixed. If the equations of motion of the matter are satisfied, the $\delta\phi$ term vanishes, and thus we obtain the conservation of energy-momentum equation. Therefore, we obtain two relations from the action principle:

(1) Einstein's equations + $(\nabla_\mu T^{\mu\nu}=0)$ $\Rightarrow$ equations of motion for matter fields;
(2) Einstein's equations + equations of motion for matter fields $\Rightarrow$ $\nabla_\mu T^{\mu\nu}=0$.

When performing a path integral for gravity, we must account for all possible gravitational configurations while eliminating the degeneracy arising from diffeomorphisms. In other words, it suffices to consider gravitational configurations within a single diffeomorphism equivalence class—analogous to gauge field theory, where the path integral only incorporates gauge equivalence classes of gauge fields. In both cases, this reduction stems from the inherent degeneracy of field configurations due to underlying symmetries. However, the path integral over spacetime configurations is significantly more intricate than its gauge theory counterpart, primarily because the diffeomorphism group is far more challenging to handle.

Let's look at the conservation of energy-momentum equation for a general ideal fluid in curved spacetime. The energy-momentum tensor of an ideal fluid can be written as

$$\widehat{T}=(\rho+p)\widehat{u}\otimes\widehat{u}+p\widehat{g}. \tag{7.162}$$

The conservation law $\nabla\cdot\widehat{T}=0$ yields

$$\begin{aligned}0&=(\nabla(\rho+p)\cdot\widehat{u})\widehat{u}+((\rho+p)\nabla\cdot\widehat{u})\widehat{u}+((\rho+p)\widehat{u})\cdot\nabla\widehat{u}+(\nabla p)\cdot\widehat{g}\\&=(\nabla_{\widehat{u}}\rho+\nabla_{\widehat{u}}p+(\rho+p)\nabla\cdot\widehat{u})\widehat{u}+(\rho+p)\nabla_{\widehat{u}}\widehat{u}+\nabla p.\end{aligned} \tag{7.163}$$

Similarly, we can project Eq. (7.163) onto the direction parallel to the 4-velocity and the direction orthogonal to it. The inner product with the 4-velocity $\widehat{u}$ gives the energy conservation equation

$$0=\widehat{u}\cdot(\nabla\cdot\widehat{T})=-\nabla_{\widehat{u}}\rho-(\rho+p)\nabla\cdot\widehat{u}. \tag{7.164}$$

While in the direction orthogonal to the 4-velocity, using the projection operator $\widehat{P}=\widehat{g}+\widehat{u}\otimes\widehat{u}$, we obtain the relativistic Euler equation in curved spacetime

$$(\rho+p)\nabla_{\widehat{u}}\widehat{u}=-\widehat{P}\cdot(\nabla p)=-(\nabla p+(\nabla_{\widehat{u}}p)\widehat{u}). \tag{7.165}$$

In component form, these become

$$\frac{\mathrm{d}\rho}{\mathrm{d}\tau} = -(\rho + p)\nabla_i u^i, \tag{7.166}$$

$$(\rho + p)\frac{\mathrm{d}u^i}{\mathrm{d}\tau} = -\nabla^i p - u^i u^j \nabla_j p. \tag{7.167}$$

In the appropriate limit, the above equations can be reduced to the equations describing the motion of a fluid in a Newtonian gravitational field. First, the energy density can be taken as $\rho = \rho_0(1+\epsilon)$, where $\rho_0 = nm_\mathrm{B}$ represents the rest mass energy of n baryons, and ϵ represents the specific internal energy of the fluid. In the Newtonian limit, the rest energy is dominant, $\epsilon \ll 1$. For the metric, we take $g_{00} = -(1+2\Phi)$, $\Phi \ll 1$, and the non-relativistic limit requires $(p/\rho_0) \ll 1$ and $v^2 \ll 1$. In these limits, we have

$$\rho_0 \left(1 + \epsilon + \frac{p}{\rho_0}\right)(u^\mu \partial_\mu u_i - \Gamma^\nu_{\mu i} u_\nu u^\mu) = -\partial_i p - u_i \frac{\mathrm{d}p}{\mathrm{d}\tau}. \tag{7.168}$$

The 4-velocity can be approximated as $u^\mu \approx (1, v^i)$, so we get

$$\rho_0 \left(\frac{\mathrm{d}v_i}{\mathrm{d}\tau} + \Gamma^0_{0i}\right) \approx -\partial_i p. \tag{7.169}$$

Since $\Gamma^0_{0i} \approx -(1/2)\partial_i g_{00} \approx \partial_i \Phi$, we obtain the fluid equation under the pressure gradient and gravitational field

$$\frac{\mathrm{d}v_i}{\mathrm{d}\tau} = -\partial_i \Phi - \frac{1}{\rho_0}\partial_i p. \tag{7.170}$$

7.7 Modifications of General Relativity

As previously discussed, Einstein's equivalence principle is neither a sacred and immutable physical law nor a rigorously absolute statement; rather, it exhibits a degree of arbitrariness in practical applications. The modern perspective holds that general relativity functions as a low-energy effective theory, implying that the equivalence principle requires modification at high-energy scales. In what follows, we delve deeper into several intriguing aspects concerning Einstein's equivalence principle.

The first consideration is the principle of covariance. According to this principle, the laws in electromagnetism must be expressed in a covariant form:

$$\partial_\mu F^{\nu\mu} = J^\nu \to \nabla_\mu F^{\nu\mu} = J^\nu. \tag{7.171}$$

However, this is not the only possible covariant formulation. As previously noted, additional terms involving $F_{\mu\nu}$ can be introduced without violating the fundamental requirements of electromagnetism. This suggests that the principle of covariance is more of a logical guideline than a stringent physical constraint.

Second, the notion that "curvature = gravity" stems from Einstein's brilliant insight. The equivalence principle guarantees that gravity is universal and cannot be detected in a small region of spacetime. However, there is no definitive experimental evidence to show that "curvature = gravity" is an absolute truth. Strictly speaking, the tests of general relativity are currently limited to the millimeter scale in short-range experiments and the scale of the solar system in large-scale observations. Consequently, it remains plausible that general relativity may require modifications in both the ultraviolet (UV) and infrared (IR) regimes. In the UV regime, potential modifications have been extensively explored in quantum gravity research, primarily due to the non-renormalizability of gravitational interactions. Meanwhile, in the IR regime, growing interest has been driven by novel cosmological observations over the past two decades, which challenge conventional gravitational models. Furthermore, recent developments in gravity/gauge duality have prompted a reassessment of gravity's fundamental nature. Some theories now propose that gravity may be an emergent phenomenon rather than a fundamental force. These advances compel us to critically re-examine the long-standing paradigm of "curvature = gravity."

Let us consider the following modified geodesic equation:

$$\frac{\mathrm{d}^2x^\mu}{\mathrm{d}\lambda^2}+\Gamma^\mu_{\ \rho\sigma}\frac{\mathrm{d}x^\rho}{\mathrm{d}\lambda}\frac{\mathrm{d}x^\sigma}{\mathrm{d}\lambda}=\alpha(\nabla_\sigma R)\frac{\mathrm{d}x^\mu}{\mathrm{d}\lambda}\frac{\mathrm{d}x^\sigma}{\mathrm{d}\lambda}, \tag{7.172}$$

where α is a dimensional coupling constant. Dimensional analysis yields:

$$\left[\frac{\mathrm{d}x^\mu}{\mathrm{d}\lambda}\right]=[g_{\mu\nu}]=L^0,\quad [\nabla_\mu]=[\Gamma^\mu_{\ \rho\sigma}]=L^{-1},\quad [R]=L^{-2}, \tag{7.173}$$

implying $[\alpha]=L^2$. Given that experimental constraints require α to be extremely small and gravity-related, a natural choice is $\alpha\sim l_{\mathrm{P}}^2$. This suggests that, if general relativity is treated as a low-energy effective field theory, such a term would arise from quantum corrections. To estimate the magnitude of this effect, consider a typical gravitational system, such as the Earth's gravity,

$$a_{\mathrm{g}}=980\ \mathrm{cm/s}^2,\quad l_{\mathrm{E}}=c^2/a_{\mathrm{g}}\sim 10^{18}\ \mathrm{cm}. \tag{7.174}$$

The ratio of the Planck length to this scale is $\dfrac{l_{\mathrm{P}}}{l_{\mathrm{E}}}\sim 10^{-51}$, which requires $\alpha\sim 10^{-102}$, an exceptionally small quantity. This analysis indicates that any potential deviations from standard geodesic motion would manifest only as a high-energy (or equivalently, quantum-gravitational) effect, far beyond current observational capabilities.

Another important question concerning Einstein's equivalence principle is whether there exists an additional long-range interaction that mimics gravity but lacks a geometric origin. Such a force would induce deviations from Einstein's equivalence principle for neutral, uncharged test particles, and is commonly termed the "fifth force." This hypothetical interaction would couple universally to matter, making it experimentally indistinguishable from conventional gravity at leading order. Possible candidates include forces mediated by massless scalar fields or gauge fields, such as those arising from couplings of the form

$$\phi \mathrm{Tr} T_{\mu\nu}, \quad A_\mu J^\mu, \tag{7.175}$$

where J^μ represents the baryon number current.

7.7.1 *Alternative theories of general relativity*

If we relax the assumption that Einstein's equivalence principle is sacrosanct and instead treat general relativity as a low-energy effective theory, modifications to the theory become possible. Historically, one of the most notable modifications involves introducing a gravitational scalar field. Such a theory can be described by the action

$$S = \int \mathrm{d}^4x\sqrt{-g}\left(f(\lambda)R - \frac{1}{2}h(\lambda)\partial_\mu\lambda\partial^\mu\lambda - V(\lambda)\right) + S_\mathrm{m}, \tag{7.176}$$

where λ is a scalar field. This action is highly general: the scalar field need not be canonical, and its coupling to gravity may be non-minimal, while S_m represents the action of other matter fields. If $f(\lambda) = \dfrac{1}{16\pi G}$, the theory reduces to standard gravity coupled to a scalar field. The scalar field mediates an attractive force, long-range if massless, or short-range if massive. More generally, if $f(\lambda) \neq \dfrac{1}{16\pi G}$, the Newtonian gravitational "constant" becomes spacetime-dependent. The gravitational field equations derived from this action are

$$G_{\mu\nu} = f^{-1}(\lambda)\left(\frac{1}{2}T^\mathrm{m}_{\mu\nu} + \frac{1}{2}T^{(\lambda)}_{\mu\nu} + \nabla_\mu\nabla_\nu f - g_{\mu\nu}\Box f\right), \tag{7.177}$$

where

$$T^{(\lambda)}_{\mu\nu} = h(\lambda)\nabla_\mu\lambda\nabla_\nu\lambda - g_{\mu\nu}\left(\frac{1}{2}h(\lambda)g^{\rho\sigma}\nabla_\rho\lambda\nabla_\sigma\lambda + V(\lambda)\right). \tag{7.178}$$

We can equivalently treat the λ field as a matter field. Note that these equations govern only the gravitational sector. The dynamics of the scalar field are instead determined by

$$h\Box\lambda + \frac{1}{2}h' g^{\mu\nu}\nabla_\mu\lambda\nabla_\nu\lambda - V' + f'R = 0, \tag{7.179}$$

where the "prime" denote the derivatives with respect to λ. For $h = 1$, this reduces to a canonical scalar field theory with an additional $f'R$ correction. Since $f(\lambda)$ governs the gravitational coupling, it is tightly constrained by experimental tests, including solar system observations and primordial nucleosynthesis, implying that its variation must be sufficiently slow. Meanwhile, the potential $V(\lambda)$ plays a key role in setting the vacuum expectation value of λ.

A prominent example of such a modified theory is the Brans–Dicke theory, characterized by the couplings

$$f(\lambda) = \frac{\lambda}{16\pi}, \quad h(\lambda) = \frac{\omega}{8\pi\lambda}, \quad V(\lambda) = 0, \tag{7.180}$$

where ω is a dimensionless parameter. In this framework, the scalar field λ dynamically influences the effective gravitational coupling. Notably, the limit $\omega \to \infty$ suppresses the kinetic term of λ, rendering the scalar field non-dynamical and recovering general relativity as a limiting case. Current experimental constraints, derived from solar system tests impose a lower bound of $\omega \geq 500$.

Through a conformal (Weyl) transformation, any general scalar-tensor theory can be recast into the form of general relativity minimally coupled to a scalar field. The transformation is defined by

$$\widetilde{g}_{\mu\nu} = 16\pi G f(\lambda) g_{\mu\nu}, \tag{7.181}$$

which effectively absorbs the non-minimal coupling $f(\lambda)R$ into the rescaled metric. In this new frame, referred to as the Einstein frame, the action reduces to the standard Einstein–Hilbert action plus a canonical scalar field action, thereby eliminating the explicit $f(\lambda)R$ dependence of the original Jordan frame.

Another well-known modified gravitational theory is extra-dimensional gravity, which traces its conceptual origins to the pioneering work of Theodor Kaluza and Oskar Klein. These theories postulate a gravitational framework in higher dimensions characterized by the metric field G_{MN}, M, $N = 0, \cdots, d+3$. The higher-dimensional metric can be decomposed into four-dimensional components through the dimensional reduction:

$$\begin{aligned} \mathrm{d}s^2 &= G_{MN}\mathrm{d}x^M\mathrm{d}x^N \\ &= g_{\mu\nu}\mathrm{d}x^\mu\mathrm{d}x^\nu + b^2(x)\gamma_{ij}(y)\mathrm{d}y^i\mathrm{d}y^j, \end{aligned} \tag{7.182}$$

where $b(x)$ is a scalar function dependent solely on the four-dimensional coordinates x^μ only, and $\gamma_{ij}(y)$ depends only on the extra dimensions. This decomposition enables a tractable analysis of the higher-dimensional theory. The corresponding gravitational dynamics is governed by the higher-dimensional Einstein–Hilbert action supplemented by matter fields:

$$S = \int \mathrm{d}^{d+4}x\sqrt{-G}\frac{1}{16\pi G_{d+4}}(R(G_{MN}) + \mathcal{L}_\mathrm{m}). \tag{7.183}$$

Since $\sqrt{-G} = b^d\sqrt{-g}\sqrt{\gamma}$ and $V_d = \int \mathrm{d}^d y\sqrt{g}$, we obtain

$$\frac{1}{16\pi G_4} = \frac{V_d}{16\pi G_{d+4}}. \tag{7.184}$$

The dimensionally reduced action then takes the form

$$\begin{aligned} S = \int \mathrm{d}^4x\sqrt{-g}\Bigg\{&\frac{1}{16\pi G_4}(b^d R[g_{\mu\nu}] + d(d-1)b^{d-2}g^{\mu\nu}\nabla_\mu b\nabla_\nu b \\ &+ d(d-1)Kb^{d-2}) + V_d b^d\mathcal{L}_\mathrm{m}\Bigg\}, \end{aligned} \tag{7.185}$$

where $K = \dfrac{R[\gamma_{ij}]}{d(d-1)}$ represents the curvature of the internal space. Remarkably, this reduction yields an effective scalar-tensor theory in four dimensions, with b emerging as a dilaton field coupling to gravity.

From the perspective of effective field theory, general relativity can be regarded as the low-energy effective theory of quantum gravity. In this framework, the gravitational action may include additional higher-order derivative terms:

$$S = \int \mathrm{d}^D x\sqrt{-g}(R + \alpha_1 R^2 + \alpha_2 R_2 + \alpha_3 g^{\mu\nu}\nabla_\mu R\nabla_n R + \cdots). \tag{7.186}$$

Gravitational theories containing higher-order derivatives are often called higher-order derivative gravity. Higher-order derivative terms can improve the ultraviolet behavior of the theory. For example, it has been proven that when there is an R^2 term, the theory is one-loop renormalizable. However, this property does not persist at higher loop orders, meaning the theory fails to be fully renormalizable in general. While most higher-order derivative extensions of general relativity are not UV-complete, certain theories may exhibit asymptotic safety, where the gravitational coupling constants flow toward a UV fixed point. Nevertheless, these theories face significant challenges: One problem with higher-order derivative gravity is that the initial value problem cannot be well-defined, as the equations of motion are not just second-order derivatives but also include higher-order derivatives; Another problem with higher-order derivative gravity is the existence of ghost fields, which casts doubt on the unitarity of the theory. These issues arise because higher-derivative terms in the equations of motion lead to additional propagating degrees of freedom, some of which may be unphysical. If we relax the requirement of diffeomorphism invariance, particularly isometry and scale invariance, it becomes possible to construct higher-derivative theories without ghosts The most famous construction of this type is the Horava–Lifshitz theory. In this theory, isotropy is broken, and time and space follow different scaling laws. The spatial derivatives can be of higher order (improving UV behavior), while time derivatives remain second-order (avoiding ghosts). This approach can lead to improved UV properties while maintaining unitarity. However, it may introduce new challenges, such as the unwanted proliferation of degrees of freedom or strong coupling issues in the infrared limit.[6,7]

If the equivalence principle holds universally, it would inherently suppress the emergence of correction terms. In string theory, the dilaton breaks Einstein's equivalence principle; however, its implications for the equivalence principle at higher-order derivatives remain incompletely understood.

7.7.2 *Infrared modifications of gravity*

Experimentally, Einstein's general relativity has only been tested within a limited range of scales: from millimeters to the size of the solar system. Without empirical confirmation, we cannot simply assume its validity at all scales, particularly on cosmological distances. Recent observations reveal that our understanding of large-scale cosmology remains highly incomplete. According to the standard ΛCDM cosmological model, the universe should contain significant amounts of dark matter and dark energy, yet their fundamental nature remains poorly understood. This provides strong motivation to consider possible modifications to general relativity on large scales, commonly referred to as infrared (IR) modifications. Various models have been proposed for such modifications, including:

(1) The Dvali–Gabadadze–Porrati (DGP) model is a braneworld scenario in which our four-dimensional universe is embedded in a

[6]Additional complications, such as the decoupling of extra degrees of freedom, may also arise in such theories.

[7]Of course, there may be other problems with this kind of theory, such as decoupling problems with extra degrees of freedom.

higher-dimensional spacetime. This framework becomes particularly compelling when considered in conjunction with particle physics, offering a promising avenue for exploring modifications to gravity at large scales.

(2) Massive gravity is a theoretical framework in which the gravitational field acquires a nonzero mass, thereby explicitly breaking diffeomorphism invariance. The most well-known realization of this idea is the Fierz–Pauli model. In recent years, substantial theoretical advances have been made in massive gravity, leading to the discovery of fully consistent nonlinear extensions. For detailed discussions, see Ref. [38].

(3) The f(R) gravity framework introduces modifications to Einstein's theory through nonlinear functions of the Ricci scalar. In particular, correction terms scaling as 1/R can dominate on cosmological scales, producing observable deviations from general relativity. However, the fundamental physical justification for these terms remains unclear, and the theory may suffer from theoretical inconsistencies.

(4) The ghost condensation scenario postulates spontaneous breaking of Lorentz symmetry at high energy scales, with the symmetry being dynamically restored at low energies. In this framework, the ghost field acquires a vacuum expectation value, leading to an effective low-energy theory that respects Lorentz invariance while maintaining modified gravitational behavior at high energies.

These modified gravity theories share a common characteristic: they introduce additional degrees of freedom that can potentially account for dark energy through dynamical scalar fields. However, solar system tests place stringent constraints on the existence of such scalar fields. Typically, each modified theory suffers from various theoretical challenges, requiring either the introduction of new mechanisms or significant fine-tuning—solutions that often compromise the theory's naturalness. In terms of both conceptual simplicity and mathematical elegance, Einstein's general relativity remains the most natural framework for gravitation.

Recent developments in gauge/gravity duality have challenged conventional notions about the fundamental nature of gravity. A particularly intriguing paradigm is emergent gravity, which posits that gravity may not be fundamental and that spacetime itself could be an emergent phenomenon. The gauge/gravity correspondence provides compelling evidence for this view, demonstrating an exact equivalence between higher-dimensional gravitational theories and lower-dimensional quantum field theories, where the additional spacetime dimensions encode the renormalization group flow of the quantum system. This remarkable correspondence suggests that higher-dimensional gravity may emerge from the collective dynamics of lower-dimensional quantum degrees of freedom. However, key challenges remain in understanding how to systematically reconstruct spacetime geometry from quantum field theory data and how to fully recover the nonlinear dynamics of general relativity in this framework.

At the Planck scale, where quantum gravitational effects become dominant, our conventional understanding of spacetime as a smooth manifold may no longer hold. This profound transition necessitates the development of fundamentally new physical frameworks and mathematical formalisms, including non-commutative geometry, matrix models, and spacetime foam approaches. Beyond the nature of spacetime itself, the fundamental role of diffeomorphism symmetry in gravitational theories merits careful examination. In certain modified gravity theories—such as Fierz–Pauli massive gravity and Horava–Lifshitz gravity—diffeomorphism invariance emerges as merely an approximate symmetry rather than a fundamental principle. The breaking of this symmetry introduces additional dynamical degrees of freedom beyond the two polarization states of massless gravitons. A critical challenge for such theories lies in naturally suppressing these extra modes while maintaining consistency with precision solar system tests of gravity.

7.8 Energy Conditions

In general relativity, the right-hand side of Einstein's equations is the energy-momentum tensor, which determines how spacetime is curved. An interesting question is what kinds of energy-momentum tensor are physically permissible or well-motivated. Another related question is, given an energy-momentum tensor that satisfies certain conditions, what is its physical impact on spacetime. Both of these questions are intimately related to the so-called "energy conditions." Since the explicit form of the energy-momentum tensor depends on the choice of coordinates, i.e., the observer's frame of reference, it is natural to require that energy conditions be coordinate-independent. This can be achieved by constructing scalar quantities from the energy-momentum tensor. If an observer's 4-velocity is ξ^μ, then the energy density of the matter he measures is $T_{\mu\nu}\xi^\mu\xi^\nu$. This expression serves as the foundation for formulating various energy conditions.

A widely studied energy condition is the weak energy condition (WEC).

Weak energy condition The WEC states that for any timelike observer with 4-velocity t^μ, the observed energy density must be non-negative:

$$T_{\mu\nu}t^\mu t^\nu \geq 0. \tag{7.187}$$

For the case of an ideal fluid, whose energy-momentum tensor takes the form $T_{\mu\nu} = (\rho+p)u_\mu u_\nu + pg_{\mu\nu}$, condition (7.187) reduces to the inequalities

$$\rho \geq 0, \quad \rho + p \geq 0. \tag{7.188}$$

The statement of the weak energy condition appears to assert that the energy density of the ideal fluid, as measured by any observers, must be non-negative. However, the preceding discussion reveals that this condition not only constrains the energy density but also imposes restrictions on the pressure.

Another energy condition closely related to the weak energy condition is the null energy condition (NEC).

Null energy condition For any null vector l^μ, the stress-energy tensor $T_{\mu\nu}$ must satisfy

$$T_{\mu\nu}l^\mu l^\nu \geq 0. \tag{7.189}$$

It is straightforward to observe that the null energy condition (NEC) modifies the weak energy condition (WEC) by replacing the timelike vector with a null vector. For an ideal fluid, the NEC implies

$$(\rho + p)(\widehat{l} \cdot \widehat{u})^2 \geq 0, \tag{7.190}$$

thus, NEC reduces to the requirement

$$\rho + p \geq 0. \tag{7.191}$$

Another important energy condition is the dominant energy condition (DEC). While the WEC simply requires non-negative energy density as measured by any observer, the DEC imposes additional constraints on both the energy density and energy flux observed locally.

Dominant energy condition For any timelike observer with 4-velocity t^μ, the stress-energy tensor must satisfy

$$T_{\mu\nu}t^\mu t^\nu \geq 0, \tag{7.192}$$

and the energy-momentum current $-T^\mu_\nu t^\nu$ must be causal and future-directed. For an ideal fluid, these conditions translate to

$$\rho \geq 0, \quad \rho \geq |p|. \tag{7.193}$$

By analogy with the dominant energy condition, we may define the null dominant energy condition (NDEC) by replacing the timelike observer's 4-velocity t^μ with a null vector l^μ. The condition then requires $T_{\mu\nu}l^\mu l^\nu \geq 0$, and the energy-momentum current $-T^\mu_\nu l^\nu$ being future-directed and causal. For an ideal fluid, this yields the modified constraint

$$\rho + p \geq 0. \tag{7.194}$$

In this case, ρ can be negative provided that the equation of state satisfies $p = -\rho$.

The last important energy condition is the strong energy condition (SEC). This condition actually originates from the proof of the singularity theorem. Physically, the SEC reflects the reasonable expectation that matter pressures should remain bounded and non-negative. Mathematically, this translates to the requirement

$$T_{\mu\nu}t^\mu t^\nu + \frac{1}{2}T \geq 0. \tag{7.195}$$

Through Einstein's field equations, this condition becomes equivalent to

$$R_{\mu\nu}t^\mu t^\nu \geq 0. \tag{7.196}$$

Similar to the previous analysis, for an ideal fluid, condition (7.196) reduces to

$$\rho + 3p \geq 0, \quad \rho + p \geq 0. \tag{7.197}$$

The most important application of SEC is to the Raychaudhuri equation, where it plays a crucial role in the proof of the singularity theorems.

The above discussion demonstrates that the DEC is stronger than the WEC, and most classical matter fields satisfy the DEC. For example, consider a massless scalar field with the stress-energy tensor

$$T_{\mu\nu} = \partial_\mu\phi\partial_\nu\phi - \frac{1}{2}g_{\mu\nu}(\partial\phi)^2. \tag{7.198}$$

It can be proved that this field satisfies the DEC. However, for a massive scalar field, the strong energy condition (SEC) is violated. When quantum effects are taken into account, all energy conditions may be broken, as quantum fields can exhibit negative energy densities. On the other hand, the conservation of energy-momentum equation $\nabla_\mu T^{\mu\nu} = 0$ actually prevents non-physical processes from occurring, such as energy propagating faster than light, and vacuum spontaneously decaying to E and $-E$.

There are two fundamental theorems in general relativity closely tied to energy conditions. The first is the conservation theorem by Hawking and Ellis.

Conservation theorem (S. Hawking and G. Ellis)
If $T_{\mu\nu}$ satisfies the DEC and vanishes in some spacelike region, then it must remain zero throughout the entire future causal domain of that region.

This theorem implies that energy cannot spontaneously appear without a source, nor can it escape outside the light cone. The other famous theorem related to energy conditions is the positive energy theorem.

Positive energy theorem The ADM energy of an asymptotically flat spacetime satisfying Einstein's equations $G_{\mu\nu} = 8\pi G T_{\mu\nu}$ is semi-positive definite.[8] Furthermore the ADM energy vanishes if and only if the spacetime is globally Minkowskian, i.e., $T_{\mu\nu} = 0$ everywhere. For the theorem to hold, the following conditions must be satisfied:

(1) Existence of a regular Cauchy surface: The spacetime must admit an initial Cauchy hypersurface free of singularities (otherwise, exotic solutions with negative mass $M < 0$ could arise as counterexamples);
(2) $T_{\mu\nu}$ must obey the DEC, ensuring causal energy flow;
(3) Additional technical assumptions. These typically involve asymptotic falloff conditions for metric components and constraints on matter fields.

The concept of Cauchy surfaces in the context of the positive energy theorem is formally introduced in Section 7.9. This theorem was initially established by Richard Schoen and Shing-Tung Yau through sophisticated geometric analysis methods. Subsequently, Edward Witten provided an alternative and physically insightful proof utilizing spinor field techniques on Riemannian manifolds. The profound significance of the positive energy theorem stems from its rigorous demonstration that Minkowski spacetime represents the stable ground state of minimal energy—effectively serving as the vacuum configuration—while all other asymptotically flat spacetimes correspond to excited states relative to this fundamental solution.

7.9 Causality

Unlike flat Minkowski spacetime, the causal structure of curved spacetime is much more complex. For simplicity, we temporarily assume a fixed spacetime background and focus solely on the evolution of matter fields and their associated causal relationships. By the equivalence principle, spacetime is

[8]The formal definition of ADM energy can be found in [64].

locally flat, ensuring that no signal can propagate faster than the speed of light. A causal curve is defined as a curve along which every point is connected by either timelike or null (lightlike) tangent vectors. Given a subset S of the spacetime manifold M, its causal future, denoted $J^+(S)$, consists of all points in M that can be reached from S via future-directed causal curves. Similarly, the chronological future of S, $I^+(S)$, is the subset of $J^+(S)$ containing only points connected to S by strictly timelike curves. Evidently, $I^+(S) \subset J^+(S)$, which also means that the spacetime points in $J^+(S)$ that do not belong to $I^+(S)$ are connected to S by null curves. A subset $S \in M$ is called achronal if no two points in S are connected by a timelike curve. A classic example is a boundaryless spacelike hypersurface in Minkowski spacetime, which is inherently achronal.

Given a closed achronal set S, the future domain of dependence $D^+(S)$ is defined as the set of all points P such that every past-directed timelike or null, inextendible curve passing through P must intersect S. Here, inextendible means that the curve can be extended indefinitely in the past direction without terminating. By definition, $S \in D^+(S)$. Analogously, the past domain of dependence $D^-(S)$ is defined by replacing the past-directed curves with future-directed curves in the above condition. The future Cauchy horizon $H^+(S)$ is then defined as the boundary of $D^+(S)$, while the past Cauchy horizon $H^-(S)$ is the boundary of $D^-(S)$. It follows directly that both $H^+(S)$ and $H^-(S)$ are null surfaces. For a visual representation, refer to Figure 7.4.

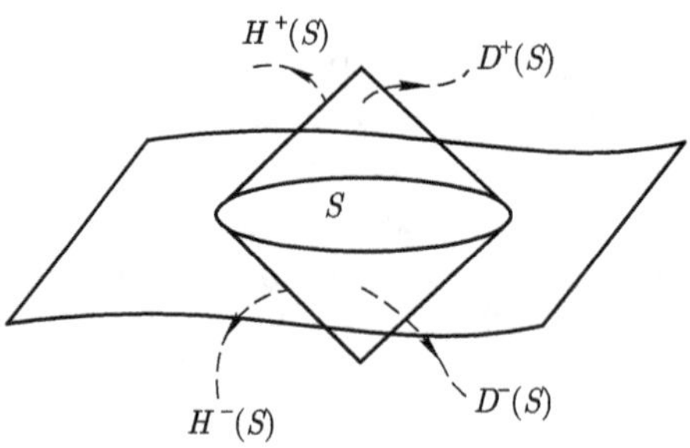

Fig. 7.4 The causal region of a non-chronal set S.

The causal domain associated with the set S denoted $D(S) = D^+(S) \cup D^-(S)$, represents the region where the physical state is fully determined by data on S. Specifically, knowledge of the conditions on S allows us to predict events in$D^+(S)$ or retrodict events in $D^-(S)$. Now consider a closed achronal surface Σ. Even if Σ is extended arbitrarily far in space, its causal domain $D(\Sigma) = D^+(\Sigma) \cup D^-(\Sigma)$ may still fail to cover the entire spacetime, a striking departure from the behavior of flat Minkowski spacetime. We call Σ a Cauchy surface if $D(\Sigma)$ encompasses the whole spacetime manifold. In such a case, specifying initial data on Σ completely determines the past and future evolution of the entire spacetime. A spacetime admitting a Cauchy surface is termed globally hyperbolic. Such spacetimes exhibit a well-behaved causal structure and a rigorously defined initial value problem, making them particularly tractable for physical and mathematical analysis.

A closed, achronal, boundaryless set is referred to as a partial Cauchy surface. Such a surface may or may not be a true Cauchy surface. If it is not, its domain of dependence $D(\Sigma)$ fails to cover the entire spacetime manifold M. This incompleteness arises in two distinct scenarios:

(1) **Ill-chosen hypersurface.**
The surface Σ itself may be poorly selected, as illustrated in Figure 7.5. In such cases, the future domain of dependence $D^+(\Sigma)$ remains confined within the past domain of dependence of some point P, preventing Σ from determining the future evolution beyond P.

(2) **Pathological spacetime structure.**
The spacetime may contain closed timelike curves (CTCs) or intrinsic singularities, both of which indicate a breakdown in causal structure. Such features render the spacetime "pathological," as they preclude the existence of a global Cauchy surface.

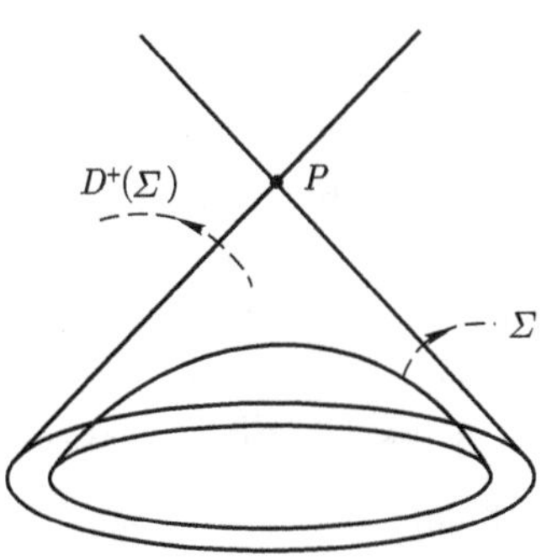

Fig. 7.5 A "bad" hypersurface.

Closed timelike curves do not exist in Newtonian gravity, where time is absolute and strictly unidirectional. In the Minkowski spacetime of special relativity, temporal evolution is further constrained: not only does time advance forward, but it is also confined within the light cone, enforcing the relativistic limit $v \leq c$. In general relativity, the trajectory of a massive particle must locally remain within its forward light cone. However, this condition is inherently local. Globally, spacetime curvature can cause light cones to tilt as one moves from one region to another. In principle, worldlines within these tilted light cones remain timelike everywhere, yet they may eventually self-intersect, forming what is known as a closed timelike curve (CTC). The existence of CTCs presents a profound challenge to the well-posedness of the initial value problem in general relativity.

Example 7.6. Misner Spacetime.

A typical example of a spacetime admitting closed timelike curves is the Misner spacetime. Consider a two-dimensional spacetime with the metric

$$\mathrm{d}s^2 = -\cos\lambda \mathrm{d}t^2 - \sin\lambda(\mathrm{d}t\mathrm{d}x + \mathrm{d}x\mathrm{d}t) + \cos\lambda \mathrm{d}x^2, \tag{7.199}$$

where the function $\lambda(t)$ is given by

$$\lambda = \pi - \cot^{-1} t = \begin{cases} 0, & t = -\infty, \\ \pi, & t = \infty. \end{cases} \tag{7.200}$$

Additionally, the spacial x-direction is compactified with periodicity, $x \sim x + 1$. While this spacetime is not an exact solution of the Einstein equations, it serves as a valuable toy model for studying the geometric and causal properties of spacetimes with closed timelike curves. The spacetime is topologically $\mathbf{R} \times \mathbf{S}^1$, forming a cylindrical geometry. For $t < 0$, the vector field ∂_t is timelike. For $t > 0$, $\cot^{-1} t < \pi/2$, $\cos\lambda < 0$, making ∂_x timelike instead. Due to the compact x-direction, any integral curve of ∂_x for $t > 0$ becomes a closed timelike curve. If we choose a spacelike hypersurface Σ at $t < 0$, the region containing the closed timelike curves lies outside its causal future $J^+(\Sigma)$. Moreover, no CTC intersects Σ, meaning the curves are causally disconnected from initial data defined on Σ. As illustrated in Figure 7.6, the tilting of light cones mediates the transition from a causally well-behaved region $t < 0$ to a CTC-allowing regime $t > 0$.

Another pathological behavior of spacetime arises from the presence of singularities. Although singularities are not part of the manifold in principle, they can be reached by geodesics within a finite affine parameter. In many cases, singularities represent points of geodesic incompleteness, where the smooth extension of spacetime breaks down. A key characteristic of these singularities is the divergence of spacetime curvature scalars, signaling an extreme gravitational regime. This divergence leads to the

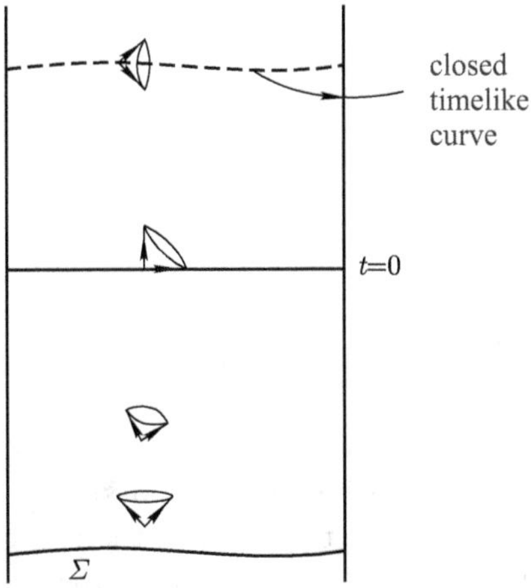

Fig. 7.6 The light cone structure of Misner spacetime.

emergence of a Cauchy horizon: a point P in the future of a singularity cannot be causally connected to any spacelike hypersurface in the past of the singularity. This is because the past-directed causal curves through P necessarily terminate at the singularity and cannot be extended further. Moreover, due to the inherently attractive nature of gravity, the formation of singularities is an unavoidable consequence in classical general relativity.

The existence of closed timelike curves and singularities has profound implications on the initial value problem in general relativity. In well-behaved spacetimes, the choice of initial hypersurface is typically straightforward–particularly when the global structure of the manifold is known and geodesically complete. However, if the initial hypersurface is improperly chosen, the resulting evolution may lead to ill-posed numerical solutions of the Einstein field equations. While spacetimes containing CTCs are inherently pathological, such configurations do not generically arise from ordinary initial value evolution. In contrast, singularities present a far more fundamental challenge. As dictated by the singularity theorems, their formation is often unavoidable under physically reasonable conditions, severely disrupting the predictability and determinism of the initial value formulation.

Exercises

1. The metric of a two-dimensional sphere is

$$ds^2 = d\theta^2 + \sin^2\theta d\phi^2. \tag{7.201}$$

On the sphere, prove that along the geodesic $\phi =$ constant, the geodesic deviation vector t_i satisfies

$$\frac{D^2 t^\theta}{ds^2} = 0, \quad \frac{D^2 t^\phi}{ds^2} = -t^\phi \left(\frac{d\theta}{ds}\right)^2. \tag{7.202}$$

Take a geodesic $\phi = \phi_0$, and the path length on it is measured from $\theta = 0$ to get $s = \theta$, and a neighboring geodesic $\phi = \phi_0 + \delta\phi_0$ also has $s = \theta$. Define $t^i(\theta)$ as the vector between the two geodesics at $s = \theta$. Prove that for all θ, $t(\theta) = (0, t^\phi)$. Further prove that if $t^\phi = 0$ when $\theta = 0$, then

$$t^\phi t_\phi = l^2 \sin^2\theta, \tag{7.203}$$

where l^2 is a constant, and the two geodesics pass through the point $\theta = \pi$.

2. In the discussion of the geodesic deviation equation, if the family of curves is not geodesics, prove that the deviation vector satisfies the equation

$$\frac{D^2 S^\mu}{d\tau^2} = R^\mu{}_{\nu\rho\sigma} T^\nu T^\rho S^\sigma + S^\nu \nabla_\nu a^\mu, \tag{7.204}$$

where $a^\mu = u^\nu \nabla_\nu u^\mu$ is the acceleration 4-vector.

3. The simplest relativistic generalization of Newtonian gravity is the so-called scalar theory, in which Newton's gravitational Poisson equation is generalized to

$$\Box\Phi = 4\pi G T^\mu_\mu. \tag{7.205}$$

For an ideal fluid, we have

$$\Box\Phi = 4\pi G(\rho - 3p). \tag{7.206}$$

Prove that in the non-relativistic limit, the equation reduces to the Poisson equation of Newtonian mechanics. How to introduce the cosmological constant into the theory?

4. The Lagrangian density of electromagnetism in curved spacetime is

$$\mathcal{L} = -\sqrt{-g}\left(\frac{1}{4} F^{\mu\nu} F_{\mu\nu} + A_\mu J^\mu\right), \tag{7.207}$$

where J^μ is the conserved current density that generates the electromagnetic field. Prove that the energy-momentum tensor of the electromagnetic field interacting with a source satisfies $\nabla_\mu T^{\mu\nu}_{\text{em}} = -F^{\nu\sigma} J_\sigma$. Thereby prove that the worldline of a particle with charge q in the electromagnetic field satisfies

$$\ddot{z}^\nu + \Gamma^\nu_{\ \sigma\mu}\dot{z}^\mu \dot{z}^\sigma = \frac{q}{m} F^\nu_{\ \sigma}\dot{z}^\sigma. \tag{7.208}$$

Explain this result physically.

5. If an additional term

$$\mathcal{L}' = \beta R^{\mu\nu} g^{\rho\sigma} F_{\mu\rho} F_{\nu\sigma}, \tag{7.209}$$

is added to the Lagrangian density in the previous question, how do Maxwell's equations change? How do Einstein's equations change? Is the current conserved?

6. Consider a conformal transformation $g_{\mu\nu} \to e^{2\omega(x)} g_{\mu\nu}$, while the electromagnetic gauge potential A_μ and the electromagnetic field strength $F_{\mu\nu}$ remain unchanged. Prove that the action of the free electromagnetic field is invariant under the conformal transformation. If in a spacetime with a metric of the form $g_{\mu\nu} = \phi^2(x)\eta_{\mu\nu}$, what changes will occur to the solutions of Maxwell's equations?

7. On the surface of the Earth, a charged object will not emit electromagnetic radiation if it remains stable, but if the object falls freely, electromagnetic radiation will occur. Does this conform to the equivalence principle?

8. Consider an ideal fluid at rest in a static spacetime, that is, the flow lines satisfy x^i = constant. Prove that the Euler equation gives

$$\frac{\partial p}{\partial x^0} = 0, \quad \frac{\partial p}{\partial x^i} = -(\rho + p)\frac{\partial \ln \sqrt{-g_{00}}}{\partial x^i}. \tag{7.210}$$

Prove that for an extremely relativistic fluid $p = \frac{1}{3}\rho$, if it is in a static equilibrium state in a gravitational field, there cannot be a free surface, that is, a surface where $\rho \to 0$.

9. Prove that if a fluid is at rest in a static metric, then its temperature satisfies the relation $T\sqrt{-g_{00}}$ = constant.

10. An ideal gas flows adiabatically and steadily in a steady gravitational field[9] with a 4-velocity $\hat{u}$. Prove that on the flow lines, there must be

$$u_0 \propto \frac{n_{\text{b}}}{\rho + p}, \tag{7.211}$$

[9]The concepts of static and steady spacetime used here will be discussed in the next chapter.

where $n_{\rm b}$ is the baryon number density. Furthermore, prove that in the case of slow speed and weak gravitational field, this equation reduces to the standard Bernoulli equation.

11. Consider a minimally coupled scalar field in 4-dimensional curved spacetime, with the action

$$S = -\frac{1}{2}\int d^4x\sqrt{-g}\partial_\mu\phi\partial^\mu\phi. \tag{7.212}$$

Assuming that under the conformal transformation $\phi \to e^{n\omega(x)}\phi$, $n \in N$, prove that the above action cannot remain invariant. On the other hand, if the action is taken as

$$S = -\int \mathrm{d}^4x\sqrt{-g}\left(\frac{1}{2}\partial_\mu\phi\partial^\mu\phi + \frac{1}{6}R\phi^2\right), \tag{7.213}$$

then under the conformal transformation $\phi \to e^{-\omega(x)}\phi$, the action remains invariant.

12. Prove that the energy-momentum tensors of the electromagnetic field and the scalar field both satisfy the dominant energy condition, and therefore also satisfy the weak, null, and dominant null energy conditions. The equations of state of these two types of matter both satisfy $\omega \geq -1$.

13. Consider a scalar field with the action

$$S = -\int p(X,\phi)\sqrt{-g}\mathrm{d}^4x, \tag{7.214}$$

where $X \equiv \frac{1}{2}g^{\mu\nu}\partial_\mu\phi\partial_\nu\phi$, and p is an arbitrary function of its arguments. Find the energy-momentum tensor of this system and prove that it can be expressed as $T^{\mu\nu} = (\rho + p)u^\mu u^\nu + pg^{\mu\nu}$, where $u_\mu = (2X)^{-1/2}\partial^\mu\phi$, $\rho = 2X(\partial p/\partial X) - p$.

Chapter 8

Spherically Symmetric Vacuum Solution

In this chapter, we examine the Schwarzschild solution—the simplest nontrivial exact solution in general relativity. This spherically symmetric solution is of great physical significance, describing both the exterior spacetime geometry of spherical astrophysical objects and serving as the fundamental model for non-rotating black holes formed through gravitational collapse.

The Einstein equations

$$R_{\mu\nu} - \frac{1}{2} g_{\mu\nu} R = 8\pi G T_{\mu\nu} \tag{8.1}$$

constitute a system of highly nonlinear differential equations. For a generic matter distribution and its associated energy-momentum tensor $T_{\mu\nu}$, obtaining exact analytical solutions is exceptionally challenging. Indeed, even numerical approaches often face significant difficulties—for instance, stable and reliable simulations of spacetime dynamics during black hole collisions were not achieved until 2005.[1] Remarkably, Einstein himself initially doubted whether exact solutions could be found at all. Yet, merely months after the formulation of general relativity, Karl Schwarzschild derived the first nontrivial exact solution.[2] Schwarzschild's approach was to seek a static, spherically symmetric solution. Due to the local nature of the Einstein equations, a static and spherically symmetric matter distribution

[1] Numerical relativity remains an active and important subfield of general relativity. Due to scope limitations and the author's expertise, this topic will not be covered in detail here; interested readers may consult references [53, 54].

[2] Karl Schwarzschild (1873–1916) was a German physicist, astronomer, and member of the Prussian Academy of Sciences. Within weeks of encountering Einstein's equations, he obtained the spherically symmetric vacuum solution while serving as an artillery officer on the Eastern Front during World War I. His result was communicated to Einstein on December 22, 1915, and published shortly afterward on January 16, 1916. Tragically, Schwarzschild succumbed to pemphigus and died months later.

naturally implies that the exterior spacetime geometry inherits these symmetries. Thus, when studying the region outside a star, we may restrict our attention to the vacuum Einstein equations. In this chapter, we will focus exclusively on the spherically symmetric vacuum solution. Notably, it can be rigorously shown that—in four dimensions—the Schwarzschild solution is the unique spherically symmetric vacuum solution to Einstein's equations.

8.1 Stationary and Static Spacetime

In classical physics, a system is said to be in a steady state if its macroscopic properties remain stable over time, even if there is persistent motion e.g., a fluid flowing at constant velocity in a pipe. In contrast, a static system is not only time-invariant but also devoid of any internal motion or evolution—all observables are stationary in a given reference frame. However, these distinctions are reference-frame-dependent. For instance, a uniformly flowing gas appears time-independent (stationary) in a static lab frame, but in a comoving frame (moving with the fluid), the same system appears static—no motion is observed. This raises a fundamental question: How can we define stationary or static spacetimes in general relativity in a coordinate-independent, observer-invariant manner? In general relativity, the distinction is made precise through Killing vector fields.

Definition A metric is called stationary if there exists a special coordinate system in which the metric is explicitly time-independent, i.e.,

$$\left.\frac{\partial g_{\mu\nu}}{\partial x^0}\right|_* = 0, \tag{8.2}$$

where x^0 is the timelike coordinate in this coordinate system. This essentially means that ∂_{x^0} is a Killing vector. In short, a spacetime is called stationary if and only if it has a timelike Killing vector field.

For a stationary spacetime, by choosing a suitable coordinate system, the metric can always be expressed in the form:

$$\mathrm{d}s^2 = g_{00}(\mathrm{d}x^0)^2 + g_{0i}\mathrm{d}x^0\mathrm{d}x^i + g_{ij}\mathrm{d}x^i\mathrm{d}x^j, \tag{8.3}$$

where all the metric components are independent of the timelike coordinate. Note that $g_{0i} \neq 0$ is allowed, indicating that the spacetime may still possess evolution. If we further impose the condition that the spacetime is static, meaning the metric remains invariant under the transformation $x^0 \to x'^0 = -x^0$, we must have

$$g_{\mu\nu} = g'_{\mu\nu} \quad \Rightarrow \quad g_{0i} = 0. \tag{8.4}$$

Thus, a general form of the metric for a static spacetime can be written as

$$\mathrm{d}s^2 = g_{00}(\mathrm{d}x^0)^2 + g_{ij}\mathrm{d}x^i\mathrm{d}x^j. \tag{8.5}$$

To provide a coordinate-independent characterization of a static spacetime, we introduce the concept of a hypersurface orthogonal vector. A hypersurface can be defined by a family of equations $f(x^\mu) = c$. At a point P on this hypersurface, the components of the normal covector are given by $n_\mu \equiv \dfrac{\partial f}{\partial x^\mu}$. A vector field $\widehat{X}$ is said to be hypersurface orthogonal if it is everywhere orthogonal to such a family of hypersurfaces, i.e.,

$$X_\mu = \lambda(x)n_\mu = \lambda f_{,\mu},$$

which leads to

$$X_\mu \partial_\nu X_\sigma = \lambda f_{,\mu}\partial_\nu(\lambda f_{,\sigma}) = \lambda f_{,\mu}\lambda_{,\nu} f_{,\sigma} + \lambda^2 f_{,\mu} f_{,\nu\sigma}. \tag{8.6}$$

Antisymmetrization this expression yields

$$X_{[\mu}\partial_\nu X_{\sigma]} = 0 \Rightarrow X_{[\mu}\nabla_\nu X_{\sigma]} = 0. \tag{8.7}$$

Any hypersurface orthogonal vector field satisfies Eq. (8.7). Conversely, any non-zero Killing vector field satisfying Eq. (8.7) must be hypersurface orthogonal.

Theorem A spacetime is static if and only if it has a hypersurface orthogonal timelike Killing vector field.

Proof For a stationary spacetime, there exists a timelike Killing vector field $X^\mu = \delta^\mu_0$:

$$X_\mu = g_{\mu\nu}X^\nu = g_{\mu 0}, \quad |X|^2 = X_\mu X^\mu = g_{\mu 0}\delta^\mu_0 = g_{00}. \tag{8.8}$$

If further, this Killing vector field is hypersurface orthogonal,

$$X_\mu = |X|^2 f_{,\mu}, \tag{8.9}$$

where $f_{,\mu}$ is required to be normalized, then we have

$$g_{0_\mu} = g_{00} f_{,\mu}. \tag{8.10}$$

For $\mu = 0$, $f_{,0} = 1$, so

$$f = x^0 + h(x^i). \tag{8.11}$$

Consider the coordinate transformation

$$x^0 \to x'^0 = x^0 + h(x^i), \quad x^i \to x'^i = x^i, \tag{8.12}$$

a vector field transforms as

$$X'^\mu = \frac{\partial x'^\mu}{\partial x^\nu} X^\nu = \delta^\mu_0, \tag{8.13}$$

and the metric tensor field transforms as

$$g'_{\mu\nu,0} = 0, \quad g'_{00} = g_{00}, \tag{8.14}$$

while

$$\begin{aligned} g'_{0i} &= \frac{\partial x^\mu}{\partial x'^0} \frac{\partial x^\nu}{\partial x'^i} g_{\mu\nu} \\ &= \frac{\partial x^0}{\partial x'^0} \left(\frac{\partial x^0}{\partial x'^i} g_{00} + \frac{\partial x^j}{\partial x'^i} g_{0i} \right) \\ &= -\partial_i h g_{00} + g_{0i}. \end{aligned} \tag{8.15}$$

We can choose $\partial_i h = g_{0i}/g_{00}$ to make $g'_{0i} = 0$. In this new coordinate system, $g_{0i} = 0$, and the metric is static.

In the preceding analysis, the static metric retains a residual coordinate freedom of the form:

$$x^0 \to x'^0 = Ax^0 + B, \quad x^i \to x'^i = h'^i(x^j), \tag{8.16}$$

where A and B are constants and h'^i is an arbitrary spatial coordinate transformation. These transformations do not change the above discussion, and the obtained metric still maintains the form of a static metric. If the boundary condition requires that $g_{00} \to 1$ at spatial infinity, then $A = 1$. In this way, we define a time coordinate called worldline time. Therefore, for a static spacetime, we regain the idea of absolute time in Newtonian gravity, that is, the spacetime can be well sliced by hypersurfaces $t =$ constant.

8.2 Birkhoff's Theorem

For the vacuum Einstein equations, the spherically symmetric solution is unique. More precisely, there is the following theorem.

Birkhoff theorem The four-dimensional Schwarzschild solution is the unique spherically symmetric solution to the vacuum Einstein equations.

Below, we outline a proof of this theorem. The term "spherical symmetry" in Birkhoff's theorem means that the solution possesses the same symmetry as a two-dimensional sphere. Specifically, a two-dimensional sphere

$\mathbf{S}^2$ admits three Killing vector fields ($V^{(1)}$, $V^{(2)}$, $V^{(3)}$), which satisfy the $so(3)$ Lie algebra:

$$\begin{aligned} [V^{(1)}, V^{(2)}] &= V^{(3)}, \\ [V^{(2)}, V^{(3)}] &= V^{(1)}, \\ [V^{(3)}, V^{(1)}] &= V^{(2)}. \end{aligned} \tag{8.17}$$

To proceed, we invoke Frobenius' theorem in differential geometry: Given a set of mutually commuting vector fields, then there exists a coordinate system in which these vector fields correspond to the partial derivatives along the coordinate directions. In other words, these vector fields generate translations in these coordinates, effectively defining a Cartesian-like coordinate system. More generally, if these vector fields do not commute but form a closed algebra, their integral curves collectively describe a submanifold. On this submanifold, the vector fields are well-defined and generate its isometry group. The submanifold's dimension is at most equal to the number of vector fields, as the latter cannot span a space of higher dimension. Crucially, since the vector fields are globally defined on the manifold, every point lies on some such submanifold. In the context of spherically symmetric spacetime, the two-dimensional sphere arises as the submanifold generated by the Killing vector fields. The full spacetime may be viewed as a foliation, with the spherical symmetry "leafing" out from this two-dimensional base.

Example 8.1. Three-dimensional Euclidean space $\mathbf{R}^3$.

Three-dimensional Euclidean space $\mathbf{R}^3$ is obviously spherically symmetric, which can be clearly seen using spherical coordinates. In spherical

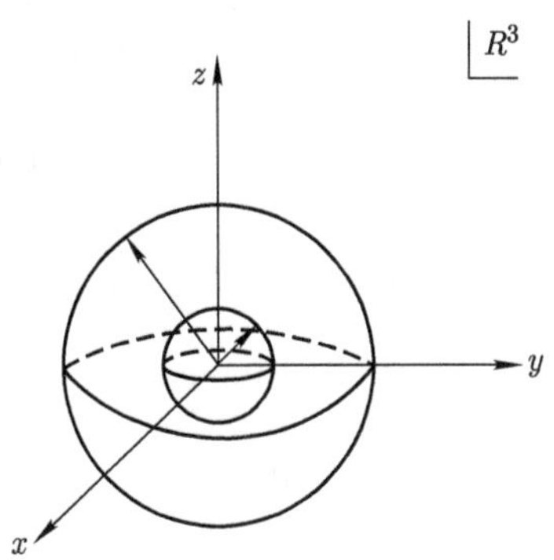

Fig. 8.1 Using spherical coordinates to foliate three-dimensional Euclidean space: at different radial positions, there is a two-dimensional sphere.

coordinates, for each fixed radial coordinate, we have a two-dimensional sphere, as shown in Figure 8.1.

Example 8.2. A wormhole with topology $\mathbf{R} \times \mathbf{S}^2$.

In this case, the spacetime exhibits spherical symmetry without a fixed origin about which the geometry rotates. Instead, the entire manifold is constructed by "attaching" the remaining dimensions to the two-dimensional sphere, as shown in Figure 8.2.

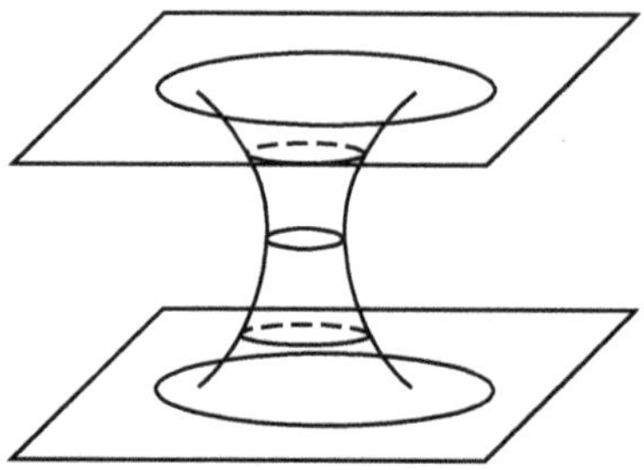

Fig. 8.2 A schematic diagram of a wormhole embedded in a higher dimension. For display convenience, each circle actually represents a two-dimensional sphere. The spacetime can be foliated along the vertical direction.

In general, consider an n-dimensional manifold foliated on an m-dimensional submanifold. We can describe such a structure using coordinate functions u^i on the m-dimensional submanifold and $n - m$ additional coordinates v^I to parametrize the transverse directions, thereby specifying the position within the foliation. The coordinate functions u^i and v^I together describe the full manifold. When the submanifold is a maximally symmetric space, such as the spherically symmetric case under consideration, a powerful simplification arises: there exists a choice of u coordinates such that the metric of the entire manifold takes the following block-diagonal form:

$$\mathrm{d}s^2 = g_{\mu\nu}\mathrm{d}x^\mu \mathrm{d}x^\nu = g_{IJ}(v)\mathrm{d}v^I \mathrm{d}v^J + f(v)\gamma_{ij}(u)\mathrm{d}u^i \mathrm{d}v^j, \tag{8.18}$$

where $\gamma_{ij}(u)$ is the metric on the maximally symmetric submanifold. Note that there are no cross terms like $\mathrm{d}v^I \mathrm{d}u^j$ in the above metric, and $g_{IJ}(v)$ and $f(v)$ are only functions of v^I, independent of u^i.

For a spherically symmetric spacetime, using coordinate transformations, we can always start from the general metric ansatz

$$\mathrm{d}s^2 = m(t,r)\mathrm{d}t^2 + n(t,r)\mathrm{d}r^2 + r^2\mathrm{d}\Omega^2 \tag{8.19}$$

to solve the vacuum Einstein equations. Here $m(t,r)$ and $n(t,r)$ are arbitrary functions, and $d\Omega^2$ is the metric on the unit 2-sphere. At this stage, the coordinates (t,r) are merely labels with no predetermined geometric interpretation. To fix the physical meaning of these coordinates, we compare (8.19) with the Minkowski spacetime metric in spherical coordinates:

$$\mathrm{d}s^2 = -\mathrm{d}t^2 + \mathrm{d}r^2 + r^2\mathrm{d}\Omega^2. \tag{8.20}$$

Far from any gravitational source (e.g., a star), we expect spacetime to approach flatness. This asymptotic flatness motivates identifying r as the radial coordinate and t as the time coordinate, with $m(t,r)$ becoming negative to preserve the timelike nature of t. For practical calculations, it is often more convenient to adopt an exponential parametrization of the metric components:

$$\mathrm{d}s^2 = -\mathrm{e}^{2\alpha(t,r)}\mathrm{d}t^2 + \mathrm{e}^{2\beta(t,r)}\mathrm{d}r^2 + r^2\mathrm{d}\Omega^2. \tag{8.21}$$

Note that this metric assumption comes entirely from the requirement of spherical symmetry and is independent of the Einstein equations, so it also applies to solving the spherically symmetric solutions of the non-vacuum Einstein equations. The non-zero Christoffel symbols corresponding to this metric are

$$\begin{aligned}
&\Gamma^0_{00} = \partial_0\alpha, \quad \Gamma^0_{01} = \partial_1\alpha, \quad \Gamma^0_{11} = \mathrm{e}^{2(\beta-\alpha)}\partial_0\beta,\\
&\Gamma^1_{00} = \mathrm{e}^{2(\alpha-\beta)}\partial_1\alpha, \quad \Gamma^1_{01} = \partial_0\beta, \quad \Gamma^1_{11} = \partial_1\beta,\\
&\Gamma^2_{12} = \frac{1}{r}, \quad \Gamma^1_{22} = -r\mathrm{e}^{-2\beta}, \quad \Gamma^3_{13} = \frac{1}{r},\\
&\Gamma^1_{33} = -r\mathrm{e}^{-2\beta}\sin^2\theta, \quad \Gamma^2_{33} = -\sin\theta\cos\theta, \quad \Gamma^3_{23} = \frac{\cos\theta}{\sin\theta},
\end{aligned}$$

where $(0,1,2,3)$ represent (t,r,θ,ϕ) respectively. From these, we can obtain the components of the Riemann tensor:

$$\begin{aligned}
R^0{}_{101} &= \mathrm{e}^{2(\beta-\alpha)}[\partial_0^2\beta + (\partial_0\beta)^2 - \partial_0\alpha\partial_0\beta]\\
&\quad + [\partial_1\alpha\partial_1\beta - \partial_1^2\alpha - (\partial_1\alpha)^2],\\
R^0{}_{202} &= -r\mathrm{e}^{-2\beta}\partial_1\alpha,\\
R^0{}_{303} &= -r\mathrm{e}^{-2\beta}\sin^2\theta\partial_1\alpha,\\
R^0{}_{212} &= -r\mathrm{e}^{-2\alpha}\partial_0\beta,\\
R^0{}_{313} &= -r\mathrm{e}^{-2\alpha}\sin^2\theta\partial_0\beta,\\
R^1{}_{212} &= r\mathrm{e}^{-2\beta}\partial_1\beta,\\
R^1{}_{313} &= r\mathrm{e}^{-2\beta}\sin^2\theta\partial_1\beta,\\
R^2{}_{323} &= (1-\mathrm{e}^{-2\beta})\sin^2\theta,
\end{aligned} \tag{8.22}$$

and the components of the Ricci tensor:

$$\begin{aligned}
R_{00} &= [\partial_0^2\beta + (\partial_0\beta)^2 - \partial_0\alpha\partial_0\beta] + \mathrm{e}^{2(\alpha-\beta)} \\
&\quad \left[\partial_1^2\alpha + (\partial_1\alpha)^2 - \partial_1\alpha\partial_1\beta + \frac{2}{r}\partial_1\alpha\right], \\
R_{11} &= -\left[\partial_1^2\alpha + (\partial_1\alpha)^2 - \partial_1\alpha\partial_1\beta - \frac{2}{r}\partial_1\beta\right] \\
&\quad + \mathrm{e}^{2(\beta-\alpha)}[\partial_0^2\beta + (\partial_0\beta)^2 - \partial_0\alpha\partial_0\beta], \\
R_{01} &= \frac{2}{r}\partial_0\beta, \\
R_{22} &= \mathrm{e}^{-2\beta}[r(\partial_1\beta - \partial_1\alpha) - 1] + 1, \\
R_{33} &= R_{22}\sin^2\theta.
\end{aligned} \tag{8.23}$$

We only need to solve the vacuum Einstein equations $R_{\mu\nu} = 0$. From $R_{01} = 0$, we find

$$\partial_0\beta = 0 \quad \Rightarrow \quad \beta = \beta(r), \tag{8.24}$$

that is, β must be a function of r. Therefore, R_{00} and R_{11} are further simplified, leading to the following two equations:

$$\begin{aligned}
\mathrm{e}^{2(\alpha-\beta)}\left[\partial_1^2\alpha + (\partial_1\alpha)^2 - \partial_1\alpha\partial_1\beta + \frac{2}{r}\partial_1\alpha\right] &= 0, \\
\partial_1^2\alpha + (\partial_1\alpha)^2 - \partial_1\alpha\partial_1\beta - \frac{2}{r}\partial_1\beta &= 0.
\end{aligned} \tag{8.25}$$

This leads to

$$\partial_1(\alpha + \beta) = 0 \quad \Rightarrow \quad \alpha + \beta = g(t), \tag{8.26}$$

where $g(t)$ is an arbitrary function of time. That is to say,

$$\alpha = -\beta(r) + g(t). \tag{8.27}$$

The metric can therefore be expressed in the form:

$$\mathrm{d}s^2 = -\mathrm{e}^{-2\beta(r)+2g(t)}\mathrm{d}t^2 + \mathrm{e}^{2\beta(r)}\mathrm{d}r^2 + r^2\mathrm{d}\Omega^2. \tag{8.28}$$

We still have the freedom to redefine the time coordinate $\mathrm{d}t \to \mathrm{e}^{-g(t)}\mathrm{d}t$ such that $g(t) = 0$. This simplifies the metric to

$$\mathrm{d}s^2 = -\mathrm{e}^{-2\beta(r)}\mathrm{d}t^2 + \mathrm{e}^{2\beta(r)}\mathrm{d}r^2 + r^2\mathrm{d}\Omega^2, \tag{8.29}$$

where all the metric components are now explicitly t-independent. In fact, we have already proven an important fact: any spherically symmetric vacuum solution necessarily admits a timelike Killing vector field ∂_t, and this metric is not only stationary but also static.

Furthermore, from $R_{22} = 0$, we obtain

$$\begin{aligned} &\mathrm{e}^{2\alpha}(2r\partial_1\alpha + 1) = 1 \\ &\Rightarrow \partial_1(r\mathrm{e}^{2\alpha}) = 1 \\ &\Rightarrow \mathrm{e}^{2\alpha} = 1 + \frac{C}{r}, \end{aligned} \tag{8.30}$$

where C is an undetermined integration constant. Finally, we obtain the metric for a spherically symmetric spacetime in vacuum

$$\mathrm{d}s^2 = -\left(1 + \frac{C}{r}\right)\mathrm{d}t^2 + \left(1 + \frac{C}{r}\right)^{-1}\mathrm{d}r^2 + r^2\mathrm{d}\Omega^2. \tag{8.31}$$

As a consistency check, we must verify whether this metric satisfies $R_{00} = 0$. Indeed, it can be confirmed that this condition holds.

As previously discussed, the spherically symmetric vacuum solution describes the spacetime geometry outside a spherically symmetric star. In the limit $r \to \infty$, where the gravitational influence of the star diminishes, we expect the metric to reduce to the Newtonian limit. Specifically, the metric components should satisfy:

$$\begin{aligned} g_{00} &= -\left(1 + \frac{C}{r}\right) \sim g_{00}|_{\text{Newton}} = -(1 + 2\Phi), \\ g_{rr} &= \left(1 + \frac{C}{r}\right)^{-1} \approx 1 - \frac{C}{r} \sim g_{rr}|_{\text{Newton}} = 1 - 2\Phi, \end{aligned} \tag{8.32}$$

where $\Phi = -GM/r$ is the Newtonian gravitational potential. By comparing these expressions, we identify the integration constant C as $C = -2GM$. Substituting this back into the metric, we arrive at the well-known Schwarzschild metric:

Schwarzschild metric

$$\mathrm{d}s^2 = -\left(1 - \frac{2GM}{r}\right)\mathrm{d}t^2 + \left(1 - \frac{2GM}{r}\right)^{-1}\mathrm{d}r^2 + r^2\mathrm{d}\Omega^2. \tag{8.33}$$

When $M \to 0$, corresponding to the absence of a central mass, the Schwarzschild metric reduces to the Minkowski spacetime, as expected for flat spacetime in special relativity. The form of the metric presented earlier assumes natural units, $c = 1$. Restoring dimensional consistency, the Schwarzschild metric including the speed of light c is given by

$$\mathrm{d}s^2 = \left(1 - \frac{2GM}{rc^2}\right)\mathrm{d}t^2 + \left(1 - \frac{2GM}{rc^2}\right)^{-1}\mathrm{d}r^2 + r^2\mathrm{d}\Omega^2. \tag{8.34}$$

In many literatures, some authors introduce the shorthand $\mu = GM/c^2$, which represents the geometric mass (with units of length). It is also common to adopt units where $G = 1$ (and often $c = 1$), simplifying the metric further while retaining physical clarity.

From the form of the Schwarzschild metric, it is evident that the metric appears singular at $r = 0$ and $r = 2GM$. However, the explicit expression of the metric is highly dependent on the choice of coordinates, or more precisely, on the observer's reference frame. These apparent singularities may not represent true physical singularities but could instead be artifacts of the coordinate system, known as coordinate singularities.

For example, consider the polar coordinate system in two-dimensional Euclidean space, where the metric is given by $\mathrm{d}s^2 = \mathrm{d}r^2 + r^2\mathrm{d}\theta^2$. At the origin $r = 0$, the metric becomes degenerate, and the component $g^{\theta\theta} = r^{-2}$ diverges. Nevertheless, the underlying geometry of the plane is everywhere flat, and no genuine singularity exists. The apparent singularity in this case arises solely from the choice of coordinates. Similarly, in the Schwarzschild metric, the singularity at $r = 2GM$is merely a coordinate singularity, as can be verified by transforming to alternative coordinate systems, where the metric remains well-behaved at this radius. The details of such coordinate transformations will be discussed in later chapter.

To determine whether a singularity is genuine (i.e., a physical singularity) rather than a mere artifact of the coordinate system, we can employ the following criterion: a true singularity occurs where curvature invariants diverge. The curvature invariants are scalar quantities constructed from the Riemann tensor, Ricci tensor, Ricci scalar, covariant derivatives, and their contractions, such as

$$R, \quad R_2 = R^{\mu\nu}R_{\mu\nu}, \quad R_4 = R^{\mu\nu\sigma\rho}R_{\mu\nu\sigma\rho}, \quad R_{\mu\nu\rho\sigma}R^{\rho\sigma\lambda\tau}R^{\mu\nu}{}_{\lambda\tau}, \cdots .$$

If any of these scalars diverges at a certain place, that place constitutes a true spacetime singularity. However, this is a sufficient but not necessary condition, as some singularities (e.g., conical singularities) may not exhibit divergent curvature scalars. Another method to identify singularities is by examining geodesic completeness—whether freely falling particles or light rays can be extended indefinitely in proper time or affine parameter. In most cases, the divergence of curvature invariants and geodesic incompleteness coincide in signaling a true singularity. For the Schwarzschild metric, since $R_{\mu\nu} = 0$, $R = 0$, $R_2 = 0$, however

$$R^{\mu\nu\rho\sigma}R_{\mu\nu\rho\sigma} = \frac{48G^2M^2}{r^6} \tag{8.35}$$

diverges as $r \to 0$. This confirms that $r = 0$ is a true curvature singularity. In contrast, at $r = 2GM$, no curvature invariants diverge, indicating that this is merely a coordinate singularity. Nevertheless, this radius is physically significant: it defines the event horizon of a Schwarzschild black hole, known as the Schwarzschild horizon or Schwarzschild radius:

$$R_{\rm s} = \frac{2GM}{c^2}. \tag{8.36}$$

From the above analysis, it is evident that our choice of the constant $C = -2GM$ is motivated by matching the solution to the Newtonian approximation in the weak-field limit. From a purely mathematical standpoint, the constant C remains arbitrary. In particular, selecting $C = 2GM$ with $M > 0$ leads to a spacetime metric that is regular everywhere except at the central singularity $r = 0$. In this case, the coordinate singularity associated with the Schwarzschild horizon vanishes, resulting in a crucial physical implication: while $r = 0$ is still a curvature singularity, the absence of an event horizon exposes this singularity, giving rise to a naked singularity. As we shall discuss later, the existence of a naked singularity is unphysical.

The Schwarzschild metric has several important properties:

(1) Asymptotically flatness.

The Schwarzschild metric is asymptotically flat: as $r \to \infty$, it converges to the Minkowski metric. This reflects the physical expectation that far from the gravitating source (e.g., a star or black hole), spacetime curvature diminishes, and the geometry approaches that of flat spacetime.

(2) Schwarzschild coordinates.

While the Schwarzschild spacetime, as a solution to the vacuum Einstein equations, is coordinate-independent, the metric (8.34) employs a specific choice of coordinates (t, r, θ, ϕ), known as Schwarzschild coordinates. These coordinates are physically meaningful: at radial infinity, they align with the frame of a static observer, consistent with the asymptotic Minkowski limit.

(3) Stationarity and staticity.

Outside a star or a black hole's event horizon, the Schwarzschild spacetime is static, as the metric components are time-independent and ∂_t is a timelike Killing vector. However, for $r < R_{\rm s}$, ∂_t becomes spacelike, although it remains a Killing vector of spacetime. This distinction has profound implications for the causal structure of black holes, which we will explore later when discussing the Schwarzschild black hole.

We have actually proven Birkhoff's theorem: the Schwarzschild metric is the unique spherically symmetric solution to the vacuum Einstein equations. Notably, our derivation imposes no prior assumptions about stationarity or staticity, nor does it depend on the nature of the source generating the spacetime. Even if the central object (e.g., a star) undergoes time-dependent spherical evolution—such as gravitational collapse—the external spacetime remains static and described by the Schwarzschild metric. In other words, no time-varying, spherically symmetric vacuum solutions exist; all such solutions must reduce to Schwarzschild.

The Schwarzschild metric exhibits remarkable uniqueness properties in general relativity, as highlighted by the following theorems.

Israel's theorem Any static vacuum solution to Einstein's field equations must be spherically symmetric and is therefore described by the Schwarzschild metric.

Thus, we have the following uniqueness theorem.[3]

Uniqueness theorem The Schwarzschild geometry is the unique static vacuum solution in general relativity.

While the Schwarzschild spacetime may appear in different forms due to coordinate choices, it is always reducible to the standard Schwarzschild metric via a suitable coordinate transformation. Historically, this has led to its "rediscovery" over twenty times in various coordinate systems.

8.3 Geodesics in Schwarzschild Spacetime

In this section, we examine the geodesic motion of particles in Schwarzschild spacetime. Our analysis encompasses both massive and massless particles, whose distinct kinematic behaviors yield different observational signatures, including those relevant to the classical tests of general relativity. Furthermore, investigating particle geodesics provides crucial insights into the fundamental properties of Schwarzschild spacetime, particularly those characterizing Schwarzschild black holes.

The geodesic equation governing the motion of a particle in spacetime

[3] A similar uniqueness theorem holds in higher dimensions under the assumptions that the theory is still Einstein's general theory of relativity and that the action has no higher derivative term. For a detailed discussion, see S. Hollands and A. Ishibashi, *Black Hole Uniqueness Theorems in Higher Dimensional Spacetime*. Class. Quant. Grav., 2012, 29: 163001.

is given by

$$\frac{\mathrm{d}^2 x^\mu}{\mathrm{d}\lambda^2} + \Gamma^\mu_{\nu\sigma} \frac{\mathrm{d}x^\nu}{\mathrm{d}\lambda} \frac{\mathrm{d}x^\sigma}{\mathrm{d}\lambda} = 0,$$

where λ is an affine parameter along the trajectory. Using the Christoffel symbols for the Schwarzschild metric

$$\begin{aligned}
&\Gamma^1_{00} = \frac{GM}{r^3}(r - R_\mathrm{s}), \quad \Gamma^1_{11} = \frac{-GM}{r(r - R_\mathrm{s})}, \quad \Gamma^0_{01} = \frac{GM}{r(r - R_\mathrm{s})}, \\
&\Gamma^2_{12} = \frac{1}{r}, \quad \Gamma^1_{22} = -(r - R_\mathrm{s}), \quad \Gamma^3_{13} = \frac{1}{r}, \\
&\Gamma^1_{33} = -(r - R_\mathrm{s}) \sin^2\theta, \quad \Gamma^2_{33} = -\sin\theta\cos\theta, \quad \Gamma^3_{23} = \frac{\cos\theta}{\sin\theta},
\end{aligned} \tag{8.37}$$

we obtain a set of geodesic equations:

$$\frac{\mathrm{d}^2 t}{\mathrm{d}\lambda^2} + \frac{R_\mathrm{s}}{r(r - R_\mathrm{s})} \frac{\mathrm{d}r}{\mathrm{d}\lambda} \frac{\mathrm{d}t}{\mathrm{d}\lambda} = 0, \tag{8.38}$$

$$\begin{aligned}
&\frac{\mathrm{d}^2 r}{\mathrm{d}\lambda^2} + \frac{GM}{r^3} r(r - R_\mathrm{s}) \left(\frac{\mathrm{d}t}{\mathrm{d}\lambda}\right)^2 - \frac{GM}{r(r - R_\mathrm{s})} \left(\frac{\mathrm{d}r}{\mathrm{d}\lambda}\right)^2 \\
&\quad - (r - R_\mathrm{s}) \left[\left(\frac{\mathrm{d}\theta}{\mathrm{d}\lambda}\right)^2 + \sin^2\theta \left(\frac{\mathrm{d}\phi}{\mathrm{d}\lambda}\right)^2\right] = 0,
\end{aligned} \tag{8.39}$$

$$\frac{\mathrm{d}^2\theta}{\mathrm{d}\lambda^2} + \frac{2}{r} \frac{\mathrm{d}\theta}{\mathrm{d}\lambda} \frac{\mathrm{d}r}{\mathrm{d}\lambda} - \sin\theta\cos\theta \left(\frac{\mathrm{d}\phi}{\mathrm{d}\lambda}\right)^2 = 0, \tag{8.40}$$

$$\frac{\mathrm{d}^2\phi}{\mathrm{d}\lambda^2} + \frac{2}{r} \frac{\mathrm{d}\phi}{\mathrm{d}\lambda} \frac{\mathrm{d}r}{\mathrm{d}\lambda} + 2\frac{\cos\theta}{\sin\theta} \frac{\mathrm{d}\theta}{\mathrm{d}\lambda} \frac{\mathrm{d}\phi}{\mathrm{d}\lambda} = 0. \tag{8.41}$$

At first glance, these coupled second-order differential equations appear highly intricate and analytically challenging. However, the spherical symmetry of the Schwarzschild spacetime allows for significant simplifications. Rather than solving them directly, we exploit the conserved quantities associated with the spacetime symmetries, namely, energy and angular momentum, to reduce the problem to a more tractable form.

The full isometry group of Schwarzschild spacetime is simply $U(1) \times SO(3)$, where the $U(1)$ factor arises from the time-translation symmetry and $SO(3)$ from spherical symmetry. The timelike Killing vector is $\widehat{\xi} = \partial_t$ (for $r > R_\mathrm{s}$), with components $\xi^\mu = (1, 0, 0, 0)$. Among the three spacelike Killing vectors, the most apparent one is $\widehat{\eta} = \partial_\phi$, representing azimuthal symmetry. Using the Killing vectors, we can define conserved quantities for a particle in geodesic motion: the inner product of the Killing vector field $\widehat{K}$ and the tangent vector $\widehat{u}$ remains constant along the geodesic:

$$K_\mu \frac{\mathrm{d}x^\mu}{\mathrm{d}\lambda} = \text{constant}, \tag{8.42}$$

where λ is the proper time for massive particles and the affine parameter for massless particles.

First, using the timelike Killing vector $\xi^\mu = (1, 0, 0, 0)$, we can define the conserved energy density of the particle. From the expressions $\xi_\mu = \left(-\left(1 - \frac{R_\mathrm{s}}{r}\right), 0, 0, 0\right)$, $u^\mu = \frac{\mathrm{d}x^\mu}{\mathrm{d}\lambda}$, we derive the conserved quantity

$$E = \left(1 - \frac{R_\mathrm{s}}{r}\right)\frac{\mathrm{d}t}{\mathrm{d}\lambda}. \tag{8.43}$$

Indeed, this satisfies $u_0 = g_{0\mu}u^\mu = -E$. For a massive particle following a timelike geodesic, as $r \to \infty$, we have $\|\widehat{\xi}\|^2 \sim -1$, and E corresponds to the total energy per unit mass as measured by a static observer at infinity. This is because such an observer would measure the particle's energy as $E_\mathrm{o} = -\widehat{u}_\mathrm{obs} \cdot \widehat{P} = mE$. Generally speaking, relative to a static observer at infinity, E represents the total energy per unit mass of a particle moving along the geodesic, incorporating both kinetic and gravitational potential energy contributions. Physically, this can be viewed as the energy required by the observer to place a unit mass particle into a specific orbital configuration, effectively, the particle's energy density. Notably, if the particle is released from rest at infinity, $E = 1$. In the case of a massless particle (e.g., a photon) following a null geodesic, E corresponds to its "total energy". However, it is important to note that due to the freedom in choosing the affine parameter for null geodesics, this "total energy" lacks a precise physical meaning—it is inherently arbitrary and does not represent a measurable physical quantity.

Secondly, the $SO(3)$ spatial rotation symmetry allows us to define the conserved angular momentum of the particle. Consider the spacelike Killing vector field $\widehat{\eta} = \partial_\phi$, or equivalently $\eta_\mu = (0, 0, 0, r^2 \sin^2\theta)$. The corresponding conserved quantity for the particle is

$$L = g^{\mu\nu}\eta_\mu u_\nu = r^2 \sin^2\theta \frac{\mathrm{d}\phi}{\mathrm{d}\lambda}. \tag{8.44}$$

For a massive particle, L represents the angular momentum per unit mass (angular momentum density) as measured by a static observer at infinity. For a massless particle, L corresponds to its "orbital angular momentum". In addition to ∂_ϕ, spherically symmetric spacetime possesses two other Killing vectors associated with rotational symmetry. The conserved quantities they define collectively correspond to the three-dimensional angular momentum vector. The invariance of this vector's direction implies that the particle's motion is confined to a fixed orbital plane. Without loss of generality, we can choose this plane to be the equatorial plane $\theta = \frac{\pi}{2}$. This can also be deduced from the geodesic equation for θ:

$$\frac{\mathrm{d}}{\mathrm{d}\lambda}\left(r^2\frac{\mathrm{d}\theta}{\mathrm{d}\lambda}\right) - \frac{L^2\cos\theta}{r^2\sin^3\theta} = 0. \tag{8.45}$$

Integrating this equation yields

$$\left(r^2\frac{\mathrm{d}\theta}{\mathrm{d}\lambda}\right)^2 = -L^2\cot^2\theta + C, \tag{8.46}$$

where C is an integration constant. Using spherical symmetry, we can always align the initial conditions such that at some $\lambda = \lambda_0$, $\theta(\lambda_0) = \frac{\pi}{2}$, $\frac{\mathrm{d}\theta}{\mathrm{d}\lambda}(\lambda_0) = 0$, meaning the particle starts in the equatorial plane with no initial polar velocity. This fixes $C = 0$. Consequently, (8.46) holds only if both sides vanish identically, leading to

$$\theta(\lambda) = \frac{\pi}{2}, \quad \frac{\mathrm{d}\theta}{\mathrm{d}\lambda}(\lambda) = 0. \tag{8.47}$$

Thus, the particle's trajectory remains confined to the equatorial plane for all time. In what follows, we restrict our analysis to $\theta = \frac{\pi}{2}$, where the angular momentum density reduces to

$$L = r^2\frac{\mathrm{d}\theta}{\mathrm{d}\lambda}. \tag{8.48}$$

Furthermore, we can utilize the normalization condition of the particle's 4-velocity to simplify our analysis. The compatibility of the metric ensures that the inner product of the particle's 4-velocity along the geodesic remains constant,

$$\epsilon = -g_{\mu\nu}\frac{\mathrm{d}x^\mu}{\mathrm{d}\lambda}\frac{\mathrm{d}x^\nu}{\mathrm{d}\lambda}, \tag{8.49}$$

where, for massive particles (parametrized by proper time) $\epsilon = 1$, and for massless particles $\epsilon = 0$. For the Schwarzschild metric, this condition gives

$$-\left(1 - \frac{R_\mathrm{s}}{r}\right)\left(\frac{\mathrm{d}t}{\mathrm{d}\lambda}\right)^2 + \left(1 - \frac{R_\mathrm{s}}{r}\right)^{-1}\left(\frac{\mathrm{d}r}{\mathrm{d}\lambda}\right)^2 + r^2\left(\frac{\mathrm{d}\phi}{\mathrm{d}\lambda}\right)^2 = -\epsilon.$$

Substituting the conserved quantities E and L, we obtain the radial equation of motion:

$$-E^2 + \left(\frac{\mathrm{d}r}{\mathrm{d}\lambda}\right)^2 + \left(1 - \frac{R_\mathrm{s}}{r}\right)\left(\frac{L^2}{r^2} + \epsilon\right) = 0. \tag{8.50}$$

This can be reformulated into a more physically intuitive form, resembling the energy equation for classical motion:

$$\frac{1}{2}\left(\frac{\mathrm{d}r}{\mathrm{d}\lambda}\right)^2 + V(r) = \frac{1}{2}E^2, \tag{8.51}$$

where the effective potential $V(r)$ is given by

$$V(r) = \frac{1}{2}\epsilon - \epsilon\frac{GM}{r} + \frac{L^2}{2r^2} - \frac{GML^2}{r^3}. \tag{8.52}$$

Equation (8.51) is analogous to the classical description of a unit-mass particle with "total energy" of $\frac{1}{2}E^2$, moving in a one-dimensional potential with potential energy $V(r)$. The effective potential energy $V(r)$ includes contributions from the particle's rest mass (ϵ-dependent term), gravitational attraction ($-GM/r$), centrifugal barrier ($L^2/2r^2$), and general relativistic correction ($-GML^2/r^3$). This equation fully determines the radial motion $r(\lambda)$ of the particle. The angular motion $\phi(\lambda)$ and coordinate time evolution $t(\lambda)$ can then be obtained straightforwardly using the conserved quantities E and L.

The effective potential not only depends on the properties of spacetime itself but also on the particle's mass and conserved angular momentum density. In the effective potential $V(r)$, the second term corresponds to the usual Newtonian gravitational potential, the third term represents the centrifugal potential, and the last term is the correction term from general relativity. The correction term becomes significant at small r. Without it, the equation reduces to the radial motion of a point particle in Newtonian gravity. The qualitative behavior of the particle's radial motion can be understood by comparing $\frac{1}{2}E^2$ with $V(r)$. Consider a particle initially moving inward from infinity. The possible scenarios are as follows:

(1) Bound motion.

Here $\frac{1}{2}E^2$ intersects $V(r)$ at two points, confining the particle's motion between them. In Newtonian mechanics, this corresponds to an elliptical orbit. If the two intersection points merge, meaning $\frac{1}{2}E^2$ coincides with a local minimum of $V(r)$, the particle undergoes stable circular motion.

(2) Unbound motion with a turning point.

If $\frac{1}{2}E^2$ intersects $V(r)$ only once, the particle reaches a turning point and reverses direction, escaping back to infinity.

(3) Captured by the star.

When $\frac{1}{2}E^2$ does not intersect $V(r)$, i.e., the particle's total energy density exceeds the potential barrier's peak, the particle continues inward until it is captured by the star. Notably, if $\frac{1}{2}E^2$ precisely matches a local maximum of $V(r)$, the particle may briefly execute unstable circular motion before either falling inward or escaping.

In the following discussion, we will treat the central celestial body as a point mass, effectively modeling the Schwarzschild spacetime as a black hole by neglecting its finite radius.

Let us first look at motions of particles in Newtonian gravity. The effective potential for a massive particle in Newtonian gravity is given by:

$$V(r) = \frac{1}{2} - \frac{GM}{r} + \frac{L^2}{2r^2}. \tag{8.53}$$

This potential admits circular orbits at the radius

$$r_{\mathrm{c}} = \frac{L^2}{GM}, \tag{8.54}$$

with a corresponding binding energy of $E = -\frac{GM}{2r_{\mathrm{c}}}$. A particle starting from rest at infinity must lose energy to settle into such an orbit. Beyond circular motion, elliptical orbits (bound trajectories) are also possible, depending on the particle's energy. For massless particles, the potential reduces to

$$V(r) = \frac{L^2}{2r^2}. \tag{8.55}$$

Unlike massive particles, photons cannot form circular or bound orbits in Newtonian gravity. A photon arriving from infinity will gradually "slow down" before being reflected back to infinity. In Newtonian gravity, massless particles are not affected by gravitational attraction, and their trajectories are straight lines.

In general relativity, the trajectories of particles deviate significantly from those predicted by Newtonian mechanics. For massive particles, the effective potential governing their motion is given by

$$V(r) = \frac{1}{2} - \frac{GM}{r} + \frac{L^2}{2r^2} - \frac{GML^2}{r^3}. \tag{8.56}$$

Notably, as $r \to 0$, the potential $V \to -\infty$, a behavior starkly contrasting with Newtonian gravity. At $r = R_{\mathrm{s}}$, the potential vanishes $V = 0$. The extremal points of the potential, corresponding to circular orbits, are determined by solving

$$\begin{aligned} 0 &= \frac{\partial V}{\partial r} = r^{-4}[GMr^2 - L^2 r + 3GML^2] \\ \Rightarrow R_{\pm} &= \frac{L^2 \pm \sqrt{L^4 - 12L^2M^2G^2}}{R_{\mathrm{s}}}. \end{aligned} \tag{8.57}$$

If the angular momentum is small, $L^2 < 12G^2M^2$, no extremal points exist. The inward-moving particle $\mathrm{d}r/\mathrm{d}\lambda < 0$ will inevitably plunge toward the central object, either colliding with a compact star or crossing the event horizon of a black hole and heading towards the center of the black hole. If the particle's angular momentum is large enough, $L^2 > 12G^2M^2$, the potential admits two critical radii: a minimum at R_+ supporting stable circular orbits; a maximum at R_-, permitting unstable circular orbits. In the limit $L \gg GM$, the stable orbit radius reduces to $R_+ \sim L^2/GM$, recovering the Newtonian result. Note that as $R_+ > 6GM$, there are no stable circular orbits smaller than $6GM$, while unstable circular orbits are confined to $3GM < R_- < 6GM$. For a given circular orbit R_c, the required angular momentum is

$$L = \left(\frac{GMR_\mathrm{c}}{1 - 3GM/R_\mathrm{c}}\right)^{\frac{1}{2}}. \tag{8.58}$$

For the circular orbit $\mathrm{d}r/\mathrm{d}\lambda = 0$, the conserved energy density of the particle satisfies

$$\frac{1}{2}E^2(R) = V(R) \Rightarrow E(R) = \frac{R - R_\mathrm{s}}{\sqrt{R(R - 3GM)}}. \tag{8.59}$$

When $R \leq 4GM$, $E \geq 1$, particles may escape to infinity if perturbed, though they risk capture by the central object. As $R \to 3GM$, then E, $L \to \infty$. If $4GM < R < 6GM$, since $E < 1$, particle remain confined to bound trajectories.

When $L^2 = 12G^2M^2$, a special case arises: the stable and unstable circular orbits coincide at a single radius, $R_+ = R_- = 6GM$. This unique orbit is known as the innermost stable circular orbit (ISCO) and plays a crucial role in astrophysics, particularly in the study of accretion disks around compact objects. In the vicinity of a massive compact star or black hole, an accretion disk forms as gas gradually spirals inward due to turbulent viscosity—often driven by magnetohydrodynamic instabilities. As the gas loses angular momentum, it migrates inward, converting gravitational potential energy into heat. Eventually, once the gas reaches the critical angular momentum $L < L_\mathrm{c}$, it can no longer perform circular motion and rapidly plunges toward the central object. The ISCO marks the final stable boundary where orbiting matter can resist this inward pull. Beyond this point, no further stable orbits exist, and the gas is inevitably captured. At the ISCO, the particle's energy density is $E^2 = 8/9$. For a particle originating from infinity where $E = 1$, the binding energy per unit mass at the ISCO is

$$E_\mathrm{B} = 1 - E = 1 - \sqrt{8/9} \approx 0.0572. \tag{8.60}$$

This corresponds to 5.7% of the particle's rest mass m_0c^2 being released before final capture. Such an energy loss is extraordinarily efficient compared to other astrophysical processes. For instance, nuclear fusion efficiency (hydrogen to helium), the latter's efficiency is only about 0.7%. The majority of this energy is expected to be emitted as gravitational waves, particularly near the ISCO, where the orbital dynamics are most extreme.

For massless particles, $\epsilon = 0$, the predictions of general relativity differ markedly from Newtonian gravity. The effective potential energy governing their motion takes the form

$$V(r) = \frac{L^2}{2r^2} - \frac{GML^2}{r^3}, \tag{8.61}$$

As with massive particles, this potential vanishes at the Schwarzschild radius $r = R_s$. For massless particles, except for $L = 0$, the potential exhibits a barrier that particles must overcome to reach the central object. The potential reaches its maximum at $r = 3GM$, corresponding to an unstable circular orbit for photons. Particles with precisely the right energy can orbit momentarily at this radius, but any perturbation will cause them to either escape to infinity or spiral inward.

8.3.1 *Free fall*

The preceding analysis primarily focused on qualitative discussions of circular orbits. We now turn our attention to the purely radial motion of particles, specifically under the assumption of zero angular momentum. Consider a massive particle in free fall from infinity, characterized by the energy and effective potential:

$$E = 1, \quad V_{\text{eff}} = \frac{1}{2} - \frac{GM}{r}. \tag{8.62}$$

The corresponding equation of motion is given by

$$\frac{1}{2}\dot{r}^2 - \frac{GM}{r} = 0 \Rightarrow \frac{\mathrm{d}r}{\mathrm{d}\tau} = \pm\sqrt{\frac{R_s}{r}}, \tag{8.63}$$

where the derivative is taken with respect to the proper time τ, and the plus and minus signs denote outward (+) or inward (−) radial motion, respectively. For our purposes, we examine the case of inward motion and thus select the negative sign. Additionally, the time-component of 4-velocity satisfies

$$\frac{\mathrm{d}t}{\mathrm{d}\tau}\left(1 - \frac{R_s}{r}\right) = 1 \Rightarrow \frac{\mathrm{d}t}{\mathrm{d}\tau} = \frac{1}{1 - \frac{R_s}{r}}. \tag{8.64}$$

Consequently, the particle's 4-velocity takes the form

$$u^\mu = \left(\left(1 - \frac{R_s}{r}\right)^{-1}, -\sqrt{\frac{R_s}{r}}, 0, 0 \right). \tag{8.65}$$

Integrating the radial equation of motion yields

$$\begin{aligned} \sqrt{r}\mathrm{d}r &= -\sqrt{R_s}\mathrm{d}\tau \quad (\text{“} - \text{”} : \text{inward motion}) \\ \Rightarrow \Delta\tau &= \frac{2}{3\sqrt{R_s}}(r_0^{3/2} - r^{3/2}), \end{aligned} \tag{8.66}$$

where $\Delta\tau$ is the proper time elapsed as the particle travels from the initial radial coordinate r_0 to the final position r. Remarkably, the proper time required for the particle to reach either the Schwarzschild radius $r = R_s$ or the central singularity $r = 0$ remains finite in both cases.

On the other hand, for a static observer at infinity, the observed time corresponds to the coordinate time t. The relationship between t and the radial motion is governed by the differential equation

$$\frac{\mathrm{d}t}{\mathrm{d}r} = \frac{\mathrm{d}t}{\mathrm{d}\tau} \Big/ \frac{\mathrm{d}r}{\mathrm{d}\tau} = -\left(\frac{R_s}{r}\right)^{-1/2} \left(1 - \frac{R_s}{r}\right)^{-1}. \tag{8.67}$$

After integration, we get

$$\begin{aligned} \Delta t = &\frac{2}{2\sqrt{R_s}}(r_0^{3/2} - r^{3/2}) + 2\sqrt{R_s}(r_0^{1/2} - r^{1/2}) \\ &+ R_s \ln \left| \left(\frac{\sqrt{r} + \sqrt{R_s}}{\sqrt{r} - \sqrt{R_s}} \right) \left(\frac{\sqrt{r_0} - \sqrt{R_s}}{\sqrt{r_0} + \sqrt{R_s}} \right) \right|. \end{aligned}$$

It gives the total coordinate time required for the particle to travel from the initial position r_0 to the final position r. As the particle approaches the Schwarzschild radius $r \to R_s$, the logarithmic term diverges, leading to $\Delta t \to \infty$. This implies that, from the perspective of a distant observer, the particle never actually reaches the Schwarzschild radius within a finite coordinate time. Remarkably, this conclusion is not exclusive to observers at infinity. Any static observer, i.e., one fixed at a constant radial position $r_o > R_s$, will similarly perceive the infalling particle as requiring an infinite time to reach R_s.

8.3.2 *Escape velocity*

We now revisit the problem of escape velocity, traditionally studied in Newtonian gravity, within the framework of general relativity. Specifically, we determine the minimum initial velocity required for a particle to overcome a star's gravitational attraction and reach infinity. For simplicity, we assume the star is non-rotating, so the most efficient escape trajectory is purely radial. To ensure the particle reaches infinity, it must possess a conserved energy density of at least $E = 1$. Consequently the radial component of its 4-velocity is given by $u^r = \sqrt{\frac{R_s}{r}}$, while the remaining components are as derived in Section 8.3.1. The initial velocity of the particle is defined relative to a static launcher positioned on the star's surface at radius R. Assuming the launcher remains at rest, its 4-velocity is

$$u^{\mu}_{\text{obs}}(R) = \left(\left(1 - \frac{R_s}{R}\right)^{-1/2}, 0, 0, 0\right). \tag{8.68}$$

The energy of the particle, as measured by the launcher, is given by $-\widehat{P} \cdot \widehat{u}_{\text{obs}}$, leading to

$$\mathcal{E}_o = -mg_{tt}u^t u^t_{\text{obs}} = m\left(1 - \frac{R_s}{R}\right)^{-1/2}. \tag{8.69}$$

In the reference frame of the launcher, this energy corresponds to the relativistic expression for a particle with velocity v, $\mathcal{E}_o = m/\sqrt{1-v^2}$, so the escape velocity of the particle is

$$v_{\text{escape}} = \left(\frac{R_s}{R}\right)^{1/2}. \tag{8.70}$$

This result coincides exactly with the escape velocity derived in Newtonian gravity.

Finally, let us examine the stable circular orbit more carefully. The radius of a stable circular orbit is given by

$$R_+ = \frac{L^2}{R_s}\left(1 + \sqrt{1 - 12\left(\frac{GM}{L}\right)^2}\right). \tag{8.71}$$

As evident from Eq. (8.71), the smaller the angular momentum, or the smaller L/GM, the smaller R_+ is. The innermost stable circular orbit (ISCO) is at

$$R_{\text{ISCO}} = 6GM. \tag{8.72}$$

For an observer at infinity, the angular velocity of a particle in a circular orbit of radius R_c is

$$\Omega = \frac{\mathrm{d}\phi}{\mathrm{d}t} = \frac{\mathrm{d}\phi/\mathrm{d}\tau}{\mathrm{d}t/\mathrm{d}\tau} = \frac{1}{R_c^2}\left(1 - \frac{R_s}{R_c}\right)\frac{L}{E}. \tag{8.73}$$

Remarkably, it satisfies the relation

$$\Omega^2 = \frac{GM}{R_c^3}, \tag{8.74}$$

which is identical to Kepler's third law for circular orbits in Newtonian mechanics. The particle's 4-velocity takes the form $u^\mu = u^t(1,0,0,\Omega)$, where

$$u^t = \left(1 - \frac{R_s}{R_c} - R_c^2\Omega^2\right)^{-1/2} = \left(1 - \frac{3GM}{R_c}\right)^{-1/2}. \tag{8.75}$$

Consider a stationary observer on the circular orbit, e.g., supported by a powerful rocket to maintain their position. Their 4-velocity is $u_{\rm H}^\mu = ((1-R_s/R_c)^{-1/2}, 0, 0, 0)$. The particle has $u_\mu = (-E, 0, 0, L)$, so the energy density γ measured by the observer is

$$\gamma = E_{\rm H} = -u_\mu u_{\rm H}^\mu = \frac{E}{(1 - R_s/R_c)^{1/2}}. \tag{8.76}$$

In special relativity, $\gamma = (1-v^2)^{-1/2}$, so the observed velocity of the particle is $v_{\rm H} = \sqrt{\dfrac{R_s/(2R_c)}{1 - R_s/R_c}}$. At $R_c = 2R_s$, this velocity coincides with the radial escape velocity derived earlier. For $R_c < 2R_s$, it exceeds the radial escape velocity, rendering the orbit unstable under perturbations. In such cases, the particle may become unbound and escape to infinity. This conclusion is consistent with our earlier analysis of orbital stability.

8.4 Classical Experimental Tests of General Relativity

In this section, we present a historical overview of key experimental tests validating general relativity. These classical experiments fundamentally rely on particle dynamics in Schwarzschild spacetime. We begin our discussion with the anomalous perihelion precession of Mercury's orbit.

8.4.1 *Precession of Mercury's perihelion*

Before the advent of general relativity, long-term astronomical observations had revealed an anomalous precession in Mercury's perihelion. Even after accounting for perturbations due to the Sun's non-spherical gravitational potential and the gravitational influences of other planets, a small but persistent discrepancy remained—one that could not be explained within the framework of Newtonian mechanics. At the time, most physicists dismissed this anomaly as insignificant, attributing it to the hypothetical presence of an undiscovered intra-Mercurial planet. Einstein himself was aware of Mercury's anomalous precession prior to formulating general relativity, though he did not consider it a primary motivation for his theory. However, in late 1915, shortly after completing the field equations of general relativity, Einstein revisited the problem. Remarkably, his theory naturally accounted for the residual precession, resolving a decades-old puzzle in celestial mechanics and providing one of the earliest empirical validations of general relativity.

In Newtonian gravity, a particle moving in a spherically symmetric gravitational potential experiences a central force given by: $\boldsymbol{F} = -m\dfrac{GM}{r^3}\boldsymbol{r}$ leading to the equation of motion $\ddot{\boldsymbol{r}} = -\dfrac{GM}{r^3}\boldsymbol{r}$. Under such a force, the particle's angular momentum $\boldsymbol{A} = \boldsymbol{r} \times m\dot{\boldsymbol{r}}$ is conserved. Defining $\boldsymbol{A} = m\boldsymbol{L}$, where $\boldsymbol{L}$ is a constant vector in both magnitude and direction, we have

$$\boldsymbol{r} \perp \boldsymbol{L}. \tag{8.77}$$

This orthogonality condition implies that the particle's motion is confined to a fixed orbital plane, allowing the trajectory to be described in polar coordinates (r, ϕ). The dynamics are governed by the following equations

$$\begin{aligned} r^2\dot{\phi} &= L, \\ \ddot{r} - r\dot{\phi}^2 &= -\frac{GM}{r^2}, \end{aligned} \tag{8.78}$$

where $L = |\boldsymbol{L}|$ is the conserved angular momentum per unit mass. The first equation expresses the conservation of angular momentum, a direct consequence of the central force; while the second equation is derived from Newton's second law in the radial direction. From these two equations, we seek to determine the particle's orbit $r(\phi)$. To simplify the analysis, we introduce the substitution $u = r^{-1}$. Differentiating $r = u^{-1}$ with respect to time and using angular momentum conservation, we have

$$\begin{aligned} \dot{r} \equiv \frac{\mathrm{d}r}{\mathrm{d}t} &= \frac{\mathrm{d}}{\mathrm{d}t}\left(\frac{1}{u}\right) = -\frac{1}{u^2}\frac{\mathrm{d}u}{\mathrm{d}\phi}\dot{\phi} \\ &= -\frac{1}{u^2}Lu^2\frac{\mathrm{d}u}{\mathrm{d}\phi} = -L\frac{\mathrm{d}u}{\mathrm{d}\phi}. \end{aligned} \tag{8.79}$$

Taking the time derivative of $\dot{r}$ yields $\ddot{r} = -L^2u^2\dfrac{\mathrm{d}^2u}{\mathrm{d}\phi^2}$. Thus, the orbit can be determined as:

$$\begin{aligned}
&\frac{\mathrm{d}^2 u}{\mathrm{d}\phi^2} + u = \frac{GM}{L^2} \quad \text{(Binet's equation)} \\
&\Rightarrow u = \frac{GM}{L^2} + C\cos(\phi - \phi_0) \quad (C, \phi_0 \text{ are constants}) \\
&\Rightarrow \frac{l}{r} = 1 + e\cos(\phi - \phi_0),
\end{aligned} \tag{8.80}$$

where

$$l = L^2/GM, \quad e = CL^2/GM. \tag{8.81}$$

This is the equation of a conic section, with l (the semi-latus rectum) determining the size of the orbit, and e known as the eccentricity, determining the shape of the orbit, while ϕ_0 determines the orientation of the orbit (relative to the x-axis). For $0 < e < 1$, the section is an ellipse, confirming that bound planetary trajectories in a spherically symmetric Newtonian potential are closed ellipses. The above relationships can be expressed equivalently as

$$r(\phi) = \frac{(1 - e^2)a}{1 + e\cos(\phi - \phi_0)}, \tag{8.82}$$

where a is the semi-major axis. Setting $\phi_0 = 0$ aligns the orbit's pericenter with the x-axis, as shown in Figure 8.3. The closed nature of Newtonian orbits arises from the inverse-square law, which exhibits a hidden symmetry: the conservation of the Laplace–Runge–Lenz (LRL) vector. This symmetry ensures that bound orbits with $e < 1$ are perfectly closed ellipses. If the central force deviates from the inverse-square law, e.g., due to higher-order corrections such as $1/r^3$ term from non-spherical mass distributions or relativistic effects as in general relativity, the LRL vector is no longer conserved, and orbits fail to close. Instead, the pericenter precesses.

In general relativity, even in the spherically symmetric Schwarzschild spacetime, a particle's motion—while capable of forming a circular orbit—does not trace a closed elliptical path as in Newtonian gravity. Instead, the orbit exhibits a characteristic precession, resulting in a rosette-like pattern. The geodesic equation derived earlier is given by

$$\dot{r}^2 + \left(1 - \frac{R_\mathrm{s}}{r}\right)\left(\frac{L^2}{r^2} + 1\right) = E^2. \tag{8.83}$$

To analyze the shape of the orbit rather than merely its radial evolution, we utilize the conserved angular momentum density. Using the chain rule,

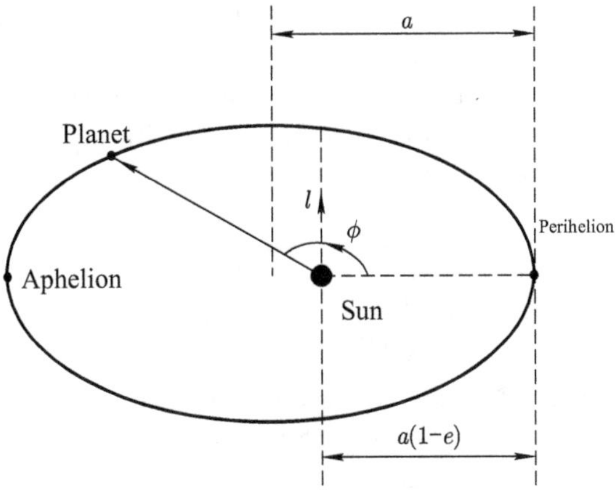

Fig. 8.3 In Newtonian mechanics, if only the two-body interaction between the sun and the planet is considered, then the closed orbit of the planet around the sun will be a strict ellipse. In the figure a is called the semi-major axis of the ellipse, the perihelion distance is $r_- = a(1-e)$, and the aphelion distance is $r_+ = a(1+e)$. By determining perihelion and aphelion, we can determine the semi-major axis and eccentricity of the orbit, and thus the entire orbit.

we find $\dot{r} = \dfrac{\mathrm{d}r}{\mathrm{d}\lambda} = \dfrac{\mathrm{d}r}{\mathrm{d}\phi}\dfrac{\mathrm{d}\phi}{\mathrm{d}\lambda} = \dfrac{\mathrm{d}r}{\mathrm{d}\phi}\dot{\phi}$, and then obtain he orbital equation

$$\left(\frac{\mathrm{d}r}{\mathrm{d}\phi}\right)^2 + \left(1 - \frac{R_\mathrm{s}}{r}\right) r^2 \left(1 + \frac{r^2}{L^2}\right) = E^2 \frac{r^4}{L^2}. \tag{8.84}$$

Introducing the dimensionless variable $x \equiv \dfrac{2L^2}{R_\mathrm{s} r}$, the equation of motion simplifies to

$$\left(\frac{\mathrm{d}x}{\mathrm{d}\phi}\right)^2 + \left(\left(\frac{2L}{R_\mathrm{s}}\right)^2 - \frac{4E^2L^2}{R_\mathrm{s}^2}\right) - 2x + x^2 - \frac{R_\mathrm{s}^2}{2L^2}x^3 = 0. \tag{8.85}$$

Here the last term on the left side of the equation represents a small relativistic perturbation to the Newtonian orbit. Differentiating this first-order equation once more with respect to $\dfrac{\mathrm{d}}{\mathrm{d}\phi}$, we have

$$\begin{aligned} &2\frac{\mathrm{d}}{\mathrm{d}\phi}\left(\frac{\mathrm{d}x}{\mathrm{d}\phi}\right)\frac{\mathrm{d}x}{\mathrm{d}\phi} - 2\frac{\mathrm{d}x}{\mathrm{d}\phi} + 2x\frac{\mathrm{d}x}{\mathrm{d}\phi} - \frac{3R_\mathrm{s}^2}{2L^2}x^2\frac{\mathrm{d}x}{\mathrm{d}\phi} = 0 \\ \Rightarrow &\left(2\frac{\mathrm{d}^2x}{\mathrm{d}\phi^2} - 2 + 2x - \frac{3R_\mathrm{s}^2}{2L^2}x^2\right)\frac{\mathrm{d}x}{\mathrm{d}\phi} = 0 \\ \Rightarrow &\frac{\mathrm{d}^2x}{\mathrm{d}\phi^2} - 1 + x = \frac{3R_\mathrm{s}^2}{4L^2}x^2. \end{aligned} \tag{8.86}$$

In the absence of the right-hand term, the final equation reduces to the normalized Binet equation from Newtonian gravity, which yields conic-section trajectories (e.g., closed ellipses). The inclusion of the right-hand term introduces a general relativistic correction, necessitating a perturbative approach. We decompose the solution as $x = x_0 + x_1$, where

$$\begin{aligned}\frac{\mathrm{d}^2 x_0}{\mathrm{d}\phi^2} - 1 + x_0 &= 0,\\ \frac{\mathrm{d}^2 x_1}{\mathrm{d}\phi^2} + x_1 &= \frac{3R_\mathrm{s}^2}{4L^2} x_0^2.\end{aligned} \tag{8.87}$$

Here x_0 describes the Newtonian elliptical orbit, while x_1 encodes the relativistic perturbation. As previously established, the zeroth-order solution is $x_0 = 1 + e\cos\phi$. Substituting this into the equation for x_1, we obtain

$$\frac{\mathrm{d}^2 x_1}{\mathrm{d}\phi^2} + x_1 = \alpha(1 + e^2\cos^2\phi + 2e\cos\phi),$$

where $\alpha = \dfrac{3R_\mathrm{s}^2}{4L^2}$. The exact solution for x_1 is

$$x_1 = \alpha\left(1 + \frac{e^2}{2} - \frac{e^2\cos 2\phi}{6} + e\phi\sin\phi\right). \tag{8.88}$$

In this solution, $1 + \dfrac{e^2}{2}$ represents a minor shift in the orbit's center, and the last term $\dfrac{e^2}{6}\cos 2\phi$ describes a small oscillation with no cumulative effect, only the term proportional to $e\phi\sin\phi$ leads to precession of the orbit. Retaining only the precession-relevant term, the combined solution becomes

$$\begin{aligned}x &= 1 + e\cos\phi + \alpha e\phi\sin\phi\\ &\approx 1 + e\cos(\phi - \alpha\phi),\end{aligned} \tag{8.89}$$

where the approximation holds for $\alpha \ll 1$. The orbit remains approximately elliptical but now exhibits a shifted period. Instead of closing after 2π, the perihelion advances by

$$\Delta\phi = 2\pi\alpha = \frac{6\pi G^2 M^2}{L^2 c^4}, \tag{8.90}$$

per revolution. This phenomenon, known as perihelion precession, manifests as a gradual rotation of the orbit's major axis (Figure 8.4). For elliptical orbits, the radial coordinate r is given by

$$r = \frac{2L^2}{R_\mathrm{s}}\frac{1}{1 + e\cos\phi} = \frac{(1 - e^2)a}{1 + e\cos\phi}, \tag{8.91}$$

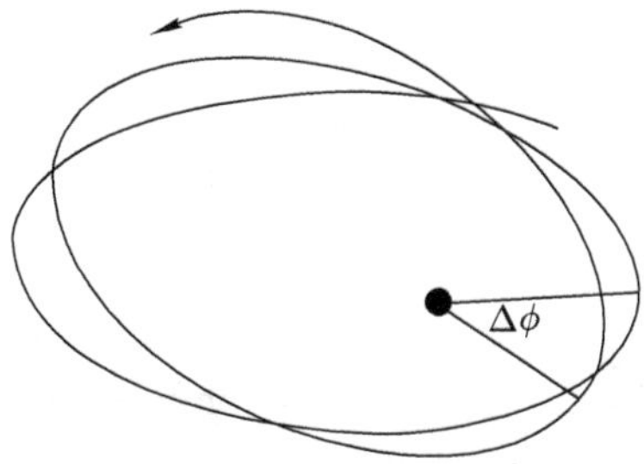

Fig. 8.4 Schematic diagram of planetary perihelion precession.

where

$$a\frac{1}{1-e^2}\frac{2L^2}{R_\mathrm{s}} \tag{8.92}$$

is the semi-major axis of the ellipse. The perihelion (closest approach) and aphelion (farthest point) of the orbit are determined by the condition $dr/d\phi = 0$. From the orbital equation, we find that the perihelion occurs at $\phi = 0$:

$$\left.\frac{\mathrm{d}r}{\mathrm{d}\phi}(r_+)\right|_{\phi=0} = 0 \Rightarrow r_+ = \frac{2L^2}{R_\mathrm{s}(1+e)} = a(1-e); \tag{8.93}$$

while the aphelion is at $\phi = \pi$:

$$\left.\frac{\mathrm{d}r}{\mathrm{d}\phi}(r_-)\right|_{\phi=\pi} = 0 \Rightarrow r_- = \frac{2L^2}{R_\mathrm{s}(1-e)} = a(1+e). \tag{8.94}$$

These relations show that the semi-major axis a can be determined observationally from the perihelion and aphelion distances $a = \frac{(r_+ + r_-)}{2}$. Using a and $L^2 = GM(1-e^2)a$, we get the precession angle of the planet's perihelion advance per orbit:

$$\Delta\phi = \frac{6\pi GM}{(1-e^2)a}. \tag{8.95}$$

For the Sun, with $GM_\odot = 1.475$ km, Mercury's orbital period is 88 days, and the precession advance per orbit is $\Delta\phi = 0.1038''$. While this per-orbit precession is minuscule, its cumulative effect over a century becomes significant due to Mercury's high orbital frequency $\Delta\phi|_\mathrm{century} = 42.98''$. Modern measurements of Mercury's perihelion advance yield $43.1'' \pm 0.5''$ per century, in striking agreement with general relativity's prediction. This precise match (within 0.3% uncertainty) stands as one of the theory's earliest and most compelling experimental validations.

Table 8.1 The perihelion precession of some planets in the solar system (unit: ″/century).

Planet	General Relativity Prediction	Experimental Observation
Venus	8.6	8.4 ± 4.8
Earth	3.8	5.0 ± 1.2
Icarus	10.3	9.8 ± 0.8

All bound orbits in the solar system exhibit this relativistic precession when modeled in the Sun's spherically symmetric spacetime. Table 8.1 summarizes theoretical and observed precession rates for select solar system bodies, highlighting the consistency of general relativity across diverse orbital configurations.

As discussed above, the precession of a planet's perihelion in general relativity arises from a correction term to Newtonian gravity. From the perspective of classical mechanics, this relativistic correction can be interpreted as the introduction of a small non-inverse-square force. In fact, various astrophysical effects can produce additional non-inverse-square forces, leading to further contributions to perihelion precession. For instance, if the Sun deviates from a perfect sphere, its gravitational potential acquires multipole moment corrections. In the equatorial plane ($\theta = \pi/2$), this potential takes the form

$$\Phi = -\frac{GM}{r} - \frac{1}{2}\frac{GMJ_2R_\odot^2}{r^3}, \tag{8.96}$$

where J_2 quantifies the quadrupole moment contribution. Comparing this with the general relativistic correction derived earlier, we obtain the quadrupole-induced precession:

$$\begin{aligned}\Delta\phi_{\text{quad}} &= \frac{6\pi GM}{(1-e^2)a}\frac{1}{2}\frac{J_2R_\odot^2}{GM(1-e^2)a} \\ &= \frac{1}{2}\frac{J_2R_\odot^2}{GM(1-e^2)a}\Delta\phi_{\text{GR}}.\end{aligned} \tag{8.97}$$

For Mercury, the accumulated contribution to the perihelion precession over a century is

$$\Delta\phi|_{\text{century}} = 42.98''(1 + 3\times 10^{-4}(J_2/10^{-7})). \tag{8.98}$$

Observational constraints indicate $J_2 \sim 10^{-7}$ for the Sun, rendering this effect negligible compared to the dominant general relativistic precession.

In fact, the gravitational perturbations from other planets exert a far more significant influence on Mercury's orbit than the relativistic or solar quadrupole effects. Through meticulous perturbation calculations in the late 19th century, astronomers derived the contributions to Mercury's perihelion precession summarized in Table 8.2.

Table 8.2 The influence of other planets in the solar system on the precession of Mercury's perihelion (unit: ″/century).

Perturbing Planet	Precession of Mercury's Perihelion
Venus	277.86 ± 0.27
Earth	90.04 ± 0.08
Mars	2.54
Jupiter	153.58
Saturn	7.30
Uranus	0.14
Neptune	0.04
Total	531.5 ± 0.3

8.4.2 *Gravitational redshift*

In Chapter 3, we demonstrated that photons experience frequency shifts when propagating through a gravitational potential. This effect was derived using the equivalence principle and simple thought experiments, yielding a redshift factor $z = \Delta\phi$ proportional to the gravitational potential difference. In this section, we present a rigorous treatment of gravitational redshift within the framework of curved spacetime, elucidating its geometric origin in general relativity.

Let us first examine a simple scenario for gravitational redshift. Consider a photon emitter fixed at coordinates $(r_{\rm E},\ \theta_{\rm E},\ \phi_{\rm E})$, and a receiver is fixed at $(r_{\rm R},\ \theta_{\rm R},\ \phi_{\rm R})$. Photons can be regarded as waves, where the distance between two successive wave crests (or troughs) determines their wavelength and frequency. The first wave crest travels from the emitter to the receiver along a null geodesic $(t_{\rm E},\ \theta_{\rm E},\ \phi_{\rm E}) \to (t_{\rm R}, \theta_{\rm R}, \phi_{\rm R})$, and the second crest follows a similar trajectory but is emitted at a slightly later time $(t_{\rm E} + \Delta t_{\rm E}) \to (t_{\rm R} + \Delta t_{\rm R})$. In Schwarzschild spacetime, the null condition

$$\mathrm{d}s^2 = 0 \quad \Rightarrow \quad \left(1 - \frac{R_{\rm s}}{r}\right)\mathrm{d}t^2 = \left(1 - \frac{R_{\rm s}}{r}\right)^{-1}\mathrm{d}r^2 + r^2\mathrm{d}\Omega^2.$$

Parameterizing the null curve with an affine parameter λ, we obtain

$$\frac{\mathrm{d}t}{\mathrm{d}\lambda} = \left(1 - \frac{R_{\rm s}}{r}\right)^{-1/2}\left(g_{ij}\frac{\mathrm{d}x^i}{\mathrm{d}\lambda}\frac{\mathrm{d}x^j}{\mathrm{d}\lambda}\right)^{1/2}$$

$$\Rightarrow t_{\rm R} - t_{\rm E} = \int_{\lambda_{\rm E}}^{\lambda_{\rm R}}\left(1 - \frac{R_{\rm s}}{r}\right)^{-1/2}\left(g_{ij}\frac{\mathrm{d}x^i}{\mathrm{d}\lambda}\frac{\mathrm{d}x^j}{\mathrm{d}\lambda}\right)^{1/2}\mathrm{d}\lambda.$$

For spatially fixed emitters and receivers, the coordinate time difference $t_{\rm R} - t_{\rm E}$ is identical for all photons. Consequently, the time intervals between successive crests satisfy $\Delta t_{\rm E} = \Delta t_{\rm R}$ for an observer at rest at infinity.

However, the proper time intervals measured by the emitter and receiver differ due to gravitational time dilation. Since $\mathrm{d}r = \mathrm{d}\theta = \mathrm{d}\phi = 0$, the proper time is given by $\mathrm{d}\tau^2 = \left(1 - \frac{R_s}{r}\right)\mathrm{d}t^2$, leading to

$$\Delta\tau_E = \left(1 - \frac{R_s}{r_E}\right)^{1/2} \Delta t_E, \quad \Delta\tau_R = \left(1 - \frac{R_s}{r_R}\right)^{1/2} \Delta t_R.$$

The ratio of these intervals is

$$\frac{\Delta\tau_R}{\Delta\tau_E} = \left(\frac{1 - R_s/r_R}{1 - R_s/r_E}\right)^{1/2}. \tag{8.99}$$

Here $\Delta\tau$ represents the period T between wave crests as measured locally. Thus, the frequency ratio observed by the receiver and emitter is

$$\frac{\nu_R}{\nu_E} = \left(\frac{1 - R_s/r_E}{1 - R_s/r_R}\right)^{1/2}. \tag{8.100}$$

If $r_R > r_E$, then $\nu_R < \nu_E$, the photon is redshifted, and the redshift factor is

$$1 + z = (\nu_R/\nu_E)^{-1} \quad \Rightarrow \quad z = \sqrt{\frac{1 - R_s/r_R}{1 - R_s/r_E}} - 1. \tag{8.101}$$

In the weak field limit, $R_s/r \ll 1$, this expression simplifies to

$$\begin{aligned} z &= \frac{1}{2}\left(\frac{R_s}{r_E} - \frac{R_s}{r_R}\right) = \frac{GM}{r_E} - \frac{GM}{r_R} \\ &= -\Phi(r_E) + \Phi(r_R) = \Delta\Phi, \end{aligned} \tag{8.102}$$

where Φ is the Newtonian gravitational potential. This matches our earlier result derived from the equivalence principle. Crucially, the agreement holds only in the weak-field regime, highlighting the necessity of general relativity for strong gravitational fields.

In fact, for a spacetime with the metric

$$\mathrm{d}s^2 = g_{00}(\boldsymbol{x})\mathrm{d}t^2 + g_{ij}(\boldsymbol{x})\mathrm{d}x^i\mathrm{d}x^j, \tag{8.103}$$

if the emitter and receiver are fixed, then

$$\frac{\nu_R}{\nu_E} = \left(\frac{g_{00}(\boldsymbol{x}_E)}{g_{00}(\boldsymbol{x}_R)}\right)^{1/2}. \tag{8.104}$$

More generally, when the emitter and receiver are not fixed in space but undergo motion, such as in free fall, how to compute the redshift? This scenario can be addressed using the general framework for measurements introduced earlier. Consider a spacetime manifold with the metric $g_{\mu\nu}$, where the emitter is located at point A and the receiver at point B. The energies of the photons detected by each are given by

$$\begin{aligned}\mathcal{E}_{\mathrm{E}}(A) &= -\widehat{P}(A)\cdot\widehat{u}_{\mathrm{E}}(A) = -P_\mu(A)u^\mu_{\mathrm{E}}(A),\\ \mathcal{E}_{\mathrm{R}}(B) &= -\widehat{P}(B)\cdot\widehat{u}_{\mathrm{R}}(B) = -P_\mu(B)u^\mu_{\mathrm{R}}(B),\end{aligned} \tag{8.105}$$

where $\widehat{P}$ is the photon's 4-momentum. Consequently, the ratio of the observed photon frequencies is

$$\frac{\nu_{\mathrm{R}}}{\nu_{\mathrm{E}}} = \frac{P_\mu(B)u^\mu_{\mathrm{R}}(B)}{P_\mu(A)u^\nu_{\mathrm{E}}(A)}. \tag{8.106}$$

If $P_\mu(A)$ is known, $P_\mu(B)$ can be determined via the null geodesic equation:

$$\frac{\mathrm{d}P_\mu}{\mathrm{d}\lambda} - \Gamma^\nu_{\mu\rho}P_\nu\frac{\mathrm{d}x^\rho}{\mathrm{d}\lambda} = 0 \Rightarrow \frac{\mathrm{d}P_\mu}{\mathrm{d}\lambda} = \Gamma^\nu_{\mu\rho}P_\nu P^\rho. \tag{8.107}$$

For a stationary spacetime, i.e., one admitting a timelike Killing vector ∂_t in some coordinate system, the time component P_0 of the photon's 4-momentum is conserved. If both the emitter and receiver are spatially fixed, their spatial components of 4-velocity are both zero:

$$u^i_{\mathrm{E}} = 0, \quad u^i_{\mathrm{R}} = 0. \tag{8.108}$$

Using the normalization condition of the 4-velocities of the emitter and receiver $g_{\mu\nu}u^\mu u^\nu = -1$, we get

$$u^0 = \frac{1}{\sqrt{-g_{00}}}. \tag{8.109}$$

Thus, the frequency ratio simplifies to

$$\frac{\nu_{\mathrm{R}}}{\nu_{\mathrm{E}}} = \frac{P_0(B)}{P_0(A)}\left(\frac{g_{00}(A)}{g_{00}(B)}\right)^{1/2} = \left(\frac{g_{00}(A)}{g_{00}(B)}\right)^{1/2}, \tag{8.110}$$

recovering our earlier result. More generally, when the emitter or receiver is in motion, the analysis becomes more intricate. We will examine this situation in detail in Section 8.6 within the context of relativistic astrophysics.

8.4.3 *Null geodesics*

Among the classical tests of general relativity, two key experiments—the deflection of light and the radar echo delay—are intimately connected to the motion of photons in Schwarzschild spacetime. To understand these phenomena, we must first carefully analyze the behavior of null geodesics in this spacetime.

Null geodesics in Schwarzschild spacetime satisfy the equation

$$-\left(1-\frac{R_s}{r}\right)^{-1}E^2+\left(1-\frac{R_s}{r}\right)^{-1}\left(\frac{\mathrm{d}r}{\mathrm{d}\lambda}\right)^2+\frac{L^2}{r^2}=0, \tag{8.111}$$

where E and L are the conserved quantities associated with the photon's "energy" and "angular momentum," respectively:

$$E=\left(1-\frac{R_s}{r}\right)\frac{\mathrm{d}t}{\mathrm{d}\lambda},\quad L=r^2\frac{\mathrm{d}\phi}{\mathrm{d}\lambda}.$$

Since the affine parameter λ is arbitrary, these quantities lack an unambiguous physical interpretation. To resolve this ambiguity, we introduce the impact parameter b, defined as

$$b^2=\frac{L^2}{E^2}, \tag{8.112}$$

which eliminates the dependence on λ. Substituting this into (8.111), we obtain

$$\frac{1}{b^2}=\frac{1}{L^2}\left(\frac{\mathrm{d}r}{\mathrm{d}\lambda}\right)^2+W_{\text{eff}}(r), \tag{8.113}$$

where the effective potential is given by:

$$W_{\text{eff}}(r)=\frac{1}{r^2}\left(1-\frac{R_s}{r}\right). \tag{8.114}$$

Equation (8.113) resembles the energy equation for a particle in a potential, with $1/b^2\sim E$ playing the role of an effective energy and W_{eff} acting as the potential. Notably, the affine parameter λ be rescaled $\lambda\to L\lambda$ to absorb L, so the equation only depends solely on b^2. While E and L are coordinate-dependent and lack direct physical meaning, the impact parameter b is a gauge-invariant quantity with a clear geometric interpretation, it characterizes the photon's trajectory relative to the central mass. We define the impact parameter as $b=|L/E|$. Working in the equatorial plane $\theta=\pi/2$, we employ polar coordinates $x=r\cos\phi$, $y=r\sin\phi$. In the asymptotic limit $r\to\infty$, we find

$$b=|L/E|=\frac{r^2\mathrm{d}\phi/\mathrm{d}\lambda}{\mathrm{d}t/\mathrm{d}\lambda}=r^2\frac{\mathrm{d}\phi}{\mathrm{d}t}. \tag{8.115}$$

At infinity, the photon's trajectory approximates a straight line with $\phi\approx\dfrac{d}{r}$, where d is the minimal distance between the light ray and the central mass, and $\dfrac{\mathrm{d}r}{\mathrm{d}t}\approx-1$ for ingoing photon. This yields

$$\frac{\mathrm{d}\phi}{\mathrm{d}t}=\frac{d}{r^2}\quad\Rightarrow\quad b=d. \tag{8.116}$$

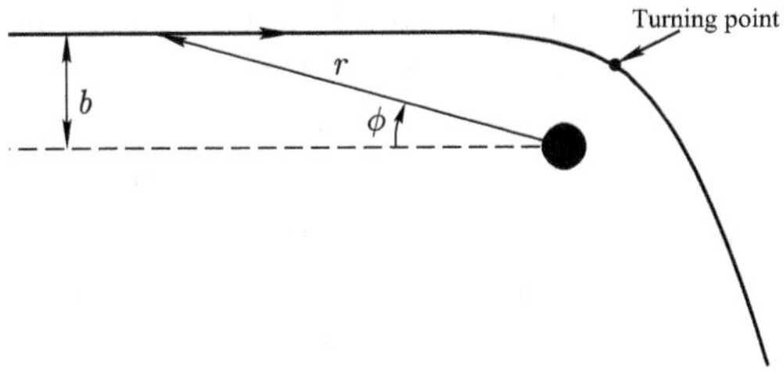

Fig. 8.5 The physical meaning of the impact parameter.

Thus, b indeed corresponds to the impact parameter: with the star's center as the origin, the light ray is approximately a straight line at asymptotic infinity, whose minimum distance from the central object is precisely b, as shown in Figure 8.5. Similar to massive particles, photon motion can be described using an effective potential $W_{\text{eff}}(r)$. This potential peaks at $r = 3GM$,

$$W_{\text{eff}}(r = 3GM) = \frac{1}{27G^2M^2}. \tag{8.117}$$

This maximum determines the critical impact parameter

$$b_c = \sqrt{27}GM. \tag{8.118}$$

The critical case $b = b_c$ corresponds to an unstable circular photon orbit. When $b > b_c$, the photon's "energy" $1/b^2$ lies below the potential barrier, resulting in a scattering trajectory with a turning point. The photon approaches from infinity, is deflected, and escapes back to infinity. When $b < b_c$, the photon overcomes the potential barrier and is captured by the central object. The corresponding capture cross-section is

$$\sigma = \pi b_c^2 = 27\pi(GM)^2. \tag{8.119}$$

For distant observers, the black hole's effective capture radius appears as $3\sqrt{3}GM = \frac{3\sqrt{3}}{2}R_s$. Any photon with impact parameter $b < 3\sqrt{3}GM$ will inevitably be captured.

We can also examine the outward photon orbit outside the event horizon, within the range $R_s < r < 3GM$. As we will later demonstrate, photonsinside the horizon $r \leq R_s$ cannot escape due to the causal structure of spacetime. In the range $R_s < r < 3GM$, photons must overcome a potential barrier to propagate to infinity. This scenario closely parallels our earlier discussion. For $b^2 < 27G^2M^2$, photons possess sufficient energy to surmount the barrier and escape to infinity. These photons have relatively small angular momentum, favoring radial motion. Conversely, for $b^2 > 27G^2M^2$, photons fail to overcome the barrier and ultimately fall back into the black hole after reaching a turning point. Here, the larger angular momentum prevents escape from the gravitational pull.

An interesting question arises. Suppose a transmitter is positioned at radial position $r = R$ in the plane $\theta = \pi/2$, emitting photons at an angle ψ relative to the radial direction. If $\psi = 0$, $b = 0$, the photons strictly move radially, and they will always escape to infinity. The question is, what is the critical emission angle ψ_c beyond which photons can no longer escape? In the transmitter's local reference frame, $\widehat{e}_0$ is its timelike 4-velocity, and the other three spacelike basis vectors are mutually orthogonal and normalized,

$$\widehat{e}_r = (0, (1 - R_s/R)^{1/2}, 0, 0), \quad \widehat{e}_0 = (0, 0, 0, 1/R). \tag{8.120}$$

The emission angle relative to the transmitter satisfies

$$\tan\psi = \frac{\widehat{u}\cdot\widehat{e}_\phi}{\widehat{u}\cdot\widehat{e}_r}, \tag{8.121}$$

where $\widehat{u}$ is the photon's 4-momentum, with components

$$\begin{aligned}
u^t &= \frac{\mathrm{d}t}{\mathrm{d}\lambda} = \frac{E}{1 - R_s/r},\\
u^r &= \frac{\mathrm{d}r}{\mathrm{d}\lambda} = L\left(\frac{1}{b^2} - W_{\text{eff}}(r)\right)^{1/2},\\
u^\theta &= \frac{\mathrm{d}\theta}{\mathrm{d}\lambda} = 0,\\
u^\phi &= \frac{\mathrm{d}\phi}{\mathrm{d}\lambda} = \frac{L}{r^2}.
\end{aligned} \tag{8.122}$$

Since $\widehat{u}\cdot\widehat{e}_\phi = g_{\phi\phi}(\widehat{e}_\phi)^\phi u^\phi = \dfrac{L}{R}$, and

$$\widehat{u}\cdot\widehat{e}_r = g_{rr}(\widehat{e}_r)^r u^r = (1 - R_s/R)^{-1/2} L\left(\frac{1}{b^2} - \frac{1}{R^2}\left(1 - \frac{R_s}{R}\right)\right)^{1/2}, \tag{8.123}$$

the critical angle ψ_c for $b^2 = 27G^2M^2$ is

$$\tan\psi_c = \frac{1}{R}(1 - R_s/R)^{1/2}\left(\frac{1}{27G^2M^2} - \frac{1}{R^2}\left(1 - \frac{R_s}{R}\right)\right)^{-1/2}.$$

When the transmitter is at $R = 3GM$, $\psi_c = \dfrac{\pi}{2}$, which corresponds to the unstable circular orbit. A slight outward perturbation allows photon escape. When the transmitter is at $R < 3GM$, ψ_c decreases, narrowing the escape cone. And when $R = R_s$, $\psi_c = 0$, meaning no escape is possible. Notably, ψ_c. Note that the angle here is actually a conical solid angle. defines a conical solid angle: any photon emitted within this angle (relative to the radial direction) can escape to infinity, regardless of its azimuthal orientation.

A closely related question is: for a stationary observer at a fixed radius $r = R$, maintained by a powerful rocket, what portion of the sky remains visible? This scenario mirrors our earlier discussion of outward photon orbits, but now we must consider inward-propagating null geodesics. The critical angle for visibility is analogous, with the key distinction being the reversed propagation direction. The critical angle satisfies

$$\sin\psi_c = (1 - R_s/R)^{1/2}\frac{b_c}{R}. \tag{8.124}$$

When the angular momentum of the light ray is too large, it cannot be seen by the observer. As the observer approaches the event horizon $R \to R_s$, the visible sky contracts to a narrow circular disk of angular radius

$$\sin\psi_c \approx (1 - R_s/R)^{1/2}\frac{\sqrt{27}}{2}. \tag{8.125}$$

The disk shrinks as $R \to R_s$, eventually vanishing entirely. On the other hand, infalling photons are increasingly blueshifted, causing the remaining visible portion of the sky to appear intensely bright.

8.4.4 *Light deflection and radar echo delay*

The phenomenon of light deflection, first predicted by Einstein's general theory of relativity, was famously confirmed during the 1919 total solar eclipse observation. This gravitational lensing effect becomes particularly significant in strong gravitational fields, making it a powerful tool in modern astrophysics. As previously discussed, when a photon's impact parameter b exceeds the critical value b_c, its trajectory features a turning point at R_0, where $\mathrm{d}r/\mathrm{d}\phi = 0$. The photon's path can then be described as originating from infinity, reaching the turning point, and returning to infinity. The orbital equation at this turning point is given by:

$$\frac{\mathrm{d}\phi}{\mathrm{d}r} = \frac{1}{r^2}\left(\frac{1}{b^2} - \frac{1}{r^3}(r - R_s)\right)^{-1/2}. \tag{8.126}$$

The turning point radius R_0 satisfies the cubic equation

$$R_0^3 - b^2(R_0 - R_s) = 0 \quad \Rightarrow \quad R_0 = \frac{2b}{\sqrt{3}}\cos\left(\frac{1}{3}\cos^{-1}\left(-\frac{b_c}{b}\right)\right). \tag{8.127}$$

The total deflection angle can be obtained by integrating Eq. (8.126) over the photon's trajectory. By symmetry, the deflection angle $\Delta\phi$ receives equal contributions from both the incoming and outgoing paths, yielding

$$\Delta\phi = 2\int_{R_0}^{\infty} \frac{\mathrm{d}r}{(r^4 b^{-2} - r(r - R_\mathrm{s}))^{1/2}}. \tag{8.128}$$

In the absence of mass, $M = 0$, we recover the expected straight-line solution

$$\Delta\phi|_{M=0} = 2\sin^{-1}(b/R_0)|_{R_0=b} = \pi. \tag{8.129}$$

For$M \neq 0$ and $\Delta\phi \neq \pi$, this indicates that due to the gravitational attraction in the Schwarzschild spacetime, the light ray is deflected. Transforming to the variable $u = 1/r$, the deflection angle can be written as

$$\Delta\phi = 2\int_{0}^{1/R_0} \frac{\mathrm{d}u}{(b^{-2} - u^2 + R_\mathrm{s}u^3)^{1/2}}. \tag{8.130}$$

The integral's complexity arises because for fixed impact parameter b, the upper limit of the integral R_0 depends on M through R_s. We address this by treating M and R_0 as independent variables, comparing deflection angles at constant R_0 rather than constant b. To the first order in M, keeping b or R_0 constant makes no difference, but it will be different at higher orders.

Using the relation $b = \sqrt{\dfrac{R_0^3}{R_0 - R_\mathrm{s}}}$, we express the deflection angle as

$$\Delta\phi(M) = 2\int_{0}^{1/R_0} \frac{\mathrm{d}u}{(R_0^{-2} - R_\mathrm{s}R_0^{-3} - u^2 + R_\mathrm{s}u^3)^{1/2}}. \tag{8.131}$$

Thus, we can perform a Taylor expansion with respect to M about $M = 0$:

$$\Delta\phi(M) = \Delta\phi(M=0) + \left.\frac{\partial(\Delta\phi)}{\partial M}\right|_{M=0} M + \cdots. \tag{8.132}$$

The first-order term evaluates to

$$\begin{aligned} \left.\frac{\partial(\Delta\phi)}{\partial M}\right|_{M=0} &= 2\int_{0}^{1/R_0} \left.\frac{G(R_0^{-3} - u^3)\mathrm{d}u}{(R_0^{-2} - R_\mathrm{s}R_0^{-3} - u^2 + R_\mathrm{s}u^3)^{1/2}}\right|_{M=0} \\ &= 2G\int_{0}^{1/b} \frac{b^{-3} - u^3}{(b^{-2} - u^2)^{3/2}}\mathrm{d}u \\ &= 4Gb^{-1}. \end{aligned} \tag{8.133}$$

Thus, the leading-order deflection is

$$\begin{aligned} \delta\phi &= \Delta\phi(M) - \Delta\phi(M=0) \approx M \left.\frac{\partial(\Delta\phi)}{\partial M}\right|_{M=0} = \frac{4GM}{b} \\ &= \frac{4GM}{bc^2}. \end{aligned} \tag{8.134}$$

The deflection angle is inversely proportional to b, the closer to the star, the greater the influence of the star, and thus the larger the deflection angle. On the other hand, it is directly proportional to M, the heavier the star, the more pronounced the light deflection. It is also directly proportional to the Newtonian gravitational constant, the larger the Newtonian gravitational coupling constant, the stronger the gravity, and therefore the larger the deflection angle.

We now present an alternative derivation of the light deflection angle, employing a perturbative method analogous to that used in calculating the perihelion precession of planetary orbits. Beginning with the null geodesic equation for light in Schwarzschild spacetime, we have:

$$\frac{\mathrm{d}^2 x}{\mathrm{d}\phi^2} + x = \alpha x^2, \tag{8.135}$$

where $\alpha = \dfrac{3R_\mathrm{s}^2}{4L^2}$ is a small parameter. In Newtonian gravity, the term on the right-hand side does not exist, the solution describes a straight-line trajectory in polar coordinates

$$x = \frac{1}{D}\sin(\phi - \phi_0). \tag{8.136}$$

or equivalently, in terms of the radial coordinate r and impact parameter b,

$$r = \frac{b}{\sin(\phi - \phi_0)}. \tag{8.137}$$

Here, we adopt the substitution $x = \dfrac{2L^2}{R_\mathrm{s} r}$ and define $D = \dfrac{R_\mathrm{s}}{2EL}$. For small deflections, we treat the right-hand side of (8.135) as a perturbation. Expanding x as $x = x_0 + \alpha x_1 + \mathcal{O}(\alpha^2)$, where $x_0 = \dfrac{\sin\phi}{D}$ (setting $\phi_0 = 0$ for simplicity), the first-order correction x_1 satisfies

$$x_1'' + x_1 = x_0^2 = \frac{\sin^2\phi}{D^2}. \tag{8.138}$$

The inhomogeneous solution to this equation is

$$x_1 = \left(1 + C\cos\phi + \frac{1}{3}\cos 2\phi\right) \Big/ 2D^2, \tag{8.139}$$

where C is an integration constant determined by boundary conditions. The smallness of the perturbation is ensured by $\alpha/D = 3R_\mathrm{s}/2b \ll 1$, consistent with weak-field observations. Finally, the full perturbative solution to first order is

$$x \approx \frac{\sin\phi}{D} + \frac{\alpha(1 + C\cos\phi + \frac{1}{3}\cos 2\phi)}{2D^2}. \tag{8.140}$$

To compute the deflection angle, we examine the asymptotic behavior as $r \to \infty$, i.e., $x \to 0$. The photon's incoming and outgoing directions correspond to the roots of $x(\phi) = 0$ at

$$\phi = -\epsilon_1, \quad \phi = \pi + \epsilon_2, \tag{8.141}$$

where $\epsilon_1, \epsilon_2 \ll 1$. Substituting these into (8.140) and linearizing yields

$$\begin{aligned} -\frac{\epsilon_1}{D} + \frac{\alpha}{2D^2}\left(\frac{4}{3} + C\right) &= 0, \\ -\frac{\epsilon_2}{D} + \frac{\alpha}{2D^2}\left(\frac{4}{3} - C\right) &= 0. \end{aligned} \tag{8.142}$$

The total deflection angle is then

$$\delta\phi = \epsilon_1 + \epsilon_2 = \frac{4\alpha}{3D} = \frac{4GM}{b}, \tag{8.143}$$

recovering the classic result for small-angle light bending. This perturbative treatment is valid only for small deflections. The above discussion only applies when the light deflection is small, and the approximate conditions are valid. For larger deflections (e.g., near photon spheres or ultracompact objects), a non-perturbative analysis of (8.135) is required, typically involving elliptic integrals or numerical methods. Such cases are critical in strong-field gravitational lensing studies, where higher-order terms in α become significant.

For a light ray grazing the solar limb, the theoretical deflection angle is obtained by substituting the solar mass $M_\odot$ and radius $R_\odot$ into (8.143), $\delta\phi \approx 1.75''$. This prediction was first put to the test during the total solar eclipse of May 29, 1919, when two expeditions led by Arthur Eddington and Frank Dyson observed star positions near the obscured Sun. The teams, stationed in Sobral, Brazil, and Príncipe Island (Gulf of Guinea), measured the apparent shifts of starlight due to the Sun's gravitational field. Their results were:

$$\begin{aligned} \delta\phi &= 1.98'' \pm 0.12'' \text{ (Sobral)}, \\ \delta\phi &= 1.60'' \pm 0.3'' \text{ (Príncipe)}. \end{aligned} \tag{8.144}$$

Eddington subsequently declared these observations a confirmation of Einstein's general relativity, marking a pivotal moment in the acceptance of the theory.[4]

The advent of radio astronomy in the late 20th century revolutionized tests of light deflection, eliminating the need for total solar eclipses. For example: every year on October 8th, the quasar 3C279 passes behind the Sun as viewed from Earth. By comparing its radio wave deflection with that of the nearby quasar 3C273 (separated by $> 10°$), astronomers could isolate the gravitational lensing effect. Similar observations of other quasars confirmed the predicted deflection to within 1% accuracy, far surpassing eclipse-based methods. Furthermore, using the Very Long Baseline Interferometer (VLBI) networks to measure the deflection of electromagnetic waves from radio quasars as they pass by the sun, changes as small as $10^{-4''}$ can be detected. The measured value of $\delta\phi$ agrees with general relativity at the level of 0.99997 ± 0.00016.

We now examine the radar echo delay experiment, first proposed by Irwin Shapiro in 1964. This experiment demonstrates the gravitational time delay effect predicted by general relativity. The experimental configuration consists of a radar signal emitted from Earth at Schwarzschild coordinate R_{E}, which passes near the Sun to reach a target planet at radius R_{P}, and then the radar signal returns to the Earth along the same path, as shown in Figure 8.6. The Sun's gravitational field causes a measurable delay in the round-trip travel time of the signal, known as the Shapiro time delay. In Schwarzschild coordinates, the time component of the photon geodesic equation yields

$$\frac{\mathrm{d}t}{\mathrm{d}r} = \left(1 - \frac{R_{\mathrm{s}}}{r}\right)^{-1} \left(1 - \left(1 - \frac{R_{\mathrm{s}}}{r}\right)\frac{b^2}{r^2}\right)^{-1/2}. \tag{8.145}$$

The total coordinate time (measured by an observer at infinity) for a one-way trip is given by

$$\frac{1}{2}\Delta t = \left(\int_{R_{\mathrm{E}}}^{R_{\mathrm{O}}} \mathrm{d}r + \int_{R_{\mathrm{O}}}^{R_{\mathrm{P}}} \mathrm{d}r\right) \left(1 - \frac{R_{\mathrm{s}}}{r}\right)^{-1} \left(1 - \left(1 - \frac{R_{\mathrm{s}}}{r}\right)\frac{b^2}{r^2}\right)^{-\frac{1}{2}}.$$

[4]However, modern historians of science doubt that the technical conditions at the time could have achieved this accuracy, especially since the number of photographs taken at the time was limited (only 2 usable images from Príncipe and 7 from Sobral). It was later found that the light deflection experiments based on total solar eclipses yield less consistent results.

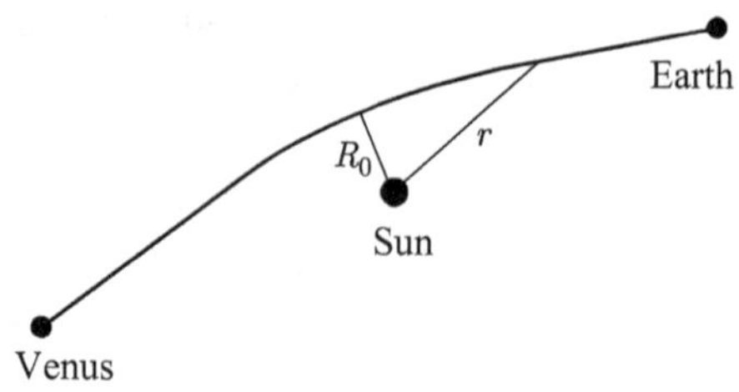

Fig. 8.6 Schematic diagram of radar echo experiment.

If there were no sun, $M = 0$ and $b = R_0$, the time taken would be

$$\begin{aligned}\frac{1}{2}\Delta t_0 &= \left(\int_{R_E}^{R_0} \mathrm{d}r + \int_{R_0}^{R_P} \mathrm{d}r\right)\left(1 - \frac{R_0^2}{r^2}\right)^{-1/2} \\ &= \left(\int_{R_E}^{R_0} + \int_{R_0}^{R_P}\right)\frac{r\mathrm{d}r}{\sqrt{r^2 - R_0^2}} \\ &= \sqrt{R_E^2 - R_0^2} + \sqrt{R_P^2 - R_0^2}.\end{aligned} \tag{8.146}$$

The delayed time is

$$\delta t = \Delta t(M) - \Delta t_0 \approx M\left(\frac{\partial \Delta t}{\partial M}\right)\bigg|_{M=0}. \tag{8.147}$$

To first order in M, the full round-trip delay is

$$\begin{aligned}\Delta t = {}& 2\left(\sqrt{R_E^2 - R_0^2} + \sqrt{R_P^2 - R_0^2}\right) + R_s\left\{2\ln\left(\frac{R_E + \sqrt{R_E^2 - R_0^2}}{R_0}\right)\right. \\ &\left. + 2\ln\left(\frac{R_P + \sqrt{R_P^2 - R_0^2}}{R_0}\right) + \left(\frac{R_E - R_0}{R_E + R_0}\right)^{1/2} + \left(\frac{R_P - R_0}{R_P + R_0}\right)^{1/2}\right\}.\end{aligned}$$

The first two terms on the right-hand side come from Δt_0. Since R_E, $R_P \gg R_0$, we obtain

$$\delta t = \frac{4GM}{c^3}\left(\ln\left(\frac{2R_E}{R_0}\right) + \ln\left(\frac{2R_P}{R_0}\right) + 1\right). \tag{8.148}$$

Note that in the above discussion, we need to vary M and compare geodesics with the same R_0 rather than b. Unlike the case of light deflection, even at the first order in M, whether to keep b or R_0 constant when M changes will affect the final result. The above discussion is for coordinate time. The observed delay on Earth accounts for gravitational time dilation

$$\delta\tau = (1 - R_s/R_E)^{1/2}\delta t. \tag{8.149}$$

In fact, it is difficult to accurately determine R_0, $R_{\rm E}$, and $R_{\rm P}$. What we can do is to obtain a formula for how $\Delta\tau$ varies with time (since the Earth and the planet are constantly moving) and treat all parameters as unknowns, then fit the data to determine these parameters.

There are two methodologies employed to measure this effect. One is "passive" observations: waiting for the right opportunity to use other planets, the sun, and the Earth to carry out radar echo experiments. The planet usually chosen is Venus. The theoretical prediction is $\delta\tau \approx 220\ \mu$s. Shapiro's experiments in 1968 and 1971 confirmed the effect within 2% accuracy. The other is "active" experiments: using spacecraft as precise reflectors to carry out radar echo experiments. It was Anderson who first carried out experiments using spacecraft in 1975. Shapiro's experiments using the Viking spacecraft in 1977 and 1979 obtained results that deviate from the theoretical prediction by only about 0.1%. In 2002, experiments using the Cassini spacecraft further reduced the deviation between theory and experiment to one in a hundred thousand.

8.4.5 *Gravitational time dilation effect*

The preceding discussion introduced the classical experimental tests of general relativity. However, numerous additional experiments have been designed to further verify the theory's predictions. In this section, we focus on the gravitational time dilation effect, which arises from differences in proper time measured by clocks carried by observers in different states of motion. Consider two initially synchronized clocks at the equator: one placed aboard a spacecraft orbiting Earth, and the other remaining stationary on the ground. Assuming $\theta = \pi/2$ and constant r, the spacetime interval is given by

$$ds^2 = -\left(1 - \frac{2GM}{rc^2}\right)c^2 dt^2 + r^2 d\phi^2 = -c^2 d\tau^2. \tag{8.150}$$

For observer A on the ground, moving with Earth's angular velocity[5] $\omega = \dfrac{d\phi}{dt}$, the proper time is

$$d\tau_A^2 = \left(1 - \frac{2GM}{Rc^2} - \frac{R^2\omega^2}{c^2}\right)dt^2, \tag{8.151}$$

[5] In this section we do not consider the effect of the rotation of the Earth on spacetime, assuming the exterior geometry is described by the Schwarzschild metric; we will later account for frame-dragging effects.

where R is Earth's radius. Given the smallness of the dimensionless quantities

$$\frac{2GM}{Rc^2} \ll 1, \quad \frac{R^2\omega^2}{c^2} \ll 1, \tag{8.152}$$

we obtain to first order

$$\delta\tau_A = \left(1 - \frac{GM}{Rc^2} - \frac{R^2\omega^2}{2c^2}\right)\delta t. \tag{8.153}$$

For clock B aboard the spacecraft at altitude $h \ll R$, moving with velocity v relative to the ground, its velocity relative to the asymptotic inertial frame is approximately $v + (R+h)\omega$. Thus,

$$\mathrm{d}\tau_B^2 = \left(1 - \frac{2GM}{(R+h)c^2} - \frac{(v+(R+h)\omega)^2}{c^2}\right)\mathrm{d}t^2. \tag{8.154}$$

Expanding to first order in h/R and neglecting $h\omega$ compared to v, we find

$$\delta\tau_B = \left(1 - \frac{GM}{(R+h)c^2} - \frac{v^2 + R^2\omega^2 + 2R\omega v}{2c^2}\right)\delta t. \tag{8.155}$$

An observer at infinity would perceive clocks A and B separating at some initial moment and reuniting after a coordinate time interval δt. The proper time difference between the clocks is

$$\delta\tau_A - \delta\tau_B = \left(-\frac{GMh}{R^2c^2} + \frac{(2R\omega + v)v}{2c^2}\right)\delta t. \tag{8.156}$$

Defining the fractional time difference as

$$\Delta \equiv \frac{\delta\tau_A - \delta\tau_B}{\delta\tau_A}, \tag{8.157}$$

we find for a spacecraft at $h = 10^4$ m with $v = 300\ \mathrm{m \cdot s^{-1}}$: If the spacecraft moves eastward, then $\Delta_\mathrm{E} = 1.0 \times 10^{-12}$; If the spacecraft moves westward, then $\Delta_\mathrm{W} = -2.1 \times 10^{-12}$. Modern cesium atomic clocks achieve precisions of 10^{-13}, making this effect measurable. Indeed, the 1970 Hafele–Keating experiment confirmed gravitational time dilation, with results deviating from theoretical predictions by approximately 10%.

8.4.6 *Parametrized post-Newtonian (PPN) limit*

From an experimental standpoint, potential modifications to general relativity can be investigated through well-studied solar system tests. For metric-based gravitational theories, the weak-field and slow-motion conditions of the solar system, where $|\Phi|/c^2 \ll 1$ and $v^2/c^2 \ll 1$, allow different theories to be compared within a unified framework. As established, general relativity predicts that a spherically symmetric vacuum solution is described by the Schwarzschild metric. However, if we relax the assumption of a specific theory and instead consider a general static, spherically symmetric spacetime, the metric takes the form

$$\mathrm{d}s^2 = -B(r)\mathrm{d}t^2 + A(r)\mathrm{d}r^2 + r^2(\sin^2\theta \mathrm{d}\phi^2 + \mathrm{d}\theta^2). \tag{8.158}$$

In the solar system, we introduce the dimensionless quantity

$$x \equiv \frac{GM_\odot}{rc^2}, \tag{8.159}$$

which characterizes the strength of the gravitational field. The metric functions $B(r)$ and $A(r)$ can then be expanded in powers of x as

$$\begin{aligned} B(r) &= 1 - 2\alpha x + 2(\beta - \gamma)x^2 + \cdots, \\ A(r) &= 1 + 2\gamma x + \cdots, \end{aligned} \tag{8.160}$$

where α, β, and γ are the post-Newtonian parameters. From the previous discussion of the Newtonian limit, we know that the (00)-component of the metric field is related to the Newtonian gravitational potential, while corrections to other components are negligible at leading order. In the weak field approximation of general relativity, these parameters take the values $\alpha = 1$, $\gamma = 1$, $\beta = 1$. Deviations from these values would indicate corrections to or extensions of general relativity. This parametrization scheme is known as the Parametrized Post-Newtonian (PPN) formalism, which provides a model-independent framework for testing gravitational theories. The expansion can, in principle, be extended to higher orders, yielding post-post-Newtonian corrections. Alternatively, the metric can be expressed in isotropic coordinates as[6]

$$\begin{aligned} \mathrm{d}s^2 = &-\left(1 - 2\alpha\frac{GM_\odot}{\rho c^2} + 2\beta\left(\frac{GM_\odot}{\rho c^2}\right)^2 + \cdots\right)\mathrm{d}t^2 \\ &+ \left(1 + 2\gamma\frac{GM_\odot}{\rho c^2}\right)(\mathrm{d}\rho^2 + \rho^2\mathrm{d}\Omega^2). \end{aligned} \tag{8.161}$$

The parametrized post-Newtonian (PPN) metric provides a powerful framework for testing and distinguishing between alternative theories of gravity. For spherically symmetric spacetimes, this approach offers a standardized parametrization that enables systematic comparisons with experimental observations. To investigate potential corrections to general

[6]This expansion was first proposed by Sir Arthur S. Eddington and Howard P. Robertson.

relativity, we may fix $\alpha = 1$, as required by the Newtonian limit, while allowing β and γ to deviate from their general relativistic values. Under the assumption that test particles follow geodesic motion, we can recompute key observational effects within this generalized framework. For the precession of the perihelion of planets, in the above spacetime, the precession advance is given by

$$\delta\phi_{\text{prec}} = \frac{1}{3}(2 + 2\gamma - \beta)\frac{6\pi GM}{c^2 a(1 - e^2)}. \tag{8.162}$$

For light deflection, if the impact parameter is still b, the deflection angle becomes

$$\delta\phi_{\text{def}} = \frac{1+\gamma}{2}\frac{4GM}{c^2 b}. \tag{8.163}$$

Finally, for Shapiro time delay, there is

$$\Delta t_{\text{delay}} = \frac{1+\gamma}{2}\frac{4GM}{c^3}\left[\ln\left(\frac{4R_{\text{E}}R_{\text{P}}}{R_0^2}\right) + 1\right]. \tag{8.164}$$

The Parametrized Post-Newtonian (PPN) formalism allows for stringent tests of general relativity through high-precision solar system experiments.

- Using the deflection of light through the sun can constrain the PPN parameter γ:

$$\gamma = 1.007 \pm 0.009. \tag{8.165}$$

- The time delay of radio signals due to gravitational potential (Shapiro effect) provides even tighter constraints. The Viking spacecraft experiment gave

$$\gamma = 1.000 \pm 0.001. \tag{8.166}$$

The Cassini spacecraft experiment between June 6 and July 7, 2002, gave

$$\gamma = 1 + (2.1 \pm 2.3) \times 10^{-5}, \tag{8.167}$$

representing the most precise measurement to date, confirming general relativity at the 10^{-5} level.

- Using the anomalous advance of Mercury's perihelion constrains the parameter β:

$$\beta = 1.000 \pm 0.003. \tag{8.168}$$

Other experiments based on satellites orbiting the Earth, such as Gravity Probe B, can also give constraints on the PPN parameters.

On the other hand, the PPN limit can serve as an approximation for spherically symmetric spacetime in weak-field regimes. In strongly relativistic systems, such as binary neutron stars or black holes, the weak-field assumption breaks down, but the PPN approach can be generalized to higher orders, e.g. 2PN, 3PN, to model phenomena like periastron advance, spin-orbit coupling, and gravitational wave emission.

8.5 Geodetic Precession

The discussion in the preceding section on experimental tests of general relativity primarily relied on traditional astronomical observations of planetary motion within the solar system. However, advances in space technology have enabled the design of more sophisticated experiments to probe the predictions of general relativity. In this section, we explore the use of gyroscopes to test gravitational effects in the framework of relativity.

In curved spacetime, when a vector is parallel-transported along a closed loop back to its initial position, its final orientation differs from the original due to spacetime curvature. This discrepancy, known as holonomy, provides a measurable manifestation of curvature. Such an effect offers a novel approach to investigating the geometric properties of spacetime. A practical application of this phenomenon involves observing the evolution of the spin vector of a rotating rigid body in a gravitational field.

A spinning probe is referred to as a gyroscope. In the absence of gravitational effects, the spin direction of a gyroscope mounted on a frictionless bracket remains invariant as the bracket moves, meaning the spin undergoes parallel transport. However, since the gyroscope is an object with a scale, the tidal force will cause a tidal torque, which will cause the spin of the gyroscope not to be strictly parallel transported and instead undergo Newtonian (tidal) precession.[7] The challenge lies in measuring the minuscule relativistic precession predicted by general relativity, as the much larger tidal precession would dominate and obscure the desired effect. To mitigate this, the gyroscope must exhibit exact spherical symmetry,

[7] A well-known example is the equinoxes precession of the Earth's spin, which results from the tidal forces of the sun and moon acting on the bulge of the Earth's equator. If the Earth were strictly spherical, this effect would disappear.

eliminating tidal torque entirely.[8] Thus, in our subsequent analysis, we idealize the gyroscope as a point-like particle with spin, neglecting its finite size to isolate purely relativistic effects.

The motion of a freely falling object is governed the geodesic equation

$$\frac{\mathrm{d}u^\alpha}{\mathrm{d}\tau} + \Gamma^\alpha_{\beta\gamma} u^\beta u^\gamma = 0. \tag{8.169}$$

Accompanying the gyroscope is its spin vector, a spacelike 4-vector $\widehat{s}(\tau)$. In the local inertial frame of the gyroscope, the spin and 4-velocity satisfy

$$\begin{cases} s^\alpha = (0, \boldsymbol{s}), \\ u^\alpha = (1, \boldsymbol{0}) \end{cases} \Rightarrow \widehat{s} \cdot \widehat{u} = 0. \tag{8.170}$$

Since this is a scalar relation, it must hold in any reference frame. The magnitude of the spin vector, $s_* = (\boldsymbol{s} \cdot \boldsymbol{s})^{1/2}$, is a constant of motion, corresponding to the spin angular momentum of the gyroscope.[9] When an object possesses a constant angular momentum magnitude, the evolution of its angular momentum direction is known as precession.

We begin by deriving the gyroscope equation, which governs the evolution of the gyroscope's spin vector. In the local inertial frame, the spin vector remains constant, satisfying: $\dfrac{\mathrm{d}s^\alpha}{\mathrm{d}\tau} = 0$. In curved spacetime, we need to replace the partial derivative with the covariant derivative: $\dfrac{\mathrm{d}}{\mathrm{d}\tau} \to \dfrac{\mathrm{D}}{\mathrm{d}\tau}$. Applying the equivalence principle, we arrive at the gyroscope equation.

Gyroscope equation

$$\frac{\mathrm{d}s^\alpha}{\mathrm{d}\tau} + \Gamma^\alpha_{\beta\gamma} s^\beta u^\gamma = 0. \tag{8.171}$$

This equation is a linear in the spin vector $\widehat{s}$, meaning it depends only on the direction of $\widehat{s}$, not its magnitude. Essentially, it describes how $\widehat{s}$ undergoes parallel transport along the geodesic. As a consequence, both the spin magnitude $\widehat{s}\cdot\widehat{s}$ and the orthogonality condition $\widehat{s}\cdot\widehat{u}$ remain constant along the geodesic.

We now examine the phenomenon of geodetic precession in a spherically symmetric spacetime. Consider a test particle following a circular orbit of

[8]For example, in the Gravity Probe B experiment, the gyroscope's deviation from sphericity was controlled to one part in a million.

[9]Note that this spin angular momentum is a classical concept, referring to the rotational angular momentum of the gyroscope itself, and should not be confused with quantum mechanical spin.

radius R in the equatorial plane $\theta = \pi/2$. The angular velocity of the orbit Ω is given by

$$u^\phi = \frac{d\phi}{d\tau} = \frac{d\phi}{dt}\frac{dt}{d\tau} = \Omega u^t, \quad \Omega^2 = \frac{GM}{R^3}. \tag{8.172}$$

The 4-velocity of the particle takes the form

$$\begin{aligned} u^\mu &= u^t(1, 0, 0, \Omega), \\ u^t &= (1 - 3GM/R)^{-1/2}. \end{aligned} \tag{8.173}$$

From the gyroscope equation, we immediately see that the polar component of the spin vector remains constant

$$\frac{ds^\theta}{d\tau} = 0. \tag{8.174}$$

If the spin vector initially points radially, $s^\theta(0) = 0$, it will remain confined to the equatorial plane

$$s^\theta = 0. \tag{8.175}$$

The orthogonality condition $\widehat{s} \cdot \widehat{u} = 0$ yields a relationship between the temporal and azimuthal components

$$s^t = R^2\Omega\left(1 - \frac{2GM}{R}\right)^{-1} s^\phi. \tag{8.176}$$

The gyroscope equation then reduces to a coupled system for the radial and azimuthal spin components:

$$\begin{aligned} \frac{ds^r}{dt} - (R - 3GM)\Omega s^\phi &= 0, \\ \frac{ds^\phi}{dt} + \frac{\Omega}{R}s^r &= 0. \end{aligned} \tag{8.177}$$

Combining these equations yields a harmonic oscillator equation for s^ϕ:

$$\frac{d^2 s^\phi}{dt^2} + \left(1 - \frac{3GM}{R}\right)\Omega^2 s^\phi = 0. \tag{8.178}$$

Using the new parameter $\Omega'^2 = \left(1 - \frac{3GM}{R}\right)\Omega^2$, the general solution to the above equation can be written as

$$\begin{aligned} s^r(t) &= s_*\left(1 - \frac{2GM}{R}\right)^{1/2} \cos(\Omega' t), \\ s^\phi(t) &= -s_*\left(1 - \frac{2GM}{R}\right)^{1/2}\left(\frac{\Omega}{\Omega' R}\right)\sin(\Omega' t). \end{aligned} \tag{8.179}$$

After one period $P = 2\pi/\Omega$, the spin vector does not return to its initial orientation. The angular deflection $\Delta\phi$ is determined by

$$\cos\Delta\phi = \left[\hat{e}_r \cdot \left(\frac{\hat{s}(t)}{s_*}\right)\right]_{t=P} = \cos\left[2\pi\left(1 - \frac{3GM}{R}\right)^{1/2}\right], \tag{8.180}$$

yielding the net precession angle

$$\Delta\phi = 2\pi[1 - (1 - 3GM/R)^{1/2}]. \tag{8.181}$$

This result gives the precession as measured by an observer at infinity. However, since the orbital velocity is purely azimuthal, a comoving observer would measure the same deflection angle.

The phenomenon of a spin vector precessing along a geodesic trajectory is known as geodetic precession. This effect was first predicted by Willem de Sitter in 1916 and is also referred to as the de Sitter–Fokker effect. De Sitter's original proposal considered the Earth-Moon system as a gravitational gyroscope in free fall within the Sun's gravitational field. The Earth-Moon system behaves as a spinning object with its spin axis aligned along the Moon's orbital angular momentum vector. The solar gravitational field induces a gradual precession of this spin axis. In the solar system, since $GM/R \ll 1$, the precession angle per orbit simplifies to

$$\Delta\phi \approx \left(\frac{3GM}{2R}\right) = \frac{3\pi GM}{c^2R}. \tag{8.182}$$

The predicted geodetic precession amounts to $0.0192''$ per year. Although this precession is very small, it can be detected through lunar laser ranging (LLR), which tracks the Moon's orbit with millimeter precision. LLR data confirmed the geodetic precession with remarkable accuracy, agreeing with theoretical predictions to within 0.7%.

The geodetic precession effect can be measured with greater precision using satellites in low Earth orbit. Consider a satellite orbiting at Earth's surface, with an orbital period $\tau = 2\pi\sqrt{\frac{R_\oplus^3}{GM}}$. The accumulated annual precession angle of the spin vector for such a satellite is

$$\Delta\phi|_{\text{year}} \approx \frac{3(GM_\oplus)^{3/2}}{2c^2}\frac{1}{R_\oplus^{5/2}}. \tag{8.183}$$

The radius of the Earth is approximately $R_\oplus = 6.38 \times 10^3$ km, and the satellite's period around the Earth is about 84.5 minutes, so the satellite can orbit the Earth 6.2×10^3 times a year, thus the annual precession angle of the spin vector is 8.4″. For a satellite at altitude h with orbital radius $R = R_\oplus + h$, the annual precession scales as

$$\Delta\phi|_{\text{year}} \approx \frac{3(GM_\oplus)^{3/2}}{2c^2} \frac{1}{R_\oplus^{5/2}} \left(\frac{R_\oplus}{R}\right)^{5/2}. \tag{8.184}$$

If the satellite's height is 650 km, the annual precession angle of the gyroscope is $\Delta\phi = 6.6''$. Due to the rotation of the Earth, this value will have a small correction. Although this value is small, it can be measured with Gravity Probe B, with a precision deviation that can reach 0.01%. The actual measurement precision deviation is about 0.28%, which is in good agreement with the theoretical prediction.

8.6 Relativistic Astrophysics

In astrophysical applications, Newtonian gravity often provides an adequate description for numerous scenarios. However, when dealing with compact objects such as neutron stars, relativistic effects become non-negligible and must be properly addressed within the framework of general relativity. In this section, we present a concise introduction to several relevant relativistic effects within the Schwarzschild spacetime.

8.6.1 *Gravitational lensing effect*

The gravitational lensing effect has very important implications in astronomy and cosmology. It provides a powerful tool for probing the physical properties of both the lensed source and the intervening lens, as well as for investigating the large-scale structure of spacetime. The underlying principle of gravitational lensing is straightforward: light undergoes deflection in a gravitational field, causing the observed image to differ from the true source position. In this process, the massive object (or galaxy) generating the gravitational field acts as a lens, bending the light rays from the background source.

In the Schwarzschild spacetime, the deflection angle of light is given by $\alpha = \dfrac{4GM}{c^2 b} = \dfrac{2R_\text{s}}{b}$, where $R_\text{s} = \dfrac{2GM}{c^2}$ is the Schwarzschild radius. When gravitational lensing occurs on cosmological scales such as when galaxies act as lenses, typical parameters are:

$$M \sim 10^{11} M_\odot, \quad R_\text{s} \sim 10^{11}\ \text{km}, \tag{8.185}$$

$$D_\text{S} \sim D_\text{L} \sim D_\text{LS} \sim 1\ \text{Gpc} \sim 3 \times 10^{22}\ \text{km}, \tag{8.186}$$

where D_S is the distance from the source to the observer, D_L is the distance from the lens to the observer, and D_{LS} is the distance from the lens to the source. The parsec (pc), the standard unit in galactic and extragalactic astronomy, is defined as the distance at which 1 astronomical unit (AU) subtends an angle of 1 arcsecond

$$1 \text{ pc} = 3.086 \times 10^{13} \text{ km}, \text{ or } 3.262 \text{ ly}. \tag{8.187}$$

The stellar distance in galaxies is a few pc, the typical size of a galaxy is on the kpc scale, the typical distance between galaxies is on the Mpc scale, and the size of the observable universe is on the Gpc scale.

In practical analyses of gravitational lensing, the condition $R_s \ll D_L$, D_S, D_{LS} allows us to adopt the thin lens approximation, which relies on three key simplifications:

(1) Straight-line propagation. Light travels along straight paths far from the lens;
(2) All deflections occur at the lens plane;
(3) Point-like objects. Both the source and lens are treated as point masses.

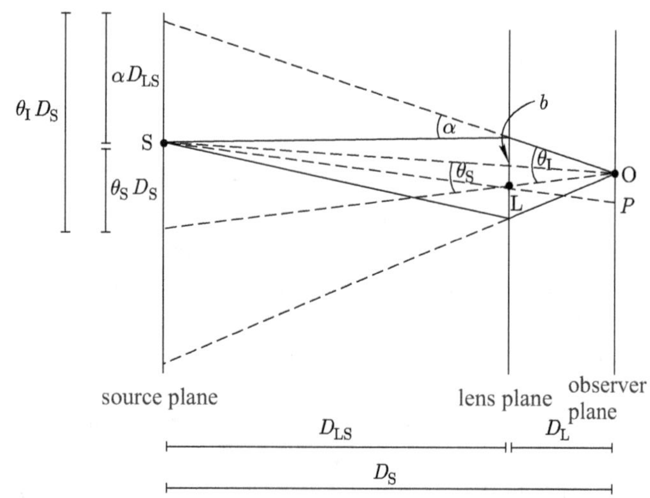

Fig. 8.7 Schematic diagram of the gravitational lensing effect under the thin lens approximation, where O represents the observer, S represents the source, and L represents the lens.

As illustrated in Figure 8.7, the angles α, θ_S, θ_I are all very small, leading to the lens equation

$$\theta_I D_S = \theta_S D_S + \alpha D_{LS}. \tag{8.188}$$

Substituting the impact parameter $b \approx \theta_{\rm I} D_{\rm L}$ and the deflection angle, we have

$$\theta_{\rm I} = \theta_{\rm S} + \frac{\theta_{\rm E}^2}{\theta_{\rm I}}, \tag{8.189}$$

where the Einstein angle $\theta_{\rm E}$ is defined as

$$\theta_{\rm E} \equiv \left(2R_{\rm s}\frac{D_{\rm LS}}{D_{\rm S}D_{\rm L}}\right)^{1/2}. \tag{8.190}$$

For a lens much closer to the observer $D_{\rm S} \approx D_{\rm LS} \gg D_{\rm L} = D$, the Einstein angle simplifies to $\theta_{\rm E}^2 \approx \dfrac{2R_{\rm s}}{D}$. A notable configuration occurs when the source, lens, and observer are perfectly aligned $\theta_{\rm S} = 0$. In this case, $\theta_{\rm I} = \theta_{\rm E}$, and the observer sees a circular image known as an Einstein ring, as shown in Figure 8.8.

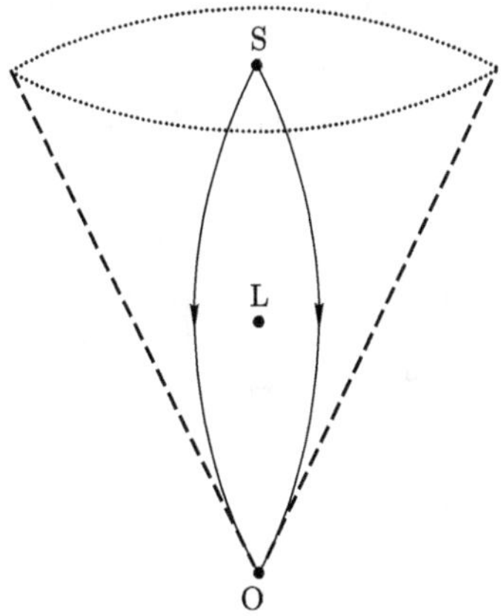

Fig. 8.8 Schematic diagram of the Einstein ring, where O represents the observer, S represents the source, and L represents the lens.

The Einstein angle $\theta_{\rm E}$ defines the characteristic angular scale of gravitational lensing phenomena. It depends solely on the distances between the source, lens, and observer, as well as the lens mass. For a lens with mass $M \sim M_\odot$, $R_{\rm s} \sim 1$ km and galactic-scale distances $D_{\rm L} \sim D_{\rm S} \sim D_{\rm LS} \sim 10$ kpc $\sim 10^{17}$ km, the Einstein angle becomes $\theta_{\rm E} \sim (10^{-3})''$, which is far below the resolution limit of conventional telescopes. This lensing phenomenon induced by solar-mass objects within the galaxy, is termed microlensing. In microlensing, although the angular separation of lensed images is unresolvable, the relative motion between the lens and source leads to time-dependent magnification of the source's brightness. This temporal variation serves as a detectable signature of microlensing events. An important application of microlensing is to search for massive astrophysical compact halo objects (MACHOs), which include white dwarfs, neutron stars, black holes, and Jupiter-sized substellar objects. These objects, with masses ranging from $10^{-2}M_{\odot}$ to several $M_{\odot}$, are candidates for explaining the dark matter in galactic halos. Since MACHOs emit negligible electromagnetic radiation, they evade direct detection. However, their gravitational influence manifested through microlensing provides a unique observational window into their existence and distribution.

For a lens with a mass $M \sim 10^{11}M_{\odot}$, the Einstein angle is typically $\theta_{\rm E} \sim 1''$, which leads to observable macrolensing phenomena. In cosmological contexts, Hubble's law provides a means to estimate the mass of a gravitational lens, such as a quasar. According to Hubble's law,

$$cz_{\rm L} = H_0 D_{\rm L}, \quad cz_{\rm S} = H_0 D_{\rm S}, \tag{8.191}$$

where H_0 is the Hubble constant, and $z_{\rm L}$, $z_{\rm S}$ denote the redshift factors of the lens and the source, respectively. From this, the lens mass can be derived as

$$M = \frac{c^3}{4GH_0}\frac{z_{\rm S}z_{\rm L}}{z_{\rm S}-z_{\rm L}}\theta_{\rm E}^2. \tag{8.192}$$

In most cases, the source, lens, and observer are not perfectly aligned. Consequently, the observer detects two distinct images, with angular separations given by

$$\theta_{\pm} = \frac{1}{2}[\theta_{\rm S} \pm (\theta_{\rm S}^2 + 4\theta_{\rm E}^2)^{1/2}]. \tag{8.193}$$

Obviously, $\theta_- < \theta_{\rm E}$ while $\theta_+ > \theta_{\rm E}$. If the lens is transparent and spatially extended, light from the source may traverse the lens directly, potentially producing an odd number of observable images. Notably, the angular positions $\theta_{\pm}$ are independent of the frequency of light, rendering the lensing phenomenon achromatic.

For lensed images with finite angular scales, their shape and brightness are the most critical observables. The angular size distribution of an image is given by

$$\Delta\theta_{\pm} = \frac{1}{2}\left(1 \pm \frac{\theta_{\rm S}}{(\theta_{\rm S}^2 + 4\theta_{\rm S}^2)^{1/2}}\right)\Delta\theta_{\rm S}, \tag{8.194}$$

where $\Delta\theta_{\rm S}$ represents the intrinsic angular size of the source. In terms of brightness, the observed intensity is proportional to the solid angle subtended by the image. The image at θ_+ is always brighter than the one at θ_-, i.e., $I_+ > I_-$. The total magnification ratio, comparing the lensed brightness to the original source brightness, is

$$\begin{aligned}\frac{I_{\text{tot}}}{I_*} &= \frac{\sin\theta_+\Delta\theta_+ + \sin\theta_-\Delta\theta_-}{\sin\theta_S\Delta\theta_S} \approx \frac{\theta_+\Delta\theta_+ + \theta_-\Delta\theta_-}{\theta_S\Delta\theta_S} \\ &= \frac{1}{2}\left(\frac{\theta_S}{(\theta_S^2 + 4\theta_E^2)^{1/2}} + \frac{(\theta_S^2 + 4\theta_E^2)^{1/2}}{\theta_S}\right) > 1.\end{aligned} \tag{8.195}$$

This demonstrates that gravitational lensing invariably enhances the apparent brightness of the source. The effect is particularly pronounced when the source is nearly aligned with the lens $\theta_S \approx 0$, leading to significant magnification.

8.6.2 *Accretion disk*

The strength of relativistic gravitational effects is characterized by the dimensionless parameter $\frac{GM}{Rc^2}$. In the case of the Sun, this quantity is 10^{-6}, indicating relatively weak relativistic effects. Nevertheless, even at this level, various observable phenomena emerge, making the solar system an important testing ground for general relativity. Indeed, solar system experiments currently provide the most precise large-scale tests of relativistic gravity. For compact objects, this parameter approaches significantly larger values: For a spherically symmetric black hole, it takes the upper limit of 0.5 at the event horizon; and for a neutron star, it is about 0.2. In these regimes, relativistic effects dominate and cannot be neglected. For simplicity, we restrict our discussion to non-rotating systems, where the spacetime geometry is well described by the Schwarzschild metric. Matter in the vicinity of such compact objects, for instance, a companion star in a binary system, experiences intense gravitational attraction, leading to accretion processes. Furthermore, in binary systems containing compact objects, the emission of gravitational waves causes continuous orbital energy loss, resulting in measurable changes in the orbital parameters over time.

In typical astrophysical scenarios, the mass of a companion star is substantially smaller than that of its compact companion (e.g., a black hole or neutron star). Consequently, the outer gaseous envelope of the star can be tidally stripped and gravitationally captured by the compact object, leading to mass transfer through accretion. However, due to angular momentum conservation, the infalling matter cannot plunge directly into the compact object. Instead, it forms a rotating accretion disk in the equatorial plane. Within this disk, complex physical processes occur, including various dissipation, turbulent transport, and radiative cooling. These mechanisms gradually extract energy and angular momentum from the orbiting material, allowing it to spiral inward. From studies of particle geodesics in strong gravity, we know an innermost stable circular orbit (ISCO) exists. Particles crossing this critical radius lose orbital stability and rapidly plunge into the event horizon. The radiative properties of accretion depend strongly on the compact object's mass. For stellar-mass black holes ($M \sim M_\odot$), the hot inner accretion disk emits predominantly in the X-ray band ($kT \sim$ keV). For supermassive black holes ($M \sim 10^6 - 10^9 M_\odot$) in galactic nuclei, the cooler disk radiates at lower energies (optical/UV). Studying these accretion processes provides crucial insights into the physics of active galactic nuclei, the nature of quasars, and black hole growth and feedback mechanisms.

In reality, the gas within the accretion disk follows a nearly circular orbital motion. However, magnetohydrodynamic (MHD) instabilities generate turbulence, and the resulting viscous forces gradually dissipate the gas's angular momentum. Consequently, the gas spirals inward, losing gravitational potential energy and heating up in the process. Eventually, the particles in the gas lose sufficient angular momentum to sustain stable motion at the ISCO, causing them to plunge rapidly into the central compact object. The radiative efficiency of this process can be estimated by considering the maximum energy release, which corresponds to the gravitational binding energy of particles at the ISCO relative to their rest mass. This is equivalent to the energy lost by a particle falling from infinity to the ISCO. For a Schwarzschild black hole, the ISCO occurs at $r = 6GM$, the ratio of the conserved energy of the particles to the rest mass is

$$\frac{E}{m_0 c^2} = \frac{2\sqrt{2}}{3} \approx 0.943. \tag{8.196}$$

Thus, the maximum radiative efficiency of the accretion disk is

$$\epsilon_{\text{acc}} \approx 1 - 0.943 = 5.7\%. \tag{8.197}$$

In comparison, nuclear processes such as hydrogen-to-helium fusion release energy with an efficiency of only

$$\epsilon_{\text{nuclear}} \approx 0.7\%. \tag{8.198}$$

Remarkably, accretion disks are nearly an order of magnitude more efficient at converting mass to energy than nuclear fusion. This extreme efficiency explains why accretion onto compact objects powers some of the most energetic phenomena in the universe.

On the other hand, studying the dynamics of accreting matter and its emitted spectra provides crucial insights into the geometry of the central compact object. For example, at a characteristic temperature $T \sim 10^7$ K of the accretion disk, heavy elements such as iron retain some bound electrons. These highly ionized atoms absorb X-rays and re-emit them as fluorescence lines. Most notably, the iron K_α line at 6.4 keV, appears in the middle of the X-ray band. However, two relativistic effects significantly distort the observed emission: one is the gravitational redshift caused by the spacetime itself, and the other is the Doppler shift caused by the motion of matter. Since the magnitude of these shifts varies across the disk, the observed emission is not a narrow spectral line but rather a broadened feature. The precise shape of this broadened line encodes information about the disk's geometry, while photons originating from the innermost orbits probe the strong-field regime of gravity, offering a unique window into extreme astrophysical environments.

We now examine two fundamental redshift effects governing photon emission from an accretion disk. For simplicity, we assume that the central black hole is non-rotating and can be described by the Schwarzschild spacetime, and the accretion disk lies entirely in the equatorial plane $\theta = \dfrac{\pi}{2}$ and is geometrically thin. As shown in Figure 8.9, consider a photon emitted at frequency $\omega_\star$ from a source at coordinates (r, ϕ) in the disk, and observed at infinity with frequency ω_∞. The 4-velocity of the source is $\widehat{u}_{\mathrm{s}}(r, \phi)$, and the 4-momentum of the emitted photons is $\widehat{P}(r, \phi)$. For the observer, its 4-velocity is $\widehat{u}_0$, and the 4-momentum of the received photons is $\widehat{P}(\infty)$. For a static observer at infinity $u_0^\mu = (1, 0, 0, 0) = \xi^\mu$, where ξ^μ is the component

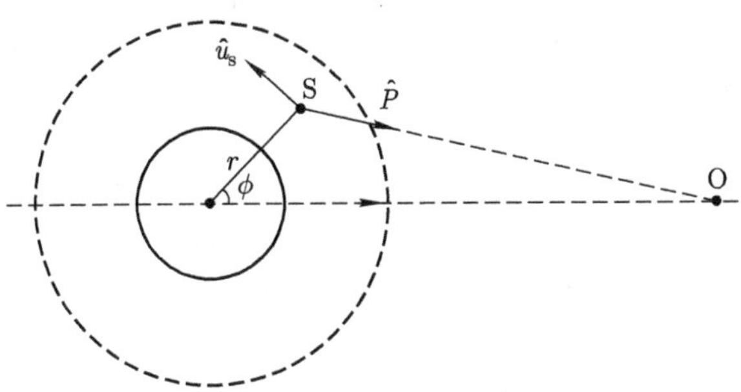

Fig. 8.9 Redshift of photon emission from accretion disk. Where O represents the observer, S represents the emission source, $\widehat{u}_{\mathrm{s}}$ represents the velocity of the source, and $\widehat{P}$ represents the momentum of the photon.

of the Killing vector ∂_t, the redshift is determined by the ratio

$$\frac{\omega_\infty}{\omega_\star} = \frac{\widehat{u}_0 \cdot \widehat{P}(\infty)}{\widehat{u}_{\rm s} \cdot \widehat{P}}. \tag{8.199}$$

The source follows a circular orbit with angular velocity

$$\Omega(r) = \frac{\mathrm{d}\phi}{\mathrm{d}t} = (GM/r^3)^{1/2}. \tag{8.200}$$

Its 4-velocity is

$$u_{\rm s}^\mu = \left(\frac{\mathrm{d}t}{\mathrm{d}\tau}, 0, 0, \frac{\mathrm{d}\phi}{\mathrm{d}\tau}\right) = u_{\rm s}^t(1, 0, 0, \Omega(r)). \tag{8.201}$$

From the normalization condition of the 4-velocity $\widehat{u} \cdot \widehat{u} = -1$, we get

$$u_{\rm s}^t = \left(1 - \frac{R_{\rm s}}{r} - r^2\Omega^2\right)^{-1/2} = \left(1 - \frac{3GM}{r}\right)^{-1/2}.$$

On the other hand, the conserved quantities of photons defined by the Killing vector are

$$E = -\widehat{P} \cdot \widehat{\xi}, \quad L = \widehat{P} \cdot \widehat{\eta}. \tag{8.202}$$

Defining the impact parameter $b = |L/E|$, we have

$$\widehat{u}_0 \cdot \widehat{P}(\infty) = \widehat{\xi} \cdot \widehat{P}(\infty) = -E, \tag{8.203}$$

$$\begin{aligned} \widehat{u}_{\rm s}\widehat{P} &= u_{\rm s}^t(\widehat{\xi} + \Omega(r)\widehat{\eta}) \cdot \widehat{P} \\ &= u_{\rm s}^t(-E + \Omega(r)L) \end{aligned} \tag{8.204}$$

$$\Rightarrow \frac{\omega_\infty}{\omega_\star} = (u_{\rm s}^t(1 \pm \Omega(r)b))^{-1}. \tag{8.205}$$

Here, "−" represents "counter-emission," meaning the emitter is moving towards the observer, while "+" represents "receding emission," meaning the emitter is moving away from the observer when emitting photons.

When the emitter radiates photons at $\phi = 0$, π, the photon's angular momentum vanishes $b = 0$ and $L = 0$. The observed redshift then simplifies to

$$\frac{\omega_\infty}{w_\star} = \left(1 - \frac{3GM}{r}\right)^{1/2}. \tag{8.206}$$

This corresponds to lateral motion of the emitter relative to the observer. Notably, the photons are always redshifted, regardless of emission position. The effect combines gravitational redshift and lateral Doppler shift, distinguishing it from pure gravitational redshift.

For emission at $\phi = \pi/2$ (receding motion), or $\phi = -\pi/2$ (counter motion), the photon's conserved quantities are

$$E = \left(1 - \frac{R_s}{r}\right)\frac{dt}{d\lambda}, \quad L = r^2 \sin^2\theta \frac{d\phi}{d\lambda}, \tag{8.207}$$

yielding the impact parameter

$$b = \frac{r^2 |P^\phi(r,\phi)|}{\left(1 - \dfrac{R_s}{r}\right) P^t(r,\phi)}. \tag{8.208}$$

Since $P^r\left(r, \pm\dfrac{\pi}{2}\right) = 0$ and $\widehat{P} \cdot \widehat{P} = 0$, we have

$$-\left(1 - \frac{R_s}{r}\right)(P^t)^2 + r^2 (P^\phi)^2 = 0 \Rightarrow b = r\left(1 - \frac{R_s}{r}\right)^{-1/2}. \tag{8.209}$$

Substituting b into (8.205), the observed frequency shift becomes

$$\frac{\omega_\infty}{\omega_\star} = \left(1 - \frac{3GM}{r}\right)^{1/2} \left(1 \pm \left(\frac{r}{GM} - 2\right)^{-1/2}\right)^{-1}. \tag{8.210}$$

In the weak field limit, we can expand (8.210) to leading orders in GM/r

$$\begin{aligned}\frac{\omega_\infty}{\omega_\star} &= 1 \mp \left(\frac{GM}{r}\right)^{1/2} - \frac{GM}{2r} + \cdots \\ &= 1 \mp v + \frac{1}{2}v^2 - \frac{GM}{r} + \cdots,\end{aligned} \tag{8.211}$$

where $v = \Omega r = (GM/r)^{1/2}$ is the local velocity of the source. The above expression includes $\mp v + \dfrac{1}{2}v^2$ from the Doppler shift and $-\dfrac{GM}{r}$ from the gravitational redshift. The net shift is a superposition of these two effects.

The observed spectrum integrates photons emitted at various radii in the accretion disk. For emissions originating specifically from the ISCO at $r = 6GM$, the frequency shifts exhibit distinct behaviors depending on the azimuthal angle ϕ of emission:

(1) Receding emission at $\phi = \pi/2$,

$$\frac{\omega_\infty}{\omega_\star} = \frac{\sqrt{2}}{3} \approx 0.47. \tag{8.212}$$

This corresponds to the lowest observed photon frequency.

(2) Approaching emission at $\phi = -\pi/2$,

$$\frac{\omega_\infty}{\omega_\star} = \sqrt{2}. \tag{8.213}$$

This is due to the fact that the Doppler blueshift dominates over gravitational redshift, resulting in a net blueshift.

(3) Lateral emission at $\phi = 0,\ \pi$,

$$\frac{\omega_\infty}{\omega_\star} = \frac{1}{\sqrt{2}} \approx 0.71. \tag{8.214}$$

In the case of receding emission, the Doppler redshift effect is the most significant, combined with the gravitational redshift effect, resulting in the largest redshift. Detection of the minimal frequency in observations serves as a signature of emission from the ISCO, confirming the central object as a black hole (where no surface exists beyond $r = 6GM$). For other compact objects (e.g., neutron stars), photon frequencies from their surfaces would be higher due to larger radii.

If the central compact star has rotation, frame-dragging reduces the ISCO radius, permitting even lower observed frequencies than the Schwarzschild case. Generally, the 6.4 keV spectral band has a minimum frequency that depends on the size, rotation, and inclination angle of the accretion disk relative to us. Furthermore, the profile of the spectral band is also affected by the relativistic beaming effect and other possible radiation sources. The relativistic beaming effect also increases the intensity of the blue end of the spectral band.

8.6.3 *Binary pulsars*

Binary pulsars represent a cornerstone in relativistic astrophysics, serving dual roles as exceptional laboratories for testing general relativity beyond the solar system and as unique probes for investigating gravitational radiation. A paradigmatic example is the Hulse–Taylor binary pulsars PSR1913+16,[10] which consists of two neutron stars in a compact 7.75-hour orbit. The system's observational significance stems from one member being a rapidly rotating pulsar exhibiting extraordinary rotational stability, effectively functioning as a precision cosmic clock:

$$P_{\rm rot} = (0.059029997929613 \pm 0.000000000000007)\ \text{s}. \tag{8.215}$$

Detected period variations revealed the pulsar's binary nature, while subsequent measurements unveiled remarkable relativistic effects. Most notably, the system demonstrates a periastron advance of

$$\delta\phi_{\rm prec} = (4.22659 \pm 0.00004)^\circ/\text{year}, \tag{8.216}$$

[10]The designation "PSR" identifies the object as a pulsar, while "B1913+16" specifies its celestial coordinates in the B1950.0 epoch.

which is about 4×10^4 times larger than Mercury's perihelion precession. Additional robust relativistic signatures include gravitational redshift, Shapiro delay, and spin-geodetic precession, collectively establishing this system as an unparalleled natural laboratory for general relativity.

We can analyze elliptical orbits within the framework of Newtonian gravity, which are determined by the period $P_{\rm b}$, eccentricity e, and semi-major axis length a. Long-term spectroscopic Doppler shift measurements reveal the orbital parameters through radial velocity variations, while post-Newtonian effects (e.g., higher-order Doppler shifts and Shapiro delay) yield precise mass determinations:

$$m_{\rm pulsar} = (1.442 \pm 0.003)M_{\odot}, \tag{8.217}$$

$$m_{\rm comp} = (1.386 \pm 0.003)M_{\odot}, \tag{8.218}$$

confirming the near-equal mass neutron star configuration. Crucially, orbital decay due to gravitational wave emission has been observed through systematic period changes, with the measured rate showing striking agreement with theoretical predictions, thus providing the first compelling indirect evidence for gravitational radiation.

Exercises

1. Consider the metric

$$\mathrm{d}s^2 = 2\mathrm{d}u\mathrm{d}v + \mathrm{d}x^2 + \mathrm{d}y^2 + H\mathrm{d}u^2 \tag{8.219}$$

where $H = H(u, x, y)$ is an arbitrary function of u, x, and y.

(1) Prove that if this metric is a solution to the vacuum Einstein equations, then it must satisfy

$$\frac{\partial^2 H}{\partial x^2} + \frac{\partial^2 H}{\partial y^2} = 0. \tag{8.220}$$

(2) Calculate the non-zero components of the Riemann tensor for this metric.

(3) Compute $R_2 \equiv R^{\mu\nu\rho\sigma}R_{\mu\nu\rho\sigma}$ for this metric.

2. For the Schwarzschild metric, calculate all non-zero Christoffel symbols. Furthermore, compute all non-zero components of the Riemann tensor.

3. By defining

$$\mathrm{d}u = \mathrm{d}t - (1 - 2GM/r)^{-1}\mathrm{d}r \tag{8.221}$$

prove that the Schwarzschild metric can be written in the Kerr–Schild form

$$\mathrm{d}s^2 = -\mathrm{d}u^2 - 2\mathrm{d}u\mathrm{d}r + r^2\mathrm{d}\Omega^2 + \frac{2GM}{r}\mathrm{d}u^2. \tag{8.222}$$

4. Consider two celestial bodies with comparable masses. Discuss their motion if we only take into account the gravitational interaction between them and neglect relativistic effects.
5. In the Schwarzschild geometry produced by a spherically symmetric star of mass M, an observer maintains a stable position at a fixed radius R. A proton emerges from the star and moves radially past the observer's laboratory, with its energy E and momentum $|\boldsymbol{p}|$ being measured.

   (1) What is the relationship between the energy E and the momentum $|\boldsymbol{p}|$?
   (2) What are the components of the proton's 4-momentum in the Schwarzschild coordinate basis? Express them in terms of E and $|\boldsymbol{p}|$.

6. Outside a black hole with mass M and external Schwarzschild geometry, a spaceship is performing a circular motion without thrust at a radius of $7GM$.

   (1) What is the orbital period of the spaceship as observed by an observer at infinity?
   (2) What is the orbital period measured by a clock on the spaceship?

7. In the Schwarzschild geometry, an observer at $r = r_1$ emits a radial light signal.

   (1) What is the instantaneous coordinate velocity $\mathrm{d}r/\mathrm{d}t$ of the signal?
   (2) If the signal is reflected back to r_1 at $r = r_2 > r_1$, what is the return time measured by an observer at infinity?
   (3) What is the return time measured by the observer at $r = r_1$?

8. In the Schwarzschild geometry, a particle with mass is emitted radially from $r = 4GM$.

   (1) What initial velocity $\mathrm{d}r/\mathrm{d}t$ must the particle have at the emission point to come to rest at $r = 10GM$?
   (2) How much proper time does the particle take to complete this journey?

9. In the Schwarzschild geometry, there are two particles, A and its antiparticle B, moving in free circular orbits with radii r_A and r_B respectively, with $r_B > r_A$. At a certain moment, A emits a photon that moves radially and is received by B. What is the ratio of the received photon frequency to the emitted photon frequency? If $r_A = r_B$, and

the two particles collide and annihilate, what is the total energy radiated relative to an observer at the collision point? Assume the rest mass of these two particles is m_0.

10. Consider the motion of a particle with mass in the Schwarzschild geometry. We can define $u = GM/r$, thus obtaining the differential equation

$$\left(\frac{\mathrm{d}u}{\mathrm{d}\phi}\right)^2 = f(u) = 2(u - u_1)(u - u_2)(u - u_3) \tag{8.223}$$

where $u_1 < u_2 < u_3$.

(1) Prove that $u_1 + u_2 + u_3 = 1/2$, and the aphelion and perihelion of the orbit are located at $r_1 = GM/u_1$ and $r_2 = GM/u_2$ respectively.

(2) Express the particle's energy density and angular momentum density in terms of u_i, $i = 1, 2, 3$.

(3) Integrating the above differential equation yields

$$u = u_1 + (u_2 - u_1)\mathrm{sn}^2(\alpha\phi|\beta) \tag{8.224}$$

where

$$\alpha = \frac{(u_3 - u_1)^{1/2}}{\sqrt{2}}, \quad \beta = \frac{(u_2 - u_1)}{(u_3 - u_1)} \tag{8.225}$$

and $\mathrm{sn}(z|\mu)$ is the Jacobian elliptic function with modulus μ. Use this to plot various trajectories of the particle.

11. Suppose a photon is emitted from a spaceship performing circular motion at radius r on the equatorial plane in Schwarzschild geometry. The frequency of the photon in the comoving reference frame of the spaceship is ν_0, and its emission direction makes an outward angle α with the tangential direction of the spaceship's motion. The motion of the photon remains in the equatorial plane. What is the frequency of the photon observed by a static observer at infinity?

12. If the Sun has a quadrupole moment, i.e., it is not perfectly spherically symmetric, then in Newtonian mechanics, there will also be a precession of Mercury's perihelion. We can approximate the gravitational potential of the Sun as (taking the unit system $G = c = 1$)

$$\Phi = -\frac{M_\odot}{r}\left(1 - J\frac{R_\odot^2}{r^2}\frac{3\cos^2\theta - 1}{2}\right) \tag{8.226}$$

where J is a constant characterizing the Sun's quadrupole moment. Prove that under this circumstance, the precession angle of Mercury after one orbit is

$$\delta\phi = \frac{6\pi M_\odot}{a(1 - e^2)} + J\frac{3\pi R_\odot^2}{a^2(1 - e^2)^2} \tag{8.227}$$

where the first term on the right-hand side is the correction from general relativity, and the second term comes from the Sun's deformation. If these two terms are comparable, what is the value of J?

13. Suppose in another theory of gravity, the metric outside a spherically symmetric star is

$$\mathrm{d}s^2 = \left(1 - \frac{2GM}{r}\right)[-\mathrm{d}t^2 + \mathrm{d}r^2 + r^2\mathrm{d}\Omega^2]. \tag{8.228}$$

Assuming that photons still move along null geodesics in this geometry, calculate the deflection angle of light by the star.

14. The gravitational deflection of light by a massive object effectively makes the object appear larger. Consider a spherical star with mass M and radius R, and light traveling from its surface to infinity. Prove that the maximum impact parameter for the light ray is

$$b = R\left(1 - \frac{2GM}{Rc^2}\right)^{-1/2}. \tag{8.229}$$

Thus, the apparent diameter of the star increases. (For the Sun, the diameter increases by 3 km.)

15. In the weak-field approximation, the Schwarzschild geometry can be approximated by the following metric:

$$\mathrm{d}s^2 = -\left(1 - \frac{2GM}{r}\right)\mathrm{d}t^2 + \left(1 + \frac{2GM}{r}\right)(\mathrm{d}x^2 + \mathrm{d}y^2 + \mathrm{d}z^2). \tag{8.230}$$

(1) Discuss the deflection of light in this metric and prove that it gives the correct deflection angle.

(2) Discuss the precession of the perihelion in this metric and prove that it gives a value of 4/3 times the correct value. This indicates that the precession of the perihelion is very sensitive to the nonlinearity of general relativity.

16. In the post-Newtonian parametrized metric, discuss respectively

(1) The precession of planetary perihelia;
(2) The deflection of light;
(3) The radar echo delay.

17. Consider the vacuum Einstein equations with a cosmological constant $G_{\mu\nu} + \Lambda g_{\mu\nu} = 0$.

(1) Find the most general spherically symmetric metric solution. Require that in the coordinates (t, r), when $\Lambda = 0$, the solution reduces to the usual Schwarzschild metric.

(2) Using the effective potential, write down the equation of motion for radial motion, discuss the effective potentials for massive and massless particles, and their possible motions.

18. For a binary system with masses m_1 and m_2, their orbits are generally elliptical around the center of mass. Prove within the framework of Newtonian mechanics that the orbital parameters are determined by the system's energy $E < 0$ and angular momentum L:
$$a = -\frac{m_1 m_2}{2E}, \quad e^2 = 1 + \frac{2EL^2(m_1+m_2)}{m_1^3 m_2^3}. \tag{8.231}$$
If the center of mass is taken as the origin, give the orbital paths of the two stars.
19. If two gyroscopes A and B both move along the same geodesic, prove that $\widehat{s}_A \cdot \widehat{s}_B$ is a constant.
20. Carefully discuss the torque produced by a finite-sized rigid body under tidal forces. Prove that in the comoving reference frame of the rigid body, the torque is
$$\tau^i = c^2 \sum_{j,k,l} \varepsilon^{ijk} R^j{}_{olo} \left(-I^{kl} + \frac{1}{2}\delta^l_k I^{rr} \right) \tag{8.232}$$
where ε^{ijk} is the three-dimensional Levi–Civita symbol, with the convention $\varepsilon^{123} = 1$, and I^{ij} is the moment of inertia tensor of the rigid body. Since the change in the spin vector is
$$\frac{\mathrm{d}s^i}{\mathrm{d}t} = \tau^i \tag{8.233}$$
prove that for a spherically symmetric rigid body, its spin vector is not affected by tidal forces. Estimate the precession effect produced by non-spherical shapes.
21. In linearized gravity, if we neglect the rotation of the source, then the spacetime metric can be written as
$$\mathrm{d}s^2 = -(1+2\Phi)\mathrm{d}t^2 + (1-2\Phi)\delta_{ij}\mathrm{d}x^i\mathrm{d}x^j. \tag{8.234}$$
In the reference frame of the gyroscope, the 4-velocity is $u^\mu = (1, 0)$, and the spin 4-vector is $s^\mu = (0, \boldsymbol{S})$. Using the gyroscope equation, prove that
$$\frac{\mathrm{d}\boldsymbol{S}}{\mathrm{d}\tau} = \boldsymbol{\Omega} \times \boldsymbol{S} \tag{8.235}$$
where
$$\boldsymbol{\Omega} = \frac{3}{2}(\nabla\Phi \times \boldsymbol{v}) + \frac{1}{2}(\boldsymbol{a} \times \boldsymbol{v}). \tag{8.236}$$
Here, $\boldsymbol{v}$ is the 3-velocity of the gyroscope, and $\boldsymbol{a}$ is its 3-acceleration (relative to the Schwarzschild observer).

22. For gyroscopes moving in stable circular orbits in Schwarzschild spacetime, which orbit shows the most pronounced geodetic precession of the gyroscopes?
23. Discuss geodetic precession in the post-Newtonian approximate metric. Prove that the geodetic precession angle in a circular orbit becomes

$$\Delta\phi \approx \left(\gamma + \frac{1}{2}\right) 2\pi \left(\frac{GM}{Rc^2}\right). \tag{8.237}$$

24. Binary star systems exhibit a significant precession of the periastron. At this time, relativistic effects are strong, and we can use the post-Newtonian approximation to discuss this issue. Surprisingly, the precession angle is still given by the formula for the precession of a test particle in Schwarzschild geometry, namely

$$\delta\phi = \frac{6\pi G(m_1 + m_2)}{c^2 a(1 - e^2)}. \tag{8.238}$$

For PSR1913+16, the masses of the binary stars are both about $1.4M_\odot$, the eccentricity $e \approx 0.617$, the semi-major axis is about 9×10^{10} cm, and the orbital period is about 2.9×10^4 s. Calculate the annual precession angle of the periastron.

Chapter 9

Schwarzschild Black Hole

The Schwarzschild geometry not only describes the spacetime outside a spherically symmetric star but also plays a crucial role in characterizing the spacetime of a spherically symmetric black hole. Strictly speaking, it represents a portion of the black hole's spacetime structure. Such a black hole could form from the gravitational collapse of an exhausted star, a process whose physical mechanism we will introduce later.

Interestingly, the concept of a black hole predates general relativity and can be traced back to Newtonian gravity. As early as 1783, the English natural philosopher John Michell first postulated the existence of such objects. Shortly thereafter, in 1796, the renowned French scientist Pierre-Simon Laplace independently arrived at the same conclusion. The reasoning behind this is straightforward: by considering the Newtonian escape velocity, $\sqrt{2GM/R}$, one finds that if the escape velocity equals the speed of light c, the celestial body must have a critical radius, now known as the Schwarzschild radius:

$$R_{\rm s} = \frac{2GM}{c^2}. \tag{9.1}$$

If a sufficiently dense celestial body contracts to a radius smaller than $R_{\rm s}$, light can no longer escape its gravitational pull. To a distant observer, such an object would appear entirely dark, earning it the historical designation of a "dark star." Incorporating special relativity further reinforces this idea: since no signal can exceed the speed of light, no information from the object can reach an outside observer. Consequently, this dark, matter-absorbing entity that emits no information came to be known as a "black hole."[1]

[1] The term "black hole" was popularized by John Wheeler in a 1967 lecture and quickly gained widespread acceptance. Prior to this, objects formed from stellar collapse were referred to as "frozen stars" or "collapsing stars."

However, within Newtonian mechanics, an observer sufficiently close to the object could still detect light emitted from its surface, since light can propagate a certain distance when emitted from the surface, meaning the object would not appear completely black. This stands in stark contrast to the general relativistic description of a black hole, where not even light can escape once within the event horizon.

For a celestial body to form a black hole, its density must be sufficiently high. For example, for a solar-mass object, $m_{\odot} \sim 10^{30}$ kg, its Schwarzschild radius $R_{\rm s}$ is about several kilometers, and for an Earth-mass object, its Schwarzschild radius is roughly the size of a ping-pong ball radius. Intriguingly, however, the density of a black hole decreases with increasing mass. This can be seen from the relation:

$$\rho = \frac{\text{volume}}{\text{mass}} = \frac{M}{\frac{4}{3}\pi R_{\rm s}^3} \propto \frac{1}{M^2}. \tag{9.2}$$

Thus, the density scales inversely with the square of the black hole's mass. For example, a solar-mass black hole has an extremely high density $\rho \sim 10^{16}\ \text{g/cm}^3$. In contrast, a supermassive black hole $M \sim 10^{11} M_{\odot}$ has a density $\rho_{\rm c} \sim 10^{-6}\ \text{g/cm}^3$.

Strictly speaking, black holes are classical objects in both Newtonian gravity and general relativity, meaning quantum effects are neglected. While quantum gravity remains an active area of research, this book focuses mainly on classical black hole physics, with black hole thermodynamics and Hawking radiation being introduced very briefly in later chapters. A natural question arises: What kinds of objects can truly be considered black holes? For example, why can't an electron be classified as one? A basic criterion is that the Compton wavelength $\lambda_{\rm c}$ of this object should be less than its Schwarzschild radius $R_{\rm s}$, otherwise its quantum behavior is too significant and it can no longer be regarded as a black hole within classical general relativity. Evaluating this condition leads to

$$\frac{R_{\rm s}}{\lambda_{\rm c}} = \frac{2Gm^2}{hc}. \tag{9.3}$$

Setting $R_{\rm s} \sim \lambda_{\rm c}$ defines a critical mass scale

$$m \geq \sqrt{\frac{hc}{2G}} = m_{\rm pl} \sim 10^{-5}\ \text{g}. \tag{9.4}$$

9.1 Schwarzschild Black Hole

Let us provide a brief overview of the fundamental properties of the Schwarzschild solution, a vacuum solution to Einstein's field equations. In

Schwarzschild coordinates, the metric takes the form:

$$ds^2 = -(1 - R_s/r)dt^2 + (1 - R_s/r)^{-1}dr^2 + r^2 d\Omega^2, \tag{9.5}$$

where $R_s = 2GM/c^2 = 2\mu$ denotes the Schwarzschild radius.[2] At large distances $r \gg R_s$, the spacetime approaches asymptotic flatness, reproducing the Newtonian gravitational potential

$$\Phi \approx -\frac{R_s}{2r}. \tag{9.6}$$

Near the Schwarzschild radius $r \to R_s$, the metric components exhibit singular behavior: $g_{tt} \to 0$ and $g_{rr} \to \infty$. However, this is merely a coordinate singularity, removable by an appropriate coordinate transformation (as we shall later demonstrate). Crucially, Schwarzschild coordinates correspond to the frame of a static observer at infinity, lending them clear physical interpretability.

Although $r = R_s$ is just a coordinate singularity, it is indeed special and has particular physical significance. First, it is called the "event horizon", the physical implications of which will be revealed in subsequent discussions. Simply put, everything that happens within the event horizon cannot be seen by an observer outside the horizon. Second, it is also an infinite redshift surface. As previously discussed, if the emitter and receiver are both fixed at a certain spatial position, the ratio of the wavelength of the received photons to the emitted wavelength is

$$\frac{\lambda_o}{\lambda_e} = \left(\frac{g_{tt}(r_o)}{g_{tt}(r_e)}\right)^{1/2}. \tag{9.7}$$

For the Schwarzschild spacetime, since $g_{tt}|_{r \to R_s} \to 0$, $r = R_s$ is an infinite redshift surface, meaning that photons emitted from this location are infinitely redshifted and cannot be detected by an observer even at finite distances outside the horizon. Moreover, the proper time interval $d\tau = \sqrt{1 - R_s/r}dt$ vanishes as $r \to R_s$, implying that clocks at the horizon appear frozen to distant observers.

Another key property of the Schwarzschild spacetime concerns its "timelike" Killing vector. By construction, the vector field ∂_t is always a Killing vector of the spacetime. However, its nature exhibits a striking transition across the horizon. When outside the horizon $r > R_s$, ∂_t remains timelike, reflecting the fact that the spacetime outside the horizon is stationary. However, within the horizon, $R_s > r > 0$, since $g_{tt} > 0$ and $g_{rr} < 0$,

---

[2] In the discussion below, we often adopt the convention $\mu \equiv GM$ and employ the natural units.

∂_t becomes a spacelike vector, while ∂_r becomes a timelike vector. More importantly, within the horizon, no timelike Killing vector exists, implying that the spacetime is no longer stationary but dynamically evolving. This loss of stationarity is intrinsic: the time-dependence of the metric for $r < R_s$ is coordinate-independent, no choice of coordinates can restore a timelike Killing vector inside the horizon. Moreover, within the horizon, the metric signature enforces a radical constraint: for timelike worldlines $ds^2 < 0$, the condition $dr^2 < 0$ must hold, meaning that a change in time must be accompanied by a change in r.

To better understand the causal structure of Schwarzschild black hole spacetime, let us analyze the behavior of radial null geodesics. Starting from the metric

$$\mathrm{d}s^2 = 0 = -\left(1 - \frac{R_s}{r}\right)\mathrm{d}t^2 + \left(1 - \frac{R_s}{r}\right)^{-1}\mathrm{d}r^2, \tag{9.8}$$

we derive the coordinate time dependence for radial light propagation

$$\frac{\mathrm{d}t}{\mathrm{d}r} = \pm\left(1 - \frac{R_s}{r}\right)^{-1} \to \begin{cases} \pm 1, & r \to \infty, \\ \pm\infty, & r \to R_s, \end{cases} \tag{9.9}$$

where the "+" sign represents outgoing motion, and the "−" sign represents ingoing motion. It appears that as $r \to R_s$, the coordinate time t diverges, implying that an infinite time is required for light to reach the horizon. However, this interpretation is misleading. The time t represents the proper time of a static observer at infinity. From this distant observer's perspective, light appears to slow down asymptotically as it approaches the horizon, never actually crossing it within finite t. From another perspective, for a distant observer, they see that light approaching the horizon moves more and more slowly, requiring an infinite amount of time to reach the horizon. This is reflected in the behavior of light cones in Schwarzschild coordinates (as shown in Figure 9.1), which appear to "close" as $r \to R_s$, creating the illusion that photons only asymptotically approach the horizon. This is not the case but rather an illusion seen by a static observer at infinity. In fact, both photons and massive particles have no problem crossing the horizon, but the distant observer cannot see this. Integrating the differential equation above yields

$$t = r + R_s \ln\left|\frac{r}{R_s} - 1\right| + \text{constant, for outgoing photons,}$$

$$t = -r - R_s \ln\left|\frac{r}{R_s} - 1\right| + \text{constant, for ingoing photons,}$$

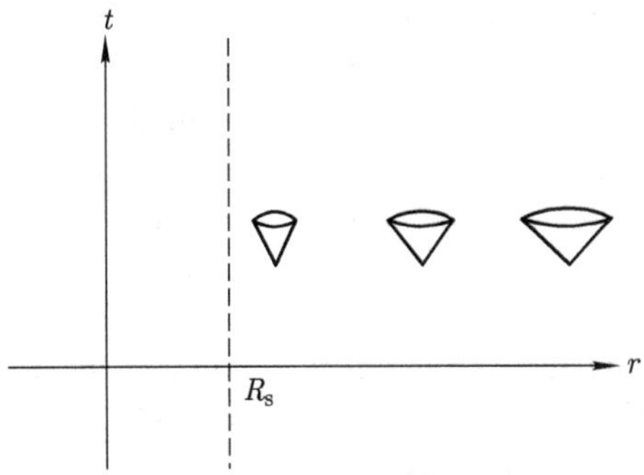

Fig. 9.1 The light cone in the external region of the Schwarzschild black hole.

which are time-reversal symmetric. For outgoing photons, outside the horizon, r increases with t, while inside the horizon, r increases as t decreases. For ingoing photons, outside the horizon, r decreases with t, whereas inside the horizon, r increases as t increases. For fixed θ and ϕ, the light cone diagram shown in Figure 9.2 reveals the following features. At large r, space-time is asymptotically flat, and light cones resemble those in Minkowski space. Near R_s, ingoing light rays require $t \to \infty$ to reach the horizon, while outgoing light rays appear to originate at $t \to -\infty$. Inside the horizon

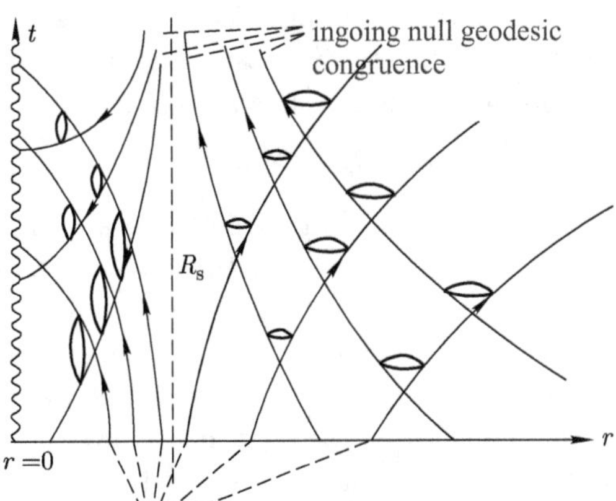

Fig. 9.2 Light cone diagram of Schwarzschild black hole. In Schwarzschild coordinates, the inward null geodesic does not seem to cross the event horizon, but only asymptotically approaches it. Inside the event horizon, the light cone pointing to the future is in the direction of decreasing r.

$r < R_s$, the roles of t and r interchanges, causing the light cone to tilt by 90°. Consequently, all photons inevitably fall onto the singularity at $r = 0$. Since the worldlines of any massive particles must lie within the light cone at each point of their worldlines, these particles also inevitably fall onto the singularity. Thus, once within the horizon, no particle or signal can avoid reaching the central singularity.

We have previously analyzed the radial motions of massive particle in detail. Consider a particle in free fall from infinity: the proper time required for it to reach the horizon from an initial radius r_0 is given by

$$\Delta\tau = \frac{1}{\sqrt{R_s}}\left(\frac{2}{3}\right)[r_0^{3/2} - (R_s)^{3/2}]. \tag{9.10}$$

For a distant observer at infinity, however, the corresponding coordinate time diverges,

$$\Delta t = R_s\left(-\frac{2}{3}\left(\frac{r}{R_s}\right)^{\frac{3}{2}} - 2\left(\frac{r}{R_s}\right)^{\frac{1}{2}} + \ln\left(\frac{\sqrt{r/R_s}+1}{\sqrt{r/R_s}-1}\right)\right)\Bigg|_{r_0}^{r}.$$

As shown in Figure 9.3, we set $t = \tau = 0$ at the initial position $r_0 = 8\mu$. In Schwarzschild coordinates, the particle's worldline exhibits an apparent singularity at the horizon, requiring infinite coordinate time to traverse from $r = 8\mu$ to $r = R_s$. Yet, from the particle's perspective, the proper time to reach the horizon remains finite, $\tau = 9.33\mu/c$. Beyond the horizon, the proper time continues to elapse until the particle reaches $r = 0$, with a total proper time of $\tau = 10.67\mu/c$. Near the horizon, the particle's motion in Schwarzschild coordinates asymptotically approaches

$$r = R_s + C\mathrm{e}^{-t/R_s}, \tag{9.11}$$

where C is a positive constant. Consequently, to a distant observer, the particle appears to linger eternally outside the horizon, $r > R_s$, never seemingly crossing it or entering the black hole. This, however, is a coordinate artifact, a limitation of the observer's frame. From the above discussion, we can see that Schwarzschild coordinates are not suitable for describing the motion of particles when $r \leq R_s$, or rather, Schwarzschild coordinates can only describe a part of the spacetime structure.

9.1.1 *Eddington–Finkelstein coordinates*

As demonstrated above, Schwarzschild coordinates exhibit significant limitations in describing the spacetime of a Schwarzschild black hole. They fail to accurately capture the motion of particles and the causal structure

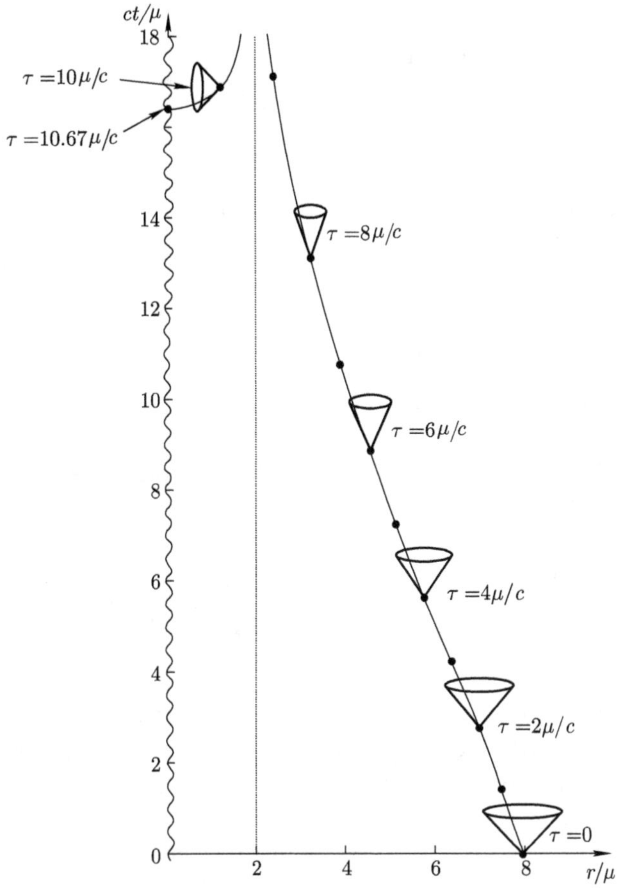

Fig. 9.3 Schematic diagram of a mass particle falling radially in Schwarzschild geometry, where $\mu = GM$.

near and within the horizon. As mentioned earlier, the spacetime manifold itself exists independently of any coordinate choice. To overcome these limitations, alternative coordinate systems can be employed.

To facilitate the understanding of particle motion, we can choose different particle worldline families to describe spacetime. While one might consider constructing coordinates based on the worldlines of massive particles, such an approach often leads to unnecessary complexity. Instead, a more practical and widely used method relies on the worldlines of massless particles (null geodesics) undergoing radial motion. For these particles, their radial motion satisfies

$$\mathrm{d}s^2 = -(1 - R_\mathrm{s}/r)\mathrm{d}t^2 + (1 - R_\mathrm{s}/r)^{-1}\mathrm{d}r^2 = 0, \tag{9.12}$$

which gives

$$\frac{\mathrm{d}t}{\mathrm{d}r} = \pm\frac{1}{1 - R_\mathrm{s}/r}. \tag{9.13}$$

To simplify the description of null geodesics, we introduce a new radial coordinate that satisfies

$$\frac{\mathrm{d}t}{\mathrm{d}r_*} = \pm 1 \Rightarrow \frac{\mathrm{d}r_*}{\mathrm{d}r} = \frac{1}{1 - R_\mathrm{s}/r}. \tag{9.14}$$

Integrating this relation gives

$$r_* = r + R_\mathrm{s}\ln\left(\frac{r}{R_\mathrm{s}} - 1\right), \tag{9.15}$$

known as the tortoise coordinate. This coordinate is well-defined for $r > R_\mathrm{s}$ but takes values from $-\infty$ to ∞. However, it can be analytically extended to the interior of the horizon $r < R_\mathrm{s}$. In terms of the tortoise coordinate, the metric becomes

$$\mathrm{d}s^2 = \left(1 - \frac{R_\mathrm{s}}{r}\right)(-\mathrm{d}t^2 + \mathrm{d}r_*^2) + r^2(r_*)\mathrm{d}\Omega^2. \tag{9.16}$$

Here, the metric components are no longer singular at the horizon, and the light cones remain open. The trade-off is that the horizon $r = R_\mathrm{s}$ is now mapped to $r_* \to -\infty$. The radial motion of massless particles can be described by the family of curves $t = \pm r_* + c$, where different values of c represent distinct geodesics, plus and minus signs representing outgoing and ingoing motion, respectively. These can be used to define null coordinates,

$$\widetilde{u} \equiv t + r_*, \quad \widetilde{v} \equiv t - r_*, \tag{9.17}$$

in which $\widetilde{u} =$ constant represent ingoing null geodesics, while $\widetilde{v} =$ constant represent outgoing null geodesics.

Using the coordinates r and $\widetilde{u}$, the Schwarzschild spacetime metric can be written as

$$\mathrm{d}s^2 = -\left(1 - \frac{R_\mathrm{s}}{r}\right)\mathrm{d}\widetilde{u}^2 + (\mathrm{d}\widetilde{u}\mathrm{d}r + \mathrm{d}r\mathrm{d}\widetilde{u}) + r^2\mathrm{d}\Omega^2. \tag{9.18}$$

This coordinate system is known as the advanced Eddington–Finkelstein (AEF) coordinates or ingoing Eddington–Finkelstein coordinates.[3] Since

[3]Arthur Eddington first discovered this coordinate transformation in 1924, but did not recognize its ability to resolve the apparent singularity at $r = R_\mathrm{s}$. In 1958, David Finkelstein rediscovered the coordinate transformation, correctly interpreting its physical significance and proposing that $r = R_\mathrm{s}$ represents the event horizon of a black hole.

the metric determinant $g = -r^4 \sin^2\theta$, the metric is non-degenerate everywhere except at $r = 0$. The radial null curves satisfy

$$\frac{\mathrm{d}\widetilde{u}}{\mathrm{d}r} = \begin{cases} 0 & \text{(ingoing)}, \\ 2\left(1 - \dfrac{R_\mathrm{s}}{r}\right)^{-1} & \text{(outgoing)}. \end{cases} \tag{9.19}$$

Thus, the ingoing null curves are characterized by $\widetilde{u} = $ constant, while the outgoing null curves are characterized by

$$\widetilde{u} = 2r + 2R_\mathrm{s} \ln\left|\frac{r}{R_\mathrm{s}} - 1\right| + \text{constant}. \tag{9.20}$$

In AEF coordinates, the light cone is well-defined even at the horizon $r = R_\mathrm{s}$, with the horizon located at a finite coordinate value. Both massless and massive particles can cross the horizon without encountering coordinate singularities. Interestingly, the light cones tilt inside the horizon such that all future-directed paths point toward decreasing r, as shown in Figure 9.4. The surface $r = R_\mathrm{s}$ is an event horizon: it acts as a one-way membrane: particles and signals can enter but never escape. An external observer cannot receive information from $r \le R_\mathrm{s}$. More precisely, $r = R_\mathrm{s}$ is both the event horizon and the Killing horizon.

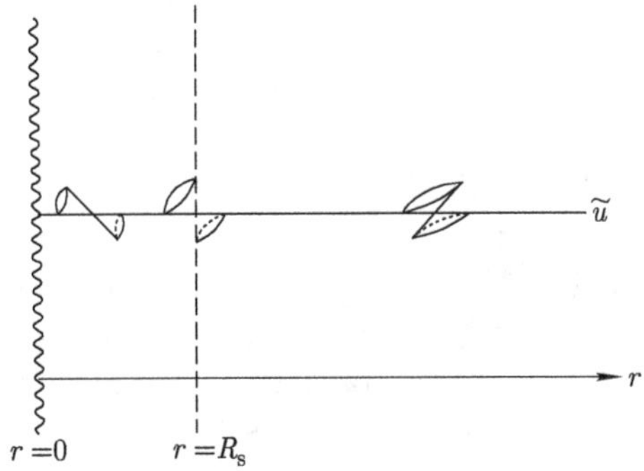

Fig. 9.4 Light cone diagram in AEF coordinates.

We can define a new time coordinate

$$t' \equiv \widetilde{u} - r \equiv t + R_\mathrm{s} \ln\left|\frac{r}{R_\mathrm{s}} - 1\right|, \tag{9.21}$$

where the set of coordinates (t', r, θ, ϕ) are also called advanced Eddington–Finkelstein coordinates. In this coordinate system, the metric takes the form

$$\mathrm{d}s^2 = -\left(1 - \frac{R_\mathrm{s}}{r}\right)\mathrm{d}t'^2 + \frac{2R_\mathrm{s}}{r}\mathrm{d}t'\mathrm{d}r + \left(1 + \frac{R_\mathrm{s}}{r}\right)\mathrm{d}r^2 + r^2\mathrm{d}\Omega^2. \tag{9.22}$$

Notably, this coordinate system is non-singular at the horizon, in contrast to the Schwarzschild coordinates. To remove the singularity at $r = R_\mathrm{s}$, we introduce the singular coordinate transformation $r \to r_*$. While this transformation is singular precisely at the horizon, it remains well-defined for $r > R_\mathrm{s}$, ensuring that the above metric still satisfies the vacuum Einstein equations and correctly describes the same spacetime outside the horizon. However, since the new coordinates extend over $0 < r < \infty$, they also allow us to explore the physics inside the horizon.

In these coordinates, the worldlines of radially moving photons are given by

$$\begin{aligned} t' &= -r + \text{constant} && \text{(ingoing)}, \\ t' &= r + 2R_\mathrm{s}\ln\left|\frac{r}{R_\mathrm{s}} - 1\right| + \text{constant} && \text{(outgoing)}. \end{aligned} \tag{9.23}$$

The Finkelstein diagram of the Schwarzschild spacetime in AEF coordinates is depicted in Figure 9.5. From Figure 9.5, it can be seen that the trajectory of an ingoing photon is smooth at the Schwarzschild radius, but the light cone structure undergoes a sharp transition at this boundary. Once an object crosses the Schwarzschild radius, its future inevitably points toward the singularity. Any particle or photon emitted from the region inside the horizon, $r < R_\mathrm{s}$) cannot escape to the region outside the horizon, $r > R_\mathrm{s}$. Thus, the Schwarzschild radius acts as an event horizon for external observers.

In AEF coordinates, the metric is not invariant under time reversal $t' \to -t'$, meaning the time reversal symmetry of the original Schwarzschild solution is lost. While the Schwarzschild metric remains a valid solution to Einstein's equations under time reversal, the AEF coordinates fail to cover the full spacetime manifold. To describe the spacetime more comprehensively, we introduce another family of coordinates based on outgoing null radial curves, parameterized by $\widetilde{v}$ instead of $\widetilde{u}$. The metric in these coordinates takes the form

$$\mathrm{d}s^2 = -\left(1 - \frac{R_\mathrm{s}}{r}\right)\mathrm{d}\widetilde{v}^2 - (\mathrm{d}\widetilde{v}\mathrm{d}r + \mathrm{d}r\mathrm{d}\widetilde{v}) + r^2\mathrm{d}\Omega^2. \tag{9.24}$$

We further define a new time coordinate

$$t^* \equiv \widetilde{v} + r, \tag{9.25}$$

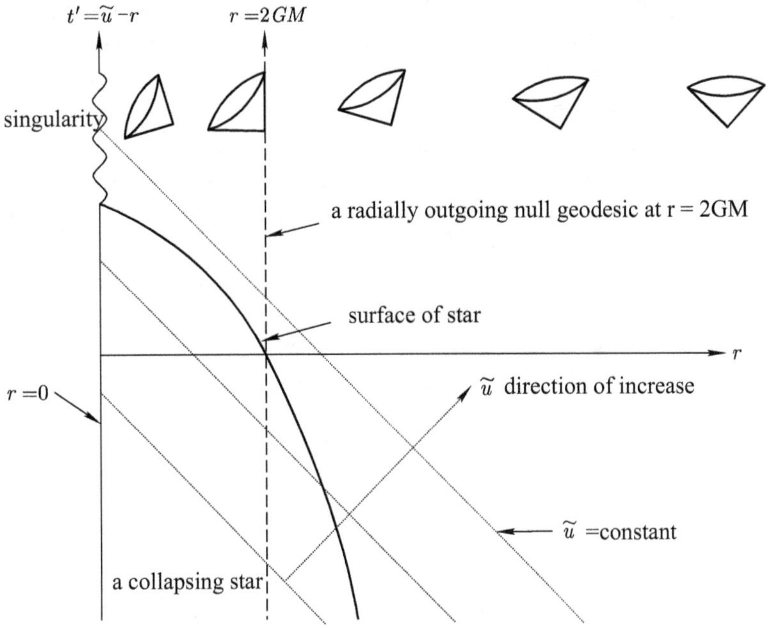

Fig. 9.5 Finkelstein diagram in AEF coordinates. The thick solid line represents the surface of the star, and the region between it and $r = 0$ represents the collapsing star.

where the set (t^*, r, θ, ϕ) constitutes the retarded Eddington–Finkelstein (REF) coordinates or outgoing Eddington–Finkelstein coordinates. The null radial curves in REF coordinates satisfy

$$\frac{\mathrm{d}\widetilde{v}}{\mathrm{d}r} = \begin{cases} 0 & \text{(outgoing)}, \\ -2(1 - R_\mathrm{s}/r)^{-1} & \text{(ingoing)}. \end{cases} \tag{9.26}$$

The corresponding worldlines for outgoing and ingoing photons are

$$\begin{aligned} t^* &= r + \text{constant} && \text{(outgoing)}, \\ t^* &= -r - 2R_\mathrm{s} \ln\left|\frac{r}{R_\mathrm{s}} - 1\right| + \text{constant} && \text{(ingoing)}. \end{aligned} \tag{9.27}$$

It is easy to see that the worldline of the outgoing photon is smooth across the Schwarzschild radius, while the ingoing one is discontinuous. Therefore, REF coordinates can well-suited for describing particles emitted from $r = 0$ and crossing outward the horizon $r = R_\mathrm{s}$. However, they provide a pathological description of infalling particles. Physically, REF coordinates describe a "white hole," a time-reversed analogue of a black hole, where matter and photons emerge from the horizon rather than being trapped by it. This scenario represents the time reversal of black hole formation. The corresponding light cone diagram is shown in Figure 9.6, and the Feinkelstein diagram is shown in Figure 9.7.

In summary, the Schwarzschild spacetime can be consistently extended across the horizon, but the direction of extension—whether future-directed or past-directed—leads to distinct spacetime regions: The original

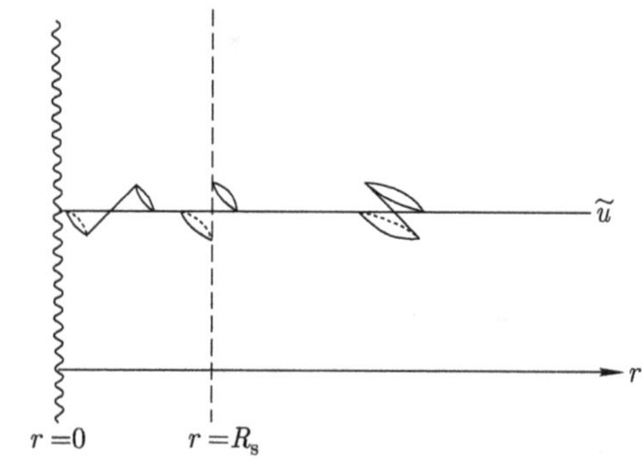

Fig. 9.6 Light cone diagram in REF coordinates.

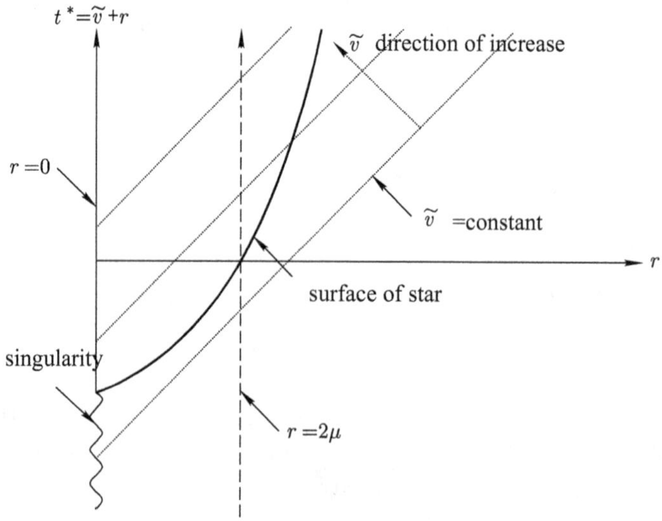

Fig. 9.7 Finkelstein diagram in REF coordinates.

Schwarzschild coordinates only cover the exterior region ($r > R_\mathrm{s}$); The AEF coordinates extend the spacetime into the black hole interior (future extension); The REF coordinates extend it into the white hole region (past extension). Thus, the AEF and REF coordinates provide complementary but distinct descriptions of the extended Schwarzschild spacetime, each revealing physically meaningful features (black hole vs. white hole dynamics) that are inaccessible in the original coordinates. Coordinate transformations between these systems are well-defined in the overlapping region ($r > R_\mathrm{s}$) but become singular elsewhere. This aligns with the standard framework of manifold theory: different coordinate charts describe different regions of the spacetime manifold, and only in the overlapping regions are the coordinate transformations smooth.

9.1.2 *Painlevé coordinates and tidal forces*

In the above discussion, we used the radial motion constant of massless particles to define our coordinate system, which well describes their geodesic motion. Alternatively, we may construct a coordinate system based on the radial motion of massive particles to better characterize their dynamics. Consider a massive particle released from rest at infinity, whose 4-velocity is given by

$$u^\mu = \left((1 - R_\mathrm{s}/r)^{-1}, -\sqrt{R_\mathrm{s}/r}, 0, 0\right). \tag{9.28}$$

The corresponding covariant components are

$$u_\mu = \left(-1, -(1 - R_\mathrm{s}/r)^{-1}\sqrt{R_\mathrm{s}/r}, 0, 0\right). \tag{9.29}$$

We now introduce a new function

$$T = t + \int^r (1 - R_\mathrm{s}/r')^{-1}\sqrt{R_\mathrm{s}/r'}\mathrm{d}r'. \tag{9.30}$$

Note that for the hypersurface $T =$ constant, $u_\mu = \partial_\mu T$ (considering the appropriate orientation). Thus, T serves as a natural time coordinate, while the normal vector of the hypersurface $T =$ constant is exactly the 4-velocity of the massive particle, meaning T defines the time direction for massive particles. From its definition, we know

$$\mathrm{d}T = \mathrm{d}t + (1 - R_\mathrm{s}/r)^{-1}\sqrt{R_\mathrm{s}/r}\mathrm{d}r, \tag{9.31}$$

yielding the metric

$$\mathrm{d}s^2 = -\mathrm{d}T^2 + \left(\mathrm{d} + \sqrt{\frac{R_\mathrm{s}}{r}}\mathrm{d}T\right)^2 + r^2\mathrm{d}\Omega^2. \tag{9.32}$$

The coordinates (T, r, θ, ϕ) are known as Painlevé coordinates. The advantage of using this metric is that $\mathrm{d}T$ is always a timelike 1-form, ensuring T remains a valid time coordinate. However, the metric's non-diagonal form is a notable drawback.[4]

The Painlevé coordinate system provides a convenient framework for analyzing tidal effects in Schwarzschild spacetime. Starting from the metric, we define an orthonormal frame field

$$\widehat{\theta^0} = \mathrm{d}T, \quad \widehat{\theta^1} = \mathrm{d}r + \sqrt{\frac{R_\mathrm{s}}{r}}\mathrm{d}T, \quad \widehat{\theta^2} = r\mathrm{d}\theta, \quad \widehat{\theta^3} = r\sin\theta\mathrm{d}\phi. \tag{9.33}$$

This frame corresponds to the local reference frame of a freely falling observer, with the metric expressed as

$$\mathrm{d}s^2 = \eta_{mn}\theta^m\theta^n, \quad m, n = 0, 1, 2, 3. \tag{9.34}$$

From the relation $\widehat{\theta^m} = e^m_\mu \mathrm{d}x^\mu$, we extract the transformation matrix e^m_μ:

$$\begin{aligned} &e^0_t = 1, \quad e^0_r = (1 - R_\mathrm{s}/r)^{-1}\sqrt{R_\mathrm{s}/r}, \\ &e^1_t = \frac{R_\mathrm{s}}{r}, \quad e^1_r = (1 - R_\mathrm{s}/r)^{-1}, \\ &e^2_\theta = r, \quad e^3_\phi = r\sin\theta. \end{aligned} \tag{9.35}$$

Using the inverse of these transformation, we compute the Riemann tensor components in Painlevé coordinates from their Schwarzschild counterparts:

$$R_{mnpq} = e^\mu_m e^\nu_n e^\rho_p e^\sigma_q R_{\mu\nu\rho\sigma}. \tag{9.36}$$

The non-zero components of the Riemann tensor in Schwarzschild spacetime are

$$\begin{aligned} R_{trtr} &= -\frac{R_\mathrm{s}}{r^3}, \quad R_{t\theta t\theta} = \frac{R_\mathrm{s}(r - R_\mathrm{s})}{2r^2}, \\ R_{t\phi t\phi} &= \frac{R_\mathrm{s}(r - R_\mathrm{s})\sin^2\theta}{2r^2}, \quad R_{r\theta r\theta} = -\frac{R_\mathrm{s}}{2(r - R_\mathrm{s})}, \\ R_{r\phi r\phi} &= -\frac{R_\mathrm{s}\sin^2\theta}{2(r - R_\mathrm{s})}, \quad R_{\theta\phi\theta\phi} = r\sin^2\theta R_\mathrm{s}. \end{aligned} \tag{9.37}$$

Transforming to the local frame yields the components

$$\begin{aligned} R_{0101} &= -\frac{R_\mathrm{s}}{r^3}, \quad R_{0202} = \frac{R_\mathrm{s}}{2r^3}, \\ R_{2323} &= \frac{R_\mathrm{s}}{r^3}, \quad R_{1212} = R_{1313} = -\frac{R_\mathrm{s}}{2r^3}. \end{aligned} \tag{9.38}$$

[4] Historically, Einstein dismissed this metric as a description of spherically symmetric spacetime precisely due to its non-diagonality.

For a family of geodesics, the deviation vector η^μ satisfies

$$\frac{\mathrm{D}^2}{\mathrm{d}\tau^2}\eta^\mu = R^\mu{}_{\nu\lambda\sigma}u^\nu u^\lambda \eta^\sigma. \tag{9.39}$$

In the local reference frame of a freely falling observer, this reduces to

$$\frac{\mathrm{D}^2}{\mathrm{d}\tau^2}\eta^m = R^m{}_{00n}\eta^n, \tag{9.40}$$

leading to the tidal acceleration equations

$$\frac{\mathrm{D}^2}{\mathrm{d}\tau^2}\eta^1 = \frac{R_\mathrm{s}}{r^3}\eta^1, \tag{9.41}$$

$$\frac{\mathrm{D}^2}{\mathrm{d}\tau^2}\eta^2 = -\frac{R_\mathrm{s}}{2r^3}\eta^2, \tag{9.42}$$

$$\frac{\mathrm{D}^2}{\mathrm{d}\tau^2}\eta^3 = -\frac{R_\mathrm{s}}{2r^3}\eta^3, \tag{9.43}$$

where η^1 represents the radial separation, while η^2 and η^3 are transverse separations. Along radial direction a positive sign indicates stretching tidal forces, while along transverse directions negative signs correspond to compression. The $1/r^3$ dependence is characteristic of four-dimensional tidal forces. Therefore, we obtain the following picture: tidal forces elongate objects radially while squeezing them transversely; Tidal effects intensify as $r \to 0$, but exhibit no divergence at the horizon. For larger black holes (with larger R_s), tidal forces at the horizon are weaker. An astronaut crossing the horizon of a supermassive black hole would experience negligible stretching or compression. The tidal forces diverge at $r = 0$, inevitably destroying any infalling object.

9.2 Maximal Extension of Schwarzschild Spacetime

The preceding discussion reveals that Schwarzschild spacetime consists of distinct regions, each described by different coordinate systems. Schwarzschild coordinates cover the region $R_\mathrm{s} < r < \infty$, while AEF and REF coordinates both span $0 < r < \infty$, but they correspond to separate spacetime regions. An interesting question is what the complete Schwarzschild spacetime manifold is. To address this, we first introduce a key definition.

Definition A manifold is maximal if every geodesic emanating from any point on the manifold can either be extended infinitely in both directions or terminates at an intrinsic singularity. In the first case, the manifold is geodesically complete.

For Schwarzschild spacetime, neither Schwarzschild coordinates nor Eddington–Finkelstein coordinates provide a maximal extension. For example, in Eddington–Finkelstein coordinates, only ingoing or outgoing null geodesics are well-defined, but not both simultaneously. So, is there a coordinate system in which both ingoing and outgoing null geodesics are well-defined? Or, what is the maximal analytic extension of the Schwarzschild solution? The maximal analytic extension of Schwarzschild spacetime is achieved via Kruskal coordinates (or Kruskal–Szekeres coordinates). Through these coordinates, we can discover the complete structure of the Schwarzschild spacetime.

As a first step, let us consider using both $\widetilde{u}$ and $\widetilde{v}$ coordinates, which respectively parametrize ingoing and outgoing null geodesics. This choice yields the Schwarzschild solution in the form

$$\mathrm{d}s^2 = -\frac{1}{2}\left(1-\frac{R_\mathrm{s}}{r}\right)(\mathrm{d}\widetilde{u}\mathrm{d}\widetilde{v}+\mathrm{d}\widetilde{v}\mathrm{d}\widetilde{u})+r^2\mathrm{d}\Omega^2, \tag{9.44}$$

where r is implicitly defined as a function of $\widetilde{u}$ and $\widetilde{v}$ through the relation

$$\frac{1}{2}(\widetilde{u}-\widetilde{v}) = r + R_\mathrm{s}\ln\left(\frac{r}{R_\mathrm{s}}-1\right). \tag{9.45}$$

For constant angular coordinates $\theta=$ constant and $\phi=$ constant, the metric simplifies to

$$\mathrm{d}s^2 = -\left(1-\frac{R_\mathrm{s}}{r}\right)\mathrm{d}\widetilde{u}\mathrm{d}\widetilde{v}. \tag{9.46}$$

By introducing coordinates

$$\widetilde{t} = \frac{1}{2}(\widetilde{u}+\widetilde{v}), \quad \widetilde{x} = \frac{1}{2}(\widetilde{u}-\widetilde{v}), \tag{9.47}$$

the metric (9.45) becomes

$$\mathrm{d}s^2 = -\left(1-\frac{R_\mathrm{s}}{r}\right)(\mathrm{d}\widetilde{t}^2-\mathrm{d}\widetilde{x}^2). \tag{9.48}$$

This form explicitly demonstrates the conformal flatness of the metric.[5] Consequently, the light cone structure in this geometry is identical to that of flat spacetime.

[5]A metric is said to be conformally flat if it differs from the metric of flat spacetime by only an overall scale function.

Proceeding further, we define new coordinates

$$u' = e^{\widetilde{u}/4\mu}, \qquad v' = -e^{-\widetilde{v}/4\mu}, \tag{9.49}$$

which relate to the original Schwarzschild coordinates via:

$$u' = \left(\frac{r}{R_s} - 1\right)^{1/2} e^{(r+t)/4\mu}, \qquad v' = -\left(\frac{r}{R_s} - 1\right)^{1/2} e^{(r-t)/4\mu}. \tag{9.50}$$

In terms of (u', v', θ, ϕ), the Schwarzschild solution becomes

$$ds^2 = -\frac{16\mu^3}{r} e^{-r/2\mu}(du'dv' + dv'du') + r^2 d\Omega^2. \tag{9.51}$$

Remarkably, this representation eliminates the coordinate singularity at the event horizon R_s. The null coordinates (u', v') can be further converted to ordinary time-like and space-like coordinates:

$$u = \frac{1}{2}(u' - v') = \left(\frac{r}{R_s} - 1\right)^{1/2} e^{r/4\mu} \cosh(t/4\mu),$$

$$v = \frac{1}{2}(u' + v') = \left(\frac{r}{R_s} - 1\right)^{1/2} e^{r/4\mu} \sinh(t/4\mu),$$

leading to

$$ds^2 = \frac{32\mu^3}{r} e^{-r/2\mu}(-dv^2 + du^2) + r^2 d\Omega^2, \tag{9.52}$$

where r is now implicitly determined by

$$(u^2 - v^2) = \left(\frac{r}{R_s} - 1\right) e^{r/R_s}. \tag{9.53}$$

These coordinates (v, u, θ, ϕ) constitute the well-known Kruskal coordinates, or Kruskal–Szekeres coordinates. In this coordinate system, the metric is singularity-free, and for fixed angular coordinates (θ, ϕ), is conformally flat.

In Kruskal coordinates, the global structure of spacetime becomes particularly transparent. First, for radial null geodesics, the trajectories in Kruskal coordinates are simply

$$v = \pm u + \text{constant}, \tag{9.54}$$

mirroring the light cone structure of Minkowski spacetime. Here, the coordinate v remains timelike everywhere, while u is always spacelike. Consequently, the tangent vectors of timelike curves are confined within a 45°

angle of the v-axis, while the tangent vectors of spacelike curves lie outside this lightcone, making an angle greater than 45° with the v-axis. Second, the event horizon, rather than lying at infinity, is located at

$$v = \pm u, \tag{9.55}$$

which defines a null surface. Meanwhile, surfaces of constant r satisfy

$$u^2 - v^2 = \text{constant}, \tag{9.56}$$

forming hyperbolas in the u-v plane, or hypersurfaces when including angular coordinates. Outside the horizon, the normal covector of these hypersurfaces is spacelike, so the hypersurfaces are timelike, while inside the horizon, the normal becomes timelike, and the surface is spacelike. This implies that the curvature singularity at $r = 0$ is a spacelike. Moreover, since the Schwarzschild spacetime is physically defined only for $r \geq 0$,[6] not all regions of the u-v plane belong to the Schwarzschild spacetime. In fact, the Schwarzschild spacetime is restricted in the region $-\infty \leq u \leq \infty$ and $v^2 < u^2 + 1$, i.e., the area bounded by the two hyperbolas at $r = 0$. The maximally extended spacetime described by these coordinates is called Kruskal spacetime. Hypersurfaces of constant Schwarzschild time t correspond to straight lines in the (u, v)-plane:

$$\frac{v}{u} = \tanh(t/4\mu) \tag{9.57}$$

all radiating from the origin with slope $\tanh(t/4\mu)$. From the preceding discussions, we obtain the Kruskal diagram shown in Figure 9.8. The Kruskal diagram encapsulates the following properties:

(1) The $r = 0$ singularity splits into two hyperbolas: a past singularity and a future singularity.

(2) The future singularity is spacelike, meaning all future-directed timelike or lightlike trajectories must intersect it.

(3) A radial timelike geodesic crosses the event horizon and eventually terminates at the future singularity.

(4) For $r < R_s$, all signals (whether photons or massive particles) are inexorably drawn toward the singularity.

(5) The horizon $r = R_s$ partitions spacetime into four distinct regions. Regions I and II can be described by AEF coordinates, where region I is the asymptotically flat region outside the horizon, describable by Schwarzschild coordinates, while region II is the black hole interior. Similarly, regions I

[6]Mathematically, the solution also holds for $r < 0$, but this corresponds to a negative-mass Schwarzschild spacetime with a naked singularity, which we disregard here.

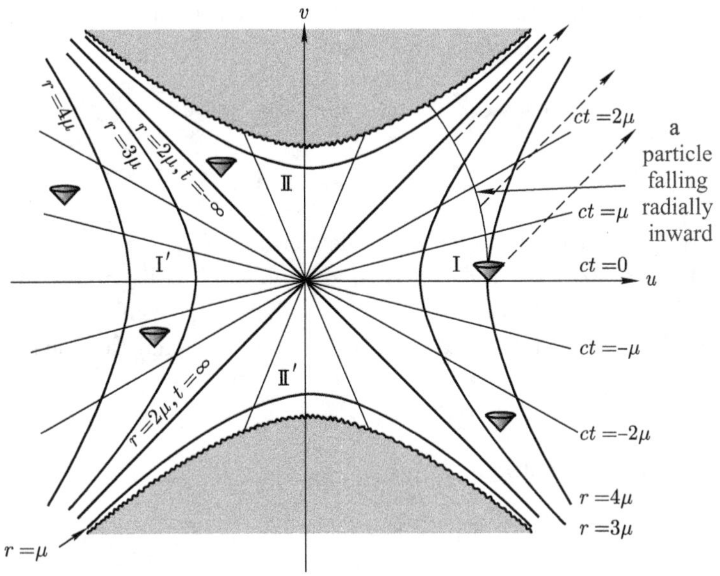

Fig. 9.8 The Schwarzschild geometry in Kruskal coordinates, where $\mu = GM$, and the shaded zigzag boundaries represent the past and future singularities at $r = 0$.

and II′ are described by REF coordinates, where region II′ corresponds to the white hole region. There is a past singularity in region II′, from which spacetime originates. Region I′ is another asymptotically flat region that we have not encountered before. Moreover, region I′ cannot be reached from region I via any timelike or lightlike path, either forward or backward, meaning two asymptotically flat regions are causally disconnected. However, a wormhole (Einstein–Rosen bridge) briefly links them, though it is non-traversable.

9.2.1 *Wormholes*

In the (u, v) coordinates system, we examine the geometry of a spacelike hypersurface defined by $v = 0$ $(t = 0)$. This hypersurface extends from $u = \infty$ to $u = -\infty$, with its line element given by

$$\mathrm{d}s^2 = \frac{32\mu^3}{r} \exp\left(-\frac{r}{2\mu}\right) \mathrm{d}u^2 + r^2(\mathrm{d}\theta^2 + \sin^2\theta^2 \mathrm{d}\phi^2). \tag{9.58}$$

To simplify the analysis, we restrict our attention to the equatorial plane ($\theta = \pi/2$), where the metric reduces to

$$\begin{aligned} ds^2 &= \frac{32\mu^3}{r} \exp\left(-\frac{r}{2\mu}\right) du^2 + r^2 d\phi^2 \\ &= \left(1 - \frac{R_s}{r}\right)^{-1} dr^2 + r^2 d\phi^2, \end{aligned} \tag{9.59}$$

where we have used the functional relationship of r on u. At $v = 0$, as u goes from ∞ to $-\infty$, r decreases monotonically to its minimum value $r = R_s$ before increasing again. To visualize this geometry, we embed this two-dimensional hypersurface into three-dimensional Euclidean space, described in cylindrical coordinates by

$$ds^2 = dz^2 + dr^2 + r^2 d\phi^2. \tag{9.60}$$

If the embedding condition

$$\left(\frac{dz}{dr}\right)^2 + 1 = \left(1 - \frac{R_s}{r}\right)^{-1}, \tag{9.61}$$

is satisfied, the induced metric matches that of the hypersurface. This requires

$$z = \sqrt{4R_s(r - R_s)} + \text{constant}, \tag{9.62}$$

revealing the geometry explicitly, as shown in Figure 9.9. This geometry is known as the Einstein–Rosen bridge, or alternatively as Schwarzschild throat, or Schwarzschild wormhole. From Figure 9.9, it can be seen that the hypersurface at $t = 0$ is actually a bridge or "wormhole" connecting two distinct asymptotically flat regions. Note that since $r > R_s$, the narrowest section of the wormhole has a radius equal to the Schwarzschild radius. Moreover, while our discussion has suppressed one angular dimension, the full geometry for fixed r or u is a two-dimensional sphere rather than a circle.

The analysis can be generalized to any $t =$ constant hypersurface by introducing isotropic coordinates (t, ρ, θ, ϕ), where the radial coordinate ρ is defined via

$$r = \left(1 + \frac{\mu}{2\rho}\right)^2 \rho. \tag{9.63}$$

In these coordinates, the metric becomes

$$ds^2 = -\left(\frac{1 - \dfrac{\mu}{2\rho}}{1 + \dfrac{\mu}{2\rho}}\right)^2 dt^2 + \left(1 + \frac{\mu}{2\rho}\right)^4 \underbrace{\left[d\rho^2 + \rho^2 d\Omega^2\right]}_{\text{three-dimensional space}}. \tag{9.64}$$

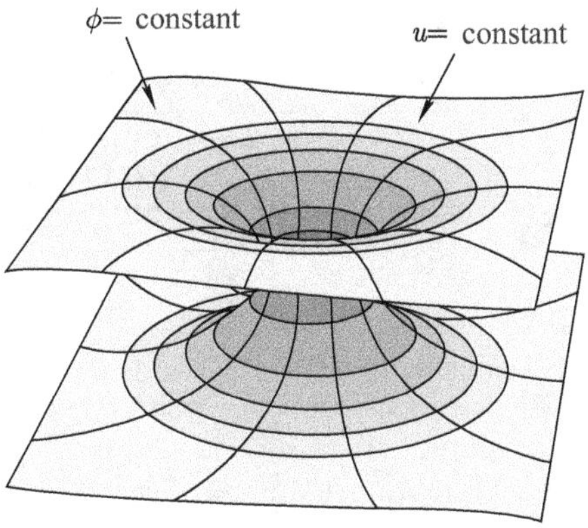

Fig. 9.9 Schematic diagram of the Einstein–Rosen bridge.

Here, the t = constant hypersurface is conformally flat. For a fixed t, there exist two distinct values of ρ, exchanged under the inversion $\rho \to (\mu)^2/(4\rho)$, with the fixed point at $\rho = \mu/2$. This transformation corresponds to $(u', v') \to (-u', -v')$ in Kruskal coordinates. Note, however, that isotropic coordinates are only valid in regions I and I$'$ of the Kruskal spacetime and are ill-defined for $r < R_s$, as shown in Figure 9.10.

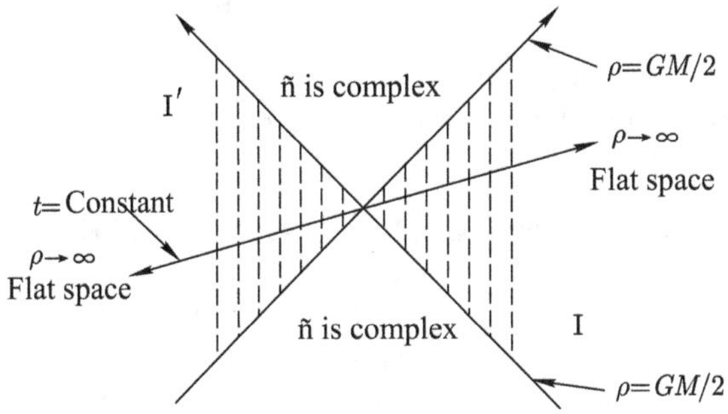

Fig. 9.10 Isotropic coordinates can only cover two asymptotically flat regions in the Kruskal spacetime.

The Schwarzschild geometry is not static in regions II and II$'$, where t becomes spacelike and r timelike. While the preceding discussion focused on $t = 0$, slicing the spacetime at different constant v values reveals a dynamical wormhole structure. As illustrated in Figures 9.11(a)–(b), the wormhole evolves as follows: at $v = -1$, it begins to open and expand;

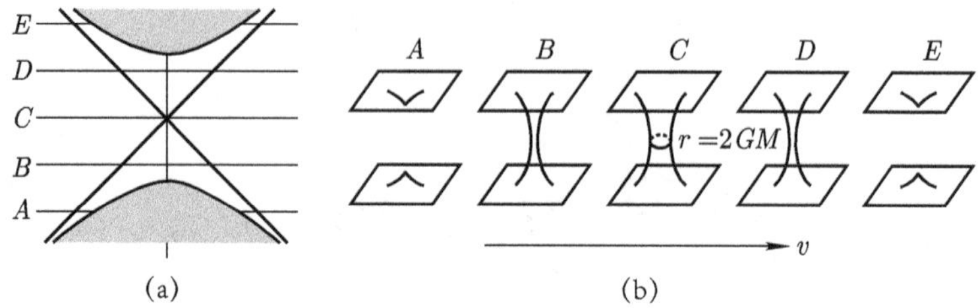

Fig. 9.11 Schematic diagram of the opening and closing of the wormhole. Corresponding to different v values, the size of the wormhole is also different.

at $v = 0$, it reaches its maximum width; and at $v = 1$, it rapidly closes. This process occurs over such a short timescale that no timelike observer can traverse the wormhole between the two asymptotically flat regions. Thus, the Schwarzschild spacetime describes a transient wormhole connecting two distinct universes, with the connection mediated by a two-dimensional spherical throat.

9.2.2 *Time translation symmetry in the Kruskal manifold*

From the Schwarzschild solution, it is evident that ∂_t is a Killing vector of the spacetime. indicating the translation $t \to t + c$ is a symmetry. In Kruskal coordinates, this symmetry manifests as

$$u' \to \mathrm{e}^{c/4\mu} u', \quad v' \to \mathrm{e}^{-c/4\mu} v', \tag{9.65}$$

which holds globally across the maximally extended Schwarzschild spacetime. The infinitesimal form of this transformation is

$$\delta u' = \frac{c}{4\mu} u', \quad \delta v' = -\frac{c}{4\mu} v'. \tag{9.66}$$

Consequently, the generator of this Killing symmetry in Kruskal coordinates is

$$\widehat{k} = \frac{1}{4\mu}\left(u' \frac{\partial}{\partial u'} - v' \frac{\partial}{\partial v'}\right). \tag{9.67}$$

In region I, this Killing vector reduce to the Killing vector ∂_t in Schwarzschild coordinates, but it is now defined throughout the manifold. This vector field exhibits the following properties:

(1) The squared norm of $\widehat{k}$ is $\widehat{k}^2 = -\left(1 - \frac{Rs}{r}\right)$, implying its causal nature depends on the region

$$\Rightarrow \begin{cases} \text{timelike,} & \text{in regions I and I}', \\ \text{spacelike,} & \text{in regions II and II}', \\ \text{null,} & \text{on the horizon, i.e., } \{u' = 0\} \cup \{v' = 0\}. \end{cases}$$

(2) The null hypersurfaces $\{u' = 0\}$ and $\{v' = 0\}$ constitute fixed point sets of the Killing vector $\widehat{k}$. On these horizons, the Killing vector takes the form

$$\begin{aligned} &\{u' = 0\}, \quad \widehat{k} = \partial/\partial\widetilde{v}, \\ &\{v' = 0\}, \quad \widehat{k} = \partial/\partial\widetilde{u}, \end{aligned} \tag{9.68}$$

where $\widetilde{u}$ and $\widetilde{v}$ are the Eddington–Finkelstein null coordinate. Along $\{u' = 0\}$, $\widetilde{v}$ serves as the natural group parameter with $-\infty < \widetilde{v} < \infty$. Similarly, along $\{v' = 0\}$, $\widetilde{u}$ plays the same role, with $-\infty < \widetilde{u} < \infty$.

(3) Every point on the Boyer–Kruskal axis $\{u' = v' = 0\}$, which geometrically is a two-dimensional sphere, is a fixed point of $\widehat{k}$. The distribution of these fixed points is illustrated in Figure 9.12.

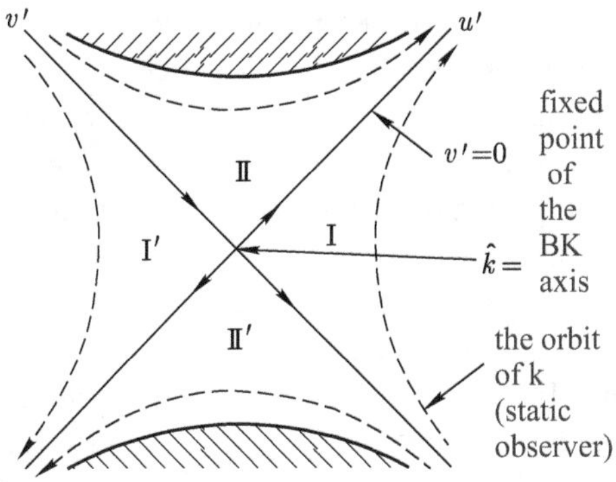

Fig. 9.12 Fixed-point set of the Kruskal spacetime under time translation.

9.3 Formation of Black Holes

In this section, we provide a concise overview of the fundamental physical processes governing black hole formation via stellar collapse.

A star maintains equilibrium through a balance between its self-gravitation and the combined pressure of gas and radiation. The radiation pressure originates from nuclear fusion, wherein light nuclei fuse into heavier ones, releasing energy. Over time, as the star radiates energy, its core contracts under gravity. This compression increases the core's temperature and density, eventually igniting successive stages of nuclear fusion—beginning with hydrogen burning (producing helium) and progressing to heavier elements such as carbon, oxygen, and silicon. During this phase, the star's outer layers expand, transforming it into a red giant. The fusion process culminates in the production of iron, the most stable atomic nucleus. Once the core consists predominantly of iron, fusion ceases, and the core begins to cool and contract under its own gravity. The subsequent fate of the star depends critically on its mass, angular momentum, and magnetic field configuration.

When nuclear fuel is exhausted, a slowly rotating star with a mass comparable to that of the Sun will collapse into a white dwarf—a dense stellar remnant supported by electron degeneracy pressure. For instance, the Sun is expected to evolve into a white dwarf in approximately 5 billion years, with a radius of about 5000 km and a density of about $10^9\ \mathrm{kg/m^3}$ in 5 billion years. White dwarfs were first identified in 1915 with the discovery of Sirius B, the faint companion of Sirius. In 1927, William A. Fowler established that white dwarfs represent an equilibrium state where electron degeneracy pressure counteracts gravitational collapse. Stars with higher masses may still become white dwarfs if significant mass loss occurs via stellar winds or supernova explosions. Additionally, due to the conservation of angular momentum, rotation plays a very important role in the collapse of stars and can in principle increase the mass threshold for white dwarf formation. However, magnetic fields may redistribute angular momentum, leading to a more symmetric collapse.

In 1930, a young Indian, Subrahmanyan Chandrasekhar, who was not yet 20 years old, had a bold idea on his long journey to UK: if a white dwarf has a greater mass, higher density, and stronger gravitational field, then the electron degeneracy pressure may not be able to resist the gravitational compression, making the white dwarf unstable and further collapse. In the following years, through careful research, he found that the mass limit for a white dwarf to be stable is $M_{\mathrm{c}} = 1.4M_{\odot}$, this critical mass is called the Chandrasekhar limit.[7] If the mass of a white dwarf is greater than M_{c}, gravity overwhelms electron degeneracy pressure, triggering further collapse.

For masses surpassing the Chandrasekhar limit, collapse proceeds until neutron degeneracy pressure halts it, forming a neutron star. This occurs via inverse β-decay:

$$\mathrm{e}^- + \mathrm{p} \rightarrow \mathrm{n} + \nu_{\mathrm{e}}, \tag{9.69}$$

where electrons and protons combine into neutrons and neutrinos, and after

[7]The 1983 Nobel Prize in Physics was shared by Subrahmanyan Chandrasekhar and William A. Fowler for their work on the evolution of stars.

the neutrinos escape, a large number of neutrons have degeneracy pressure. The resulting neutron degeneracy pressure, far stronger than its electron counterpart, can stabilize the star if its mass remains below $\sim 3M_\odot$ and $4M_\odot$, called the Oppenheimer–Volkoff limit. Neutron stars typically exhibit radii of $\sim$ 30 km and densities of $\sim 10^{16}$ kg/m^3, with relativistic effects becoming significant. A supernova explosion may leave behind a rapidly rotating neutron star with highly regular radiation pulses, known as a pulsar.

For stellar remnants exceeding the Oppenheimer–Volkoff limit, no known mechanism can prevent complete gravitational collapse, leading to black hole formation. In the idealized case of spherical symmetry, this results in a Schwarzschild black hole. However, a key question arises: is the collapse process stable against perturbations? Is it possible that small asymmetric perturbations in the collapse are amplified and destroy the formation of the event horizon? In the 1960s, Roger Penrose addressed this using topological methods, proving the singularity theorem. This theorem establishes that if a closed trapped surface forms during collapse, a singularity must inevitably arise within it—regardless of asymmetric perturbations. This result provides robust theoretical evidence that black holes are natural outcomes of stellar collapse in our universe.[8]

In an idealized scenario, we model the star as a spherically symmetric object with uniform density and negligible internal pressure—effectively approximating it as a pressureless dust cloud (i.e., an ideal fluid). In the absence of a pressure gradient, particles on the star's outer surface follow purely radial geodesics, moving inward under gravitational collapse. For simplicity, we assume the initial configuration consists of a mass layer situated at infinity and at rest. Alternatively, one may consider the collapse to begin at a finite radius $r = r_0$ with an inward initial velocity. The observed dynamics of the collapse differ for two distinct classes of static observers:

(1) Co-moving observer on the stellar surface. This observer remains attached to the star's surface throughout the collapse, eventually reaching $r = 0$.

(2) Distant observer at infinity or sufficiently far away. As shown in Figure 9.13, this observer never perceives the star's surface crossing the event horizon $r = R_\mathrm{s}$. Instead, photons emitted from the surface exhibit an increasing redshift as the star approaches R_s, with the redshift factor, $z \to$

[8]Roger Penrose won 2020 Nobel Prize for the discovery that black hole formation is a robust prediction o the general theory of relativity.

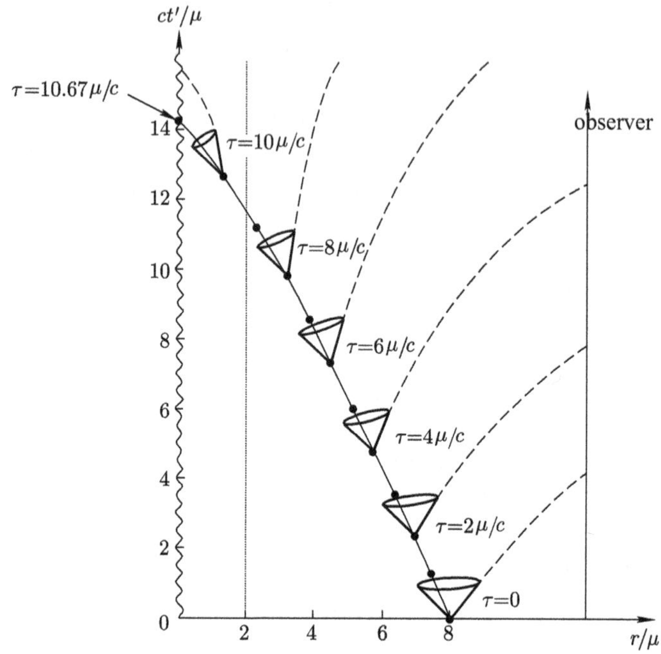

Fig. 9.13 The collapse of the surface of a pressureless star to form a black hole in AEF coordinates, where $\mu = GM$. We have already assumed that the surface of the star is released from rest at infinity, and when $r = 8\mu$, $\tau = t' = 0$ is taken.

∞. Consequently, the star's apparent luminosity diminishes asymptotically, eventually fading to invisibility.

In summary, while the co-moving observer experiences the full collapse to a singularity, the distant observer witnesses an apparent slowdown of the collapse, with the star asymptotically freezing near $R_{\rm s}$. The emitted radiation becomes infinitely redshifted, rendering the star dark and undetectable in finite observation time.

We can quantify the above picture. Consider a particle on the surface of a star emitting a radially outgoing photon at coordinates $(t_{\rm E}, r_{\rm E})$, which is later received by a distant observer at $(t_{\rm R}, r_{\rm R})$. For a freely falling emitter approaching the event horizon, the radial coordinate evolves as

$$r_{\rm E}(t_{\rm E}) \approx R_{\rm s} + a \exp\left(-\frac{t_{\rm E}}{R_{\rm s}}\right), \tag{9.70}$$

where a is a positive constant dependent on M, whose exact form is irrelevant for our discussion. Since $a > 0$, the emitter only asymptotically reaches the event horizon in infinite coordinate time. On the other hand, the star's apparent brightness decays on a characteristic timescale R_s/c, meaning it dims rapidly as the surface nears the horizon. Hence, for any approximately free-falling collapse, the approach to the black hole is effectively instantaneous. We can calculate the redshift factor. The redshift factor is determined by the frequency ratio of emitted and received photons

$$\frac{\nu_R}{\nu_E} = \frac{u^\mu_R p_\mu(R)}{u^\mu_E p_\mu(E)}. \tag{9.71}$$

For the freely falling emitter, its 4-velocity is

$$u^\mu_E = \left[\left(1 - \frac{R_s}{r}\right)^{-1}, -\left(\frac{R_s}{r}\right)^{1/2}, 0, 0\right], \tag{9.72}$$

and for a sufficiently distant stationary receiver, its 4-velocity is approximately

$$u^\mu_R = (1, 0, 0, 0). \tag{9.73}$$

Thus, the frequency ratio becomes

$$\frac{\nu_R}{\nu_E} = \frac{p_0(R)}{u^0_E p_0(E) + u^1_E p_1(E)} = \left(u^0_E + \frac{p_1(E)}{p_0(E)} u^1_E\right)^{-1}. \tag{9.74}$$

Since the photon's 4-momentum is a null vector, we have

$$p_1 = -\left(1 - \frac{R_s}{r}\right)^{-1} p_0, \tag{9.75}$$

and

$$\frac{\nu_R}{\nu_E} = \left(1 - \frac{R_s}{r_E}\right)\left(1 + \sqrt{\frac{R_s}{r_E}}\right)^{-1} = 1 - \sqrt{\frac{R_s}{r_E}}. \tag{9.76}$$

As $r_E \to R_s$, the received frequency vanishes $v_R \to 0$:

$$\frac{\nu_R}{\nu_E} \approx \frac{r_E - R_s}{2R_s} \propto \exp\left(-\frac{t_R}{2R_s}\right). \tag{9.77}$$

For the distant observer, the star's luminosity decays exponentially, $\exp\left(-\frac{t_R}{R_s}\right)$. The collapse's characteristic timescale is $2R_s/c$. Given that

$$\frac{\mu}{c^3} = 5 \times 10^{-6} \left(\frac{M}{M_\odot}\right) \text{s}, \tag{9.78}$$

this timescale is astronomically brief. Thus, to an external observer, the star effectively "turns black" almost instantaneously.

For an observer on the surface of the star, crossing the event horizon is uneventful. However, once inside, outward-directed photons can no longer escape; instead, all signals inevitably propagate toward $r = 0$. This is because, within the event horizon, r becomes a timelike coordinate, and its decrease defines the future direction. Therefore, once the star crosses the event horizon, its collapse to the singularity is inevitable. No matter how extreme the internal pressure or density becomes, even if the collapse deviates from spherical symmetry, the star's fate is sealed: it must terminate in a singularity of infinite density.

9.4 Spherically Symmetric Geometry Inside a Star

The Schwarzschild spacetime, which we discussed in detail earlier, represents a solution to the vacuum Einstein equations. It describes both the geometry of a spherically symmetric black hole formed by stellar collapse and the spacetime exterior to any static, spherically symmetric star. In this section, we extend our analysis to the spacetime geometry within a spherically symmetric star. Unlike the vacuum case, the stellar interior contains matter characterized by a non-vanishing energy-momentum tensor, which modifies the spacetime geometry. Maintaining spherical symmetry, we adopt the metric ansatz:

$$\mathrm{d}s^2 = -\mathrm{e}^{2\alpha(r)}\mathrm{d}t^2 + \mathrm{e}^{2\beta(r)}\mathrm{d}r^2 + r^2\mathrm{d}\Omega^2. \tag{9.79}$$

We need to solve the Einstein equations with an energy-momentum tensor $G_{\mu\nu} = 8\pi G T_{\mu\nu}$. For the above metric, the components of the Einstein tensor are given by:

$$\begin{aligned}
G_{tt} &= \frac{\mathrm{e}^{2(\alpha-\beta)}}{r^2}(2r\partial_r\beta - 1 + \mathrm{e}^{2\beta}),\\
G_{rr} &= \frac{1}{r^2}(2r\partial_r\alpha + 1 - \mathrm{e}^{2\beta}),\\
G_{\theta\theta} &= r^2\mathrm{e}^{-2\beta}\left(\partial_r^2 + (\partial_r\alpha)^2 - \partial_r\alpha\partial_r\beta - \frac{1}{r}\partial_r(\alpha-\beta)\right),\\
G_{\phi\phi} &= \sin^2\theta G_{\theta\theta}.
\end{aligned} \tag{9.80}$$

The matter content is well-modeled as an ideal fluid, with an energy-momentum tensor of the form

$$T_{\mu\nu} = (\rho + p)u_\mu u_\nu + p g_{\mu\nu}, \tag{9.81}$$

where u_μ denotes the 4-velocity of the fluid in the comoving frame, and $\rho(r)$ and $p(r)$ are the proper energy density and isotropic pressure, respectively.

Due to spherical symmetry, $\rho(r)$ and $p(r)$ are only functions of r. A natural question is whether all forms of matter can lead to spherically symmetric solutions, and whether such solutions impose restrictions on the equation of state of matter. To address this, note that the Einstein equations require $G_{ti} = 0$, implying $u_i u_0 = 0$. However, the 4-velocity normalization $u_\mu u^\mu = -1$ ensures $u_0 \neq 0$, forcing $u_i \equiv 0$. Thus, the 4-velocity simplifies to $u_\mu = (\mathrm{e}^\alpha, 0, 0, 0)$, indicating that the fluid is stationary relative to a distant observer. Consequently, the energy-momentum tensor of the fluid reduces to

$$\begin{aligned} T_{tt} &= \mathrm{e}^{2\alpha}, \quad T_{rr} = \mathrm{e}^{2\beta} p, \\ T_{\theta\theta} &= r^2 p, \quad T_{\phi\phi} = (r^2 \sin^2\theta) p. \end{aligned} \tag{9.82}$$

The Einstein equations become

$$8\pi G\rho(r) = \frac{\mathrm{e}^{-2\beta}}{r^2}(2r\partial_r\beta - 1 + \mathrm{e}^{2\beta}), \tag{9.83}$$

$$8\pi G p(r) = \frac{\mathrm{e}^{-2\beta}}{r^2}(2r\partial_r\alpha + 1 - \mathrm{e}^{2\beta}), \tag{9.84}$$

$$8\pi G p(r) = \mathrm{e}^{-2\beta}\left(\partial_r^2\alpha + (\partial_r\alpha)^2 - \partial_r\alpha\partial_r\beta - \frac{1}{r}\partial_r(\alpha - \beta)\right). \tag{9.85}$$

We define

$$\mathrm{e}^{2\beta} = \left(1 - \frac{2Gm(r)}{r}\right)^{-1}. \tag{9.86}$$

Substituting this into the (tt)-component of the Einstein equations yields

$$\frac{\mathrm{d}m(r)}{\mathrm{d}r} = 4\pi r^2 \rho(r), \tag{9.87}$$

which determines the mass function $m(r)$ upon integration. However, this mass function is not merely the integral of the matter density $\rho(r)$. First, note that in the above curved spacetime, the proper spatial volume element is

$$\sqrt{\gamma}\mathrm{d}^3x = \mathrm{e}^\beta r^2 \sin\theta \mathrm{d}r\mathrm{d}\theta\mathrm{d}\phi, \tag{9.88}$$

leading to a proper integrated "mass" (bare mass)

$$\overline{M} = \int_0^R \left(1 - \frac{2Gm(r')}{r'}\right)^{-1/2} 4\pi(r')^2\mathrm{d}r'\rho(r'). \tag{9.89}$$

On the other hand, $m(r)$ appears in the metric, which must match the exterior Schwarzschild metric at the stellar surface R, where $m(R) = M$ is the total mass observed asymptotically:

$$m(r) = 4\pi\int_0^r r'^2\rho(r')\mathrm{d}r', \quad m(R) = M. \tag{9.90}$$

The mass function $m(r)$ can be regarded as the total mass (energy) within radius r. The proper mass given earlier is different from M. In fact, the difference $\overline{M} - M = E_{\text{binding}} > 0$ represents the gravitational binding energy—the work required to disperse the star's matter to infinity.

In Newtonian gravity, $m(r)$ unambiguously denotes the mass within radius r, but in general relativity, it encodes additional contributions:

$$m(r) = m_0(r) + U(r) + \Omega(r), \tag{9.91}$$

where $m_0(r)$ is the rest mass energy, $U(r)$ is the internal energy, and $\Omega(r)$ is the gravitational potential energy. More precisely, from the proper volume, we get the total rest mass within radius r

$$m_0 = \int_0^r \mu_0 n \sqrt{\gamma} \mathrm{d}^3 x = \int_0^r \left(1 - \frac{2Gm(r')}{r'}\right)^{-1/2} 4\pi (r')^2 \mu_0 n \mathrm{d}r', \tag{9.92}$$

where $\mu_0 n$ is the internal energy that gives the rest mass. For example, for dust, μ_0 is the mass of a single particle, n is the particle number density, and the internal energy is $U(r) = \int_0^r (\rho - \mu_0 n)\sqrt{\gamma}\mathrm{d}^3 x$, coming from the interactions between the particles that make up the matter. The gravitational potential, that is, the binding energy,

$$\begin{aligned} \Omega &= -\int_0^r \rho \left(\left(1 - \frac{2Gm(r')}{r'}\right)^{-1/2} - 1 \right) 4\pi (r')^2 \mathrm{d}r' \\ &\approx -G \int_0^r (\rho m(r')/r') 4\pi (r')^2 \mathrm{d}r' \quad (Gm(r)/r \ll 1). \end{aligned} \tag{9.93}$$

Let us consider an idealized case: the star is composed of baryons with number density $n(r)$. The total baryon number is

$$B = \int_0^R 4\pi r^2 \left(1 - \frac{2Gm(r)}{r}\right)^{-1/2} n(r)\mathrm{d}r. \tag{9.94}$$

If the mass of a single baryon is m_{N}, the bare mass is $\overline{M} = m_{\mathrm{N}} B$. Assuming the baryon number density is constant $n(r) = n$ such that the energy density is also constant, $\rho(r) = \rho_*$, we have

$$m(r) = \frac{4\pi}{3}\rho_* r^3, \tag{9.95}$$

and the baryon number

$$\begin{aligned} B &= 4\pi n \int_0^R (1 - Ar^2)^{-1/2} r^2 \mathrm{d}r \\ &= \frac{4\pi n R^3}{3} f(x), \end{aligned} \tag{9.96}$$

where

$$A = \frac{8\pi G\rho_*}{3c^2},$$
$$f(x) = \frac{3}{2}\frac{\sin^{-1}x - x\sqrt{1-x^2}}{x^3}, \quad x = R\sqrt{A}. \tag{9.97}$$

For weakly relativistic stars, $x \ll 1$,

$$f(x) \approx 1 + \frac{3}{5}\frac{GM}{R} + \cdots, \tag{9.98}$$

yielding the binding energy in the Newtonian limit:

$$\Delta M = \overline{M} - M = M(f(x) - 1) = \frac{3}{5}\frac{GM^2}{Rc^2}. \tag{9.99}$$

From the (rr)-component of the Einstein equations, we get

$$\frac{\mathrm{d}\alpha}{\mathrm{d}r} = \frac{Gm(r) + 4\pi Gr^3 p(r)}{r^2(1 - 2Gm(r)/r)}, \tag{9.100}$$

which, upon integration, gives the metric component g_{tt}. However, this equation depends on the pressure profile $p(r)$, necessitating an analysis of the energy-momentum conservation law $\nabla_\mu T^{\mu\nu} = 0$. This leads to

$$(p + \rho)\frac{\mathrm{d}\alpha}{\mathrm{d}r} = -\frac{\mathrm{d}p}{\mathrm{d}r}$$
$$\Rightarrow \frac{\mathrm{d}p}{\mathrm{d}r} = -\frac{(\rho(r) + p(r))[Gm(r) + 4\pi Gr^3 p(r)]}{r(r - 2Gm(r))}. \tag{9.101}$$

Equations (9.87) and (9.101) constitute the Tolman–Oppenheimer–Volkoff (TOV) equations, also known as the relativistic hydrostatic equilibrium equations. Equation (9.87) indicates that, given appropriate boundary conditions, the mass function $m(r)$ is determined solely by $\rho(r)$, while Eq. (9.101) establishes a relationship between $p(r)$ and $\rho(r)$. One might question why the $(\theta\theta)$- and $(\phi\phi)$-components of the Einstein equations are not explicitly required. First, these two components are not independent-due to spherical symmetry, only one is functionally distinct. Second, the conservation of the energy-momentum tensor $\nabla_\mu T^{\mu\nu} = 0$ has already been imposed, rendering these additional equations redundant.

The left-hand side of Eq. (9.101) represents the pressure gradient. At the stellar surface, where $p = 0$, the condition $\frac{\mathrm{d}p}{\mathrm{d}r} < 0$ implies that the pressure decreases monotonically with increasing radius. Conversely, as we move inward, the pressure rises due to the increasing weight of the overlying matter. The central pressure is particularly enhanced because it must support the cumulative gravitational force exerted by the outer layers. Moreover, the larger the energy density, the larger this gradient, reflecting that to counteract gravity, the closer to the center, the faster the pressure increases. Notably, the pressure gradient depends not only on the energy density $\rho(r)$ but also on the pressure $p(r)$ itself—a feature absent in Newtonian gravity. This reflects a key distinction between general relativity and Newtonian theory: in the relativistic regime, pressure itself contributes to the gravitational field. In the non-relativistic limit, where $r \gg R_s$ and $\rho \gg p$, Eq. (9.101) reduces to

$$\frac{\mathrm{d}p}{\mathrm{d}r} = -\frac{4\pi G}{3}\rho^2 r. \tag{9.102}$$

This is precisely the result in Newtonian gravity: the pressure gradient is independent of pressure and only related to the energy density.

To solve this set of TOV equations, we must supplement them with an equation of state (EOS) relating pressure and energy density, $p(\rho)$. For many astrophysical systems, matter obeys a polytropic EOS of the form $p = K\rho^\gamma$, where $\gamma = 1 + \frac{1}{n}$, with K and the polytropic index n being constants. In this case, the TOV equations form a pair of coupled first-order differential equations. With appropriate boundary conditions, they uniquely determine the spacetime geometry, energy density profile $\rho(r)$, and pressure distribution $p(r)$. A physically meaningful solution requires two boundary conditions:

(1) Mass conditions: $m(0) = 0$ and $m(R) = M$;

(2) Pressure condition: $p(R) = 0$.

The central pressure $p(r = 0) = p_c$ then defines the stellar model. For a given EOS, the family of spherically symmetric solutions forms a one-parameter sequence, with the central density ρ_c as the free parameter. Numerical integration yields the well-known Lane–Emden solutions for such systems.

For analytical insight, consider a star with uniform density:

$$\rho(r) = \begin{cases} \rho_*, & r \leq R, \\ 0, & r > R. \end{cases} \tag{9.103}$$

This unphysical assumption corresponds to an incompressible fluid, implying an infinite speed of sound

$$c_s = \left(\frac{\mathrm{d}p}{\mathrm{d}\rho}\right)^{1/2}, \tag{9.104}$$

which violates relativity. Nevertheless, it serves as a reasonable approximation for the high-density interior of neutron stars. For this model, the mass function $m(r)$ takes a simple form

$$m(r)=\begin{cases}\frac{4}{3}\pi r^3\rho_*, & r\le R,\\ \frac{4}{3}\pi R^3\rho_*\equiv M, & r>R.\end{cases} \tag{9.105}$$

The second TOV equation reduces to

$$\frac{\mathrm{d}p}{\mathrm{d}r}=-\frac{4}{3}\pi r\frac{(\rho_*+p)(\rho_*+3p)}{1-8\pi r^2\rho_*/3}, \tag{9.106}$$

which integrates to

$$\frac{\rho_*+3p}{\rho_*+p}=\frac{\rho_*+3p_c}{\rho_*+p_c}\left(1-\frac{2\mu}{r}\right)^{1/2}. \tag{9.107}$$

Imposing the surface condition $p(r=R)=0$, the central pressure is

$$p_{\rm c}=\rho_*\frac{1-(1-R_{\rm s}/R)^{1/2}}{3(1-R_{\rm s}/R)^{1/2}-1}. \tag{9.108}$$

From the above relations, we can see that for a star of fixed mass, the central pressure diverges $p_{\rm c}\to\infty$ as R approaches $9R_{\rm s}/8$. Equivalently, a star of fixed radius R has a maximum mass $M_{\max}=\frac{4R}{9G}$, beyond which no equilibrium solution exists and gravitational collapse into a black hole becomes inevitable.

For the constant-density model, the spacetime metric inside the star is

$$\mathrm{d}s^2=-\left(\frac{3}{2}(1-AR^2)^{1/2}-\frac{1}{2}(1-Ar^2)^{1/2}\right)^2\mathrm{d}t^2+\frac{\mathrm{d}r^2}{1-Ar^2}+r^2\mathrm{d}\Omega^2, \tag{9.109}$$

where R is the radius of the star, and

$$A=\frac{2GM}{R^3}. \tag{9.110}$$

It is easy to see that this metric is consistent with the Schwarzschild spacetime outside the star.

In the constant-density model discussed earlier, we find that when the stellar mass exceeds the critical value $M>M_{\max}=\frac{4R}{9G}$, the central pressure diverges, indicating that the star no longer remain stable. In fact, a more general theorem applies, independent of the equation of state.

Buchdahl theorem No stable stellar configuration can exist with a radius smaller than $\frac{9GM}{4}$.

This theorem imposes an upper limit on the mass of a star with a given radius. Consider a star with a radius $\frac{9GM}{4}$. If a small, spherically symmetric inward force is applied, or equivalently, if the mass is slightly increased while holding the radius fixed, the hydrostatic equilibrium will be disrupted, triggering gravitational collapse. The ultimate fate of such a collapse is the formation of a Schwarzschild black hole. This limit is readily attainable in astrophysical scenarios. For instance, a neutron star with a density of $\sim 10^{16}$ kg·m^{-3} must satisfy $M < 7 \times 10^{31}$ kg, or approximately less than 35 solar masses, a mass range typical for many stars in our galaxy.

Another analytically tractable stellar model assumes an equation of state of the form $\rho = 12(p_* p)^{1/2} - 5p$, where p_* is a constant. It preserves causality, provided $c_\mathrm{s} = \sqrt{\frac{\mathrm{d}p}{\mathrm{d}\rho}} < 1$, which requires $p < p_*$ or equivalently $\rho < 7p_*$. In the low-pressure limit $p \ll p_*$, the equation of state reduces to $\rho = 12\sqrt{p_* p}$, corresponding to the $n = 1$ polytropic model in Newtonian stellar theory.

9.5 Charged Spherically Symmetric Black Hole

In this section, we explore a charged generalization of the Schwarzschild black hole, namely the Reissner–Nordström (RN) spacetime. As a spherically symmetric solution to the Einstein–Maxwell equations, the RN spacetime shares certain similarities with the Schwarzschild solution but introduces several distinctive features. It exhibits multiple horizons, a timelike singularity, and a maximum extension of spacetime comprising an infinite sequence of distinct regions. These properties make the RN solution a rich and structurally complex extension of the Schwarzschild geometry.

9.5.1 *Reissner–Nordström spacetime*

In electrodynamics, a spherically symmetric charge distribution necessarily implies a spherically symmetric energy-momentum tensor. This observation motivates the search for spacetime solutions describing the exterior of a charged, spherically symmetric star, where both gravitational and electromagnetic interactions must be treated self-consistently. For electromagnetic sector, the field strength tensor is defined as $F_{\mu\nu} = \partial_\mu A_\nu - \partial_\nu A_\mu =$

$\nabla_\mu A_\nu - \nabla_\nu A_\mu$, where the differential form representation highlights that $F_{\mu\nu}F^{\mu\nu}$ is independent of the affine connection. The complete action of the theory, in addition to the usual Einstein–Hilbert term, also includes the Maxwell action

$$S = \frac{1}{16\pi G_D}\int \mathrm{d}^D x\sqrt{-g}\left(R - \frac{g_0^2}{4\pi}F^{\mu\nu}F_{\mu\nu}\right). \tag{9.111}$$

Varying this action with respect to the metric yields the gravitational Einstein equations

$$R_{\mu\nu} - \frac{1}{2}g_{\mu\nu}R = 8\pi G_D g_0^2 T^{(F)}_{\mu\nu}, \tag{9.112}$$

where the electromagnetic energy-momentum tensor is

$$T^{(F)}_{\mu\nu} = \frac{1}{4\pi}(F_{\mu\rho}F^{\rho}_{\ \nu} - \frac{1}{4}g_{\mu\nu}F_{\rho\sigma}F^{\rho\sigma}). \tag{9.113}$$

The gauge field dynamics follow from the variation with respect to A_μ, giving the generalized Maxwell equations

$$\nabla_\mu F^{\mu\nu} = \frac{1}{\sqrt{-g}}\partial_\mu(\sqrt{-g}F^{\mu\nu}) = 0. \tag{9.114}$$

We wish to find a 4-dimensional spherically symmetric static solution characterized by the Killing vectors ∂_t and ∂_ϕ. Following the approach for the Schwarzschild solution, we can assume that the metric preserving spherical symmetry is

$$\mathrm{d}s^2 = -\mathrm{e}^{2\alpha(r)}\mathrm{d}t^2 + \mathrm{e}^{2\beta(r)}\mathrm{d}r^2 + r^2\mathrm{d}\Omega^2. \tag{9.115}$$

The spherically symmetric gauge field strength can be postulated as[9]

$$\widehat{F}_2 = F_{tr}(r)\mathrm{d}t\wedge\mathrm{d}r + F_{\theta\phi}(\theta)\mathrm{d}\theta\wedge\mathrm{d}\phi, \tag{9.116}$$

satisfying $\mathrm{d}\widehat{F}_2 = 0$. We first handle the equation of motion of the gauge field. The source-free Maxwell equations $\nabla_\mu F^{\mu\nu} = 0$ can be elegantly solved using exterior calculus. In terms of the Hodge dual, the equation becomes $\mathrm{d}*\widehat{F} = 0$. The Hodge dual of the field strength is given by

$$(*F)_{\mu\nu} = \frac{1}{2}\varepsilon_{\mu\nu\lambda\sigma}F^{\lambda\sigma} = \frac{\sqrt{-g}}{2}\widetilde{\varepsilon}_{\mu\nu\lambda\sigma}g^{\lambda\alpha}g^{\sigma\beta}F_{\alpha\beta}. \tag{9.117}$$

[9]This ansatz is motivated by the form of the field strength for spherically symmetric electric and magnetic charges in flat spacetime. Alternatively, it can be derived by requiring invariance of the field strength tensor under the Lie derivative along the Killing vectors.

Using the assumed form of the metric, we have

$$\begin{aligned}(*F)_{tr} &= \widetilde{\varepsilon}_{tr\theta\phi}(\mathrm{e}^{\alpha+\beta}r^2\sin\theta)\frac{1}{r^2}\frac{1}{r^2\sin^2\theta}F_{\theta\phi} \\ &= \frac{\mathrm{e}^{\alpha+\beta}}{r^2\sin\theta}F_{\theta\phi}(\theta),\end{aligned} \tag{9.118}$$

$$\begin{aligned}(*F)_{\theta\phi} &= -\widetilde{\varepsilon}_{\theta\phi tr}(\mathrm{e}^{\alpha+\beta}r^2\sin\theta)\mathrm{e}^{-2\alpha}\mathrm{e}^{-2\beta}F_{tr} \\ &= -\mathrm{e}^{-(\alpha+\beta)}r^2\sin\theta F_{tr}.\end{aligned} \tag{9.119}$$

Thus, the dual field strength takes the form

$$*\widehat{F} = -\mathrm{e}^{-(\alpha+\beta)}r^2\sin\theta F_{tr}\mathrm{d}\theta\wedge\mathrm{d}\phi + \frac{\mathrm{e}^{\alpha+\beta}}{r^2\sin\theta}F_{\theta\phi}(\theta)\mathrm{d}t\wedge\mathrm{d}r. \tag{9.120}$$

The condition $\mathrm{d}*\widehat{F}=0$ then yields

$$\begin{aligned}\mathrm{d}(*\widehat{F}) &= -\partial_r\left(\mathrm{e}^{-(\alpha+\beta)}r^2F_{tr}\right)\sin\theta\mathrm{d}r\wedge\mathrm{d}\theta\wedge\mathrm{d}\phi \\ &\quad + \partial_\theta((\sin\theta)^{-1}F_{\theta\phi}(\theta))r^{-2}\mathrm{e}^{\alpha+\beta}\mathrm{d}\theta\wedge\mathrm{d}t\wedge\mathrm{d}r,\end{aligned}$$

which requires

$$\partial_r\left(\mathrm{e}^{-(\alpha+\beta)}r^2F_{tr}\right)=0, \quad \partial_\theta((\sin\theta)^{-1}F_{\theta\phi}(\theta))=0. \tag{9.121}$$

The solutions to (9.121) are

$$\mathrm{e}^{-(\alpha+\beta)}r^2F_{tr}(r)=\text{constant}, \quad (\sin\theta)^{-1}F_{\theta\phi}(\theta)=\text{constant}. \tag{9.122}$$

These constants correspond to the electric and magnetic charges, respectively:

$$F_{tr}(r)=\frac{Q}{4\pi g_0}\frac{\mathrm{e}^{\alpha+\beta}}{r^2}, \quad F_{\theta\phi}=\frac{P}{4\pi g_0}\sin\theta, \tag{9.123}$$

where Q is the electric charge and P is the magnetic charge.

Furthermore, we can solve the Einstein equations and find that the metric functions satisfy $\alpha+\beta=0$. Following similar reasoning as in the Schwarzschild case, we solve the complete set of Einstein equations to obtain the Reissner–Nordström (RN) solution:

$$\mathrm{d}s^2_{\mathrm{RN}} = -\Delta(r)\mathrm{d}t^2+\Delta^{-1}(r)\mathrm{d}r^2+r^2\mathrm{d}\Omega_2^2, \tag{9.124}$$

where the metric function $\Delta(r)$ is given by

$$\Delta(r)=1-\frac{2\mu}{r}+\frac{Q^2+P^2}{r^2}. \tag{9.125}$$

When interpreting this solution as describing a black hole formed from the collapse of a charged star, we identify the horizons as the roots of

$$\begin{aligned}&\Delta = 0\\&\Rightarrow r^2 - (2\mu)r + (Q^2 + P^2) = 0\\&\Rightarrow r_\pm = \mu \pm \sqrt{\mu^2 - (Q^2 + P^2)}.\end{aligned} \tag{9.126}$$

This reveals two distinct horizons, the outer one at r_+ and the inner one at r_-. When the two horizons coincide, the black hole becomes extremal, occurring when

$$r_+ = r_- = \mu = \sqrt{Q^2 + P^2}. \tag{9.127}$$

The RN solution exhibits the following properties:

(1) The solution is asymptotically flat. The electromagnetic field components F_{tr} and $F_{\theta\phi}$ correspond to the electric field of a point charge Q and the magnetic field of a point magnetic charge P, respectively. At infinity, both fields yield a vanishing energy-momentum tensor. From the explicit form of the solution, it is evident that the electromagnetic contribution decays more rapidly than the mass term. Consequently, the RN metric asymptotically approaches the flat Minkowski spacetime.

(2) When both the electric and magnetic charges are zero, that is, $Q = 0 = P$, the RN solution reduces to the Schwarzschild metric.

(3) The RN solution shares the same spacetime symmetry as the Schwarzschild solution. Outside the outer horizon r_+, it possesses a timelike Killing vector and exhibits spherical symmetry. Moreover, the solution is static in this region.

(4) Similar to the Schwarzschild solution, the RN solution is unique; it is the only spherically symmetric solution of the Einstein–Maxwell theory.

While the mathematical form of the RN solution permits arbitrary values of (M, Q, P), physical considerations impose restrictions. For simplicity, we restrict our discussion to the electrically charged RN spacetime, i.e. $P = 0$. For $Q^2 > \mu^2$, the metric components are given by

$$g_{00} = -(g_{11})^{-1} = \Delta(r) = \Sigma/r^2, \tag{9.128}$$

where $\Sigma = r^2 - 2\mu r + Q^2$. In this case, the discriminant of the quadratic form Σ satisfies $\mu^2 - Q^2 < 0$, ensuring $\Sigma > 0$ for all $r \neq 0$. Thus, the metric is regular everywhere except at $r = 0$. The coordinate t is always timelike, and the coordinate r is always spacelike, devoid of coordinate singularities. However, $r = 0$ is still an intrinsic singularity, as the Kretschmann scalar $R_4 = R^{\mu\nu\sigma\rho}R_{\mu\nu\sigma\rho}$ diverges at $r = 0$. Notably, in this case, there is no event horizon, exposing the singularity at $r = 0$ as a naked singularity. Geodesic analysis reveals that this singularity exhibits a repulsive character: timelike geodesics cannot reach $r = 0$ but are instead deflected away after approaching it. Although the spacetime is asymptotically flat with a conventional causal structure at infinity, the presence of a naked singularity violates the cosmic censorship hypothesis. In fact, a black hole formed via gravitational collapse cannot satisfy $\mu^2 < Q^2$. If this is the case, the total mass of the black hole is less than the energy of the electromagnetic field. In other words, matter carrying electric or magnetic charges must have negative energy to satisfy this inequality. This scenario is manifestly unphysical. Hence, solutions with $Q^2 > \mu^2$ are deemed nonphysical.

We now focus exclusively on the physically admissible RN black hole, where $Q^2 < \mu^2$. In this case, there are two horizons located at $r_\pm = \mu \pm \sqrt{\mu^2 - Q^2}$. The spacetime can be partitioned into three distinct regions:

(1) Region I: $r_+ < r < \infty$, the asymptotically flat exterior region;
(2) Region II: $r_- < r < r_+$, a non-static, dynamical region;
(3) Region III: $0 < r < r_-$, the innermost region, containing the singularity.

The radii $r_\pm$ represent coordinate singularities, analogous to the Schwarzschild horizon $r = R_s$ in the Schwarzschild solution. Outside r_+, the spacetime is asymptotically flat. In regions I and III, the coordinate t is timelike, and r is spacelike, rendering these two regions static. In region II, t becomes spacelike, while r assumes a timelike role, indicating a dynamical spacetime.

To examine the null geodesics in this spacetime, we can choose the tortoise coordinates, as in the Schwarzschild case. The trajectories of null geodesics are shown in Figure 9.14. For the radially ingoing null geodesics, the trajectories follow 45° straight lines, while the radially outgoing photons exhibit curved trajectories. The light cone structure reveals that $r = r_+$ is an event horizon. The light cones in region II are tilted inward, compelling all particles to move toward $r = r_-$. In region III, the light cones revert to a standard orientation, suggesting particles are no longer forced toward the singularity. This spacetime diagram can be somewhat misleading: the diagram suggests that outgoing particles in region III can only asymptotically approach the inner horizon, analytical continuation of the RN solution demonstrates that particles can traverse $r = r_-$ in a finite proper time.

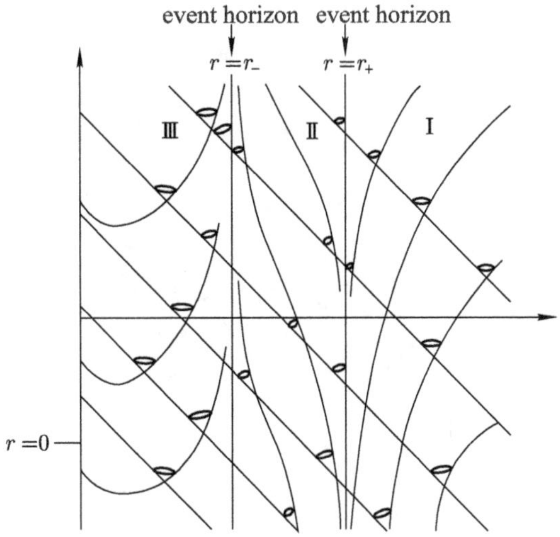

Fig. 9.14 The trajectory of photons in the RN spacetime in AEF coordinates.

9.5.2 *Motion of particles*

Next, we briefly analyze the dynamics of neutral massive particles in RN spacetime. Since the particles are electrically neutral, their interaction with the electromagnetic field can be neglected. For simplicity, we restrict our analysis to radial motion. The 4-velocity of such a particle is given by $u^\mu = (u^0, u^1, 0, 0) = (\dot{t}, \dot{r}, 0, 0)$, and its geodesic equation takes the form

$$\dot{u}_\sigma = \frac{1}{2}(\partial_\sigma g_{\mu\nu})u^\mu u^\nu. \tag{9.129}$$

Since ∂_t is a Killing vector, the quantity $u_0 = g_{00}\dot{t} =$ is conserved. And from the normalization condition

$$g_{\mu\nu}u^\mu u^\nu = -1 \Rightarrow g_{00}(u^0)^2 + g_{11}(u^1)^2 = -1, \tag{9.130}$$

we obtain the radial equation of motion

$$\dot{r}^2 + c^2\Delta = u_0^2/c^2. \tag{9.131}$$

This effectively reduces the problem to a one-dimensional dynamical system, where Δ plays the role of an effective potential, as shown in Figure 9.15.

For such a one-dimensional problem, the trajectory of motion depends critically on the constant of motion u_0. If $u_0 = c^2$, this corresponds to a

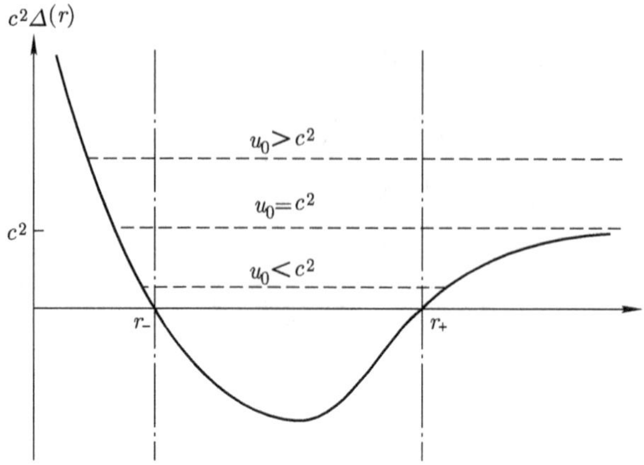

Fig. 9.15 The effective potential for the motion of a massive particle in the RN spacetime.

particle being freely released from infinity. When $u_0 > c^2$, the motion is unbound, the particle originating from infinity, reaching a minimal radius and escaping back to infinity. And when $u_0 < c^2$, the motion is bound, confined within a finite radial range. In all cases, the particle encounters a turning point in the radial direction and cannot reach the central singularity at $r = 0$; instead, it is repelled outward upon approaching this limit. At first glance, this behavior appears counterintuitive, as a radially infalling particle crossing the event horizon would classically be unable to escape. The resolution lies in considering the maximal analytic extension of the RN spacetime. A particle falling inward crosses r_+ and r_-, but instead of re-emerging in the same spacetime, it passes through another region, effectively crossing a white hole, before re-entering an asymptotically flat spacetime (for a detailed discussion, refer to [64]).

Consider a spherical shell composed of particles with a charge-to-mass ratio

$$\gamma = \frac{Q}{GM}, \quad |\gamma| < 1, \tag{9.132}$$

where Q is the total charge and M is the total mass. Each particle on the shell moves radially inward, and its worldline obeys (see Problem 15 of this chapter)

$$\left(\frac{\mathrm{d}r}{\mathrm{d}\tau}\right)^2 = E^2 - V_{\text{eff}} \quad (E < 1), \tag{9.133}$$

where the effective potential is given by

$$V_{\text{eff}} = 1 - (1 - E\gamma^2)\frac{2\mu}{r} + (1 - \gamma^2)\frac{Q^2}{r^2}. \tag{9.134}$$

The minimum point of this effective potential is at

$$r_0 = \frac{(1 - \gamma^2)Q^2}{(1 - E\gamma^2)\mu} = \frac{\gamma^2(1 - \gamma^2)}{1 - E\gamma^2}\mu. \tag{9.135}$$

This implies that the gravitational collapse is halted by electrostatic repulsion. Thus, even for charged particles, the singularity remains inaccessible; instead, after crossing both horizons, the particles are expelled outward, effectively forming a white hole.

To better visualize this process, we analyze the particle trajectories using advanced Eddington–Finkelstein (AEF) coordinates $(\widetilde{u}, r)$, in which the metric takes the form

$$\mathrm{d}s^2 = -\Delta \mathrm{d}\widetilde{u}^2 + 2\mathrm{d}\widetilde{u}\mathrm{d}r + r^2\mathrm{d}\Omega^2. \tag{9.136}$$

For a radially infalling particle, the 4-velocity components are

$$\dot{r}_{\text{in}} = -(u_0^2 - \Delta)^{1/2}, \tag{9.137}$$

$$\dot{\widetilde{u}}_{\text{in}} = \frac{u_0 - (u_0^2 - \Delta)^{1/2}}{\Delta}. \tag{9.138}$$

If the particle enters the inner horizon r_- and subsequently reverses direction, its outgoing 4-velocity components become

$$\dot{r}_{\text{out}} = (u_0^2 - \Delta)^{1/2}, \tag{9.139}$$

$$\dot{\widetilde{u}}_{\text{out}} = \frac{u_0 + (u_0^2 - \Delta)^{1/2}}{\Delta}. \tag{9.140}$$

As the particle approaches a horizon, $\Delta \to 0$, for inward motion, the $\widetilde{u}$-component remains finite,

$$\dot{\widetilde{u}}_{\text{in}} \approx (2u_0)^{-1}, \tag{9.141}$$

confirming that horizon crossing is well-defined; for outward motion, however, the $\widetilde{u}$-component diverges,

$$\dot{\widetilde{u}}_{\text{out}} \approx \frac{2u_0}{\Delta} \to \infty. \tag{9.142}$$

This indicates a coordinate singularity in the AEF framework, necessitating the introduction of alternative coordinates for a complete description of the outgoing trajectory.

For outward-moving particles, the retarded Eddington–Finkelstein (REF) coordinates provide a more appropriate description. In these coordinates, the metric takes the form

$$\mathrm{d}s^2 = -\Delta \mathrm{d}\widetilde{v}^2 - 2\mathrm{d}\widetilde{v}\mathrm{d}r + r^2\mathrm{d}\Omega^2. \tag{9.143}$$

The 4-velocity components for an outward-moving particle are given by

$$\dot{r}_{\text{out}} = (u_0^2 - \Delta)^{1/2}, \tag{9.144}$$

$$\dot{\widetilde{v}}_{\text{out}} = \frac{u_0 - (u_0^2 - \Delta)^{1/2}}{\Delta}. \tag{9.145}$$

Crucially, as the particle approaches the horizon ($\Delta \to 0$), the $\widetilde{v}$-component remains finite,

$$\dot{\widetilde{v}}_{\text{out}} \approx (2u_0)^{-1}, \tag{9.146}$$

demonstrating that no coordinate singularity occurs for outward motion through the horizon. This confirms that the REF coordinates, which effectively describe a white hole geometry, provide a complete and regular description of the particle's trajectory.

More careful analysis reveals that different asymptotically flat regions in Reissner–Nordström spacetime are connected through what may be termed a black hole tunnel. In this scenario, particles can enter a black hole from one asymptotically flat region, traverse through the tunnel, and then emerge into another asymptotically flat region. It is particularly noteworthy that the inability of particles to reach the central singularity does not stem from angular momentum effects (and the associated centrifugal barrier), but rather from the intrinsic geometry of the spacetime itself. Specifically, inside the inner horizon, the gravitational field exhibits repulsive characteristics, and this repulsive nature ultimately leads to the black hole tunnel structure. While the existence of the black hole tunnel in the RN spacetime is undeniable mathematically, its physical realizability is highly constrained. Its existence is contingent upon perfect staticity and spherical symmetry, any perturbation within the black hole interior destabilizes the tunnel structure. As shown in [56], even minor deviations from ideal conditions cause the tunnel to collapse. Consequently, while the black hole tunnel represents a fascinating theoretical feature of the exact RN solution, it does not provide a viable mechanism for traversing to other universes or distant spacetime regions in realistic astrophysical scenarios.

9.5.3 *Extreme RN black hole*

Finally, we briefly discuss the case of the extreme RN black hole. In this limit, where $Q^2 = G^2M^2$, the two horizons coincide, leaving only a single event horizon, while the intrinsic singularity at $r = 0$ is still timelike. The form of the metric becomes

$$\mathrm{d}s^2 = -\left(1-\frac{\mu}{r}\right)^2 \mathrm{d}t^2 + \frac{\mathrm{d}r^2}{\left(1-\frac{\mu}{r}\right)^2} + r^2\mathrm{d}\Omega^2, \tag{9.147}$$

which is singular at $r = \mu$. To resolve this singularity, we introduce the Regge–Wheeler coordinate

$$r^* = r + 2\mu \ln\left|\frac{r-\mu}{\mu}\right| - \frac{(\mu)^2}{r-\mu} \quad \Rightarrow \quad \mathrm{d}r^* = \frac{\mathrm{d}r}{1-\frac{\mu}{r}}, \tag{9.148}$$

and adopt AEF coordinates as before. The metric then becomes

$$\mathrm{d}s^2 = -\left(1-\frac{\mu}{r}\right)^2 \mathrm{d}\tilde{u}^2 + 2\mathrm{d}\tilde{u}\mathrm{d}r + r^2\mathrm{d}\Omega^2. \tag{9.149}$$

which is now regular at the null hypersurface $r = \mu$.

To analyze the asymptotic behavior near $r \to \mu$, we introduce new coordinates $r = \mu(1+\lambda)$ and keep only the leading-order terms in λ. This yields

$$F \sim \mathrm{d}\lambda \wedge \mathrm{d}t, \tag{9.150}$$

$$\mathrm{d}s^2 \sim \underbrace{(-\lambda^2\mathrm{d}t^2 + \mu^2\lambda^2\mathrm{d}\lambda^2)}_{\mathrm{AdS}_2} + \mu^2\mathrm{d}\Omega^2. \tag{9.151}$$

The first part of the metric corresponds to an AdS_2 spacetime, while the second part is a 2-dimensional sphere of radius μ. Thus, the near-horizon geometry is $\mathrm{AdS}_2 \times \mathbf{S}^2$, a Robinson–Bertotti spacetime. This configuration represents a Kaluza–Klein vacuum, where two dimensions are compactified, resulting in an effective 2-dimensional AdS_2 spacetime with constant negative curvature.

In the extreme RN black hole, the gravitational attraction and electromagnetic repulsion precisely balance. This equilibrium enables the construction of multi-centered extreme black holes. Specifically, two extreme RN black holes with different masses and like charges experience mutual gravitational attraction and electromagnetic repulsion that exactly cancel. The exact multi-centered solution is given by

$$\mathrm{d}s^2 = -H^{-2}(\boldsymbol{x})\mathrm{d}t^2 + H^2(\boldsymbol{x})[\mathrm{d}x^2 + \mathrm{d}y^2 + \mathrm{d}z^2], \tag{9.152}$$

where H satisfies the Laplace equation

$$\nabla^2 H = (\partial_x^2 + \partial_y^2 + \partial_z^2)H = 0. \tag{9.153}$$

A general solution is

$$H = 1 + \sum_{i=1}^{N} \frac{\mu_i}{|\boldsymbol{x} - \boldsymbol{x}_i|}, \tag{9.154}$$

describing a black hole with N distinct centers. This is the well-known the Majumdar–Papapetrou solution.

9.6 Observation of Black Holes

Although black holes have been widely accepted in astrophysics for nearly half a century, their observational confirmation remains a subject of debate. The fundamental challenge lies in their nature: black holes do not emit detectable radiation, making them invisible to traditional observation methods. For isolated black holes formed from the gravitational collapse of massive stars, direct detection is particularly difficult. A promising approach involves studying binary star systems, which account for nearly two-thirds of stellar populations in the universe. Among these, X-ray binaries serve as key candidates for identifying black holes. In such systems, a compact object, either a black hole or a neutron star, accretes matter from its companion star, forming a luminous accretion disk. The presence of a compact object at the disk's center can then be inferred through observational signatures. Distinguishing between black holes and neutron stars in X-ray binaries relies on key diagnostics such as mass and pulse frequency. X-ray emissions from these systems are among the brightest astrophysical sources. Unusual temporal variability and spectral features in the X-ray emission can indicate an orbiting compact companion. If the accreting object exhibits stable pulsations, it is likely a pulsar (a rapidly rotating neutron star). However, if no such pulsations are detected, dynamical measurements, such as the companion star's orbital velocity and radius, allow estimation of the compact object's mass via spectroscopic analysis. Since neutron stars cannot exceed approximately $3M_\odot$ (the Tolman–Oppenheimer–Volkoff limit), a more massive compact object must be a black hole.

In recent years, gravitational wave astronomy has opened a new observational window into black holes and neutron stars. The detection of merging binary systems, including black hole-black hole (BH-BH) and black hole-neutron star (BH-NS) mergers, has provided unprecedented insights into their physical properties. Notably, the first observed gravitational wave signal (GW150914) originated from the coalescence of two black holes, each tens of solar masses in size. Subsequent detections have revealed a population of black holes with masses exceeding standard theoretical predictions, challenging existing models of stellar evolution and black hole formation. Understanding the origins of these unexpectedly massive black holes has since emerged as a forefront topic in astrophysical research.

On the other hand, modern astrophysics strongly supports the existence of a supermassive black hole (SMBH) at the center of every galaxy, with masses ranging from $10^6 M_\odot$ to $10^9 M_\odot$. The closest such object is Sagittarius A* (Sgr A*), the SMBH at the Milky Way's center, with a mass of $4.3 \times 10^6 M_\odot$. Since SMBHs do not emit detectable radiation, their presence must be inferred indirectly—primarily through their gravitational influence on nearby stars, as revealed by infrared observations and stellar orbit tracking. Substantial evidence confirms their existence. By meticulously analyzing the motion of stars orbiting Sgr A* over decades, researchers have constrained its mass and properties with remarkable precision.[10] However, there are still some key questions about SMBH formation and growth remain unresolved:

(1) Did SMBHs form concurrently with their host galaxies, or did they emerge later through gradual processes?

(2) Were they born as massive seeds ($\sim 10^6 M_\odot$) or as smaller black holes that grew via accretion and mergers?

(3) Did their mass increase primarily through gas accretion, stellar consumption, or mergers with other black holes?

Current models suggest that massive black holes in galactic nuclei serve as the progenitors of SMBHs, growing over cosmic time via gas accretion, tidal disruption of stars, and hierarchical mergers. There are active galactic nuclei (AGN) or quasars at the centers of galaxies. Quasars, in particular, are among the brightest objects in the universe, with luminosities reaching 10^{48} erg/s, outshining their host galaxies by factors of 100–1000. Moreover, quasars are very compact, with a small size of only a few parsecs or even smaller, appear like stars in optical images, and exhibit broad, non-thermal emission spectra. Many also produce relativistic radio jets, further supporting the black hole accretion model. British astrophysicist Donald

[10]Half of the 2020 Nobel Prize in Physics was awarded to Reinhard Genzel and Andrea Ghez for their pioneering work in demonstrating the presence of a supermassive, compact object at the Galactic Center—now widely accepted as a black hole.

Lynden-Bell (1969) proposed that quasar activity arises from gas accretion onto SMBHs—a theory now widely validated by observations.

In recent years, a new tool for the study of supermassive black holes has emerged. In 2017, the Event Horizon Telescope (EHT) collaboration achieved a milestone by capturing the first image of the black hole in M87*, using a global network of radio telescopes. The image revealed a dark "shadow" encircled by a bright ring, the photon capture region of the black hole's accretion disk. This breakthrough not only confirmed general relativity in strong gravitational fields but also opened new avenues for studying accretion physics and jet formation.

Another possible observation of black holes comes from the laboratory. Over the past decade or so, theoretical studies have explored whether microscopic black holes could be produced in particle colliders like the Large Hadron Collider (LHC), and if so, how to observe their signals. While such scenarios raise dramatic questions, e.g., could they grow uncontrollably or endanger Earth?, calculations suggest they pose no threat. First, it is highly improbable to be produced under current energy scales. Second, even if formed, Hawking radiation would cause near-instantaneous evaporation, potentially leaving detectable quantum signatures.

Exercises

1. Consider a massive particle outside a Schwarzschild black hole that falls radially towards the event horizon. Assume its geodesic has a conserved energy density $E = 0.95$.

   (1) Calculate the proper time it takes to travel from $r = 3\mu$ to $r = R_s$;
   (2) Calculate the proper time it takes to travel from $r = R_s$ to $r = 0$;
   (3) In the Schwarzschild coordinate basis, find its 4-velocity at $r = 2.001\mu$;
   (4) When it passes $r = 2.001\mu$, it emits a radially outward photon to an observer at a fixed radial position far away. Calculate the redshift of the photon.

2. In the Schwarzschild geometry, consider the motion of a massive particle.

   (1) For a particle with a definite angular momentum density L, how much energy density must it have to reach the Schwarzschild radius?

(2) If the above condition is changed to the impact parameter b, what is this value?

(3) Conversely, for a definite impact parameter b, what is the maximum value of the angular momentum density for the particle to reach the Schwarzschild radius?

3. Introduce the coordinate

$$r = \rho\left(1 + \frac{\mu}{2\rho}\right)^2, \tag{9.155}$$

prove that the line element of the Schwarzschild geometry can be written as

$$\begin{aligned} ds^2 = &-\left(1 - \frac{\mu}{2\rho}\right)^2\left(1 + \frac{\mu}{2\rho}\right)^{-2} dt^2 \\ &+\left(1 + \frac{\mu}{2\rho}\right)^4 (d\rho^2 + \rho^2 d\theta^2 + \rho^2 \sin^2\theta d\phi^2), \end{aligned} \tag{9.156}$$

and prove that in the weak field limit $\mu \ll \rho$, $g_{00} \approx 1 - R_s/\rho$.

4. Consider an astronaut falling radially from infinity, who emits a radially outward light signal during the fall, which is received by a spaceship fixed at a certain radial position r_o.

(1) Assuming the photon is emitted at the radial position r_e, calculate the redshift factor of the received photon;

(2) Prove that for an observer at infinity, when the emitter approaches the event horizon, the times of photon emission and reception are

$$t_e = -2\mu \ln\left(1 - \frac{R_s}{r_e}\right) + \text{constant}, \tag{9.157}$$

$$t_o = -4\mu \ln\left(1 - \frac{R_s}{r_e}\right) + \text{constant}, \tag{9.158}$$

thereby proving that the change in the wavelength of the photon is

$$\frac{\lambda_o}{\lambda_e} \propto \exp(t_o/4\mu). \tag{9.159}$$

5. A spherical dust matter composed of a star with a coordinate radius R and a total mass M collapses inward from rest due to its own gravity. Prove that when the collapse occurs, the coordinate radius r of the star's surface and the proper time τ passed by an observer on the surface satisfy the relationship

$$\tau(r) = -\frac{1}{(R_s)^{1/2}} \int_R^r \left(\frac{r}{1 - r/R}\right)^{1/2} dr. \tag{9.160}$$

By making a coordinate transformation $r = R\cos^2(\psi/2)$, prove that the above relationship can be expressed as

$$r = \frac{R}{2}(1+\cos\psi), \quad \tau = \frac{R}{2}\left(\frac{R}{R_s}\right)^{1/2}(\psi + \sin\psi), \tag{9.161}$$

and calculate the proper time passed by the observer before the star collapses into a point.

6. Consider the motion of a massive particle inside a Schwarzschild black hole with mass M (not necessarily geodesic motion). Using Schwarzschild coordinates (t, r, θ, ϕ), prove that the r-coordinate of this particle must decrease and the rate of decrease satisfies

$$\left|\frac{\mathrm{d}r}{\mathrm{d}\tau}\right| \geq \sqrt{\frac{R_s}{r} - 1}, \tag{9.162}$$

where τ is the proper time of the particle. From this, further calculate the maximum time this particle can reach from the event horizon to the singularity, and find the trajectory of the particle that reaches this time.

7. If an astronaut is in circular motion at a radius r outside a Schwarzschild black hole, calculate the tidal force he feels.
8. In the text, we discussed the transformation from Schwarzschild coordinates to Kruskal coordinates, which is only correct for $r > R_s$. Please give the transformation from Schwarzschild coordinates to Kruskal coordinates for $r < R_s$.
9. In Kruskal coordinates, prove that the Schwarzschild black hole is locally flat at $u = v = 0$.
10. Consider the Schwarzschild geometry in Kruskal coordinates. When $v = 0$, we found the Einstein–Rosen bridge. Please consider the geometry when $0 < v < 1$, $v = 1$, and $v > 1$, and thereby explain that the Einstein–Rosen bridge narrows as v increases, the two asymptotic regions are connected at the singularity when $v = 1$, and the two asymptotically flat regions are completely separated when $v > 1$.
11. Consider a spherically symmetric star with a single density ρ.

    (1) Prove that

$$R^2 < \frac{c^2}{3\pi G\rho}, \quad M^2 < \frac{16c^6}{243\pi\rho G^3}, \tag{9.163}$$

    where R is the radius of the star and M is the mass of the star.

(2) Prove that if we require the star not to be within its Schwarzschild radius, then we must have

$$M^2 < \frac{3c^6}{32\pi\rho G^3}. \tag{9.164}$$

(3) Prove that if a photon is emitted from the surface of the star and received by a static observer at infinity, then the observed redshift must satisfy $z < 2$.

(4) Prove that if the photon is emitted from the center of the star, then the observed redshift can be arbitrarily large.

12. Please calculate the tidal force felt by an observer falling radially in the RN black hole spacetime.
13. In the coordinates (t, r, θ, ϕ), please give the worldline of a radially moving photon in the RN spacetime geometry. Furthermore, by introducing AEF coordinates

$$t' = t - \frac{r_-^2}{r_+ - r_-} \ln\left|\frac{r}{r_-} - 1\right| + \frac{r_+^2}{r_+ - r_-} \ln\left|\frac{r}{r_+} - 1\right|, \tag{9.165}$$

give the form of the RN line element, and prove that in the AEF coordinate system, the worldlines of radially moving photons are given by the following equations:

$$t' + r = \text{constant} \quad \text{(inward)}, \tag{9.166}$$

$$\frac{dt'}{dr} = \frac{2 - \Delta}{\Delta} \quad \text{(outward)}. \tag{9.167}$$

14. A particle with mass m and charge q has the action

$$S = \int d\lambda \left(\frac{1}{2} e^{-1} \dot{x}^\mu \dot{x}^\nu g_{\mu\nu}(x) - \frac{1}{2} m^2 e - q\dot{x}^\mu A_\mu(x) \right), \tag{9.168}$$

where A_μ is the electromagnetic 4-potential. Prove that if for the Killing vector $\widehat{\xi}$,

$$(\mathcal{L}_{\widehat{\xi}} A)_\mu = \xi^\nu \partial_\nu A_\mu + (\partial_\mu \xi^\nu) A_\nu = 0, \tag{9.169}$$

then the action is invariant under the transformation $x^\mu \to x^\mu + \epsilon\xi^\mu(x)$ (keeping the first order in ξ). Prove that the corresponding Noether charge

$$\mathcal{Q} = -\xi^\mu (m u_\mu - q A_\mu) \tag{9.170}$$

is a constant of motion, where u^μ is the 4-momentum of the particle.

15. Continuing from the previous problem, consider a massive particle moving in the RN spacetime, prove that for $\hat{\xi} = \partial_t$ we have $\mathcal{L}_{\hat{\xi}} A = 0$, and thus derive

$$\left(1 - \frac{2\mu}{r} + \frac{Q^2}{r^2}\right)\frac{\mathrm{d}t}{\mathrm{d}\tau} = E - \frac{q}{m}\frac{Q}{r}, \tag{9.171}$$

where E is the energy density of the particle. Prove that for a massive particle with no angular momentum, its worldline $r(\tau)$ satisfies

$$\left(\frac{\mathrm{d}t}{\mathrm{d}\tau}\right)^2 = (E^2 - 1) + \left(1 - E\frac{qQ}{m\mu}\right)\frac{2\mu}{r} + \left(\left(\frac{q}{m}\right)^2 - 1\right)\frac{Q^2}{r^2}. \tag{9.172}$$

Consider the case when $q^2 = m^2$, $qQ = m\mu$, $E = 1$, and discuss the motion of the particle.

16. In the RN spacetime geometry, what is the magnetic field component measured by an observer in circular motion at a coordinate radius $r = R$?

17. Solve for the spherically symmetric solution with charge in a spacetime with a cosmological constant.

18. Suppose the metric and gauge potential have the following forms:

$$\begin{aligned} \mathrm{d}s^2 &= -H^{-2}(x)\mathrm{d}t^2 + H^2(x)(\mathrm{d}x^2 + \mathrm{d}y^2 + \mathrm{d}z^2), \\ A_\mu \mathrm{d}x^\mu &= \mp H^{-1}(x)\mathrm{d}t, \end{aligned} \tag{9.173}$$

prove that if the function H satisfies the Laplace equation $\nabla^2 H = 0$, then they are solutions to the Einstein–Maxwell equations.

19. Prove that in isotropic coordinates, the metric of the extreme RN black hole can be written as

$$\mathrm{d}s^2 = -\left(1 + \frac{\mu}{\rho}\right)^{-2}\mathrm{d}t^2 + \left(1 + \frac{\mu}{\rho}\right)^2(\mathrm{d}\rho^2 + \rho^2\mathrm{d}\Omega^2). \tag{9.174}$$

(1) Prove that along any curve with constant t, the proper distance from $\rho = 0$ to any finite ρ is infinite.

(2) Prove that along any timelike or null curve, as $\rho \to 0$, $|t| \to \infty$, but the affine parameter of a timelike or null radial geodesic reaching $\rho = 0$ is finite.

(3) Introduce a null coordinate to prove that $\rho = 0$ is only a coordinate singularity.

(4) What happens when a particle reaches $\rho = 0$?

Chapter 10

Kerr Black Hole

In the preceding chapters, we have explored the properties of spherically symmetric spacetimes. Such spacetimes, particularly the Schwarzschild spacetime, serve as excellent approximations for the exterior gravitational field of non-rotating stars. Indeed, in the absence of rotation, massive celestial bodies would assume an almost perfectly spherical shape, as gravitational forces act to suppress any deviations from spherical symmetry. This is especially true for black holes: in the idealized case where they possess neither rotation nor electric charge, their spacetime geometry is exactly described by the Schwarzschild metric. However, in reality, all astrophysical objects exhibit some degree of rotation. During their formation, conservation of angular momentum ensures that even initially slowly rotating matter acquires significant spin. For example, the Sun, with a rotational period of approximately 27 days, exhibits a fractional deviation from spherical symmetry on the order of 10^{-4}. While seemingly negligible, such deviations can perturb the Newtonian gravitational potential, leading to observable effects such as the precession of planetary orbits.

The spacetime geometry surrounding a rotating object depends not only on its mass M but also on its angular momentum J. In this chapter, we will study the geometry of rotating spacetimes and their physical implications. We begin by analyzing the effects of rotation on the spacetime geometry in the weak field approximation, before introducing the Kerr solution, which is a vacuum solution to Einstein's field equations that describes a stationary, axisymmetric black hole with angular momentum. Within the Kerr spacetime, novel phenomena arise, including frame-dragging effects and the Penrose process, which we shall explore in detail.

10.1 Spacetime Geometry of Rotating Objects in the Weak Field Approximation

The external spacetime geometry of a rotating body is determined not only by its mass but also by its rotational motion. For small angular velocities Ω, the deviation of the rotating spacetime from spherical symmetry is of order $\mathcal{O}(\Omega)$. Since the centrifugal acceleration scales as $\mathcal{O}(\Omega^2)$, the object's shape remains unchanged to first order in the angular velocity. In other words, under slow rotation, the deformation of the body is negligible, allowing it to be approximated as spherical. Nevertheless, the external spacetime geometry is still influenced by rotation, meaning the curvature depends not only on the mass distribution but also on the object's state of motion. The leading-order contribution from rotation is of magnitude $\mathcal{O}\left(\frac{v}{c}\frac{GM}{Rc^2}\right)$, which leads to the gravitomagnetic effect.[1]

10.1.1 *Weak field approximation and electrodynamics*

When an object rotates at non-relativistic speeds, the external spacetime geometry can be analyzed using the weak-field approximation. In this regime, the Einstein field equations simplify to

$$\Box\bar{h}_{\mu\nu} = -16\pi G T_{\mu\nu}, \tag{10.1}$$

where $\bar{h}_{\mu\nu}$ is the trace-reversed perturbation, defined as

$$\bar{h}_{\mu\nu} = h_{\mu\nu} - \frac{1}{2}\eta_{\mu\nu}h. \tag{10.2}$$

In general relativity, matter can often be well approximated as an ideal fluid, with an energy-momentum tensor given by

$$T_{\mu\nu} = (\rho + p)u_\mu u_\nu + p g_{\mu\nu}. \tag{10.3}$$

For non-relativistic matter (e.g., dust), the pressure is negligible ($p \approx 0$), and the fluid velocity is much smaller than the speed of light, with the 4-velocity taking the form $u^\mu = (1, v^i/c)$, where $v^i \ll c$. Consequently, the components of $T_{\mu\nu}$ reduce to

$$T_{00} = \rho, \quad T_{0i} = -\rho v_i/c, \quad T_{ij} \propto v_i v_j/c^2. \tag{10.4}$$

Since T_{ij} is second-order in v/c, it can be neglected, leaving only T_{00} and T_{0i} as the dominant contributions. Under these approximations, the spatial

[1]To clarify the origin and physical meaning of the gravitomagnetic field, the speed of light c is retained explicitly in the theoretical discussion of this chapter's first two sections.

components of the trace-reversed perturbation vanish, $\bar{h}_{ij} = 0$, while the remaining components satisfy

$$\begin{aligned} \Box \bar{h}_{00} &= -\frac{16\pi G}{c^2}\rho, \\ \Box \bar{h}_{0i} &= -\frac{16\pi G}{c^2}T_{0i}. \end{aligned} \tag{10.5}$$

Here, $\bar{h}_{0i} = h_{0i}$ is suppressed by a factor of v/c compared to h_{00}. Noting that $\bar{h} = -\bar{h}_{00}$, we obtain

$$h_{00} = h_{ii} = \frac{1}{2}\bar{h}_{00}. \tag{10.6}$$

For clarity, we define the perturbations in terms of a scalar potential Φ and a vector potential A_i:

$$h_{00} = -2\Phi/c^2, \quad \bar{h}_{0i} = h_{0i} = A_i/c. \tag{10.7}$$

Assuming a stationary matter distribution, $\partial_t \rho = 0$, we compute the non-vanishing Christoffel symbols:

$$\begin{aligned} \Gamma^0_{i0} &= \frac{1}{c^2}\nabla_i \Phi, \\ \Gamma^i_{00} &= \frac{1}{c^2}\nabla_i \Phi + \frac{1}{c}\partial_t A_i, \\ \Gamma^k_{i0} &= \frac{1}{2c}(A_{k,i} - A_{i,k}), \\ \Gamma^k_{im} &= -\frac{1}{c^2}(-\delta_{im}\nabla_k \Phi + \delta^k_i \nabla_m \Phi + \delta^k_m \nabla_i \Phi), \\ \Gamma^0_{im} &= -\frac{1}{2c}(A_{i,m} + A_{m,i}). \end{aligned} \tag{10.8}$$

Consider the motion of a particle in the weak field. The geodesic equation for the particle is

$$\frac{du_\mu}{d\tau} + (h_{\mu\alpha,\beta}u^\alpha u^\beta - \frac{1}{2}h_{\alpha\beta,\mu}u^\alpha u^\beta) = 0. \tag{10.9}$$

If the background spacetime admits a timelike Killing vector, $\partial_0 h_{\alpha\beta} = 0$, the particle's energy is conserved. In the non-relativistic limit, the particle's spatial velocity satisfies $v_i = \dfrac{dx^i}{dt} \ll 1$, and the proper time τ approximately coincides with coordinate time t, $\dfrac{dt}{d\tau} \approx 1$. Consequently, terms of order v^2 can be neglected. However, cross terms of the form $h \times v$, which scale as $\mathcal{O}(GM/r \cdot \sqrt{GM/r})$, are larger than v^2 terms and may dominate in certain regimes. At the lowest order (Newtonian approximation) both v^2 and $h \times v$

terms are negligible, reducing the dynamics to classical gravity. Retaining the leading-order relativistic corrections ($h \times v$ terms) while neglecting $\mathcal{O}(v^2)$ terms, the geodesic equation simplifies to

$$\frac{\mathrm{d}u_\mu}{\mathrm{d}t} + (h_{\mu 0,l} - h_{0l,\mu})v^l + \frac{1}{2}(2h_{0\mu,0} - h_{00,\mu}) = 0. \tag{10.10}$$

For the spatial components, this yields the equation of motion

$$\frac{\mathrm{d}v_i}{\mathrm{d}t} = -(\partial_i \Phi + \partial_t A_i) + (\partial_i A_l - \partial_l A_i)v_l. \tag{10.11}$$

Introducing the gravitoelectric field $\boldsymbol{g}$ and gravitomagnetic field $\boldsymbol{b}$ as

$$\begin{aligned} \boldsymbol{g} &= -\nabla\Phi - \partial_t \boldsymbol{A}, \\ \boldsymbol{b} &= \nabla \times \boldsymbol{A}, \end{aligned} \tag{10.12}$$

the geodesic equation takes the form

$$\frac{\mathrm{d}\boldsymbol{v}}{\mathrm{d}t} = \boldsymbol{g} + \boldsymbol{v} \times \boldsymbol{b}. \tag{10.13}$$

This resembles the Lorentz force law in electrodynamics: $\boldsymbol{g}$ corresponds to the Newtonian gravitational field, analogous to the electrostatic field; $\boldsymbol{b}$, analogous to the magnetic field, arises from the object's motion and induces velocity-dependent forces on test particles.

The gravitoelectric $\boldsymbol{g}$ and gravitomagnetic $\boldsymbol{b}$ fields satisfy field equations remarkably similar to Maxwell's equations. From their definitions in Eq. (10.12), we immediately obtain two source-free equations:

$$\begin{cases} \nabla \cdot \boldsymbol{b} = 0, \\ \nabla \times \boldsymbol{g} + \dfrac{\partial \boldsymbol{b}}{\partial_t} = \boldsymbol{0}, \end{cases} \tag{10.14}$$

which are exact analogues of Maxwell's equations for magnetism and Faraday's law of induction. The sourced equations emerge from the weak-field Einstein equations. Starting from

$$\begin{aligned} \nabla^2 h^{00} &= -8\pi G T^{00}, \\ \nabla^2 h^{0i} &= -16\pi G T^{0i}, \end{aligned} \tag{10.15}$$

we derive the gravitoelectromagnetic field equations with sources:

$$\begin{aligned} \nabla \cdot \boldsymbol{g} &= -4\pi G\rho, \\ \nabla \times \boldsymbol{b} &= -\frac{16\pi G}{c^2}\boldsymbol{j}, \end{aligned} \tag{10.16}$$

where $\boldsymbol{j} = \rho \boldsymbol{v}$ is the mass-current density, playing the role analogous to the electric current density in electromagnetism.

There are two key differences between gravito-electromagnetic field equations and standard Maxwell equations. First, the gravitational "charge" density appears with a negative sign in Gauss's law (10.16), reflecting gravity's attractive nature between like charges. Second, the gravitomagnetic equation contains an extra factor of 4 in front of the flux $\boldsymbol{j}$ compared to Ampère's law, originating from the spin-2 nature of the gravitational field (versus spin-1 for electromagnetism).

This gravito-electromagnetic framework holds strictly for the conditions that the perturbed spacetime must be stationary, the matter is non-relativistic, and the weak-field approximation holds. When these conditions are violated, the electromagnetic analogy becomes less precise. For instance, the time-dependent metric perturbations introduce additional terms. Particularly, in strongly dynamical spacetimes, non-linear gravitational effects dominate and the weak field approximation breaks down.

In summary, in the weak field approximation, considering a stationary spacetime and non-relativistic matter sources, we find that the gravitational field is remarkably similar to the electromagnetic field. In fact, there is a correspondence with electromagnetic gauge potential, in which the electrostatic potential is related to the Newtonian gravitational potential, and the vector potential to the $(0i)$-components of the metric tensor:

$$\Phi_{\mathrm{e}} \leftrightarrow \Phi_{\mathrm{g}} = -\frac{1}{2}h_{00}, \quad A_{\mathrm{e}i} \leftrightarrow A_i. \tag{10.17}$$

10.1.2 *External spacetime of slowly rotating objects*

We now investigate how matter rotation influences the external spacetime. The perturbation described by Eq. (10.72) satisfies a linear wave equation, which admits a solution via the Green's function method:

$$\bar{h}_{\mu\nu}(t, \boldsymbol{x}) = \frac{4G}{c^4}\int \frac{T_{\mu\nu}(t - |\boldsymbol{x} - \boldsymbol{y}|, \boldsymbol{y})}{|\boldsymbol{x} - \boldsymbol{y}|}\mathrm{d}^3 y, \tag{10.18}$$

where the integral over y spans the matter distribution region. For a stationary spacetime, the time dependence vanishes, simplifying the perturbation to:

$$\bar{h}_{\mu\nu}(\boldsymbol{x}) = \frac{4G}{c^4}\int \frac{T_{\mu\nu}(\boldsymbol{y})}{|\boldsymbol{x} - \boldsymbol{y}|}\mathrm{d}^3 y. \tag{10.19}$$

This leads to the following expressions for the gravitational potential and gravitomagnetic vector potential

$$\Phi(\boldsymbol{x}) = -G\int \frac{\rho(\boldsymbol{y})}{|\boldsymbol{x} - \boldsymbol{y}|}\mathrm{d}^3 y, \tag{10.20}$$

$$A_i(\boldsymbol{x}) = 4G\int \frac{T_{0i}(\boldsymbol{y})}{|\boldsymbol{x} - \boldsymbol{y}|}\mathrm{d}^3 y. \tag{10.21}$$

The corresponding spacetime line element takes the form

$$\mathrm{d}s^2 = -(1+2\Phi)\mathrm{d}t^2 + 2A_i \mathrm{d}t\mathrm{d}x^i + (1-2\Phi)(\mathrm{d}x^2+\mathrm{d}y^2+\mathrm{d}z^2). \quad (10.22)$$

At large distances from the source, we employ the multipole expansion

$$\frac{1}{|\boldsymbol{x}-\boldsymbol{y}|} \approx \frac{1}{r} + \frac{x_i}{r^2}\frac{y^i}{r} + \cdots . \quad (10.23)$$

In the center-of-mass coordinate frame, this yields

$$\Phi = -\frac{GM}{r}, \quad M = \int \rho \mathrm{d}^3 y, \quad (10.24)$$

and for the gravitomagnetic potential

$$\begin{aligned} A_i &= 4G\int T_{0i}\left(\frac{1}{r} + \frac{x_j}{r^2}\frac{y^j}{r} + \cdots\right)\mathrm{d}^3 y \\ &\approx 4G\int \frac{T_{0i}x_j y^j}{r^3}\mathrm{d}^3 y \\ &= \frac{4G}{r^3}x_j\int T_{0i}y^j\mathrm{d}^3 y. \end{aligned} \quad (10.25)$$

Introducing the total angular momentum of the source,

$$\boldsymbol{J} = \int(\boldsymbol{y}\times\boldsymbol{p})\mathrm{d}^3 y, \quad (10.26)$$

where $p_i = -T_{0i} = \rho v_i$, we find that the gravitomagnetic potential simplifies to

$$\begin{aligned} A_i &= -\frac{2G}{r^3}(\boldsymbol{J}\times\boldsymbol{x})_i \\ &= \frac{2G}{r^3}\varepsilon_{ijk}x^j J^k. \end{aligned} \quad (10.27)$$

(1) Spacetime geometry outside a cylindrically symmetric matter distribution.

We examine the spacetime geometry generated by a massive object rotating about the z-axis with a cylindrically symmetric matter distribution. To describe the system, we adopt coordinates X^i, where $X^3 = z$, and introduce polar coordinates in the X^1-X^2 plane for convenience. In this coordinate system, the components of the stress-energy tensor associated with the rotating matter are given by

$$T_{01} = -\frac{\rho v_x}{c} = -\frac{\rho v}{c}\sin\alpha, \quad T_{02} = -\frac{\rho v_y}{c} = -\frac{\rho v}{c}\cos\alpha, \quad (10.28)$$

where v_x, v_y denote the velocity components, and α is the azimuthal angle in the X^1-X^2 plane. From symmetry considerations, several integrals vanish:

$\int T_{01}\mathrm{d}^3X = 0$, $\int X^1 \frac{\mathrm{d}X^1}{dt} = 0$, $\int X^3 \frac{\mathrm{d}X^1}{dt} = 0$, while the cross term is non-zero, $\int X^2 \frac{\mathrm{d}X^1}{dt} \neq 0$. The metric perturbations $\bar{h}_{01}, \bar{h}_{02}$ in the far-field regime take the form

$$\bar{h}_{01} = \frac{4G}{c^2}\frac{y}{r^3}\int X^2 T_{01}\mathrm{d}^3X = -\frac{4G}{c^2}\frac{y}{r^3}\int X^2 T^{01}\mathrm{d}^3X, \tag{10.29}$$

$$\bar{h}_{02} = -\frac{4G}{c^2}\frac{x}{r^3}\int X^1 T^{02}\mathrm{d}^3X. \tag{10.30}$$

The total angular momentum of the matter distribution is given by

$$J^3 = \int (X^1P^2 - X^2P^1)\mathrm{d}^3X = c\int (X^1T^{02} - X^2T^{01})\mathrm{d}^3X, \tag{10.31}$$

where $P^i = \rho v^i$ is the momentum density. For a cylindrically symmetric system, the perturbations simplify to

$$\begin{aligned} \bar{h}_{01} &= \frac{2G}{c^3}\frac{y}{r^3}J^3, \\ \bar{h}_{02} &= -\frac{2G}{c^3}\frac{x}{r^3}J^3. \end{aligned} \tag{10.32}$$

(2) Spacetime geometry outside a spherically symmetric matter distribution.

For a spherically symmetric matter distribution, the metric perturbation due to rotation takes the form

$$\bar{h}_{0i} = \frac{2G}{c^3r^3}\varepsilon_{ikm}x^kJ^m, \tag{10.33}$$

where ε_{ikm} is the three-dimensional Levi–Civita symbol, and J^m represents the angular momentum vector. In the linear approximation, the components of the metric are given by

$$\begin{aligned} g_{00} &= -\left(1 - \frac{2GM}{rc^2}\right), \\ g_{ik} &= \left(1 + \frac{2GM}{rc^2}\right)\delta_{ik}, \\ g_{0i} &= \frac{2G}{r^3c^3}\varepsilon_{ikm}x^kJ^m. \end{aligned} \tag{10.34}$$

Here, the angular momentum is assumed to be aligned along the z-axis. Note that this solution is not exact but rather a linearized approximation to the Einstein equations. Therefore, for a slowly rotating spherically symmetric object, the external spacetime geometry can be expressed as:

$$\mathrm{d}s^2 = \mathrm{d}s^2_{\mathrm{Sch}} - \frac{4GJ}{c^3 r^2}\sin^2\theta(r\mathrm{d}\phi)(c\mathrm{d}t) + O(J^2), \tag{10.35}$$

where $\mathrm{d}s^2_{\mathrm{Sch}}$ is the Schwarzschild metric, and the second term represents the leading-order rotational correction. The angular momentum of the object can be estimated as

$$J \sim I\Omega \sim MR^2\Omega \sim MRv, \tag{10.36}$$

where I is the moment of inertia, Ω is the angular velocity, R is the characteristic radius, and v is the tangential velocity. Substituting this into the metric perturbation yields

$$\frac{4GJ}{c^3R^2} \sim \frac{GM}{Rc^2}\frac{v}{c}. \tag{10.37}$$

This term leads to a gravitomagnetic effect, analogous to the magnetic field in electrodynamics. Just as electric currents generate magnetic fields, moving matter (i.e., mass currents) generates a gravitomagnetic field. The first term in (10.35) corresponds to the Schwarzschild metric, where the weak-field expansion includes the Newtonian gravitational potential, analogous to the electrostatic potential, while the second term is suppressed by a factor of v/c, making it a relativistic correction akin to magnetism in electromagnetism. Thus, the gravitomagnetic field arises from the motion of mass, mirroring the relationship between electric currents and magnetic fields in Maxwell's theory.

10.2 Physical Effects of the Gravitomagnetic Field

From the preceding discussion, it is clear that the gravitomagnetic field exhibits a profound analogy with the magnetic field in electromagnetism. Building upon this correspondence, this section delves deeper into the physical manifestations of the gravitomagnetic field. In electrodynamics, a well-established coupling exists between magnetic fields and magnetic dipoles possessing non-zero magnetic moments. Similarly, in quantum mechanics, electrons—endowed with intrinsic spin—experience coupling with magnetic fields, giving rise to phenomena such as the anomalous Zeeman effect. A closer examination further reveals that the electron's spin undergoes Thomas precession, as previously elucidated. Extending this parallelism to general relativity, objects with spin interact with the gravitomagnetic field. This interaction induces various spin precession effects, mirroring the behavior observed in electromagnetic systems while highlighting the unique implications of gravitomagnetism in relativistic gravity.

10.2.1 *Lense–Thirring precession*

A prominent example of spin precession induced by the gravitomagnetic field is the Lense–Thirring precession. To illustrate this phenomenon, we begin with a simplified scenario: a gyroscope moving in the spacetime described by the line element (10.35). Our focus lies in how the spacetime's rotation influences the gyroscope's spin orientation.

In the absence of rotation, i.e., when $J = 0$, the spin vector $\widehat{s}$ remains constant if the gyroscope undergoes a radial motion, due to the $\phi \to -\phi$ symmetry. This behavior contrasts with the geodetic precession discussed earlier. For analytical convenience, we switch to Cartesian coordinates, where the metric takes the form:

$$\mathrm{d}s^2 = \mathrm{d}s^2_{\mathrm{Sch}} + \frac{4GJ}{c^3 r^2}(c\mathrm{d}t)\left(\frac{x\mathrm{d}y - y\mathrm{d}x}{r}\right) + O(J^2). \tag{10.38}$$

Since the leading-order gravitomagnetic correction scales as $\sim \frac{1}{c^3}$, the spherically symmetric contribution from the mass M encoded in the Schwarzschild part ds^2_{Sch} does not interfere at this order. Thus, we may set $M = 0$ to isolate the rotational effect. We further simplify the analysis by assuming the gyroscope moves along the z-axis, with its spin confined to the x-y plane,

$$u^\alpha = (u^t, 0, 0, u^z), \quad s^\alpha = (0, s^x, s^y, 0). \tag{10.39}$$

The relevant connection coefficients governing the spin evolution are

$$(\Gamma^x_{ty})_z = \frac{2GJ}{c^2 z^3} = -(\Gamma^y_{tx})_z. \tag{10.40}$$

The gyroscope equation then becomes

$$\frac{\mathrm{d}s^x}{\mathrm{d}t} = -\frac{2GJ}{c^2 z^3} s^y,$$
$$\frac{\mathrm{d}s^y}{\mathrm{d}t} = \frac{2GJ}{c^2 z^3} s^x.$$

This set of equations describe a Lense–Thirring of the spin vector in the x-y plane, with a precession rate $\Omega_{\mathrm{LT}} = \frac{2GJ}{c^2 z^3}$. Notably, although the gyroscope's motion is purely along the z-axis, its spin undergoes precession due to the frame-dragging effect generated by the spacetime's rotation. In Newtonian mechanics, where only the gravitational potential governs dynamics, inertial frames are not dragged by rotating masses. Thus, the Lense–Thirring precession is a purely relativistic signature of gravitomagnetism.

10.2.2 *Gravitomagnetic clock effect*

Another fascinating manifestation of the gravitomagnetic field is the gravitomagnetic clock effect. To illustrate this phenomenon, consider two clocks initially synchronized on Earth's equatorial plane, each placed on separate spaceships in circular orbits moving in opposite directions. In the absence of Earth's rotation, $\boldsymbol{J} = 0$, the clocks would remain synchronized upon completing a full orbit and returning to their starting point. However, when the Earth's rotation is taken into account, the clocks exhibit a measurable time discrepancy due to the gravitomagnetic field. Assume that the Earth rotates about the z-axis, $\boldsymbol{J} = J\boldsymbol{e}_z$. The gravitomagnetic vector potential components in this scenario are

$$A_x = \frac{2G}{r^3c^3}yJ, \quad A_y = -\frac{2G}{r^3c^3}xJ, \quad A_z = 0. \tag{10.41}$$

The spacetime metric outside the Earth is then given by

$$\begin{aligned} \mathrm{d}s^2 = &-\left(1 - \frac{2GM}{rc^2}\right)c^2\mathrm{d}t^2 + \frac{4GJ}{r^3c^2}(y\mathrm{d}x - x\mathrm{d}y)\mathrm{d}t \\ &+\left(1 + \frac{2GM}{rc^2}\right)(\mathrm{d}x^2 + \mathrm{d}y^2 + \mathrm{d}z^2). \end{aligned} \tag{10.42}$$

For circular motion in the equatorial plane, $\mathrm{d}\phi = \omega\mathrm{d}t$, the geodesic equation yields the orbital angular velocities:

$$\omega^2 + \frac{2GJ}{r^3c^2}\omega - \frac{GM}{r^3} = 0. \tag{10.43}$$

Solving this, we find the angular frequencies for prograde (ω_+) and retrograde (ω_-) orbits:

$$\omega_\pm = \sqrt{\left(\frac{GJ}{r^3c^2}\right)^2 + \frac{GM}{r^3}} \pm \frac{GJ}{r^3c^2}. \tag{10.44}$$

From the metric condition $\mathrm{d}s^2 = -c^2\mathrm{d}\tau^2$, the proper time for a clock in orbit is approximated by

$$\mathrm{d}\tau^2 \approx \frac{1}{\omega^2}\left(1 - \frac{3GM}{rc^2} + \frac{6GJ\omega}{rc^4}\right)\mathrm{d}\phi^2, \tag{10.45}$$

or equivalently

$$\mathrm{d}\tau \approx \frac{1}{\omega}\left(1 - \frac{3GM}{2rc^2} + \frac{3GJ\omega}{rc^4}\right)\mathrm{d}\phi. \tag{10.46}$$

Integrating over one full revolution, $\phi : 0 \to 2\pi$, the orbital period in proper time is

$$T = \frac{2\pi}{\omega}\left(1 - \frac{3GM}{2rc^2}\right) + \frac{6\pi GJ}{rc^4}. \tag{10.47}$$

The key result emerges when comparing the periods for prograde (T_+) and retrograde (T_-) orbits. The time difference is

$$\begin{aligned} T_+ - T_- &= \left(\frac{1}{\omega_+} - \frac{1}{\omega_-}\right) 2\pi \left(1 - \frac{3GM}{2rc^2}\right) \\ &= \frac{2J}{Mc^2} 2\pi \left(1 - \frac{3GM}{2rc^2}\right) \\ &\approx \frac{4\pi J}{Mc^2}. \end{aligned} \tag{10.48}$$

Remarkably, the leading-order term is independent of the orbital radius r and the Newtonian gravitational constant G, depending only on the Earth's angular momentum J and mass M!

For Earth, with $J = \frac{2}{5}MR^2\omega$, radius $R \approx 6.4 \times 10^6$ m, and angular frequency $\omega \approx 7.3 \times 10^{-5}$ s^{-1}, the time discrepancy is

$$T_+ - T_- \approx 1.7 \times 10^{-7} \text{ s per orbit.} \tag{10.49}$$

This subtle yet measurable effect is a prime target for precision experiments such as the proposed Gravity Probe C(lock) mission, which aims to directly observe the gravitomagnetic influence on timekeeping in orbiting systems.

10.2.3 ***General discussion on spin precession**

We have already discussed two distinct types of gyroscope precession: geodetic precession and Lense–Thirring precession. The former arises independently of rotational effects, while the latter stems from the dragging of inertial frames due to spacetime rotation. For simplicity, our earlier discussions treated these phenomena separately. In this section, we present a comprehensive analysis of gyroscope precession.

The gyroscope's spin 4-vector is perpendicular to its 4-velocity, i.e., $\widehat{s} \cdot \widehat{u} = 0$. This yields

$$s_0 = -\frac{1}{c}\frac{\mathrm{d}x^i}{\mathrm{d}t} s_i. \tag{10.50}$$

Furthermore, the parallel transport of the gyroscope's spin is governed by the gyroscope equation

$$\frac{\mathrm{D}\widehat{s}}{\mathrm{d}\tau} = 0 \quad \Rightarrow \quad \frac{\mathrm{d}s_\mu}{\mathrm{d}\tau} = \Gamma^\lambda_{\mu\nu} s_\lambda \frac{\mathrm{d}x^\nu}{\mathrm{d}\tau}. \tag{10.51}$$

From this, we derive

$$\begin{aligned} \frac{\mathrm{d}s_i}{\mathrm{d}t} &= (\Gamma^0_{i\nu} s_0 + \Gamma^k_{i\nu} s_k) \frac{\mathrm{d}x^\nu}{\mathrm{d}t} \\ &= \left(-\Gamma^0_{i0}\frac{\mathrm{d}x^k}{\mathrm{d}t} - \frac{1}{c}\Gamma^0_{im}\frac{\mathrm{d}x^m}{\mathrm{d}t}\frac{\mathrm{d}x^k}{\mathrm{d}t} + c\Gamma^k_{i0} + \Gamma^k_{im}\frac{\mathrm{d}x^m}{\mathrm{d}t}\right) s_k. \end{aligned} \tag{10.52}$$

The equation of motion for the spin vector is then

$$\begin{aligned}\frac{\mathrm{d}\boldsymbol{s}}{\mathrm{d}t} &= \frac{2}{c^2}(\boldsymbol{v}\cdot\boldsymbol{s})\nabla\Phi - \frac{1}{c^2}(\nabla\Phi\cdot\boldsymbol{s})\boldsymbol{v} + \frac{1}{c^2}(\nabla\Phi\cdot\boldsymbol{v})\boldsymbol{s} \\ &\quad + \frac{c}{2}[\boldsymbol{s}\times(\nabla\times\boldsymbol{A})].\end{aligned} \tag{10.53}$$

The final term in this equation is velocity-independent and depends solely on the angular momentum of the rotating body, leading to the Lense–Thirring effect. To solve the above equation, we employ the following techniques. First, the magnitude of the spin 4-vector is constant, so

$$\begin{aligned}\frac{\mathrm{d}}{\mathrm{d}t}(g^{\mu\nu}s_\mu s_\nu) = 0 &\Rightarrow g^{00}(s_0)^2 + g^{ik}s_i s_k \approx \text{constant} \\ &\Rightarrow \boldsymbol{s}^2 - \frac{2\Phi}{c^2}\boldsymbol{s}^2 - \frac{1}{c^2}(\boldsymbol{v}\cdot\boldsymbol{s}) \approx \text{constant}.\end{aligned} \tag{10.54}$$

Second, we may redefine the spin vector

$$\boldsymbol{s} = \left(1+\frac{\Phi}{c^2}\right)\boldsymbol{\sigma} + \frac{1}{2c^2}\boldsymbol{v}(\boldsymbol{v}\cdot\boldsymbol{\sigma}). \tag{10.55}$$

Retaining leading-order terms, we find

$$\boldsymbol{s}^2 = \left(1+\frac{2\Phi}{c^2}\right)\boldsymbol{\sigma}^2 + \frac{1}{c^2}(\boldsymbol{v}\cdot\boldsymbol{\sigma}). \tag{10.56}$$

From Eq. (10.54), it follows that $\boldsymbol{\sigma}^2$ is a constant, neglecting the sub-leading-order effects, and we may express

$$\boldsymbol{\sigma} = \left(1+\frac{\Phi}{c^2}\right)\boldsymbol{s} - \frac{1}{2c^2}(\boldsymbol{v}\cdot\boldsymbol{s}). \tag{10.57}$$

Using $\boldsymbol{\sigma}$ instead of the original spin vector $\boldsymbol{s}$ simplifies subsequent calculations. Differentiating Eq. (10.57) yields

$$\frac{\mathrm{d}\boldsymbol{\sigma}}{\mathrm{d}t} = \frac{\mathrm{d}\boldsymbol{s}}{\mathrm{d}t} - \frac{1}{c^2}\frac{\mathrm{d}\Phi}{\mathrm{d}t}\boldsymbol{s} - \frac{1}{2c^2}\frac{\mathrm{d}\boldsymbol{v}}{\mathrm{d}t}(\boldsymbol{v}\cdot\boldsymbol{s}) - \frac{1}{2c^2}\boldsymbol{v}\left(\frac{\mathrm{d}\boldsymbol{v}}{\mathrm{d}t}\cdot\boldsymbol{s}\right). \tag{10.58}$$

Substituting the relations

$$\begin{aligned}\frac{\mathrm{d}\Phi}{\mathrm{d}t} &= \frac{\partial\Phi}{\partial t} + \nabla\Phi\cdot\boldsymbol{v} = \nabla\Phi\cdot\boldsymbol{v}, \\ \frac{\mathrm{d}\boldsymbol{v}}{\mathrm{d}t} &= \nabla\Phi,\end{aligned} \tag{10.59}$$

where we assume that the Newtonian potential is time-independent, and combining with Eq. (10.53), we obtain

$$\frac{\mathrm{d}\boldsymbol{\sigma}}{\mathrm{d}t} = \boldsymbol{\Omega}\times\boldsymbol{\sigma}, \tag{10.60}$$

where

$$\boldsymbol{\Omega} = -\frac{c}{2}\nabla \times \boldsymbol{A} + \frac{3}{2c^2}\boldsymbol{v} \times \nabla\Phi. \tag{10.61}$$

For a spherically symmetric distribution of matter, the potentials are

$$\Phi = -\frac{GM}{r}, \quad \boldsymbol{A} = \frac{2G}{r^3c^3}\boldsymbol{r} \times \boldsymbol{J}, \tag{10.62}$$

leading to

$$\begin{aligned}\boldsymbol{\Omega} &= -\frac{c}{2}\nabla \times \left(\frac{2G}{r^3c^3}\boldsymbol{r} \times \boldsymbol{J}\right) + \frac{3GM}{2c^2}\boldsymbol{v} \times \nabla\left(\frac{1}{r}\right) \\ &= \frac{2G}{r^3c^2}\left(\frac{3(\boldsymbol{J}\cdot\boldsymbol{r})\boldsymbol{r}}{r^2} - \boldsymbol{J}\right) + \frac{3GM}{2c^2r^3}\boldsymbol{r} \times \boldsymbol{v}.\end{aligned} \tag{10.63}$$

The first term in Eq. (10.63), which depends only on the body's angular momentum, generates the Lense–Thirring effect. The second term, governed by the mass M and the gyroscope's motion, produces the de Sitter–Fokker effect, or geodesic precession. Expressing the angular momentum as $\boldsymbol{J} = I\boldsymbol{\omega}$, where I is the moment of inertia, we may write

$$\begin{aligned}\boldsymbol{\Omega} &= \boldsymbol{\Omega}_{\text{LT}} + \boldsymbol{\Omega}_{\text{dS}}, \\ \boldsymbol{\Omega}_{\text{LT}} &= \frac{2GI}{r^3c^2}\left(\frac{3(\boldsymbol{\omega}\cdot\boldsymbol{r})\boldsymbol{r}}{r^2} - \boldsymbol{J}\right), \\ \boldsymbol{\Omega}_{\text{dS}} &= \frac{3GM}{2c^2r^3}\boldsymbol{r} \times \boldsymbol{v}.\end{aligned} \tag{10.64}$$

From the expression for the Lense–Thirring effect, it is evident that the gravitomagnetic field exhibits a striking analogy to the magnetic field generated by a magnetic dipole in electrodynamics. In electrodynamics, a charged particle moves around another particle with a magnetic dipole moment $\boldsymbol{m}$ will experience a precession of its angular momentum. The magnetic field produced by such a magnetic dipole is

$$\boldsymbol{B} = \frac{3\boldsymbol{e}_r(\boldsymbol{e}_r \cdot \boldsymbol{m}) - \boldsymbol{m}}{r^3}, \tag{10.65}$$

where $\boldsymbol{e}_r = \dfrac{\boldsymbol{r}}{r}$ is the unit radial vector. The coupling between this field and the particle's own magnetic moment $\boldsymbol{m}'$ is described by the torque

$$\boldsymbol{\tau} = \boldsymbol{m}' \times \boldsymbol{B}. \tag{10.66}$$

Similarly, in the gravitational case, a spherically symmetric rotating object produces a gravitomagnetic field

$$\boldsymbol{b} = -4G\frac{3\boldsymbol{e}_r(\boldsymbol{e}_r \cdot \boldsymbol{J}) - \boldsymbol{J}}{2r^3}. \tag{10.67}$$

The coupling of this field with a gyroscope's spin $\boldsymbol{s}$ results in a gravitational torque:

$$\boldsymbol{\tau}_{\mathrm{g}} = \frac{\boldsymbol{s}}{2} \times \boldsymbol{b}. \tag{10.68}$$

Both the Lense–Thirring effect and the de Sitter–Fokker effect describe spin precession phenomena in curved spacetime, where the gyroscope follows a geodesic trajectory. However, in quantum mechanics, the electron's spin exhibits Thomas precession, a distinct effect arising from non-geodesic motion. This suggests that if a gyroscope experiences additional interactions causing deviation from geodesic motion, it should similarly display Thomas precession. Consider a gyroscope moving in flat spacetime subject to external interactions, with worldline $x^\mu(\tau)$. In its comoving frame, the spin 4-vector is purely spacelike $\widehat{s} = (0, \boldsymbol{s}(\tau))$, perpendicular to the 4-velocity. The spin evolution obeys

$$\frac{\mathrm{d}s^\mu}{\mathrm{d}\tau} = k u^\mu. \tag{10.69}$$

This is because along the worldline $\widehat{s} \cdot \widehat{s} = \text{constant}$, so its rate of change must be perpendicular to the original spin 4-vector, and thus parallel to the 4-velocity. The coefficient k is determined by the constraint $\widehat{s} \cdot \widehat{u} = 0$, yielding $k = \widehat{s} \cdot \widehat{a}$, i.e.,

$$\frac{\mathrm{d}s^\mu}{\mathrm{d}\tau} = u^\mu (\widehat{s} \cdot \widehat{a}). \tag{10.70}$$

Thomas precession emerges from the non-commutativity of Lorentz transformations—successive boosts induce an additional rotation. In the laboratory reference frame S_{o}, the spin components are

$$s^\mu|_{S_{\mathrm{o}}} = \left(\frac{\gamma}{c} \boldsymbol{v} \cdot \boldsymbol{s}, \boldsymbol{s} + \frac{\gamma^2}{c^2(\gamma+1)} (\boldsymbol{v} \cdot \boldsymbol{s}) \boldsymbol{v} \right), \tag{10.71}$$

with 4-velocity and 4-acceleration

$$\begin{aligned} u^\mu &= \left(\gamma, \frac{\gamma}{c} \boldsymbol{v} \right), \\ a^\mu &= \left(\dot{\gamma}, \frac{\dot{\gamma}}{c} \boldsymbol{v} + \frac{\gamma}{c} \dot{\boldsymbol{v}} \right). \end{aligned} \tag{10.72}$$

The spin-acceleration coupling gives

$$\widehat{s} \cdot \widehat{a} = \frac{\gamma}{c} \left(\dot{\boldsymbol{v}} \cdot \boldsymbol{s} + \frac{\gamma^2}{c^2(\gamma+1)} (\dot{\boldsymbol{v}} \cdot \boldsymbol{v})(\boldsymbol{v} \cdot \boldsymbol{s}) \right). \tag{10.73}$$

After calculation, we obtain the precession equation

$$\frac{\mathrm{d}\boldsymbol{s}}{\mathrm{d}\tau} = \boldsymbol{\Omega}_{\mathrm{T}} \times \boldsymbol{s}, \tag{10.74}$$

with Thomas frequency

$$\boldsymbol{\Omega}_{\mathrm{T}} = \frac{\gamma^2}{\gamma + 1}\frac{\boldsymbol{v} \times \boldsymbol{a}}{c^2}. \tag{10.75}$$

To leading order, this simplifies to

$$\boldsymbol{\Omega}_{\mathrm{T}} \approx \frac{1}{2c^2}(\boldsymbol{v} \times \boldsymbol{a}) = \frac{1}{2mc^2}(\boldsymbol{v} \times \boldsymbol{F}). \tag{10.76}$$

Note that this discussion concerns the precession of a gyroscope in flat spacetime within the framework of special relativity, where the force $\boldsymbol{F}$ it experiences can be the usual gravitational force or any other force. If it is the gravitational force and the source of gravity is spherically symmetric, then

$$\boldsymbol{F}_{\mathrm{g}} = -\frac{GMm}{r^3}\boldsymbol{r}, \tag{10.77}$$

which can be compared with the de Sitter–Fokker precession discussed earlier, but the numerical values differ by a factor of three. This discrepancy arises because Thomas precession is purely a special relativistic effect, while the de Sitter–Fokker precession incorporates general relativistic curvature effects.

The evolution of a gyroscope's spin vector along its worldline is universally described by Eq. (10.70), which can be expressed covariantly as

$$\frac{\mathrm{D}\widehat{\mathrm{s}}}{\mathrm{d}\tau} = \widehat{u}(\widehat{s} \cdot \widehat{a}). \tag{10.78}$$

This fundamental equation governs spin dynamics in both flat and curved spacetime. When the gyroscope experiences no external interactions, $\widehat{a} = 0$, we recover the standard gyroscope equation of parallel transport. This case describes pure gravitational precession effects in curved spacetime. When additional interactions are present, the right-hand side of (10.78) becomes non-zero, leading to Thomas precession. This demonstrates that any non-gravitational acceleration induces additional spin evolution. The complete precession dynamics can be summarized by

$$\frac{\mathrm{d}\boldsymbol{s}}{\mathrm{d}t} = \boldsymbol{\Omega} \times \boldsymbol{s}, \tag{10.79}$$

where the total precession frequency combines all contributing effects

$$\boldsymbol{\Omega} = \boldsymbol{\Omega}_{\mathrm{T}} + \boldsymbol{\Omega}_{\mathrm{dS}} + \boldsymbol{\Omega}_{\mathrm{LT}}, \tag{10.80}$$

where $\boldsymbol{\Omega}_{\mathrm{T}}$ represents Thomas precession from non-gravitational accelerations, $\boldsymbol{\Omega}_{\mathrm{dS}}$ corresponds to geodetic (de Sitter–Fokker) precession, and $\boldsymbol{\Omega}_{\mathrm{LT}}$ accounts for frame-dragging (Lense–Thirring) effects.

The separation of de Sitter precession from Lense–Thirring precession in experimental measurements relies on their distinct characteristic behaviors. The precession frequency $\boldsymbol{\Omega}_{\mathrm{dS}}$ depends solely on orbital radius and is independent of Earth's rotation

$$\boldsymbol{\Omega}_{\mathrm{dS}} = \frac{3GM}{2c^2r^3}\boldsymbol{r} \times \boldsymbol{v} \tag{10.81}$$

For circular orbits, $\boldsymbol{\Omega}_{\mathrm{dS}}$ remains constant in magnitude regardless of orbital inclination (equatorial, polar, or inclined). Defining the orbital angular direction $\boldsymbol{h}$, unit vector normal to the orbital plane, via $\boldsymbol{r} \times \boldsymbol{v} = |\boldsymbol{r} \times \boldsymbol{v}|\boldsymbol{h}$. The de Sitter precession then satisfies $\boldsymbol{\Omega}_{\mathrm{dS}} \propto \boldsymbol{h}$ and $\boldsymbol{\delta s}_{\mathrm{dS}} \propto \boldsymbol{h} \times \boldsymbol{s}$. Maximum observable effect occurs when the spin vector $\boldsymbol{s}$ lies entirely within the orbital plane. Unlike the de Sitter effect, $\boldsymbol{\Omega}_{\mathrm{LT}}$ depends explicitly on Earth's angular momentum vector $\boldsymbol{J}$, and the frame-dragging effect varies with orbital inclination relative to Earth's rotation axis.

If the circular orbit lies exactly in the equatorial plane and the spin vector also resides in this plane, then $\boldsymbol{s} \perp \boldsymbol{\omega}$, meaning the gyroscope's spin vector is perpendicular to the Earth's spin vector. Additionally, $\boldsymbol{h}//\boldsymbol{\omega}$. Consequently,

$$\boldsymbol{\omega} \cdot \boldsymbol{r} = 0 \Rightarrow \boldsymbol{\Omega}_{\mathrm{LT}} \propto \boldsymbol{\omega}, \quad \boldsymbol{\delta s}_{\mathrm{LT}} \propto \boldsymbol{\omega} \times \boldsymbol{s}. \tag{10.82}$$

Here $\boldsymbol{\delta s}_{\mathrm{LT}}//\boldsymbol{\delta s}_{\mathrm{dS}}$, indicating that the two precessions superimpose and cannot be distinguished experimentally.

For a circular orbit passing through the poles, the Lense–Thirring precession comprises two contributions:

$$\begin{aligned} \boldsymbol{\delta s}_{\mathrm{LT}}(\mathrm{i}) &\propto \boldsymbol{r} \times \boldsymbol{s}, \\ \boldsymbol{\delta s}_{\mathrm{LT}}(\mathrm{ii}) &\propto \boldsymbol{\omega} \times \boldsymbol{s}. \end{aligned} \tag{10.83}$$

To maximize $\boldsymbol{\delta s}_{\mathrm{LT}}(\mathrm{ii})$, we make $\boldsymbol{s}$ perpendicular to the Earth's spin vector, keeping it fixed in direction, parallel to the equatorial plane, as the gyroscope traverses the poles. Under this configuration, $\boldsymbol{\delta s}_{\mathrm{LT}}(\mathrm{ii})//\boldsymbol{h}$, causing the gyroscope's spin vector to deviate from the orbital plane. In contrast, $\boldsymbol{\delta s}_{\mathrm{dS}} \sim \boldsymbol{s} \times \boldsymbol{h}$, meaning the geodetic precession confines the spin vector's evolution to the orbital plane. Thus, the two precessions produce distinct and separable effects, enabling independent measurement.

In fact, since $\boldsymbol{\omega} \cdot \boldsymbol{r}$ varies continuously along the orbit in the Lense–Thirring precession, we compute its average over one orbital period:

$$\langle \boldsymbol{\Omega}_{\mathrm{LT}} \rangle = \frac{GI}{cr^3}\left\langle \frac{3(\boldsymbol{\omega} \cdot \boldsymbol{r})\boldsymbol{r}}{r^2} - \boldsymbol{\omega} \right\rangle. \tag{10.84}$$

Assuming the Earth's rotation vector aligns with z-direction, $\boldsymbol{\omega} = \omega \boldsymbol{k}$, we find

$$\langle \boldsymbol{r}(\boldsymbol{\omega} \cdot \boldsymbol{r}) \rangle = \frac{\omega r^2}{2} \boldsymbol{k}. \tag{10.85}$$

Substituting the Earth's moment of inertia $I = \frac{2}{5} M R^2$, we obtain

$$\langle \boldsymbol{\Omega}_{\rm LT} \rangle = \frac{G M R^2 \omega}{5 c^2 r^3} = 0.065'' \left(\frac{R}{r} \right)^3 \Big/ \text{year}. \tag{10.86}$$

For a satellite at an altitude of 650 km, this yields

$$\langle \boldsymbol{\Omega}_{\rm LT} \rangle = 0.048''/\text{year}. \tag{10.87}$$

The Lense–Thirring precession arises from frame dragging, where a rotating massive object (e.g., the Earth) induces a slow rotation of the surrounding inertial frame. This effect can be visualized as the inertial frame behaving like a fluid dragged by the rotating body. The precession rate varies depending on the orbital configuration:

$$\begin{aligned} (\boldsymbol{\Omega}_{\rm LT})_{\rm pole} &= \frac{2GI}{c^2 r^3} \boldsymbol{\omega}, \\ (\boldsymbol{\Omega}_{\rm LT})_{\rm eq} &= -\frac{GI}{c^2 r^3} \boldsymbol{\omega}. \end{aligned} \tag{10.88}$$

At the Earth's surface $r = R_\oplus$, these expressions reduce to

$$\begin{aligned} (\boldsymbol{\Omega}_{\rm LT})_{\rm pole} &= \frac{4GM}{5c^2 R_\oplus} \boldsymbol{\omega} = 5.52 \times 10^{-10} \boldsymbol{\omega}, \\ (\boldsymbol{\Omega}_{\rm LT})_{\rm eq} &= -\frac{2GM}{5c^2 R_\oplus} \boldsymbol{\omega} = -2.76 \times 10^{-10} \boldsymbol{\omega}. \end{aligned} \tag{10.89}$$

The gyroscope precession effects have been tested using satellite-based experiments, with a long history of missions designed to probe general relativity. From 1964 to 2004, a collaboration between NASA and Stanford University, Gravity Probe B aimed to measure the de Sitter–Fokker effect and the Lense–Thirring effect. After meticulous data analysis, the final results was published in 2013, obtaining $\delta\phi_{\rm dS} = (6.6018 \pm 0.018)''/\text{year}$, with a deviation of 0.28%, and $\delta\phi_{\rm LT} = (0.0372 \pm 0.0072)''/\text{year}$, with a deviation of 19%. Other experimental results regarding gyroscope precession are as follows:

(1) Lunar laser ranging experiments: Measured the de Sitter geodetic precession with a precision of 0.7%.

(2) Laser ranging of the LAGEOS and LAGEOS II satellites: Detected the Lense–Thirring frame-dragging effect with deviations ranging from 10% to 30%.

Figure 10.1 illustrates the Gravity Probe B experiment, including theoretical predictions of the precession rates.

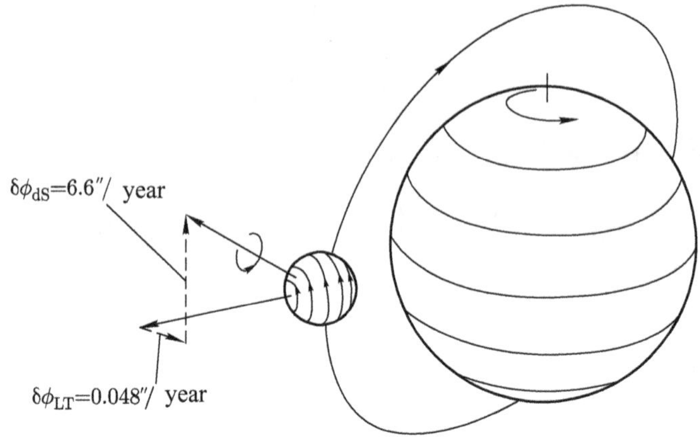

Fig. 10.1 Various spin precession phenomena can be examined using the gyroscope on the satellite. In order for a spherical gyroscope to be discussed approximately as a particle, it must be highly spherical, with deviations from perfect sphericity limited to less than one part per million.

10.2.4 *Lense–Thirring effect and the equivalence principle*

The Lense–Thirring precession arises from the influence of a rotating object on the spacetime metric, specifically due to the gravitomagnetic field generated by its rotation. An interesting question is whether this phenomenon can be interpreted within the framework of the equivalence principle. Consider a particle in an inertial frame with velocity $\boldsymbol{v}'$ and acceleration $\boldsymbol{a}'$. When transitioning to a reference frame rotating with angular velocity ω, and neglecting centrifugal effects, the observed acceleration becomes

$$\boldsymbol{a} = \boldsymbol{a}' - 2\boldsymbol{v}' \times \boldsymbol{\omega}, \tag{10.90}$$

where the second term represents the Coriolis force. Comparing this with the geodesic equation, we have

$$a^i = \frac{\mathrm{d}^2 x^i}{\mathrm{d}t^2} = -\nabla_i \Phi_\mathrm{g} - c\frac{\partial A^i}{\partial t} - c(A_{i,k} - A_{k,i})v^k, \tag{10.91}$$

or equivalently,

$$\boldsymbol{a} = \boldsymbol{g} + c\boldsymbol{v} \times (\nabla \times \boldsymbol{A}), \tag{10.92}$$

where $\boldsymbol{g} = -\nabla\Phi_\mathrm{g} - c\dfrac{\partial \boldsymbol{A}}{\partial t}$. If we identify $\nabla \times \boldsymbol{A} = -2\boldsymbol{\omega}$, then the two descriptions are consistent. This implies that $\nabla \times \boldsymbol{A}$ corresponds to the frame-dragging effect induced by the rotating reference frame

$$\begin{aligned}\boldsymbol{\omega} &= -\frac{1}{2}\nabla\times\boldsymbol{A} \\ &= \frac{G}{c^3}\nabla\times\left(\frac{\boldsymbol{J}\times\boldsymbol{r}}{r^3}\right) \\ &= \frac{G}{c^3 r^5}(3\boldsymbol{r}(\boldsymbol{J}\cdot\boldsymbol{r}) - r^2\boldsymbol{J}) \\ &= \boldsymbol{\Omega}_{\mathrm{LT}}.\end{aligned} \tag{10.93}$$

Thus, the rotational effects of massive objects can indeed be interpreted as manifestations of a rotating reference frame. We may compare this with Mach's principle. Suppose an observer resides in a rotating frame within flat spacetime, while the surrounding matter rotates relative to them. In this frame, fictitious forces such as the Coriolis and centrifugal forces emerge—artifacts of the non-inertial perspective. Conversely, if the observer is inertial and the rest of the universe rotates with an opposing angular velocity, could the collective motion of distant matter produce analogous inertial effects? Affirmative answers would suggest a relativistic notion of inertia, where accelerating a reference frame is dynamically equivalent to imparting an opposite acceleration to the universe. While such a correspondence holds in the linearized approximation of general relativity, it notably fails in the full nonlinear theory.

10.3 Kerr Geometry

In the preceding analysis, we restricted our discussion to slowly rotating objects, where perturbations to the spacetime metric remain small. However, astrophysical systems often involve rapidly rotating bodies (e.g., neutron stars, Kerr black holes), necessitating a more rigorous treatment of rotational effects on spacetime geometry. When rotation is present, spacetime is no longer static, but it may remain stationary if the rotation is steady. In such cases, the spacetime admits a timelike Killing vector ∂_t, reflecting time-translation symmetry. Additionally, if the system is axisymmetric, e.g., rotation about the z-axis, there exists a spacelike Killing vector ∂_ϕ associated with rotational invariance. Note that for spherically symmetric spacetime, in addition to the timelike Killing vector, the rotational symmetry form a group $SO(3)$. For axisymmetric rotating spacetimes, this reduces to $U(1)$, preserving only azimuthal symmetry. Using these Killing symmetries, the metric components $g_{\mu\nu}$ depend solely on the remaining two coordinates (x_1, x_2), i.e., $g_{\mu\nu}(x^1, x^2)$. Further imposing invariance under simultaneous time reversal $t \to -t$ and rotation reversal $\phi \to -\phi$ eliminates certain off-diagonal metric terms:

$$g_{01} = g_{02} = g_{13} = g_{23} = 0. \tag{10.94}$$

The line element then simplifies to

$$\mathrm{d}s^2 = g_{00}\mathrm{d}t^2 + 2g_{03}\mathrm{d}t\mathrm{d}\phi + g_{33}\mathrm{d}\phi^2 + [g_{11}(\mathrm{d}x^1)^2 + 2g_{12}\mathrm{d}x^1\mathrm{d}x^2 + g_{22}(\mathrm{d}x^2)^2].$$

In this line element, the part in the square brackets describe a two-dimensional Riemannian manifold, which is always conformally flat and its metric can be rewritten as $g_{ab} = \Omega^2(x)\eta_{ab}$ in terms of appropriate coordinates. Adopting polar coordinates (r, θ) and regrouping terms, the metric can be expressed in a more intuitive form

$$\mathrm{d}s^2 = -A\mathrm{d}t^2 + B(\mathrm{d}\phi - \omega\mathrm{d}t)^2 + C\mathrm{d}r^2 + D\mathrm{d}\theta^2, \tag{10.95}$$

where A, B, C, D, and ω are functions of r and θ. The coordinates (t, r, θ, ϕ) are defined relative to a distant inertial observer. This line element may describe the spacetime exterior to a stationary rotating object. The functions A, B, C and D describe the warping of spacetime due to mass and rotation, and ω encodes the frame-dragging effect, representing the angular velocity of inertial frames induced by the source's rotation.

In this spacetime, consider photons emitted in the $\pm\phi$ directions at a fixed coordinates (r, θ). The null paths of these photons satisfy the equation

$$\begin{aligned} &\mathrm{d}s^2 = 0 \Rightarrow g_{tt}\mathrm{d}t^2 + 2g_{t\phi}\mathrm{d}t\mathrm{d}\phi + g_{\phi\phi}\mathrm{d}\phi^2 = 0 \\ &\Rightarrow \frac{\mathrm{d}\phi}{\mathrm{d}t} = -\frac{g_{t\phi}}{g_{\phi\phi}} \pm \left(\left(\frac{g_{t\phi}}{g_{\phi\phi}}\right)^2 - \frac{g_{tt}}{g_{\phi\phi}}\right)^{1/2}. \end{aligned} \tag{10.96}$$

When $g_{tt} < 0$, the derivative $\dfrac{\mathrm{d}\phi}{\mathrm{d}t}$ take either positive or negative values, indicating that photons may propagate freely in either the $+\phi$ or $-\phi$ direction, either aligned with or opposed to the spacetime's rotation. However, if $g_{tt} > 0$, $\dfrac{\mathrm{d}\phi}{\mathrm{d}t}$ has the same sign as $-g_{t\phi}$, implying that photon motion is constrained to follow the direction of spacetime rotation. At the surface defined by $g_{tt} = 0$, we obtain

$$\frac{\mathrm{d}\phi}{\mathrm{d}t} = \begin{cases} -\dfrac{2g_{t\phi}}{g_{\phi\phi}} = 2\omega, \\ 0. \end{cases} \tag{10.97}$$

From Eq. (10.97), it can be seen that for an observer at infinity, photons emitted in the direction of spacetime rotation exhibit an enhanced higher angular velocity, while those emitted in the opposite direction appear to propagate radially at a fixed angle. The surface defined by $g_{tt} = 0$ is termed the stationary limit surface. Within the stationary limit surface, $g_{tt} > 0$, no particle can remain static at a fixed (r, θ, ϕ) position; instead, it must co-rotate with the spacetime. This can also be seen by considering the motion of a massive particle. Suppose such a particle is stationary, with a 4-velocity $u^\mu = (u^t, 0, 0, 0)$. Its normalization condition,

$$\widehat{u} \cdot \widehat{u} = -1, \tag{10.98}$$

cannot be satisfied when $g_{tt} > 0$, further justifying the name stationary limit surface.

Moreover, the surface defined by $g_{tt} = 0$ also constitutes an infinite redshift surface. As previously discussed, for a photon emitted by a static transmitter at point A and received by a static receiver at point B, the frequency shift is given by

$$\frac{\nu_{\mathrm{R}}}{\nu_{\mathrm{E}}} = \left(\frac{g_{tt}(A)}{g_{tt}(B)}\right)^{1/2}. \tag{10.99}$$

As $g_{tt}(A) \to 0$, $\nu_{\mathrm{R}} \to 0$, indicating that the received photon undergoes infinite redshift. In Schwarzschild spacetime, the condition $g_{tt} = 0$ gives $r = R_{\mathrm{s}}$, which coincides exactly with the event horizon. However, in stationary spacetime, the infinite redshift surface generally differs from the event horizon.

The defining property of an event horizon is that it must be a null surface, a hypersurface whose normal vector at every point is null. To formalize this, consider a surface defined by the equation $f(x^\mu) = 0$. The corresponding normal covector is given by $n_\mu = \nabla_\mu f = \partial_\mu f$. For this surface to be null, its normal must satisfy $g^{\mu\nu} n_\mu n_\nu = 0$. Since a particle or photon pointing to the future can only cross a 3-dimensional null surface in one direction, this null surface may constitute an event horizon. In a stationary spacetime, the null surface condition simplifies significantly. If we assume the surface depends only on r and θ, i.e., $f(r, \theta) = 0$, then the null surface condition becomes

$$\begin{aligned} & g^{\mu\nu} \partial_\mu f \partial_\nu f = 0 \\ \Rightarrow\ & g^{rr} (\partial_r f)^2 + g^{\theta\theta} (\partial_\theta f)^2 = 0. \end{aligned} \tag{10.100}$$

Choosing coordinates such that $f(r, \theta) = f(r)$, the null condition reduces to $g^{rr} (\partial_r f)^2 = 0$. Since $\partial_r f \neq 0$, otherwise the surface would be degenerate, the event horizon is determined by $g^{rr} = 0$ or equivalently $g_{rr} = \infty$. In the

Schwarzschild metric, this condition yields the well-known Schwarzschild radius $r = R_s$.

The metric (10.95) is not required to satisfy the Einstein equations but is instead restricted by symmetry considerations. If we further impose the vacuum Einstein equations $R_{\mu\nu} = 0$, we find that the equations alone are insufficient to uniquely determine all metric coefficients, a key distinction from the spherically symmetric case. This arises because axial symmetry, being significantly weaker than spherical symmetry, does not provide enough constraints to fully fix the metric. For example, a cosmic string with axial symmetry also satisfies the Einstein equations. Consequently, additional physical conditions must be introduced:

(1) Asymptotic flatness. The spacetime must approaches Minkowski metric as $r \to \infty$;

(2) Regularity of the horizon. There exists a smooth, convex event horizon, with a nonsingular geometry outside it.

Under these two requirements, it can be rigorously proven that the axisymmetric spacetime is uniquely described by the Kerr geometry.

Carter–Robinson theorem If (M, g) is an asymptotically flat, stationary, and axisymmetric vacuum spacetime that is non-singular on and outside an event horizon, then (M, g) belongs to the Kerr family of solutions, characterized by just two parameters: mass and angular momentum.

Notably, the assumption of axial symmetry in this theorem is actually redundant. Subsequent work by S. Hawking and R. Wald demonstrated that stationarity alone, under reasonable physical conditions, implies axial symmetry for black holes. Since stationarity corresponds to an equilibrium configuration, we may expect that the end state of gravitational collapse settles into a stationary spacetime. The uniqueness theorem thus implies that if gravitational collapse results in a black hole, its final state is entirely determined by its mass and angular momentum.[2] This result has profound implications: in the process of black hole formation, all higher-order multipole moments beyond the monopole and dipole are radiated away. The monopole moment corresponds to the total mass, while the dipole moment associated with angular momentum persists due to the spin-2 nature of gravitational radiation.

[2]When electromagnetic interactions are included, the uniqueness theorem generalizes further, yielding the Kerr–Newman solution, which is characterized by mass, angular momentum, and electric (or magnetic) charge.

In the Boyer–Lindquist coordinates, the line element of the Kerr spacetime geometry takes the form

$$\begin{aligned} ds^2 = &-dt^2 + \frac{\rho^2}{\Delta}dr^2 + \rho^2 d\theta^2 + (r^2+a^2)\sin^2\theta d\phi^2 \\ &+ \frac{2GMr}{\rho^2}(a\sin^2\theta d\phi - dt)^2, \end{aligned} \tag{10.101}$$

where

$$\Delta(r) = r^2 - 2GMr + a^2, \quad \rho^2(r,\theta) = r^2 + a^2\cos^2\theta. \tag{10.102}$$

The parameters in the metric (10.101) have the following physical interpretations: M represents the mass of the rotating object, and a is the Kerr parameter, defined as

$$a = J/M, \tag{10.103}$$

characterizing the angular momentum per unit mass and thus the rotation rate.[3] By introducing the quantity $\Sigma^2 \equiv (r^2+a^2)^2 - a^2\Delta\sin^2\theta$, the metric (10.101) can be rewritten in an alternative form

$$ds^2 = -\frac{\rho^2\Delta}{\Sigma^2}dt^2 + \frac{\Sigma^2\sin^2\theta}{\rho^2}(d\phi - \omega dt)^2 + \frac{\rho^2}{\Delta}dr^2 + \rho^2 d\theta^2, \tag{10.104}$$

where $\omega = 2GMra/\Sigma^2$. This representation explicitly reveals the rotational nature of the spacetime around the ϕ-axis. The corresponding components of the inverse metric are given by

$$g^{rr} = \frac{\Delta}{\rho^2}, \quad g^{\theta\theta} = \frac{1}{\rho^2}, \quad g^{tt} = -\frac{\Sigma^2}{\rho^2\Delta},$$

$$g^{\phi t} = -\frac{2GMar}{\rho^2\Delta}, \quad g^{\phi\phi} = -\frac{a^2\sin^2\theta - \Delta}{\rho^2\Delta\sin^2\theta}.$$

Replacing Δ in (10.101) with

$$\Delta(r) = r^2 - 2GMr + a^2 + e^2, \tag{10.105}$$

where

$$e = \sqrt{Q^2+P^2}, \tag{10.106}$$

and Q and P denote the electric and magnetic charges, respectively, yields the line element describing the Kerr–Newman spacetime. For this solution, the Maxwell 1-form field is

$$A = \frac{Qr(dt - a\sin^2\theta d\phi) - P\cos\theta[adt - (r^2+a^2)d\phi]}{\rho}. \tag{10.107}$$

[3]This vacuum solution was first discovered by Roy Kerr in 1963, originally expressed in the so-called Kerr–Schild coordinates.

The Kerr–Newman solution is a three-parameter family, characterized by M, J, and e. In the limit $a \to 0$, it reduces to the Reissner–Nordström solution. Similar to the Kerr solution, it exhibits a discrete symmetry

$$t \to -t, \quad \phi \to -\phi. \tag{10.108}$$

The transformation $\phi \to -\phi$ is equivalent to reversing the sign of a, thereby changing the direction of the black hole's rotation. Without loss of generality, we may assume $a > 0$. Since astrophysical objects are typically electrically neutral, the Kerr solution holds greater physical significance. In the subsequent discussion, we focus on the properties of the Kerr spacetime. It should be noted that, unlike spherically symmetric spacetime, no axisymmetric interior solution for a star has yet been found that smoothly matches the exterior Kerr vacuum solution of Einstein's equations.

From the explicit form of the line element, the two Killing vectors of the spacetime can be immediately identified as the generators of time translation and axial symmetry

$$\widehat{\xi} = \partial_t, \quad \widehat{\eta} = \partial_\phi. \tag{10.109}$$

The vector ∂_t is not orthogonal to any hypersurface of constant t. This confirms that the Kerr metric is stationary but not static. Beyond the Killing vectors, the Kerr spacetime also admits Killing tensors, which satisfy $\nabla_{(\sigma}\xi_{\mu_1\cdots\mu_n)} = 0$. While the simplest Killing tensors are the metric tensor and tensor product of Killing vectors, the Kerr geometry possesses a non-trivial rank-2 Killing tensor. This $(0,2)$-type Killing tensor is given by

$$\xi_{\mu\nu} = 2\rho^2 l_{(\mu}n_{\nu)} + r^2 g_{\mu\nu}, \tag{10.110}$$

where l_μ and n_μ are null vectors defined as

$$\begin{aligned} l^\mu &= \frac{1}{\Delta}(r^2 + a^2, \Delta, 0, a), \\ n^\mu &= \frac{1}{2\rho^2}(r^2 + a^2, -\Delta, 0, a), \end{aligned} \tag{10.111}$$

normalized such that $l^\mu n_\mu = -1$. These null vectors form part of the Newman–Penrose tetrad, a powerful framework for analyzing gravitational radiation, geodesic motion, and other physical properties of the spacetime. The tetrad simplifies calculations in problems involving curvature invariants, Petrov classification, and integrability conditions.

We now examine the behavior of the Kerr metric in different limits. First, in the slow rotation limit, we can retain only the first-order terms in a, obtaining the metric

$$\mathrm{d}s^2 = \mathrm{d}s^2_{\mathrm{Sch}} - \frac{4GJ}{r}\sin^2\theta \mathrm{d}\phi \mathrm{d}t. \tag{10.112}$$

This is precisely the metric used earlier to discuss the Lense–Thirring precession, where rotational deformation is negligible, and the central object can be approximated as spherical. This metric is very useful in dealing with many astrophysical problems. If relativistic effects are small, or that gravity is weak, the metric reduces to

$$\mathrm{d}s^2 = -(1 - 2GM/r)\mathrm{d}t^2 + (1 + 2GM/r)(\mathrm{d}r^2 + r^2\mathrm{d}\Omega^2) - \frac{4GJ}{r}\sin^2\theta\mathrm{d}\phi\mathrm{d}t,$$

or equivalently, in Cartesian coordinates

$$\begin{aligned}\mathrm{d}s^2 &= -(1 - R_\mathrm{s}/r)\mathrm{d}t^2 + (1 + R_\mathrm{s}/r)(\mathrm{d}x^2 + \mathrm{d}y^2 + \mathrm{d}z^2)\\ &\quad - \frac{4GJ}{r^3}(x\mathrm{d}y - y\mathrm{d}x)\mathrm{d}t,\end{aligned} \tag{10.113}$$

where $r = \sqrt{x^2 + y^2 + z^2}$.

We next investigate the behavior of Kerr spacetime in the limit $M \to 0$. To analyze this limit, we adopt the Kerr–Schild coordinates, defined by

$$x + \mathrm{i}y = (r + \mathrm{i}a)\sin\theta\exp\left[\mathrm{i}\int\left(\mathrm{d}\phi + \frac{a}{\Delta}\mathrm{d}r\right)\right], \tag{10.114}$$

$$z = r\cos\theta, \tag{10.115}$$

$$\tilde{t} = \int\left(\mathrm{d}t + \frac{r^2 + a^2}{\Delta}\mathrm{d}r\right) - r. \tag{10.116}$$

Here, the radial function $r = r(x, y, z)$ satisfies

$$r^4 - (x^2 + y^2 + z^2 - a^2)r^2 - a^2z^2 = 0. \tag{10.117}$$

In these coordinates, the metric becomes

$$\begin{aligned}\mathrm{d}s^2 &= -\mathrm{d}\tilde{t}^2 + \mathrm{d}x^2 + \mathrm{d}y^2 + \mathrm{d}z^2\\ &\quad + \frac{2Mr^3}{r^4 + a^2z^2}\left[\frac{r(x\mathrm{d}x + y\mathrm{d}y) - a(x\mathrm{d}y - y\mathrm{d}x)}{r^2 + a^2} + \frac{z\mathrm{d}z}{r} + \mathrm{d}\tilde{t}\right]^2.\end{aligned} \tag{10.118}$$

As expected, for $M \to 0$, spacetime reduces to Minkowski spacetime. When constant r, the coordinates x, y, z relates to r, θ, ϕ via

$$\begin{aligned}x &= (r^2 + a^2)^{1/2}\sin\theta\cos\phi,\\ y &= (r^2 + a^2)^{1/2}\sin\theta\sin\phi,\\ z &= r\cos\theta.\end{aligned} \tag{10.119}$$

The resulting hypersurface satisfies

$$\frac{x^2 + y^2}{r^2 + a^2} + \frac{z^2}{r^2} = 1, \tag{10.120}$$

describing an ellipsoid rotating about the z-axis. At $r = 0$, we have $z = 0$, i.e., the equatorial plane, and

$$x^2 + y^2 = a^2 \sin^2 \theta. \tag{10.121}$$

This represents a disk (not a circle) of radius a, as θ varies. For constant θ, the hypersurface becomes a hyperboloid

$$\frac{x^2 + y^2}{a^2 \sin^2 \theta} - \frac{z^2}{a^2 \cos^2 \theta} = 1. \tag{10.122}$$

Similar to the Schwarzschild solution, the Kerr metric describes the spacetime geometry of a rotating black hole. It turns out that an intrinsic singularity occurs when $\rho = 0$, where

$$\rho^2 = r^2 + a^2 \cos^2 \theta = 0. \tag{10.123}$$

This requires both

$$r = 0, \quad \theta = \frac{\pi}{2}. \tag{10.124}$$

As discussed earlier, $r = 0$ corresponds to a disk of coordinate radius a in the equatorial plane, and $\theta = \pi/2$ is the edge of this disk. Thus, the singularity is not a point but rather a ring lying entirely in the equatorial plane, defined by: $x^2 + y^2 = a^2$, $z = 0$, as shown in Figure 10.2.

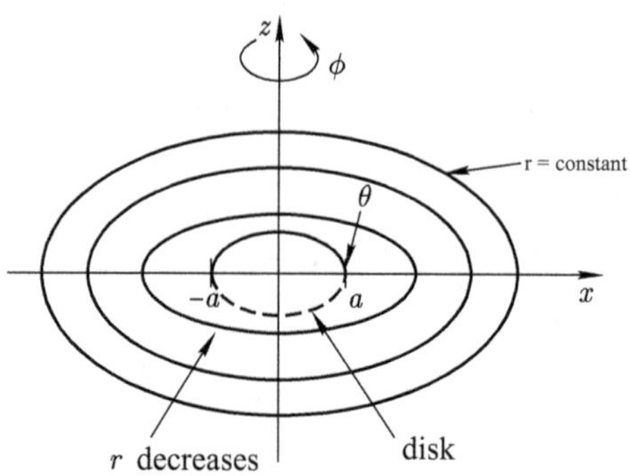

Fig. 10.2 The singularity of Kerr spacetime is a ring.

10.3.1 *Event horizon*

The event horizon of a black hole is determined by the condition $g^{rr} = 0$ or, equivalently, $g_{rr} = -\rho^2/\Delta = \infty$. Solving $\Delta = 0$, we obtain the radii of the event horizons:

$$r_{\pm} = \mu \pm (\mu^2 - a^2)^{1/2}, \tag{10.125}$$

where $\mu = GM$. Thus, the Kerr black hole possesses two distinct event horizons, $r_\pm$. At the outer horizon $r = r_+$, there exists a null vector

$$l^\alpha = (1, 0, 0, \Omega_\mathrm{H}), \tag{10.126}$$

where Ω_H is the angular velocity of the black hole as measured by a distant observer at infinity. This angular velocity is given by

$$\Omega_\mathrm{H} = \omega|_{r=r_+} = \frac{a}{2\mu r_+}. \tag{10.127}$$

The null vector $\widehat{l}$ corresponds to the tangent vector of null geodesics that generate the event horizon. From the geodesic equation, it follows that light rays propagating along these directions remain confined to the event horizon. These light rays, which generate the event horizon, rotate with an angular velocity Ω_H relative to a static observer at infinity.

Consider the metric at constant t and $r = r_\pm$. Under these conditions, the metric simplifies to

$$\mathrm{d}\sigma^2 = \rho_\pm^2 \mathrm{d}\theta^2 + \left(\frac{2\mu r_\pm}{\rho_\pm}\right)^2 \sin^2\theta \mathrm{d}\phi^2. \tag{10.128}$$

This geometry does not describe a perfect sphere but rather resembles an axisymmetric ellipsoid flattened along the rotation axis. Intuitively, this distortion can be attributed to centrifugal forces induced by the black hole's rotation. The presence of the outer horizon r_+ confirms the existence of a black hole, with a horizon area given by

$$A = 8\pi\mu r_+ = 8\pi\mu(\mu + \sqrt{\mu^2 - a^2}). \tag{10.129}$$

The spacetime can be partitioned into three distinct regions based on the horizons $r_\pm$:

$$\mathrm{I} : r_+ < r < \infty, \quad \mathrm{II} : r_- < r < r_+, \quad \mathrm{III} : 0 < r < r_-. \tag{10.130}$$

However, not all parameter combinations (μ, a) corresponds to a black hole. Event horizons exist only when the parameters satisfy the following condition:

$$a^2 < \mu^2 \Rightarrow \left(\frac{J}{M}\right)^2 < (\mu)^2. \tag{10.131}$$

For $a^2 < \mu^2$, the singularity at $\rho = 0$ is concealed within the horizon, consistent with the cosmic censorship hypothesis. In the extremal case $a^2 = \mu^2$, the horizons coincide $r_+ = r_- = \mu$, resulting in an extremal black hole. Notably, astrophysical observations have identified near-extremal black holes, with $a \approx 0.998\mu$. For $a^2 > \mu^2$, no event horizons exist, exposing a naked singularity, a scenario analogous to that of the Reissner–Nordström black hole.

For the Kerr black hole, the infinite redshift surface is determined by the condition

$$\begin{aligned} g_{tt} = 0 &\Rightarrow r^2 - 2\mu r + a^2\cos^2\theta = 0 \\ &\Rightarrow r_{S\pm} = \mu \pm (\mu^2 - a^2\cos^2\theta)^{1/2}. \end{aligned} \tag{10.132}$$

This redshift surface is axisymmetric. When $r = r_{S\pm}$ and t is held constant, the induced metric takes the form

$$\mathrm{d}\sigma^2 = \rho_{S\pm}^2\mathrm{d}\theta^2 \left(\frac{2\mu r_{S\pm}(2\mu r_{S\pm} + 2a^2\sin^2\theta)}{\rho_{S\pm}^2}\right) sin^2\theta\mathrm{d}\phi^2.$$

Embedding this metric in three-dimensional Euclidean space reveals an axisymmetric ellipsoid flattened along the rotation axis. In the limit $a \to 0$, the Schwarzschild solution is recovered:

$$r_{S+} \to r = R_\mathrm{s}, \quad r_{S-} \to 0. \tag{10.133}$$

Finally, let us discuss the Killing horizon briefly. The Killing horizon is a null hypersurface orthogonal to the Killing vector field. In Schwarzschild spacetime, the "timelike" Killing vector $\xi^\mu = \partial_t$ becomes null at the event horizon and spacelike inside it, causing the Killing horizon to coincide with the event horizon. In Kerr spacetime, however, the norm of the "timelike" Killing vector $\xi^\mu = \partial_t$ is given by $\xi^\mu\xi_\mu = -\frac{1}{\rho^2}(\Delta - a^2\sin^2\theta)$. At the outer event horizon, this becomes $\xi^\mu\xi_\mu = \frac{a^2}{\rho^2}\sin^2\theta \geq 0$, indicating that the Killing vector is spacelike everywhere on the outer horizon except at the poles, where it is null. The Killing horizon associated with $\widehat{\xi}$ is defined by $\xi^\mu\xi_\mu = 0$, leading to

$$(r-\mu)^2 = \mu^2 - a^2\cos^2\theta, \tag{10.134}$$

which coincides precisely with the infinite redshift surface $r_{S\pm}$. Thus, the stationary limit or infinite redshift surface constitutes a Killing horizon. This does not, however, imply that the event horizon of the Kerr black hole fails to be a Killing horizon. In fact, the outer event horizon is also a Killing horizon, but its associated Killing vector is $\partial_t + \Omega_\mathrm{H}\partial_\phi$.

To examine the Kerr black hole's behavior near the horizons $r = r_\pm$, we introduce the Kerr coordinates $\widetilde{u}$ and χ, analogous to advanced Eddington–Finkelstein coordinates:

$$\mathrm{d}\widetilde{u} = \mathrm{d}t + \frac{(r^2 - a^2)}{\Delta}\mathrm{d}r, \tag{10.135}$$

$$\mathrm{d}\chi = \mathrm{d}\phi + \frac{a}{\Delta}\mathrm{d}r. \tag{10.136}$$

In these coordinates, the Kerr metric assumes the regular form

$$\begin{aligned}\mathrm{d}s^2 = &-\frac{(\Delta - a^2\sin^2\theta)}{\rho}\mathrm{d}\widetilde{u}^2 + 2\mathrm{d}\widetilde{u}\mathrm{d}r - \frac{2a\sin^2\theta(r^2+a^2-\Delta)}{\rho}\mathrm{d}\widetilde{u}\mathrm{d}\chi \\ &- 2a\sin^2\theta\mathrm{d}\chi\mathrm{d}r + \frac{\left[(r^2+a^2)^2 - \Delta a^2\sin^2\theta\right]}{\rho}\sin^2\theta\mathrm{d}\chi^2 + \rho\mathrm{d}\theta^2.\end{aligned} \tag{10.137}$$

The absence of singularities at $r = r_\pm$ confirms that these are merely coordinate singularities, not intrinsic to the spacetime geometry.

10.3.2 *Observers*

To fully characterize the Kerr black hole's geometry, it is instructive to analyze the behavior of three distinct classes of observers. Each provides unique insights into the spacetime's rotational dynamics and frame-dragging effects.

The first class is zero-angular-momentum observers (ZAMOs). They are defined by the condition that their angular momentum density vanishes, $L = \widehat{u}\cdot\widehat{\eta} = 0$, where $\widehat{u}$ is the observer's 4-velocity and $\eta = \partial_\phi$ is the axial Killing vector. In terms of the metric components, this implies

$$g_{\phi t}\dot{t} + g_{\phi\phi}\dot{\phi} = 0, \tag{10.138}$$

with "$\cdot$" denoting differentiation with respect to proper time. For the Kerr metric, solving (10.138) yields the ZAMO angular velocity:

$$\Omega \equiv \frac{\mathrm{d}\phi}{\mathrm{d}t} = \omega = -\frac{g_{t\phi}}{g_{\phi\phi}}. \tag{10.139}$$

Remarkably, despite having zero angular momentum, ZAMOs must rotate with angular velocity ω relative to infinity. As $r \to r_+$, ω approaches the black hole's angular velocity Ω_H, forcing ZAMOs into corotation. At large distances $r \gg GM$, $\omega \approx 2J/r^3$, demonstrating the decay of frame-dragging effects far from the source. This phenomenon arises from the black hole's spin and exemplifies the dragging of inertial frames, a generic feature of rotating spacetimes.

The second type of observers in Kerr spacetime is the static observer. Such an observer has a 4-velocity proportional to the Killing vector $\widehat{\xi}$, given by

$$u^\mu = \gamma(1,0,0,0), \tag{10.140}$$

where $\gamma = (-g_{00})^{-1/2}$ is a normalization factor ensuring proper time parametrization. Unlike freely falling observers, static observers require an external force (e.g., a rocket) to maintain their position, meaning their trajectories are non-geodesic. However, static observers cannot be defined everywhere in Kerr spacetime. Since $\hat{\xi}$ is timelike only outside the stationary limit surface, these observers cease to exist in regions where $\hat{\xi}$ becomes spacelike.

Next, we examine stationary observers, which rotate with a fixed angular velocity Ω in the ϕ direction. Their 4-velocity is a linear combination of the two Killing vectors $\hat{\xi}$ and $\hat{\eta}$:

$$u^\mu = \gamma(\xi^\mu + \Omega\eta^\mu), \tag{10.141}$$

where γ is again a normalization factor. The timelike condition for $\hat{\xi} + \Omega\hat{\eta}$ imposes a constraint on γ:

$$\begin{aligned}\gamma^{-2} &= -g_{\mu\nu}(\xi^\mu + \Omega\eta^\mu)(\xi^\nu + \Omega\eta^\nu) \\ &= -g_{tt} - 2\Omega g_{t\phi} - \Omega^2 g_{\phi\phi} \\ &= -g_{\phi\phi}(\Omega^2 - 2\omega\Omega + g_{tt}/g_{\phi\phi}),\end{aligned} \tag{10.142}$$

with $\omega = -g_{t\phi}/g_{tt}$ denoting the frame-dragging angular velocity of the black hole as measured at infinity. Stationary observers also cannot exist everywhere in Kerr spacetime. For the observer to remain timelike, i.e., $\hat{\xi} + \Omega\hat{\eta}$ be timelike, $\gamma^{-2} > 0$, leading to the permissible range for Ω:

$$\Omega_- < \Omega < \Omega_+, \quad \Omega_\pm = \omega \pm \sqrt{\omega^2 - g_{tt}/g_{\phi\phi}}. \tag{10.143}$$

Using the explicit form of the Kerr metric, this becomes

$$\Omega_\pm = \omega \pm \frac{\Delta^{1/2}\rho^2}{\Sigma^2 \sin\theta}. \tag{10.144}$$

Static observers with $\Omega = 0$ are a special case of stationary observers, restricted to regions outside the stationary-limit surface. As an observer approaches the event horizon, the allowed range of Ω narrows, with Ω_- increasing while Ω_+ decreasing, until $\Omega_- = \Omega_+$, which means $\Omega = \omega$, forcing the observer to corotate with the black hole at the angular velocity of the horizon:

$$\Omega_{\mathrm{H}} \equiv \omega(r_+) = \frac{a}{r_+^2 + a^2}. \tag{10.145}$$

10.3.3 *Geodesic motion*

The study of geodesic motion in Kerr spacetime can be significantly simplified by employing constants of motion. Unlike spherically symmetric spacetimes, the absence of spherical symmetry in Kerr geometry generally prevents the guarantee of planar motion for particles. At first glance, only two constants of motion appear to be available, seemingly insufficient to reduce the geodesic equations to one-dimensional form. However, the existence of hidden symmetries allows for the introduction of an additional constant, the Carter constant, which facilitates further simplification of the analysis. Here, we do not intend to provide a complete discussion of geodesics in Kerr spacetime but focus only on geodesic motion confined to the equatorial plane $\theta = \pi/2$. It can be shown that a particle's trajectory remains restricted to the equatorial plane if its initial conditions satisfy $\theta = \pi/2$ and $u^\theta = 0$. Under these conditions, the motion is invariant under the transformation $\theta \to \pi \to \theta$, ensuring $u^\theta = 0$ for all proper time τ. Consequently, the particle's dynamics can be described by the simplified metric:

$$\mathrm{d}s^2 = -(1 - 2\mu/r)\mathrm{d}t^2 - \frac{4a\mu}{r}\mathrm{d}t\mathrm{d}\phi + \frac{r^2}{\Delta}\mathrm{d}r^2 + (r^2 + a^2 + 2\mu a^2/r)\mathrm{d}\phi^2.$$

Exploiting the Killing symmetries of the spacetime, we define the conserved energy per unit mass and angular momentum density of the particle:

$$\begin{aligned} E &= -\widehat{\xi} \cdot \widetilde{u} = -(g_{tt}u^t + g_{t\phi}u^\phi), \\ L &= \widehat{\eta} \cdot \widehat{u} = g_{\phi t}u^t + g_{\phi\phi}u^\phi. \end{aligned}$$

The corresponding 4-velocity components are then expressed as

$$\begin{aligned} u^t &= \frac{\mathrm{d}t}{\mathrm{d}\tau} = \frac{1}{\Delta}\left((r^2 + a^2 + 2\mu a^2/r)E - \frac{2\mu a}{r}L\right), \\ u^\phi &= \frac{\mathrm{d}\phi}{\mathrm{d}\tau} = \frac{1}{\Delta}\left((1 - 2\mu/r)L + \frac{2\mu a}{r}E\right). \end{aligned}$$

Using the normalization condition of the 4-velocity and $u^\theta = 0$, we finally obtain the radial equation of motion

$$\frac{E^2 - 1}{2} = \frac{1}{2}\left(\frac{\mathrm{d}r}{\mathrm{d}\tau}\right)^2 + V_{\text{eff}}(r, E, L), \tag{10.146}$$

where the effective potential is given by

$$V_{\text{eff}}(r, E, L) = -\frac{\mu}{r} + \frac{L^2 - a^2(E^2 - 1)}{2r^2} - \frac{\mu(L - aE)^2}{r^3}.$$

Thus, the equatorial geodesic motion reduces to an equivalent one-dimensional problem governed by the effective potential V_{eff}, analogous to classical central-force dynamics.

The analysis of photon motion in Kerr spacetime follows a similar framework to that of massive particles, but with key distinctions arising from the null nature of photon trajectories. In this case, the 4-velocity of the photon is a null vector $\widehat{u} \cdot \widehat{u} = 0$, and its motion depends on the impact parameter $b = |L/E|$. An important feature of photon motion in Kerr spacetime is its dependence on the direction of propagation relative to the black hole's rotation, co-rotating ($\sigma = \text{sign}(L) > 0$) or counter-rotating ($\sigma = \text{sign}(L) < 0$). This distinction arises from frame-dragging, an effect of the black hole's rotation that breaks the symmetry between prograde and retrograde orbits. The radial motion of photons can be described by the equation

$$\frac{1}{L^2}\left(\frac{\mathrm{d}r}{\mathrm{d}\lambda}\right)^2 = \frac{1}{b^2} - V_{\text{eff}}(r, b, \sigma), \tag{10.147}$$

where λ is an affine parameter, and the effective potential takes the form

$$V_{\text{eff}} = \frac{1}{r^2}\left(1 - \left(\frac{a}{b}\right)^2 - \frac{2\mu}{r}\left(1 - \sigma\frac{a}{b}\right)^2\right). \tag{10.148}$$

When $a = 0$, we return to the Schwarzschild spacetime. Clearly, due to the σ-dependent term, co-rotating ($\sigma = 1$) and counter-rotating ($\sigma = -1$) photons experience different effective potentials, leading to distinct orbital behaviors. The term $(a/b)^2$ introduces a spin-dependent modification, implying that photons with small b are more strongly influenced by rotation.

The Kerr spacetime allows for the analysis of diverse geodesic motions, including circular orbits, the radii of unstable photon circular orbits, light deflection, and the morphology of bound orbits. As an illustrative example, consider an extremal black hole with spin parameter $a = \mu$. The radius of the unstable photon circular orbit depends on the photon's rotation relative to the black hole: (1) If co-rotating, the unstable circular orbit is at $r = \mu$; (2) If counter-rotating, the unstable circular orbit is at $r = 4\mu$.

For massive particles, the innermost stable circular orbit (ISCO) is determined by the following conditions. First, the vanishing radial velocity implies

$$\left.\frac{\mathrm{d}r}{\mathrm{d}\tau}\right|_{r=R} = 0 \Rightarrow \frac{E^2 - 1}{2} = V_{\text{eff}}(R, E, L). \tag{10.149}$$

Second, the absence of radial acceleration requires

$$\left.\frac{\partial V_{\text{eff}}}{\partial r}\right|_{r=R} = 0, \tag{10.150}$$

while stability further demands

$$\left.\frac{\partial^2 V_{\text{eff}}}{\partial r^2}\right|_{r=R} \geq 0. \tag{10.151}$$

In the non-rotating limit $a = 0$, the ISCO reduces to $R = 6\mu$, recovering the Schwarzschild result. However, for an extremal Kerr black hole $\frac{a}{\mu} = 1$, the ISCO radii bifurcate based on the direction of orbital motion:

$$\begin{cases} R = \mu, & \text{(co-rotating)}, \\ R = 9\mu, & \text{(counter-rotating)}. \end{cases} \tag{10.152}$$

For an extremal Kerr black hole, the co-rotating innermost stable circular orbit (ISCO) imposes specific constraints on the energy and angular momentum of a test particle,

$$E = \frac{1}{\sqrt{3}}, \quad L = \frac{2\mu}{\sqrt{3}}, \quad R_{\text{ISCO}} = \mu. \tag{10.153}$$

The binding energy density of the particle is defined as the difference between its energy density at infinity, where $E_\infty = 1$ for a particle at rest, and its energy density on the ISCO. Consequently, the binding energy per unit rest mass is given by $1 - E$, which takes a maximum value of $1 - \frac{1}{\sqrt{3}} \approx 42\%$ on the co-rotating ISCO orbit. This implies that if a particle is released from infinity and settles into the co-rotating ISCO orbit, up to 42% of its rest mass must be released, an extraordinary efficiency far surpassing that of nuclear fusion. While astrophysical black holes are not perfectly extremal, near-extremal black holes with a slightly less than GM still exhibit remarkably high energy extraction efficiencies for particles spiraling into the ISCO.

Using the geodesic motion in Kerr spacetime, we can test the cosmic censorship hypothesis. The cosmic censorship hypothesis asserts that gravitational collapse cannot produce naked singularities, as singularities must always be hidden behind an event horizon. For Kerr black holes, this implies that the spin parameter must satisfy $a \leq \mu$. Solutions with $a > \mu$, so-called "overspinning" Kerr metrics, exhibit a naked singularity but are deemed unphysical under the cosmic censorship hypothesis. Consider an extremal Kerr black hole. Suppose a test particle with specific energy density E, angular momentum density L, and mass $m \ll M$ falls into the black hole. The changes in the black hole's mass and angular momentum are

$$\left.\begin{aligned} \delta M &= mE \\ \delta J &= mL \end{aligned}\right\} \Rightarrow \delta a \approx \frac{m}{M}(L - aE). \tag{10.154}$$

If the particle has $L > 2EGM$, then

$$\delta a > \frac{m}{M}(2EGM - aE) = mE\left(2G - \frac{a}{M}\right) \geq G\delta M. \tag{10.155}$$

This suggests that $(a + \delta a) > G(M + \delta M)$, thereby violating the cosmic censorship hypothesis. What is the problem with the above discussion? The problem is that when the particle's angular velocity is large enough, it may not be captured by the black hole. This can be demonstrated by a quantitative analysis. The radial motion is governed by an effective potential, where capture requires the maximum of the effective potential V_{eff} is smaller than $\frac{E^2-1}{2}$, otherwise the particle cannot be captured and will instead be scattered back to infinity. For $L = 2GME$, the potential has a maximum at $r = GM$ with $V_{\text{eff}}(r = GM) = \frac{E^2-1}{2}$. Here, the particle undergoes marginally bound circular motion, and any perturbation either ejects it to infinity or allows absorption without overspinning the black hole. For $L > 2GME$, the particle encounters a potential barrier, preventing capture. Instead, it scatters back to infinity, leaving $a \leq \mu$ intact. Thus, the cosmic censorship hypothesis cannot be violated via test-particle accretion in Kerr spacetime.

Before concluding the discussion on geodesic motion, we would like to point out that closed timelike curves (CTCs) exist in the Kerr black hole. The intrinsic singularity of the Kerr black hole has a ring-like topology, enclosing a disk-shaped region. If a particle traverses the interior of this ring, it enters a distinct asymptotically flat universe described by the Kerr metric with $r < 0$. In this region, there is no event horizon since Δ can not be zero for $r < 0$, and near the singularity ring there exist closed timelike curves. Consider the motion restricted to the equatorial plane with fixed t and r. The line element reduces to

$$\mathrm{d}s^2 = (r^2 + a^2 + 2\mu a^2/r)\mathrm{d}\phi^2. \tag{10.156}$$

For $r < 0$ and $|r| \ll \mu$, the dominant term $2\mu a^2/r$ becomes negative, potentially rendering $ds^2 < 0$. Since ϕ is periodic, such paths represent closed timelike curves, violating causality by allowing observers to return to their own past.

10.3.4 *Penrose process*

In the Kerr black hole, the infinite redshift surface r_{S+} and the event horizon r_+ do not coincide, giving rise to a distinct intermediate region known as the ergoregion, as shown in Figure 10.3. Within this region, the metric component $g_{tt} > 0$, indicating that the Killing vector ∂_t ceases to be timelike. An observer inside the ergoregion cannot remain static with respect

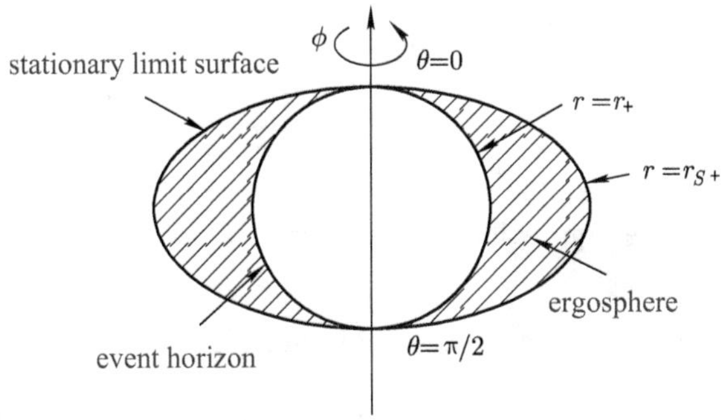

Fig. 10.3 Kerr space-time has an energy layer between the stationary limit surface and the event horizon.

to the (r, θ, ϕ) coordinates; instead, they are inevitably dragged into co-rotation with the black hole due to frame-dragging effects. However, unlike within the event horizon, the observer has not yet crossed into a region of no return; they retain the freedom to either descend further toward the horizon, or escape outward and re-emerge across the stationary limit surface. Remarkably, as Roger Penrose demonstrated, the ergoregion permits the extraction of rotational energy from the Kerr black hole—a process now known as the Penrose mechanism. This energy-harvesting property also lends the region its name.[4]

Penrose proposed the following mechanism to extract rotational energy from a Kerr black hole. Suppose an observer at infinity sends a particle A into the ergoregion of the Kerr black hole. For this observer, the energy of A is given by

$$E^{(A)} = -\widehat{P}^{(A)}(\mathcal{E}) \cdot \widehat{u}_{\text{obs}} = -P_t^{(A)}(\mathcal{E}), \tag{10.157}$$

where $u^{\mu}_{\text{obs}} = (1, 0, 0, 0)$ is the observer's 4-velocity. Suppose that at a point $\mathcal{D}$ within the ergoregion, particle A undergoes a decay into two daughter particles, B and C. By Einstein's equivalence principle, the 4-momentum is locally conserved: $\widehat{P}^{(A)}(\mathcal{D}) = \widehat{P}^{(B)}(\mathcal{D}) + \widehat{P}^{(C)}(\mathcal{D})$. If particle C escapes back to infinity, its energy as measured by the distant observer is

$$E^{(C)} = -P_t^{(C)}(\mathcal{R}) = -P_t^{(C)}(\mathcal{D}), \tag{10.158}$$

where the equality holds because the observer's 4-velocity coincides with the timelike Killing vector of the spacetime. Similarly, since $P_t^{(A)}$ is conversed

[4]The term ergoregion derives from the Greek "ergon" (meaning "work"), reflecting its role as a zone where energy extraction becomes possible. In the context of Kerr black holes, the ergoregion is sometimes alternatively referred to as the ergosphere.

along geodesics, $P_t^{(A)}(\mathcal{D}) = P_t^{(A)}(\mathcal{E})$. Substituting into the conservation equation yields

$$E^{(C)} = P_t^{(B)}(\mathcal{D}) + E^{(A)}. \tag{10.159}$$

The critical feature enabling energy extraction arises from the spacelike nature of ∂_t in the ergoregion. On the other hand, $P_t^{(B)} = \widehat{e}_t \cdot \widehat{P}^{(B)}$ and $\widehat{e}_t \cdot \widehat{e}_t = g_{tt}$. If B escapes from r_{S+}, then

$$g_{tt} < 0 \Rightarrow \widehat{e}_t \text{ is timelike,}$$

thus,

$$-P_t^{(B)} = E^{(B)} > 0 \Rightarrow E^{(C)} < E^{(A)}.$$

This implies that the energy of the outgoing particle C is less than the energy of the incoming particle A. However, if B does not escape from r_{S+} and instead falls into the event horizon, within the ergoregion $g_{tt} > 0$, $\widehat{e}_t$ is a spacelike vector, such that $P_t^{(B)}$ can be positive or negative. If $P_t^{(B)} > 0$, then

$$E^{(C)} > E^{(A)}, \tag{10.160}$$

meaning the energy of the escaping particle carries more energy than the original particle. This mechanism, illustrated in Figure 10.4, is the celebrated Penrose process.

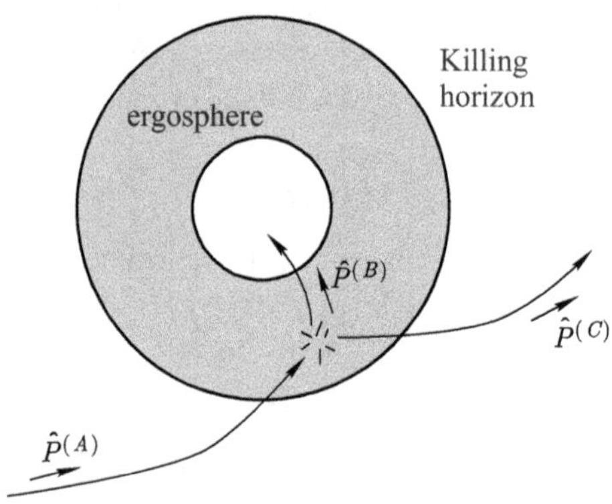

Fig. 10.4 Penrose process.

Due to energy conservation, the excess energy cannot be created out of nothing. In the Penrose process, particles effectively extracts energy from a Kerr black hole. More specifically, this process harnesses rotational energy by reducing the black hole's angular momentum. When particle B crosses the event horizon, it alters the black hole's mass and angular momentum according to

$$M \to M - P_t^{(B)}/c^2, \quad J \to J + P_\phi^{(B)}. \tag{10.161}$$

Here, the conditions $P_t^{(B)} > 0$ and $P_\phi^{(B)} < 0$ imply that particle B simultaneously decreases both the mass and angular momentum of the black hole. To demonstrate why $P_\phi^{(B)} < 0$, consider an observer fixed at r and θ within the ergoregion, with 4-velocity

$$u^\mu = u^t(1, 0, 0, \Omega), \quad \Omega = \mathrm{d}\phi/\mathrm{d}t. \tag{10.162}$$

For such an observer, particle B's energy must appear positive

$$\begin{aligned} &E^{(B)} = -P_\mu^{(B)} u^\mu = -u^t(P_t^{(B)} + P_\phi^{(B)}\Omega) > 0 \\ &\Rightarrow P_t^{(B)} + P_\phi^{(B)}\Omega < 0 \\ &\Rightarrow P_\phi^{(B)} < -P_t^{(B)}/\Omega. \end{aligned} \tag{10.163}$$

Given that $P_t^{(B)} > 0$, this necessarily requires $P_\phi^{(B)} < 0$. Thus, any particle entering the black hole with $P_t^{(B)} > 0$ must carry negative angular momentum, thereby reducing the black hole's angular momentum.

Through repeated Penrose processes, the black hole's rotational energy can be progressively extracted. This continues until complete depletion of angular momentum, ultimately transforming the Kerr black hole into a Schwarzschild black hole. The angular momentum constraint derived above applies universally to all observers within the ergoregion. The maximum angular velocity occurs at the outer horizon $r = r_+$, where $\Omega_{\max} = \Omega_{\mathrm{H}}$. This gives the fundamental limitation

$$\delta J \le \frac{\delta M}{\Omega_{\mathrm{H}}}, \quad \text{or} \quad \delta M \ge \delta J \Omega_{\mathrm{H}}, \tag{10.164}$$

where

$$\delta M = -P_t^{(B)}, \quad \delta J = P_\phi^{(B)}. \tag{10.165}$$

Notably, inequality (10.164) remains valid regardless of $P_t^{(B)}$'s sign. When $P_t^{(B)}$ is negative, the particle's angular momentum can be positive or negative, but its magnitude remains bounded by this inequality. For the Penrose process specifically, both δJ and δM are negative. This relationship fundamentally constrains the achievable reduction in the black hole's angular momentum, with the optimal case occurring when $\delta J = \delta M/\Omega_{\mathrm{H}}$.

10.4 Black Hole Thermodynamics

In fact, the area of the black hole event horizon always increases, which is known as the second law of black hole thermodynamics. This is because black holes always attract matter into the horizon, and classically, no matter can escape from the event horizon of a black hole. For a Schwarzschild black hole, the horizon area is proportional to the square of the black hole mass. Thus, when the black hole absorbs external matter, its mass increases, and the horizon area necessarily grows as well. The case of a rotating Kerr black hole is more subtle. At first glance, the Penrose process appears to allow a decrease in the black hole's mass and, consequently, its horizon area.[5] However, a deeper analysis reveals that this is not the case.

While the Penrose process extracts rotational energy (reducing the black hole's mass and angular momentum), it does so in a way that still preserves the non-decreasing nature of the horizon area. For the Kerr black hole, the area of the outer horizon depends on both its mass and angular momentum

$$A = 4\pi(r_+^2 + a^2) = 8\pi\mu(\mu + \sqrt{\mu^2 - a^2}). \tag{10.166}$$

When the black hole absorbs a particle, the resulting change in horizon area is given by

$$\delta A = \frac{8\pi}{\kappa}(\delta M - \Omega_{\rm H}\delta J), \tag{10.167}$$

where

$$\kappa = \frac{\sqrt{\mu^2 - a^2}}{2\mu r_+} \tag{10.168}$$

denotes the surface gravity, and $\Omega_{\rm H} = \dfrac{a}{2\mu r_+}$ is the angular velocity at the horizon. As established in the preceding discussion, the inequality $\delta M \geq \Omega_{\rm H}\delta J$ holds, ensuring that the horizon area never decreases: $\delta A \geq 0$. This observation motivates the definition of an irreducible mass, proportional to

[5] We will only discuss the classical physics of black holes here, ignoring the quantum effects of gravity for the time being. Although the Penrose process involves the decay of particles, it has nothing to do with the quantum behavior of particles in the black hole. It remains a classical mechanism, distinct from quantum phenomena such as Hawking radiation.

the horizon area:

$$\begin{aligned} M_{\rm irr}^2 &= \frac{A}{16\pi G^2} \\ &= \frac{1}{4G^2}(r_+^2 + a^2) \\ &= \frac{1}{2}\left(M^2 + \sqrt{M^4 - (Ma/G)^2}\right) \\ &= \frac{1}{2}\left(M^2 + \sqrt{M^4 - (J/G)^2}\right). \end{aligned} \tag{10.169}$$

The irreducible mass relates to the black hole's total mass M via:

$$M^2 = M_{\rm irr}^2 + \frac{J^2}{4M_{\rm irr}^2}. \tag{10.170}$$

Clearly, this irreducible mass is non-decreasing,

$$\delta M_{\rm irr} \geq 0. \tag{10.171}$$

The Penrose process can reduce the black hole's mass and angular momentum but cannot reduce the irreducible mass. The maximum extractable energy from a Kerr black hole via the Penrose process occurs when the black hole's angular momentum is entirely depleted, reducing it to a Schwarzschild black hole. Since the final mass cannot be less than the initial irreducible mass, the maximum energy release is

$$M - M_{\rm irr} = M - \frac{1}{\sqrt{2}}\left(M^2 + \sqrt{M^4 - (J/G)^2}\right)^{1/2}. \tag{10.172}$$

For an extremal Kerr black hole, $J = GM^2$, this process can extract up to approximately 29% of the black hole's original energy.

In the context of the Penrose process for Kerr black holes, the change in the horizon area due to the absorption of a particle is given by:

$$\delta A = \frac{8\pi}{\kappa}(\delta M - \Omega_{\rm H}\delta J), \tag{10.173}$$

where κ is the surface gravity (constant over horizon). Rearranging (10.173), we obtain

$$\delta M = \frac{\kappa}{8\pi}\delta A + \Omega_{\rm H}\delta J. \tag{10.174}$$

This equation bears a striking resemblance to the first law of thermodynamics, and is thus recognized as the first law of black hole mechanics or black hole thermodynamics.

The second law of black hole thermodynamics, which states that the total horizon area of a system of black holes cannot decrease, has two important implications: One is on the possible energy loss in black hole collisions, and the other is the restriction on black hole fission.

(1) Energy conversion rate in black hole collisions.

Consider two black holes with masses M_1 and M_2 colliding to form a black hole with mass M_3. The energy radiated away in gravitational waves is $M_1 + M_2 - M_3$, and the efficiency is

$$\eta = \frac{M_1 + M_2 - M_3}{M_1 + M_2} = 1 - \frac{M_3}{M_1 + M_2}. \tag{10.175}$$

Assuming the initial black holes are spherically symmetric, $A_1 = 16\pi M_1^2$ and $A_2 = 16\pi M_2^2$. The area non-decrease theorem requires

$$A_3 \geq 16\pi(M_1^2 + M_2^2). \tag{10.176}$$

Since the minimal area of a black hole of mass M_3 is$16\pi M_3^2$, achieved for a Schwarzschild black hole, we have

$$16\pi M_3^2 \geq A_3 \geq 16\pi(M_1^2 + M_2^2),$$

which implies:

$$M_3 \geq \sqrt{M_1^2 + M_2^2}. \tag{10.177}$$

Substituting into (10.175), the efficiency is bounded by

$$\eta \leq 1 - \frac{\sqrt{M_1^2 + M_2^2}}{M_1 + M_2} \leq 1 - \frac{1}{\sqrt{2}}. \tag{10.178}$$

The radiated energy can be used to do work, so the area non-decrease theorem limits the energy that can be extracted from black holes, similar to how the second law of thermodynamics limits the efficiency of heat engines.

(2) Black holes cannot split.

Suppose a black hole of mass M_3 could spontaneously split into two black holes with masses M_1 and M_2, $M_3 \to M_1 + M_2$. The area non-decrease theorem requires

$$M_3 \leq \sqrt{M_1^2 + M_2^2} \leq M_1 + M_2. \tag{10.179}$$

However, energy conservation demands $M_3 \geq M_1 + M_2$ (with $M_3 - M_1 - M_2$ being the mass radiated away). Contradiction! Thus, black hole fission is forbidden by the second law of black hole thermodynamics.

The first law of black hole thermodynamics bears a striking resemblance to the standard first law of thermodynamics,

$$\mathrm{d}U = T\mathrm{d}S + \text{work terms}. \tag{10.180}$$

This suggests a profound analogy between the two sets of laws, with the following natural identifications:

$$A \sim S, \quad \kappa \sim T. \tag{10.181}$$

Extending this correspondence, we can formulate the four laws of black hole thermodynamics in direct parallel with the classical thermodynamic laws.

The zeroth law Temperature T is constant in thermal equilibrium $\leftrightarrow$ surface gravity κ is constant over the entire event horizon for a stationary black hole.

The first law $\mathrm{d}U = T\mathrm{d}S + \text{work} \leftrightarrow \delta M = \frac{\kappa}{8\pi}\delta A + \Omega_{\mathrm{H}}\delta J$.

The second law Entropy increase principle $\delta S > 0 \leftrightarrow \delta A > 0$, the horizon area never decrease.

The third law It is impossible to reach absolute zero temperature $T = 0$ through a finite number of physical processes $\leftrightarrow$ it is impossible to reach $\kappa = 0$ through any finite number of physical processes.

In the third law, $\kappa = 0$ corresponds to an extremal black hole, where a naked singularity would appear. Thus, the third law is deeply tied to the cosmic censorship hypothesis.

One of the most remarkable insights from the above discussion is that black holes possess entropy and temperature, a profound and initially counterintuitive discovery. But is this entropy physically meaningful, or is it merely a mathematical artifact? To illustrate the puzzle, consider a simple thermodynamic process: mixing a cup of hot coffee with a cup of cold coffee increases the entropy of the universe. Yet, if we were to pour this mixture into a black hole, thereby erasing any record of the entropy increase, how would the total entropy of the system evolve? This very question was raised in 1970 by John Wheeler during a conversation with the young Jacob Bekenstein. Wheeler noted that an external observer could, in principle, discard entropy-carrying matter into a black hole, effectively decreasing the entropy outside the horizon. At first glance, this scenario appears to violate the second law of thermodynamics—unless black holes themselves contribute to the total entropy. Bekenstein's groundbreaking resolution was to postulate that black holes must indeed carry entropy. By accounting for the black hole's entropy, the total entropy of the system (black hole plus exterior) always increases. Crucially, he found that this entropy scales with the horizon area:

$$S_{\mathrm{BH}} \propto A. \tag{10.182}$$

Later, Stephen Hawking refined this result by determining the precise proportionality constant, culminating in the celebrated Bekenstein–Hawking entropy formula:

$$S = \frac{\kappa_{\mathrm{B}} A}{4\hbar G}, \tag{10.183}$$

where κ_{B} denotes the Boltzmann constant.

The appearance of the Planck constant $\hbar$ in the Bekenstein–Hawking entropy formula indicates that the black hole's entropy is inherently a quantum mechanical phenomenon. In the classical limit $\hbar \to 0$, this entropy diverges, underscoring that its origin lies beyond classical gravity. Even more striking is the area-law scaling of black hole entropy rather than the conventional volume-law behavior observed in most physical systems. This departure from locality carries profound implications. First, in systems governed by local interactions (e.g., gases or solids), entropy scales with volume. The area-law suggests that quantum gravity cannot be described by any straightforward local field theory. Second, the area law of black hole entropy actually reflects the holographic nature of quantum gravity, which suggests that the information content of a gravitational system is encoded on its boundary. In 1993, Nobel laureate Gerard 't Hooft proposed the holographic principle, conjecturing that a gravitational system's degrees of freedom could be represented by a lower-dimensional boundary theory. In 1994, Leonard Susskind further developed this principle.[6] In 1997, Juan Maldacena proposed the AdS/CFT correspondence [59, 60], which specifically realized the holographic principle of quantum gravity. In short, the AdS/CFT correspondence tells us that a quantum gravity theory in AdS spacetime is equivalent to a conformally invariant quantum field theory on the AdS boundary. This correspondence not only provides a concrete realization of holography but also revolutionizes our approach to quantum gravity and strongly coupled quantum field theories. By mapping intractable problems in one domain to more tractable ones in the other, AdS/CFT has become one of the most fertile frontiers in theoretical physics over the past two decades, offering insights into quantum gravity, black holes, quark-gluon plasmas, quantum many body systems, and even quantum chaos.

10.4.1 *Hawking radiation and information loss*

In the study of black holes, Hawking not only discovered that black holes have entropy but further pointed out that black holes have temperature

$$T = \frac{\hbar\kappa}{2\pi k_{\rm B}}, \tag{10.184}$$

and emit thermal radiation akin to blackbody radiation, now known as Hawking radiation. A comprehensive understanding of this phenomenon

[6] For a review of the holographic principle, see Ref. [58].

requires the framework of quantum field theory in curved spacetime. However, for the purposes of this discussion, we restrict ourselves to a simplified qualitative analysis.

According to quantum field theory, the vacuum in flat spacetime is subject to continuous quantum fluctuations, giving rise to virtual particle-antiparticle pairs. Ordinarily, energy conservation prevents these virtual pairs from materializing as real particles. However, in the presence of an external field, these fluctuations may extract energy, enabling pair production. A well-known example is the Schwinger effect, where a strong electric field generates real particle pairs from the vacuum. Similarly, in curved spacetime, virtual pairs can extract energy from the gravitational field, leading to the creation of real particles, a mechanism underlying Hawking radiation. To illustrate this, consider a Schwarzschild black hole. The Killing vector field $\widehat{\xi} = (1, 0, 0, 0)$ has a norm given by

$$\widehat{\xi} \cdot \widehat{\xi} = -(1 - R_{\mathrm{s}}/r), \tag{10.185}$$

indicating that $\widehat{\xi}$ is timelike outside the horizon ($r > R_{\mathrm{s}}$) and spacelike inside the horizon ($r < R_{\mathrm{s}}$). This distinction is crucial for Hawking radiation. Consider a particle with 4-momentum $\widehat{P}$. Outside the horizon ($r > R_{\mathrm{s}}$), the inner product with $\widehat{\xi}$ yields

$$-\widehat{\xi} \cdot \widehat{P} = E > 0, \quad \text{(for the particle)},$$

$$-\widehat{\xi} \cdot \widehat{\bar{P}} = E > 0, \quad \text{(for the antiparticle)}.$$

Since both the particle and antiparticle have positive energy, vacuum particle pair production is forbidden by energy conservation. However, inside the horizon ($r < R_{\mathrm{s}}$), $-\widehat{\xi} \cdot \widehat{p}$ represents a spatial momentum component and may assume negative values. In such cases, particle-antiparticle pairs can be produced without violating energy conservation. Near the horizon, local 4-momentum conservation permits a scenario where the antiparticle falls into the black hole while the particle escapes to infinity. From the perspective of a distant observer, this process manifests as a net outflow of energy, reducing the black hole's mass, effectively, the black hole radiates. This is the essence of Hawking radiation.

We can estimate the Hawking radiation rate of the black hole $\dfrac{\mathrm{d}M}{\mathrm{d}t}$. First, Hawking radiation is a quantum mechanical process, so its rate should naturally depend on the Planck constant $\hbar \approx 2.62 \times 10^{-66}\ \mathrm{cm}^2$. Moreover, the emission is most efficient for massless particles (e.g., photons), leading to a generic form for the mass-loss rate:

$$\frac{\mathrm{d}M}{\mathrm{d}t} = -\nu \frac{\hbar}{M^2}, \tag{10.186}$$

where the constant ν is to be determined. From quantum field theory in curved spacetime, Hawking radiation is thermal, with the black hole emitting as a blackbody at temperature

$$k_{\rm B}T = \frac{\hbar}{8\pi M} = \frac{c^3\hbar}{8\pi GM}. \tag{10.187}$$

The power radiated per unit area by a blackbody follows the Stefan–Boltzmann law, Flux $= \sigma T^4$, where $\sigma = \pi^2 k_{\rm B}^4/(60\hbar^3 c^2)$ is the Stefan–Boltzmann constant. Thus, the energy radiated through the black hole horizon is

$$16\pi M^2\pi^2 \frac{c^{12}\hbar^4}{(8\pi GM)^4 \times 60\hbar^3 c^2} = \frac{c^{10}}{4\times 64\times 60\pi G^4}\frac{\hbar}{M^2} \Rightarrow \nu = \frac{1}{15360\pi}.$$

The Hawking radiation of a black hole can be estimated as follows. The temperature of a black hole is given by $T = 6.2\times 10^{-8}\left(\frac{M_\odot}{M}\right)$ K, where $M_\odot$ denotes the solar mass. This expression reveals an inverse relationship between the black hole's mass and temperature: as M increases, T decreases, whereas a smaller black hole exhibits a higher temperature. The time evolution of the black hole mass is governed by

$$M(t) = [3\nu\hbar(t_* - t)]^{1/3}, \tag{10.188}$$

from which we derive the total evaporation time

$$\tau_{\rm H} \approx \frac{1}{3\nu}\frac{M^3}{\hbar} \approx 8.3\times 10^{-26}\left(\frac{M}{1\ {\rm g}}\right)^3 \ {\rm s}. \tag{10.189}$$

For a solar-mass black hole $M \sim M_\odot$, the required time $\tau_{\rm H} \gg 14$ billion years, far longer than the age of the universe. In contrast, primordial black holes formed in the early universe with masses on the order of

$$10^{14}\ {\rm g} \sim 10^{-19} M_\odot, \tag{10.190}$$

comparable to a terrestrial mountain, it would take $\tau_{\rm H} = 8.3\times 10^{16}$ s to fully evaporate. Thus, any such black holes formed in the early universe would still be radiating today. Notably, the rate of Hawking radiation accelerates significantly in the late stages of evaporation.

The preceding analysis neglects the backreaction of Hawking radiation, which is negligible initially but becomes critical in the final stages. Although this treatment is semi-classical, Hawking radiation has profound physical consequences, most notably the black hole information paradox. If Hawking radiation is indeed thermal, a black hole formed from gravitational collapse would eventually evaporate, leaving behind a spacetime devoid of an event horizon. Since thermal radiation carries no information, this process would irreversibly erase the quantum information of infalling matter, violating unitarity. A resolution to this paradox is expected within a complete theory of quantum gravity. The AdS/CFT correspondence offers a promising avenue: if quantum gravity is dual to a unitary quantum field theory, information loss should not occur. However, the precise mechanism of information recovery remains unclear. For further discussion, see Refs. [61–63].

Exercises

1. For a stationary source, prove that $\partial_0 T^{0i} = 0$, and hence prove that

$$\int_V (T^{0i} y^j + T^{0j} y^i) \mathrm{d}^3 y = 0, \tag{10.191}$$

where the spatial volume V encloses the source.

2. For a non-relativistic stationary source, prove that in the center-of-mass coordinates,

$$\begin{aligned} \bar{h}^{00}(\boldsymbol{x}) &= -\frac{4GM}{|\boldsymbol{x}|} + O\left(\frac{1}{|\boldsymbol{x}|^2}\right), \\ \bar{h}^{0i}(\boldsymbol{x}) &= \frac{2G}{|\boldsymbol{x}|^3} x_j J^{ij} + O\left(\frac{1}{|\boldsymbol{x}|^3}\right), \\ \bar{h}^{ij}(\boldsymbol{x}) &= 0, \end{aligned} \tag{10.192}$$

where M and J^{ij} are given by

$$M = \int_V \rho(\boldsymbol{y}) \mathrm{d}^3 y, \quad J^{ij} = \int_V [y^i p^j(\boldsymbol{y}) - y^j p^i(\boldsymbol{y})) \mathrm{d}^3 y, \tag{10.193}$$

and $\rho(\boldsymbol{y})$ is the energy density distribution of the source, $p^j(\boldsymbol{y}) = \rho(\boldsymbol{y}) u^i(\boldsymbol{y})$ is the momentum density distribution of the source. Further, prove that the gravitational scalar potential and vector potential are given to leading order by

$$\Phi_{\mathrm{g}}(\boldsymbol{x}) = -\frac{GM}{|\boldsymbol{x}|}, \quad \boldsymbol{A}_{\mathrm{g}}(\boldsymbol{x}) = -\frac{2G}{|\boldsymbol{x}|^3} \boldsymbol{J} \times \boldsymbol{x}, \tag{10.194}$$

where $\boldsymbol{J} = \int (\boldsymbol{y} \times \boldsymbol{p}) \mathrm{d}^3 y$ is the total angular momentum of the source. Prove that these relations are exact for a spherically symmetric source in the linear approximation. Hence, for a spherically symmetric distribution of the source rotating around the z-axis, the line element can be written as

$$\begin{aligned}\mathrm{d}s^2 = &-\left(1-\frac{2GM}{r}\right)\mathrm{d}t^2 - \frac{4GJ}{r^3}(x\mathrm{d}y - y\mathrm{d}x)\mathrm{d}t \\ &+\left(1+\frac{2GM}{r}\right)(\mathrm{d}x^2+\mathrm{d}y^2+\mathrm{d}z^2),\end{aligned} \tag{10.195}$$

where $r = |\boldsymbol{x}|$.

3. Prove that in linearized gravity, for a spherically symmetric object with mass M rotating steadily with angular momentum $\boldsymbol{J}$, the "gravitoelectric" and "gravitomagnetic" fields are given by

$$\boldsymbol{g}(\boldsymbol{x}) = -\frac{GM}{|\boldsymbol{x}|^2}\boldsymbol{e}_x, \quad \boldsymbol{b}(\boldsymbol{x}) = \frac{2G}{|\boldsymbol{x}|^3}[\boldsymbol{J} - 3(\boldsymbol{J}\cdot\boldsymbol{e}_x)\boldsymbol{e}_x], \tag{10.196}$$

where $\boldsymbol{e}_x$ is the unit vector in the $\boldsymbol{x}$ direction.

4. In the slow-rotation limit, the spacetime outside the Earth can be described by the following metric:

$$\begin{aligned}\mathrm{d}s^2 = &-\left(1-\frac{2GM}{rc^2}\right)c^2\mathrm{d}t^2 + \frac{4GJ}{r^3c^3}(y\mathrm{d}x - x\mathrm{d}y)\mathrm{d}t \\ &+\left(1+\frac{2GM}{rc^2}\right)(\mathrm{d}x^2+\mathrm{d}y^2+\mathrm{d}z^2).\end{aligned} \tag{10.197}$$

Consider circular motion in the equatorial plane, using $\mathrm{d}\phi = \omega\mathrm{d}t$, prove that for geodesic motion,

$$\omega^2 + \frac{2GJ}{r^3c^2}\omega - \frac{GM}{r^3} = 0. \tag{10.198}$$

5. Consider a stationary axisymmetric spacetime that is invariant under translations along the axis and reflections along the axis of symmetry. Prove that the line element can be written as

$$\mathrm{d}s^2 = -A(\rho)\mathrm{d}t^2 + \mathrm{d}\rho^2 + B(\rho)\mathrm{d}\phi^2 + C(\rho)\mathrm{d}z^2, \tag{10.199}$$

where A, B, and C are arbitrary functions of ρ. Calculate the non-zero Christoffel symbols and the non-zero components of the Ricci tensor for this line element.

6. Consider a stationary, infinitely long, cylindrically symmetric distribution of matter with constant radius, invariant under Lorentz transformations along the axis of symmetry, known as a cosmic string. Prove that outside this matter distribution, the spacetime line element can be written as

$$\mathrm{d}s^2 = -\mathrm{d}t^2 + \mathrm{d}\rho^2 + (\alpha+\beta\rho)^2\mathrm{d}\phi^2 + \mathrm{d}z^2, \tag{10.200}$$

where α and β are constants. For $\alpha = 0$, consider the spacelike surface defined by $t =$ constant and $z =$ constant. Calculate the circumference of the circle defined by constant coordinate radius ρ on this surface. Hence, prove that for $\beta < 1$, the geometry of the spacelike surface is a two-dimensional cone embedded in three-dimensional Euclidean space.

7. Continuing from the previous problem, if we choose cylindrical coordinates (t, r, ϕ, z), prove that the Einstein field equations require the energy-momentum tensor to have the form

$$T^{\mu\nu} = \mathrm{diag}(\rho, 0, 0, -\rho). \tag{10.201}$$

Thus, there is a negative pressure (or tension) along the cosmic string. On the string, the line element is of the form

$$\mathrm{d}s^2 = -\mathrm{d}t^2 + \mathrm{d}r^2 + B(r)\mathrm{d}\phi^2 + \mathrm{d}z^2, \tag{10.202}$$

where $B(r)$ satisfies

$$\frac{B''}{2B} - \frac{(B')^2}{4B^2} = -8\pi G\rho. \tag{10.203}$$

By introducing $b(r) = \sqrt{B(r)}$, we have

$$b'' = -8\pi G\rho b. \tag{10.204}$$

Furthermore, assume the energy density is uniformly distributed,

$$\rho(r) = \begin{cases} \rho_0, & r \le r_0, \\ 0, & r > r_0. \end{cases} \tag{10.205}$$

If the spacetime geometry is non-singular at the axis $r \to 0$, then $g_{\phi\phi} \to r^2$ as $r \to r$. Hence, prove that for $r \le r_0$, the line element is

$$\mathrm{d}s^2 = -\mathrm{d}t^2 + \mathrm{d}r^2 + \left(\frac{\sin\lambda r}{\lambda r}\right)^2 \mathrm{d}\phi^2 + \mathrm{d}z^2, \tag{10.206}$$

where $\lambda = \sqrt{8\pi G\rho_0}$. By requiring the metric and its derivative to be continuous at $r = r_0$, prove that for $r > r_0$, the line element is

$$\mathrm{d}s^2 = -\mathrm{d}t^2 + \mathrm{d}r^2 + \left(\frac{\sin\lambda r_0}{\lambda r} + (r - r_0)\cos\lambda r_0\right)^2 \mathrm{d}\phi^2 + \mathrm{d}z^2. \tag{10.207}$$

For our case of interest, $\lambda r_0 \ll 1$, prove that when $r \gg r_0$, the line element takes the form

$$\mathrm{d}s^2 = -\mathrm{d}t^2 + \mathrm{d}r^2 + \left(1 - \frac{8G\mu}{c^2}\right) r^2\mathrm{d}\phi^2 + \mathrm{d}z^2, \tag{10.208}$$

where $\mu = \pi r_0^2\rho_0$ is the mass per unit length of the string. Interpret the physical meaning of this line element.

8. Calculate the inverse of the Kerr metric in Boyer–Lindquist coordinates.
9. In the Kerr spacetime geometry, consider the surfaces defined by $t =$ constant and $r = r_{\pm}$. Prove that for each surface, the circumference of a circle around the poles is less than that around the equator. Prove that this fact also holds for surfaces defined by $t =$ constant and $r = r_{S_{\pm}}$.
10. Calculate the area of the outer horizon of a Kerr black hole. Prove that for any two Kerr black holes with masses M_1 and M_2 and angular momenta J_1 and J_2, the total horizon area is less than that of a Kerr black hole with mass $M_1 + M_2$ and angular momentum $J_1 + J_2$.
11. Prove that in the equatorial plane of an extremal Kerr black hole with mass M, there exists a photon performing circular motion in the same direction as the black hole at Boyer–Lindquist radius $r = \mu$, and a photon performing circular motion in the opposite direction at Boyer–Lindquist radius $r = 4\mu$.
12. Consider the geodesic motion of particles in the equatorial plane of the Kerr spacetime geometry in Boyer–Lindquist coordinates. Prove that the geodesic equations are

$$\begin{aligned} \dot{t} &= \frac{1}{\Delta}\left(\left(r^2 + a^2 + \frac{2\mu a^2}{r}\right)E - \frac{2\mu a}{r}L\right), \\ \dot{\phi} &= \frac{1}{\Delta}\left(\frac{2\mu a}{r}E + \left(1 - \frac{2\mu}{r}\right)L\right), \\ \dot{r}^2 &= (E^2 - \epsilon^2) + \frac{2\epsilon^2\mu}{r} + \frac{a^2(E^2 - \epsilon^2) - L^2}{r^2} + \frac{2\mu(L - aE)^2}{r^3}, \end{aligned} \tag{10.209}$$

where $\epsilon^2 = 1$ for massive particles and $\epsilon^2 = 0$ for massless particles. Furthermore, prove that these equations reduce to the geodesic motion of particles in Schwarzschild geometry in the limit $a \to 0$.
13. Consider the geodesic circular motion of particles in the equatorial plane of the Kerr spacetime geometry in Boyer–Lindquist coordinates. Prove that the coordinate angular velocity $\Omega = \mathrm{d}\phi/\mathrm{dt}$ satisfies

$$\Omega = \frac{\mu^{1/2}}{a\mu^{1/2} \pm r^{3/2}}, \tag{10.210}$$

where the plus sign corresponds to motion in the same direction as the spacetime rotation, and the minus sign corresponds to motion in the opposite direction.
14. Prove that if a particle starts in the equatorial plane of the Kerr black hole, it will remain in the equatorial plane.

15. If an observer performs circular motion in the equatorial plane of the Kerr spacetime, with angular velocity $\Omega = \mathrm{d}\phi/\mathrm{d}t$ relative to a distant observer, express the 4-velocity components u^t, u^ϕ, u_t, and u_ϕ in terms of Ω, r, μ, and a.
16. Consider the Kerr–Newman spacetime geometry in Boyer–Lindquist coordinates.

    (1) Find the event horizon radius r_+ and the angular velocity Ω_H.
    (2) Prove that the vector field

    $$\widehat{l} = \frac{r^2+a^2}{\Delta}\partial_t - \partial_r + \frac{a}{\Delta}\partial_\phi \tag{10.211}$$

    is tangent to the ingoing null geodesic congruence, and prove that

    $$\widetilde{u} \equiv + \int \frac{r^2+a^2}{\Delta}\mathrm{d}r, \quad \psi \equiv \phi + \int \frac{a}{\Delta}\mathrm{d}r \tag{10.212}$$

    are constant along each line of the congruence.
    (3) Prove that the area of the Kerr–Newman black hole horizon is

    $$A = 8\pi\left[\mu^2 - \frac{e^2}{2} + \sqrt{\mu^4 - e^2\mu^2 - J^2}\right]. \tag{10.213}$$

17. For the Kerr–Newman spacetime, the corotating electric potential can be defined as

    $$\Phi = \xi^\mu A_\mu. \tag{10.214}$$

    Prove that this potential is constant on the future event horizon. Specifically, prove that if the electrostatic potential $\Phi = 0$ at infinity, then

    $$\Phi_\mathrm{H} = \frac{Qr_+}{r_+^2 + a^2}, \tag{10.215}$$

    where $r_+ = \mu + \sqrt{\mu^2 - Q^2 - a^2}$.
18. Let l^μ be a null geodesic vector field in flat spacetime, and use it along with an arbitrary scalar function H to construct a new metric

    $$g_{\mu\nu} = \eta_{\mu\nu} + Hl_\mu l_\nu, \tag{10.216}$$

    where $l_\mu = \eta_{\mu\nu}l^\nu$. This metric is called the Kerr–Schild metric.

    (1) Prove that $\widehat{l}$ is null with respect to both metrics $g_{\mu\nu}$ and $\eta_{\mu\nu}$.
    (2) Give the expression for $g^{\mu\nu}$.
    (3) Prove that $l_\mu = g_{\mu\nu}l^\nu$ and $l^\mu = g^{\mu\nu}l_\nu$, i.e., the index raising and lowering of the null vector can be done using either $g_{\mu\nu}$ or $\eta_{\mu\nu}$.

(4) Calculate the Christoffel symbols for $g_{\mu\nu}$, and prove that they satisfy the relations

$$\lambda_\mu \Gamma^\mu_{\alpha\beta} = -\frac{1}{2}\dot{H} l_\alpha l_\beta, \quad l^\mu \Gamma^\alpha_{\mu\beta} = \frac{1}{2}\dot{H} l^\alpha l_\beta, \tag{10.217}$$

where $\dot{H} \equiv l^\mu \partial_\mu H$.

(5) Prove that $l^\nu \nabla_\nu l^\mu = 0$, i.e., $\hat{l}$ is a geodesic vector field for both metrics $g_{\mu\nu}$ and $\eta_{\mu\nu}$.

(6) Prove that for the Ricci tensor, $R_{\mu\nu} l^\mu l^\nu = 0$.

Chapter 11

Gravitational Waves

In this chapter, we explore one of general relativity's most remarkable predictions—gravitational waves. As previously noted, gravity is the weakest fundamental interaction in nature, characterized by an exceptionally small Newtonian coupling constant. On the one hand, this makes the detection of gravitational waves extraordinarily difficult. Indeed, mankind's first direct detection only occurred in late 2015, when[1] LIGO measured signals from merging black holes. On the other hand, the weak coupling ensures minimal interaction with intervening matter, allowing gravitational waves to traverse cosmic distances virtually undisturbed. Consequently, they preserve pristine information about their sources, offering an unparalleled window into relativistic astrophysics and cosmology. With gravitational wave astronomy now a reality, this field is poised to revolutionize our understanding of compact objects and the early universe.

Our analysis employs the weak-field approximation, where the metric decomposes into a flat background and its perturbation:

$$g_{\mu\nu} = \eta_{\mu\nu} + h_{\mu\nu}, \tag{11.1}$$

where $h_{\mu\nu}$ can be regarded as a symmetric rank-2 tensor field in Minkowski spacetime, possessing gauge symmetry

$$h_{\mu\nu} \to h_{\mu\nu} + \partial_\mu \xi_\nu + \partial_\nu \xi_\mu. \tag{11.2}$$

Through the trace-reversed perturbation, the Einstein equations simplify to a wave equation

$$\Box \bar{h}_{\mu\nu} = -16\pi G T_{\mu\nu}. \tag{11.3}$$

This formalism enables us to systematically investigate the basic properties of gravitational waves, the generation mechanism of gravitational waves,

[1]LIGO: Laser Interferometer Gravitaional-Wave Observatory.

and the detection of gravitational waves. For a more in-depth discussion of gravitational wave physics, we refer to [49, 51].

11.1 Gravitational Waves in Vacuum

The weak-field approximation provides a convenient framework for analyzing gravitational waves. Given that Earth is sufficiently distant from potential gravitational wave sources, this approximation proves particularly effective for theoretical considerations. We begin by examining the propagation of gravitational waves in vacuum, assuming they have already been generated. In vacuum, the "trace-reversed" perturbation satisfies the wave equation $\Box\bar{h}_{\mu\nu} = 0$, where $\Box = -\partial_t^2 + \nabla^2$ is the d'Alembert operator. This equation admits plane wave solutions of the form

$$\bar{h}_{\mu\nu} = C_{\mu\nu}\mathrm{e}^{\mathrm{i}k_\sigma x^\sigma}, \tag{11.4}$$

where $C_{\mu\nu}$ is a constant symmetric (0, 2) tensor, and k^σ is the wave vector characterizing the plane wave's propagation direction and frequency. Substituting the plane wave solution into the wave equation yields

$$\begin{aligned} 0 = \Box\bar{h}_{\mu\nu} &= \eta^{\rho\sigma}\partial_\rho\partial_\sigma\bar{h}_{\mu\nu} = \eta^{\rho\sigma}\partial_\rho(\mathrm{i}k_\sigma\bar{h}_{\mu\nu}) = -\eta^{\rho\sigma}k_\rho k_\sigma\bar{h}_{\mu\nu} \\ &= -k_\sigma k^\sigma\bar{h}_{\mu\nu} \\ &\Rightarrow k_\sigma k^\sigma = 0, \end{aligned} \tag{11.5}$$

implying that the wave vector is null. This null condition confirms that gravitational waves propagate at the speed of light, consistent with the interpretation of the graviton as a massless particle analogous to the photon. Expressing the wave vector as $k^\mu = (\omega, k^1, k^2, k^3)$, the null condition further requires $\omega^2 = \delta_{ij}k^ik^j$. Additionally, since the wave equation is both relativistic and linear, the principle of superposition holds: any linear combination of solutions remains a valid solution.

11.1.1 *Physical degrees of freedom*

The propagation degrees of freedom of gravitational waves in four-dimensional spacetime can be determined by carefully analyzing the gauge freedom inherent in the theory. First, since $C_{\mu\nu}$ is a symmetric tensor, it has 10 independent components, most of which are gauge degrees of freedom rather than physical propagation modes. Imposing the harmonic gauge condition,

$$0 = \partial_\mu\bar{h}^{\mu\nu} = \partial_\mu(C^{\mu\nu}\mathrm{e}^{\mathrm{i}k_\sigma x^\sigma}) = \mathrm{i}C^{\mu\nu}k_\mu\mathrm{e}^{\mathrm{i}k_\sigma x^\sigma}, \tag{11.6}$$

we obtain the transversality condition

$$k_\mu C^{\mu\nu} = 0. \tag{11.7}$$

This relation suggests that gravitational waves are transverse waves, perpendicular to the direction of propagation of the wave vector. Since (11.7) represents four independent equations, the number of independent components in $C_{\mu\nu}$ reduces from 10 to 6. However, the harmonic gauge condition

$$\Box x^\mu = 0, \tag{11.8}$$

does not fully fix the coordinate system. A residual gauge freedom remains: any coordinate transformation $x^\mu \to x^\mu + \zeta^\mu$, where $\Box\zeta^\mu = 0$, preserves the harmonic gauge. Note that $\Box\zeta^\mu = 0$ itself is a wave equation, and once a solution is chosen, all gauge degrees of freedom are fixed. Let us choose the solution

$$\zeta_\mu = B_\mu e^{\mathrm{i}k_\sigma x^\sigma}, \tag{11.9}$$

where k_σ is the wave vector, and B_μ is an undetermined constant coefficient. Under the coordinate function shift $x^\mu \to x^\mu + \zeta^\mu$, the metric perturbation changes as

$$h_{\mu\nu}^{(\mathrm{new})} = h_{\mu\nu}^{(\mathrm{old})} - \partial_\mu\zeta_\nu - \partial_\nu\zeta_\mu. \tag{11.10}$$

Consequently, the trace-reversed perturbation transforms as

$$\begin{aligned} \bar{h}_{\mu\nu}^{(\mathrm{new})} &= h_{\mu\nu}^{(\mathrm{new})} - \frac{1}{2}\eta_{\mu\nu}h^{(\mathrm{new})} \\ &= h_{\mu\nu}^{(\mathrm{old})} - \partial_\mu\zeta_\nu - \partial_\nu\zeta_\mu - \frac{1}{2}\eta_{\mu\nu}(h^{(\mathrm{old})} - 2\partial_\lambda\zeta^\lambda) \\ &= \bar{h}_{\mu\nu}^{(\mathrm{old})} - \partial_\mu\zeta_\nu - \partial_\nu\zeta_\mu + \eta_{\mu\nu}\partial_\lambda\zeta^\lambda. \end{aligned} \tag{11.11}$$

Using the plane wave solution of the perturbation and Eq. (11.9), we have

$$C_{\mu\nu}^{(\mathrm{new})} = C_{\mu\nu}^{(\mathrm{old})} - \mathrm{i}k_\mu B_\nu - \mathrm{i}k_\nu B_\mu + \mathrm{i}\eta_{\mu\nu}k_\lambda B^\lambda. \tag{11.12}$$

The four free parameters in B_μ allow us to impose additional conditions on $C_{\mu\nu}$, further reducing the independent components.

Proposition For any gravitational wave amplitude, there always exists a choice of coefficients B_μ such that the transformed gravitational wave amplitude satisfies both

$$C_\mu^{(\mathrm{new})\mu} = 0 \text{ (traceless condition)}, \tag{11.13}$$

and

$$C_{0\nu}^{(\mathrm{new})} = 0. \tag{11.14}$$

Proof The traceless condition (11.13) requires

$$0 = C_{\mu}^{(\mathrm{old})\mu} + 2\mathrm{i}k_{\lambda}B^{\lambda}, \tag{11.15}$$

which yields the constraint

$$k_{\lambda}B^{\lambda} = \frac{1}{2}C_{\mu}^{(\mathrm{old})\mu}. \tag{11.16}$$

Next, the $\nu = 0$ component of Eq. (11.14) requires

$$\begin{aligned} 0 &= C_{00}^{(\mathrm{old})} - 2\mathrm{i}k_0 B_0 - \mathrm{i}k_{\lambda}B_{\lambda}, \\ &= C_{00}^{(\mathrm{old})} - 2\mathrm{i}k_0 B_0 + \frac{1}{2}C_{\mu}^{(\mathrm{old})\mu}, \end{aligned} \tag{11.17}$$

$$B_0 = -\frac{\mathrm{i}}{2k_0}\left(C_{00}^{(\mathrm{old})} + \frac{1}{2}C_{\mu}^{(\mathrm{old})\mu}\right). \tag{11.18}$$

The $\nu = j$ component of Eq. (11.14) requires

$$\begin{aligned} 0 &= C_{0j}^{(\mathrm{old})} - \mathrm{i}k_0 B_j - \mathrm{i}k_j B_0 \\ &= C_{0j}^{(\mathrm{old})} - \mathrm{i}k_0 B_j - \mathrm{i}k_j\left[\frac{-\mathrm{i}}{2k_0}\left(C_{00}^{(\mathrm{old})} + \frac{1}{2}C_{\mu}^{(\mathrm{old})\mu}\right)\right]. \end{aligned} \tag{11.19}$$

Or

$$B_j = \frac{\mathrm{i}}{2(k_0)^2}\left[-2k_0 C_{0j}^{(\mathrm{old})} + k_j\left(C_{00}^{(\mathrm{old})} + \frac{1}{2}C_{\mu}^{(\mathrm{old})\mu}\right)\right].$$

One can readily verify that these solutions are mutually consistent.

In four dimensions, gravitational waves possess only two physical degrees of freedom, as evidenced by the following constraints:

(1) The transverse condition $k_{\mu}C^{\mu\nu} = 0$ provides 4 constraints.
(2) The traceless condition $C_{\mu}^{\mu} = 0$ gives 1 constraint.
(3) The temporal gauge condition $C_{0\nu} = 0$ imposes 4 constraints.
(4) Note that $C_{0\nu} = 0$ already implies $k_{\mu}C^{\mu 0} = 0$, meaning only the spatial components ($\nu = j$) of the transverse condition are independent. Thus, we have a total of 8 independent constraints, leaving exactly 2 physical degrees of freedom.

The situation bears close resemblance to the case of electromagnetic waves. The electromagnetic potential can be expressed in a similar plane-wave form $A_{\mu} = C_{\mu}\mathrm{e}^{\mathrm{i}k^{\sigma}x_{\sigma}}$, where k^{σ} is the wave vector. The relativistic wave equation imposes the transversality condition $C_{\mu}k^{\mu} = 0$, which constitutes one constraint on the polarization components. Gauge symmetry further allows us to impose additional conditions. For instance, in the

Lorenz gauge $\partial^\mu A_\mu = 0$, we have $k^\mu A_\mu = 0$, which is already satisfied by the transversality condition. To fully fix the gauge, we might also require $A_0 = 0$ or similar, reducing the remaining freedom. Ultimately, just as in the gravitational case, these constraints leave the electromagnetic field with only two physical degrees of freedom in four-dimensional spacetime, corresponding to the two transverse polarization states of light.

11.1.2 *TT gauge*

From the earlier discussion, we have seen that gravitational waves can be made traceless and transverse by choosing an appropriate gauge. Let us consider a wave propagating along the z-direction, for which the wave vector takes the form

$$k^\mu = (\omega, 0, 0, k^3) = (\omega, 0, 0, \omega), \tag{11.20}$$

where $k^3 = \omega$ follows from the null nature of k^μ. The gauge conditions $k^\mu C_{\mu\nu} = 0$ and $C_{0\nu} = 0$ further imply

$$C_{3\nu} = 0. \tag{11.21}$$

Thus, the only non-vanishing components of $C_{\mu\nu}$ are C_{11}, C_{12}, C_{21}, and C_{22}. Imposing the tracelessness and symmetry conditions, we may express $C_{\mu\nu}$ as

$$C_{\mu\nu} = \begin{pmatrix} 0 & 0 & 0 & 0 \\ 0 & C_{11} & C_{12} & 0 \\ 0 & C_{12} & -C_{11} & 0 \\ 0 & 0 & 0 & 0 \end{pmatrix}. \tag{11.22}$$

This choice of subgauge within the harmonic gauge is known as the transverse traceless (TT) gauge, also referred to as the radiation gauge, where the metric perturbation is both transverse to the wave vector and trace-free. Note that the preceding discussion pertains to the trace-reversed perturbation $\bar{h}_{\mu\nu}$ rather than $h_{\mu\nu}$ itself. However, in the TT gauge, the two coincide

$$\bar{h}^{\mathrm{TT}}_{\mu\nu} = h^{\mathrm{TT}}_{\mu\nu}. \tag{11.23}$$

Thus, in this gauge, the descriptions of $\bar{h}_{\mu\nu}$ and $h_{\mu\nu}$ become equivalent.

While other gauges can be used to analyze gravitational waves, the transverse traceless (TT) gauge is particularly advantageous for studying gravitational radiation. In fact, for radiation effects, the following fundamental theorem holds:[2]

---

[2]For a detailed proof, see Section 2.2 of arXiv:gr-qc/0501041.

Theorem For any solution $h_{\mu\nu}$ of the linearized Einstein equation, only the transverse-traceless part $h_{ij}^{\rm TT}$ of its spatial components carries physical information about gravitational radiation. This property is gauge-invariant.

Here, the transverse and traceless conditions are explicitly given by

$$\partial_i h_{ij}^{\rm TT} = 0, \quad \delta_{ij} h_{ij}^{\rm TT} = 0, \quad \bar{h}_{ij}^{\rm TT} = h_{ij}^{\rm TT}. \tag{11.24}$$

Even if the plane-wave solution is initially derived in a different gauge, it can always be transformed into the TT gauge. To achieve this, we introduce the projection operator $P_{\mu\nu} = \eta_{\mu\nu} - n_\mu n_\nu$, which projects vectors onto the plane orthogonal to the unit spacelike vector n_μ. For gravitational waves propagating along n_j, we set $n_0 = 0$ and $n_j = k_j/\omega$. The transverse component of the perturbation is then obtained by applying $P_{\mu\nu}$ twice: $P_\mu^\rho P_\nu^\sigma h_{\rho\sigma}$. Finally, subtracting the trace yields the transverse-traceless perturbation:

$$h_{\mu\nu}^{\rm TT} = P_\mu^\rho P_\nu^\sigma h_{\rho\sigma} - \frac{1}{2} P_{\mu\nu} P^{\rho\sigma} h_{\rho\sigma}. \tag{11.25}$$

11.1.3 *Effects of gravitational waves on point particles*

To detect gravitational waves, it is essential to understand their observable physical effects. Since these effects must be coordinate-independent, we focus on the relative motion of adjacent particles. According to Einstein's equivalence principle, the curvature of spacetime and thus the presence of gravity cannot be detected within a sufficiently small region. Consequently, to observe spacetime perturbations induced by gravitational waves, we must examine the geodesic deviation of neighboring particles. Consider a bundle of adjacent particles in a local reference frame, where one particle has a 4-velocity $u^\mu(x)$ and a deviation vector S^μ. Their relative motion is governed by the geodesic deviation equation:

$$\frac{\mathrm{D}^2}{\mathrm{d}\tau^2} S^\mu = R^\mu{}_{\nu\rho\sigma} u^\nu u^\rho S^\sigma. \tag{11.26}$$

To first order in the metric perturbation $h_{\mu\nu}$, and assuming non-relativistic particle motion $u^\nu \simeq (1, 0, 0, 0)$, the Riemann tensor $R^\mu{}_{\nu\rho\sigma}$ contributes dominantly through $R^\mu{}_{00\sigma}$. For the linearized theory, we have

$$R_{\mu 00\sigma} = \frac{1}{2}(\partial_0\partial_0 h_{\mu\sigma} + \partial_\sigma\partial_\mu h_{00} - \partial_\sigma\partial_0 h_{\mu 0} - \partial_\mu\partial_0 h_{\sigma 0}). \tag{11.27}$$

Imposing the transverse-traceless (TT) gauge condition $h_{\mu 0} = 0$, this simplifies to

$$R_{\mu 00\sigma} = \frac{1}{2}\partial_0\partial_0 h_{\mu\sigma}. \tag{11.28}$$

For slow-moving particles, the proper time coincides with coordinate time, $\tau = x^0 = t$, reducing the geodesic deviation equation to

$$\frac{\partial^2}{\partial t^2} S^\mu = \frac{1}{2} S^\sigma \frac{\partial^2}{\partial t^2} h^\mu_\sigma. \tag{11.29}$$

If the gravitational wave propagates along the x^3 direction, only the transverse components S^1 and S^2 are perturbed, demonstrating that gravitational waves are transverse, much like electromagnetic waves, where the oscillating fields are orthogonal to the propagation direction.

Gravitational waves are characterized by two independent polarization amplitudes, denoted $C_+ = C_{11}$ and $C_\times = C_{12}$, which produce distinct physical effects. We analyze these components separately. First, consider the case where $C_\times = 0$. The geodesic deviation equations for the separation vectors reduce to

$$\frac{\partial^2}{\partial t^2} S^1 = \frac{1}{2} S^1 \frac{\partial^2}{\partial t^2} (C_+ \mathrm{e}^{\mathrm{i}k_\sigma x^\sigma}), \quad \frac{\partial^2}{\partial t^2} S^2 = -\frac{1}{2} S^2 \frac{\partial^2}{\partial t^2} (C_+ \mathrm{e}^{\mathrm{i}k_\sigma x^\sigma}).$$

To leading order, the solutions are

$$S^1 = \left(1 + \frac{1}{2} C_+ \mathrm{e}^{\mathrm{i}k_\sigma x^\sigma}\right) S^1(0), \quad S^2 = \left(1 - \frac{1}{2} C_+ \mathrm{e}^{\mathrm{i}k_\sigma x^\sigma}\right) S^2(0).$$

The physical picture of this solution is as follows: Particles initially separated along the x^1-axis oscillate back and forth along the x^1 direction; similarly, particles separated along the x^2-axis oscillate back and forth along the x^2 direction as well, but in anti-phase with the x^1-motion. If a ring of test particles is initially at rest in the x^1-x^2 plane, a gravitational wave propagating along the x^3 deforms the ring into an oscillating "+" shape, as shown in Figure 11.1. Now, consider the case where $C_+ = 0$ but $C_\times \neq 0$. The solutions to the geodesic deviation equations are

$$S^1 = S^1(0) + \frac{1}{2} C_\times \mathrm{e}^{\mathrm{i}k_\sigma x^\sigma} S^2(0), \tag{11.30}$$

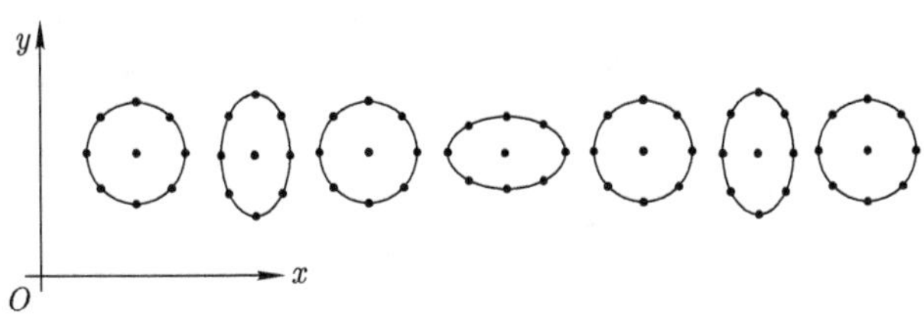

Fig. 11.1 Response of test particles under gravitational wave perturbation C_+.

$$S^2 = S^2(0) + \frac{1}{2} C_\times e^{ik_\sigma x^\sigma} S^1(0). \tag{11.31}$$

In this case, a circular ring of test particles would deform into an oscillating "×" pattern, rotated by 45° relative to the C_+ polarization case, as shown in Figure 11.2.

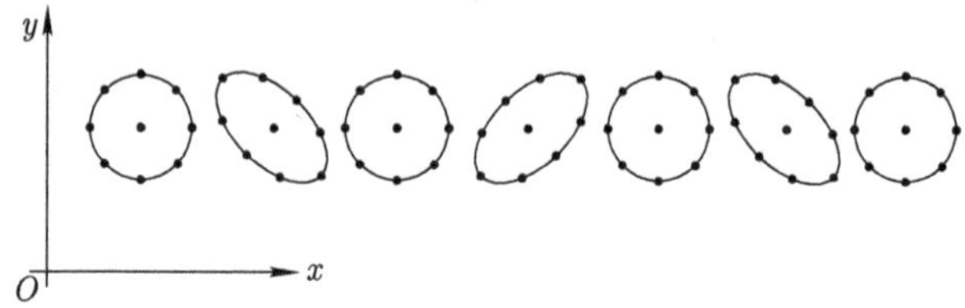

Fig. 11.2 Response of test particles under gravitational wave perturbation $C_\times$.

In analogy with electromagnetic waves, we can define two circular polarization modes for gravitational waves through the linear combinations:

$$\begin{aligned} C_R &= \frac{1}{\sqrt{2}}(C_+ + iC_\times), \\ C_L &= \frac{1}{\sqrt{2}}(C_+ - iC_\times), \end{aligned} \tag{11.32}$$

where C_R and C_L represent right- and left-circularly polarized waves, respectively. A purely left-circularly polarized wave ($C_L \neq 0, C_R = 0$) produces a characteristic rotational pattern in the test particle displacements. Specifically, a ring of particles in the transverse plane will exhibit counter-clockwise rotation as the wave passes through, as illustrated in Figure 11.3. Conversely, a right-circularly polarized wave would induce clockwise rotation.

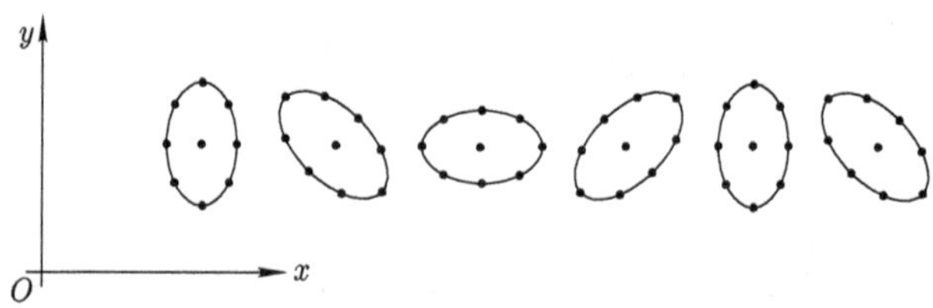

Fig. 11.3 Response of test particles under the perturbation of a circularly polarized gravitational wave.

Both electromagnetic waves and gravitational waves share fundamental characteristics as transverse waves with two propagating physical degrees of freedom. These manifest as distinct polarization states for each wave type. For electromagnetic waves propagation along the z direction, the two polarization states lie in the transverse x-y plane, and each polarization state remains unchanged under a 2π rotation in the plane. Upon quantization, this corresponds to photons with spin 1. For massless fermions, such as neutrinos treated as massless particles, they transform differently. Under a 2π rotation, the fermion field acquires a minus sign. Only after a complete 4π rotation does the field return to its original state. This behavior characterizes particles with spin $\frac{1}{2}$. More generally, for any field of spin S, if the polarization mode remains unchanged after a rotation by an angle θ, then the spin is $S = 2\pi/\theta$. For the gravitational field, as established, gravitational waves represent massless excitations of the gravitational field. Their polarization states exhibit invariance under a π rotation, which implies that the spin of the associated quantum, the graviton, must carry spin 2.

11.2 Gravitational Radiation

The "trace-reversed" perturbations satisfy the linear relativistic wave equation $\Box\bar{h}_{\mu\nu} = -16\pi G T_{\mu\nu}$, where the energy-momentum tensor acts as the source of the gravitational wave. This equation bears a strong resemblance to the wave equation governing electromagnetic waves in electrodynamics, differing primarily in the nature of the source term. Consequently, we may solve it using the well-established Green's function method. The Green's function $G(x^\sigma - y^\sigma)$ is defined by

$$\Box_x G(x^\sigma - y^\sigma) = \delta^4(x^\sigma - y^\sigma), \tag{11.33}$$

where $\Box_x$ denotes the d'Alembertian operator with respect to the coordinate x^σ. The general solution to the linearized Einstein equation then takes the form

$$\bar{h}_{\mu\nu}(x^\sigma) = -16\pi G \int G(x^\sigma - y^\sigma) T_{\mu\nu}(y^\sigma) \mathrm{d}^4 y. \tag{11.34}$$

The choice of Green's function, either retarded or advanced, depends on the causal structure of the problem: the retarded Green's function describes waves propagating forward in time, while the advanced Green's function corresponds to backward propagation. For physical scenarios involving gravitational waves emitted by a source in the past and detected at a later time, the retarded Green's function is the appropriate choice, as it captures the cumulative effect of signals arriving from the past. The retarded Green's function is explicitly given by

$$G(x^\sigma - y^\sigma) = -\frac{1}{4\pi|\boldsymbol{x}-\boldsymbol{y}|}\delta[|\boldsymbol{x}-\boldsymbol{y}| - (x^0 - y^0)]\theta(x^0 - y^0), \tag{11.35}$$

where $\theta(x^0 - y^0)$ is the Heaviside step function,

$$\theta(x^0 - y^0) = \begin{cases} 1, & x^0 > y^0, \\ 0, & x^0 < y^0. \end{cases} \tag{11.36}$$

Substituting this into the general solution yields the disturbance

$$\bar{h}_{\mu\nu}(t, \boldsymbol{x}) = 4G \int \frac{1}{|\boldsymbol{x}-\boldsymbol{y}|} T_{\mu\nu}(t - |\boldsymbol{x}-\boldsymbol{y}|, y)\mathrm{d}^3 y. \tag{11.37}$$

Where $t = x^0$. Here, we define the retarded time: $t_r = t - |\boldsymbol{x}-\boldsymbol{y}|$, which reflects the time delay due to the finite propagation speed of the gravitational wave. The physical interpretation of this solution is that the gravitational perturbation observed at $(t, \boldsymbol{x})$ originates from the energy-momentum distribution $T_{\mu\nu}$ evaluated on the past light cone of the observation point, i.e., at $(t_{\mathrm{r}}, \boldsymbol{y})$. This causal relationship is illustrated in Figure 11.4.

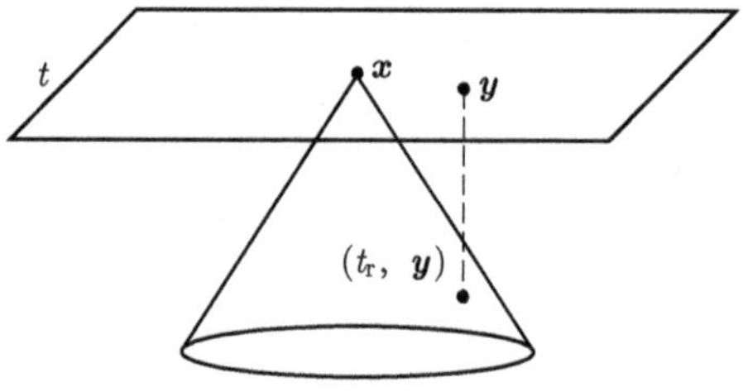

Fig. 11.4 Retarded time.

11.2.1 *Gravitational radiation from isolated sources*

In our universe, astrophysical sources of gravitational radiation are typically located at vast distances from Earth. If we exclude primordial gravitational waves and focus solely on those generated by mass distributions, these sources can be treated as isolated systems with non-relativistic bulk motion. To analyze the emitted gravitational waves, we first perform a Fourier transform in the time domain

$$\begin{aligned} \widetilde{\phi}(\omega, \boldsymbol{x}) &= \frac{1}{\sqrt{2\pi}} \int \mathrm{d}t \mathrm{e}^{\mathrm{i}\omega t} \phi(t, \boldsymbol{x}), \\ \phi(t, \boldsymbol{x}) &= \frac{1}{\sqrt{2\pi}} \int \mathrm{d}\omega \mathrm{e}^{-\mathrm{i}\omega t} \widetilde{\phi}(\omega, \boldsymbol{x}). \end{aligned} \tag{11.38}$$

Applying this transformation to the transverse-traceless perturbation, we obtain

$$\begin{aligned}\tilde{\bar{h}}_{\mu\nu}(\omega, \boldsymbol{x}) &= \frac{1}{\sqrt{2\pi}} \int \mathrm{d}t \mathrm{e}^{\mathrm{i}\omega t} \bar{h}_{\mu\nu}(t, \boldsymbol{x}) \\ &= \frac{4G}{\sqrt{2\pi}} \int \mathrm{d}t \mathrm{d}^3 y \mathrm{e}^{\mathrm{i}\omega t} \frac{T_{\mu\nu}(t - |\boldsymbol{x} - \boldsymbol{y}|, \boldsymbol{y})}{|\boldsymbol{x} - \boldsymbol{y}|} \\ &= \frac{4G}{\sqrt{2\pi}} \int \mathrm{d}t_{\mathrm{r}} \mathrm{d}^3 y \mathrm{e}^{\mathrm{i}\omega t_{\mathrm{r}}} \mathrm{e}^{\mathrm{i}\omega |\boldsymbol{x} - \boldsymbol{y}|} \frac{T_{\mu\nu}(t_{\mathrm{r}}, \boldsymbol{y})}{|\boldsymbol{x} - \boldsymbol{y}|} \\ &= 4G \int \mathrm{d}^3 y \mathrm{e}^{\mathrm{i}\omega |\boldsymbol{x} - \boldsymbol{y}|} \frac{\widetilde{T}_{\mu\nu}(\omega, \boldsymbol{y})}{|\boldsymbol{x} - \boldsymbol{y}|},\end{aligned} \tag{11.39}$$

where $\widetilde{T}_{\mu\nu}(\omega, \boldsymbol{y})$ is the Fourier transform of the energy-momentum tensor. Suppose the source's center of mass is located at a distance R from the observer, with deviations δR due to the finite size of the source distribution. Since the source is isolated and distant, the size of the source distribution relative to the distance to us is a small quantity, i.e., $\delta R \ll R$, as shown in Figure 11.5. Assuming the source itself does not change violently and the frequency of the gravitational waves it produces is low, then $\delta R \ll \omega^{-1}$ and the phase factor $\mathrm{e}^{\mathrm{i}\omega|\boldsymbol{x}-\boldsymbol{y}|}$varies slowly across the source. Under these approximation, we may simplify the integral by evaluating the geometric factor at the leading order:

$$\mathrm{e}^{\mathrm{i}\omega|\boldsymbol{x}-\boldsymbol{y}|}/|\boldsymbol{x} - \boldsymbol{y}| \sim \mathrm{e}^{\mathrm{i}\omega R}/R. \tag{11.40}$$

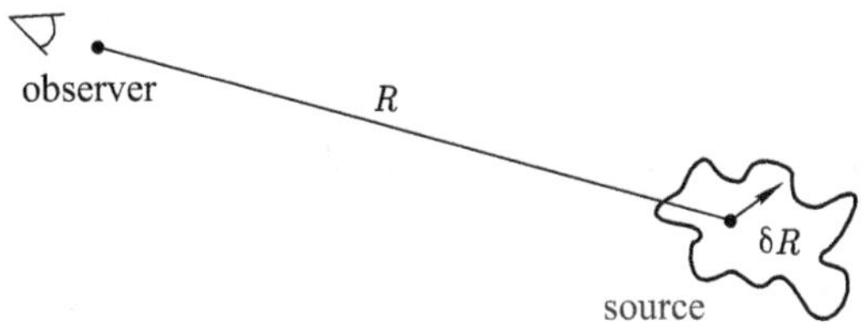

Fig. 11.5 Schematic diagram of isolated source.

Under the distant-source approximation, the Fourier-transformed perturbation simplifies to

$$\tilde{\bar{h}}_{\mu\nu}(\omega, \boldsymbol{x}) = 4G \frac{\mathrm{e}^{\mathrm{i}\omega R}}{R} \int \mathrm{d}^3 y \widetilde{T}_{\mu\nu}(\omega, \boldsymbol{y}). \tag{11.41}$$

The harmonic gauge condition

$$\partial_\mu \bar{h}^{\mu\nu}(t, \boldsymbol{x}) = 0 \Rightarrow \tilde{\bar{h}}^{0\nu} = -\frac{\mathrm{i}}{\omega} \partial_i \tilde{\bar{h}}^{i\nu}. \tag{11.42}$$

This reveals that knowing $\tilde{\bar{h}}^{ij}$ allows us to determine $\tilde{\bar{h}}^{0j}$ and then $\tilde{\bar{h}}^{00}$. Thus, we focus on calculating $\tilde{\bar{h}}_{ij}(\omega, \boldsymbol{x})$. To evaluate the spatial integral of $\widetilde{T}_{\mu\nu}(\omega, \boldsymbol{y})$, we employ integration by parts and exploit the conservation of the energy-momentum tensor $\partial_\mu T^{\mu\nu} = 0$, which in Fourier space becomes

$$\partial_k \widetilde{T}^{k\mu} = \mathrm{i}\omega \widetilde{T}^{0\mu}. \tag{11.43}$$

Applying these constraints, we derive

$$\begin{aligned} \int \mathrm{d}^3y \widetilde{T}^{ij}(\omega, \boldsymbol{y}) &= -\mathrm{i}\omega \int y^i \widetilde{T}^{0j} \mathrm{d}^3y \\ &= -\frac{\mathrm{i}\omega}{2} \int (y^i \widetilde{T}^{0j} + y^i \widetilde{T}^{0i}) \mathrm{d}^3y \\ &= -\frac{\mathrm{i}\omega}{2} \int \left[\partial_l (y^i y^j \widetilde{T}^{0l}) - y^i y^j (\partial_l \widetilde{T}^{0l})\right] \mathrm{d}^3y \\ &= -\frac{\omega^2}{2} \int y^i y^j \widetilde{T}^{00} \mathrm{d}^3y. \end{aligned} \tag{11.44}$$

This demonstrates that, in the slow-motion approximation, the gravitational wave emission is governed by the energy density distribution $\widetilde{T}^{00}$, encapsulated in the quadrupole moment:

$$q_{ij}(t) = \int y^i y^j \widetilde{T}^{00}(t, \boldsymbol{y}) \mathrm{d}^3y. \tag{11.45}$$

Substituting the quadrupole moment into the perturbation, we obtain its frequency-domain representation

$$\tilde{\bar{h}}_{ij}(\omega, \boldsymbol{x}) = -2G\omega^2 \frac{\mathrm{e}^{\mathrm{i}\omega R}}{R} \widetilde{q}_{ij}(\omega). \tag{11.46}$$

Transforming back to the time domain yields the celebrated quadrupole formula:

$$\begin{aligned} \bar{h}_{ij}(t, \boldsymbol{x}) &= -\frac{1}{\sqrt{2\pi}} \frac{2G}{R} \int \mathrm{d}\omega \mathrm{e}^{-\mathrm{i}\omega(t-R)} \omega^2 \widetilde{q}_{ij}(\omega) \\ &= \frac{1}{\sqrt{2\pi}} \frac{2G}{R} \frac{\mathrm{d}^2}{\mathrm{d}t^2} \int \mathrm{d}\omega \mathrm{e}^{-\mathrm{i}\omega t_\mathrm{r}} \widetilde{q}_{ij}(\omega) \\ &= \frac{2G}{R} \frac{\mathrm{d}^2 q_{ij}}{\mathrm{d}t^2}(t_\mathrm{r}), \end{aligned} \tag{11.47}$$

where $t_\mathrm{r} = t - R$ is the retarded time, accounting for the propagation delay from the source to the observer.

Example 11.1. Binary star system.

Consider a binary system consisting of two stars, A and B, with equal masses M orbiting in the x^1-x^2 plane, separated by a distance $2r$, as shown in Figure 11.6. In the Newtonian approximation, their circular orbits are maintained by the balance between gravitational attraction and centrifugal force

$$\frac{GM^2}{(2r)^2} = \frac{Mv^2}{r} \quad \Rightarrow \quad v = \left(\frac{GM}{4r}\right)^{1/2}, \tag{11.48}$$

where v is the orbital velocity. The orbital period is $T = \frac{2\pi r}{v}$, and the angular frequency is $\Omega = \frac{2\pi}{T} = \left(\frac{GM}{4r^3}\right)^{1/2}$. The stars' positions evolve as

$$\begin{aligned} &\text{Star A: } x_A^1 = r\cos\Omega t, \qquad x_A^2 = r\sin\Omega t \\ &\text{Star B: } x_B^1 = -r\cos\Omega t, \qquad x_B^2 = -r\sin\Omega t. \end{aligned} \tag{11.49}$$

The system's energy density distribution is

$$\begin{aligned} T^{00}(t,\boldsymbol{x}) = M\delta(x^3)[&\delta(x^1 - r\cos\Omega t)\delta(x^2 - r\sin\Omega t) \\ &+ \delta(x^1 + r\cos\Omega t)\delta(x^2 + r\sin\Omega t)]. \end{aligned}$$

The non-vanishing components of the quadrupole moment tensor are

$$\begin{aligned} q_{11} &= 2Mr^2\cos^2\Omega t = Mr^2(1+\cos^2\Omega t), \\ q_{22} &= 2Mr^2\sin^2\Omega t = Mr^2(1-\cos^2\Omega t), \\ q_{12} &= q_{21} = 2Mr^2(\cos\Omega t)(\sin\Omega t) = Mr^2\sin^2\Omega t, \\ q_{i3} &= 0. \end{aligned} \tag{11.50}$$

The transverse-traceless metric perturbation at distance R is

$$\bar{h}_{ij}(t,\boldsymbol{x}) = \frac{8GM}{R}\Omega^2 r^2 \begin{pmatrix} -\cos 2\Omega t_{\mathrm{r}} & -\sin 2\Omega t_{\mathrm{r}} & 0 \\ -\sin 2\Omega t_{\mathrm{r}} & \cos 2\Omega t_{\mathrm{r}} & 0 \\ 0 & 0 & 0 \end{pmatrix}. \tag{11.51}$$

The remaining components can be determined using the harmonic gauge condition.

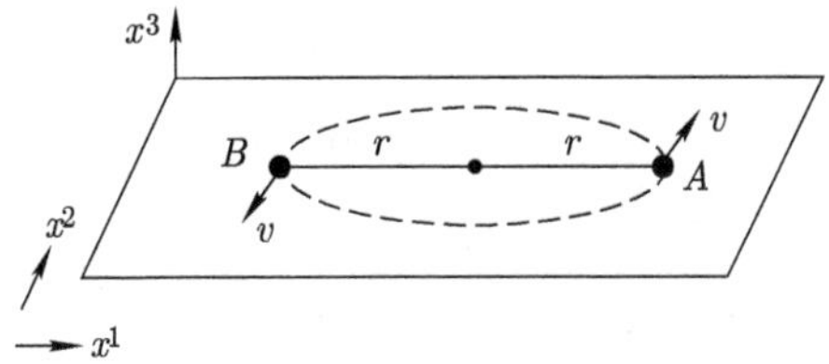

Fig. 11.6 Two stars moving in a circle on a plane.

11.2.2 *Multipole expansion of gravitational radiation*

For general gravitational wave sources, a qualitative discussion can be conducted using multipole expansion. As introduced earlier, the general solution for the gravitational perturbation is given by

$$\bar{h}_{\mu\nu}(t, \boldsymbol{x}) = 4G \int \frac{1}{|\boldsymbol{x} - \boldsymbol{y}|} T_{\mu\nu}(t - |\boldsymbol{x} - \boldsymbol{y}|, \boldsymbol{y}) \mathrm{d}^3 y. \tag{11.52}$$

By placing the center of mass of the source at the origin and denoting $r \equiv |\boldsymbol{x}|$ as the distance to the far field, we employ the multipole expansion:

$$\frac{1}{|\boldsymbol{x} - \boldsymbol{y}|} = \frac{1}{r} + y^i \frac{x_i}{r^3} + y^i y^j \left(\frac{3x_i x_j - r^2 \delta_{ij}}{r^5} \right) + \cdots, \tag{11.53}$$

which leads to the perturbation expressed term by term as

$$\begin{aligned} \bar{h}^{\mu\nu} &= 4G \left[\frac{1}{r} \int T^{\mu\nu}(t_{\mathrm{r}}, \boldsymbol{y}) \mathrm{d}^3 y + \frac{x_i}{r^3} \int T^{\mu\nu}(t_{\mathrm{r}}, \boldsymbol{y}) y^i \mathrm{d}^3 y + \cdots \right] \\ &= 4G \sum_{l=0}^{\infty} \frac{(-1)^l}{l!} M^{\mu\nu i_1 \cdots i_l}(t_{\mathrm{r}}) \partial_{i_1} \cdots \partial_{i_l} \left(\frac{1}{r} \right), \\ M^{\mu\nu i_1 \cdots i_l}(t) &= \int T^{\mu\nu}(t, \boldsymbol{y}) y^{i_1} \cdots y^{i_l} \mathrm{d}^3 y. \end{aligned} \tag{11.54}$$

Hence, the amplitude of the l-th multipole moment scales as $\frac{1}{r^{l+1}}$, approximately inversely proportional to the $(l+1)$-th power of the distance. Higher multipole moments therefore contribute less significantly to the total signal.

For compact sources, the dominant contribution comes from $l = 0$, the monopole moment ($l = 0$):

$$\bar{h}^{\mu\nu} = \frac{4G}{c^4 r} \int T^{\mu\nu}(t - r, \boldsymbol{y}) \mathrm{d}^3 y, \tag{11.55}$$

where $\int T^{00} \mathrm{d}^3 y \equiv Mc^2$ represents the total energy, $\int T^{0i} \mathrm{d}^3 y \equiv P^i c$ corresponds to the linear momentum, and $\int T^{ij} \mathrm{d}^3 y$ denotes the integrated stress tensor related to the internal shear stresses of the source. For an isolated source, energy-momentum conservation $\partial_\mu T^{\mu\nu} = 0$ implies that M and P^i are conserved quantities. In the center-of-mass frame and in the absence of rotation, we have $P^i = 0$, which leads to

$$\bar{h}^{00} = \frac{4GM}{c^2 r}, \quad \bar{h}^{i0} = \bar{h}^{0i} = 0. \tag{11.56}$$

It is illuminating to compare with electromagnetic radiation. In electromagnetism, the gauge potential satisfies the equation $\Box A_\mu = -4\pi J_\mu$. In the Lorenz gauge $\partial^\mu A_\mu = 0$, the general solution takes the form

$$A_\mu(\boldsymbol{x}) = \int \mathrm{d}^3x' \frac{\mathrm{e}^{\mathrm{i}k|\boldsymbol{x}-\boldsymbol{x}'|}}{|\boldsymbol{x}-\boldsymbol{x}'|} J_\mu(x'), \tag{11.57}$$

where $A_\mu(\boldsymbol{x},t) = A_\mu(\boldsymbol{x})\mathrm{e}^{-\mathrm{i}\omega t}$, $J_\mu(\boldsymbol{x},t) = J_\mu(\boldsymbol{x})\mathrm{e}^{-\mathrm{i}\omega t}$, and $k = |\boldsymbol{k}| = \omega$. In the far-field region $r \gg 1/k$, we have $|\boldsymbol{x}-\boldsymbol{x}'| \approx r - \widehat{\boldsymbol{n}}\cdot\boldsymbol{x}'$, yielding

$$\begin{aligned} A_\mu(\boldsymbol{x}) &= \frac{\mathrm{e}^{\mathrm{i}kr}}{r}\int \mathrm{d}^3x' J_\mu(x')\mathrm{e}^{-\mathrm{i}k\widehat{\boldsymbol{n}}\cdot\boldsymbol{x}'} \\ &= \frac{\mathrm{e}^{\mathrm{i}kr}}{r}\sum_b \frac{(-\mathrm{i}k)^b}{b!}\int \mathrm{d}^3x' J_\mu(x')(\widehat{\boldsymbol{n}}\cdot\boldsymbol{x}')^b. \end{aligned} \tag{11.58}$$

At the lowest order $b = 0$, the gauge potential components are

$$\begin{aligned} A_0(\boldsymbol{x},t) &= \int \mathrm{d}^3x' \frac{J_0(\boldsymbol{x}', t-|\boldsymbol{x}-\boldsymbol{x}'|)}{|\boldsymbol{x}-\boldsymbol{x}'|} \\ &\approx \int \mathrm{d}^3x' \frac{J_0(\boldsymbol{x}', t-r)}{r} = -\frac{1}{r}\int \mathrm{d}^3x' \rho(\boldsymbol{x}', t-r) \\ &= -\frac{q(t_\mathrm{r})}{r} = -\frac{q}{r}, \\ \boldsymbol{A}(\boldsymbol{x},t) &= \frac{\mathrm{e}^{\mathrm{i}(kr-\omega t)}}{r}\int \mathrm{d}^3x' \boldsymbol{J}(\boldsymbol{x}'). \end{aligned} \tag{11.59}$$

This shows that the electrostatic potential in the far field reduces to that of the total charge. Using the identity $\int \boldsymbol{x}'(\nabla\cdot\boldsymbol{J})\mathrm{d}^3x' = -\int \mathrm{d}^3x' J(\boldsymbol{x}')$ and the charge conservation condition $\nabla\cdot\boldsymbol{J} = \mathrm{i}\omega\rho$, we obtain

$$\begin{aligned} \boldsymbol{A}(\boldsymbol{x},t) &= -\frac{\mathrm{i}\omega\mathrm{e}^{\mathrm{i}(kr-\omega t)}}{r}\int \mathrm{d}^3x' \boldsymbol{x}'\rho(\boldsymbol{x}') = -\frac{\mathrm{i}\omega\mathrm{e}^{\mathrm{i}(kr-\omega t)}}{r}\boldsymbol{P} \\ &= -\frac{\mathrm{i}\omega\boldsymbol{P}(t_\mathrm{r})}{r} = \frac{1}{r}\left(\frac{\mathrm{d}\boldsymbol{P}}{\mathrm{d}t}\right)_{t_\mathrm{r}}, \end{aligned} \tag{11.60}$$

where $\boldsymbol{P}$ is the electric dipole moment. For higher multipole orders, the gauge potential behaves approximately as $A \approx \frac{1}{r}\sum_l \frac{\mathrm{d}^l}{\mathrm{d}t^l}(E_l + M_l)$ with the electric and magnetic multipole moments defined by

$$E_l \propto \int (x')^l\rho(x')\mathrm{d}^3x', \quad M_l \propto \int (x')^l v(x')\rho(x')\mathrm{d}^3x'. \tag{11.61}$$

Consider a source confined to a region L with total charge Q, characteristic internal velocity v, and periodic time variation. Then

$$E_l \propto QL^l \Rightarrow \frac{\mathrm{d}^l E_l}{\mathrm{d}t^l} \propto Q\left(\frac{v}{c}\right)^l,$$
$$M_l \propto QvL^l \Rightarrow \frac{\mathrm{d}^l M_l}{\mathrm{d}t^l} \propto Q\left(\frac{v}{c}\right)^{l+1}. \tag{11.62}$$

Each successive multipole order is suppressed by an additional factor of $\frac{v}{c}$. Importantly, the monopole moment ($l = 0$) does not contribute to radiation due to charge conservation.

For gravitational radiation, the field equation is $\Box h_{\mu\nu} = -16\pi G T_{\mu\nu}$. Analogous to the electromagnetic case, the radiation field can be expressed as

$$h_l^M \propto \frac{1}{r}\frac{\mathrm{d}^l}{\mathrm{d}t^l}M_l, \quad h_l^S \propto \frac{1}{r}\frac{\mathrm{d}^l}{\mathrm{d}t^l}S_l, \tag{11.63}$$

where M_l and S_l denote the mass multipole moment and current multipole moments, respectively:

$$M_l \propto \int (x')^l \rho(x')\mathrm{d}^3x', \quad S_l \propto \int (x')^l v(x')\rho(x')\mathrm{d}^3x'. \tag{11.64}$$

Since the metric perturbation is dimensionless, it is necessary to restore factors of G and c in the physical quantities: mass $\to G/c^2$, time $\to c$, velocity $\to 1/c$. Thus, the multipole contributions become

$$h_l^M \propto \frac{1}{r}\frac{G}{c^{l+2}}\frac{\mathrm{d}^l}{\mathrm{d}t^l}M_l, \quad h_l^S \propto \frac{1}{r}\frac{G}{c^{l+3}}\frac{\mathrm{d}^l}{\mathrm{d}t^l}S_l. \tag{11.65}$$

Each higher multipole order is suppressed by an additional factor of $\frac{v}{c}$. The leading contribution arises from the mass monopole moment ($l = 0$):

$$h_0^M = \frac{G}{c^2}\frac{M_0}{r}, \quad \text{with } M_0 = \int \mathrm{d}^3x' \rho(x') \equiv M_{\text{system}},$$

which corresponds to the Newtonian gravitational potential. Due to mass-energy conservation, M_0 is constant in time and does not generate gravitational radiation. The next term is the mass dipole moment ($l = 1$):

$$h_1^M \propto \frac{1}{r}\frac{G}{c^3}\frac{\mathrm{d}M_1}{\mathrm{d}t}, \quad M_1 = \int \mathrm{d}^3x' x' \rho(x'). \tag{11.66}$$

Since $\frac{\mathrm{d}M_1}{dt}$ is proportional to the total momentum, which is conserved, this term is pure gauge and does not contribute to radiation. The current dipole moment ($l = 1$) gives

$$h_1^S \propto \frac{1}{r}\frac{G}{c^4}\frac{\mathrm{d}S_1}{\mathrm{d}t}, \quad S_1 = \int \mathrm{d}^3x' x' v(x')\rho(x'). \tag{11.67}$$

This is related to the angular momentum J, which is also conserved; hence, it does not produce radiation. The first non-vanishing radiative contribution comes from the mass quadrupole moment ($l = 2$):

$$h_2^M \propto \frac{1}{r}\frac{G}{c^4}\frac{\mathrm{d}^2 M_2}{\mathrm{d}t^2}, \quad M_2 = \int \mathrm{d}^3x' x'_1 x'_2 \rho(x'). \tag{11.68}$$

This is the dominant source of gravitational radiation. To estimate its amplitude, note that $\frac{\mathrm{d}x'}{\mathrm{d}t} \propto v_{\text{int}}$, which implies

$$h_2^M \propto \frac{1}{r}\frac{G}{c^4} M_{\text{tot}} v_{\text{int}}^2 \sim \frac{1}{r}\frac{G}{c^4}(KE)_{\text{int}}, \tag{11.69}$$

where $(KE)_{\text{int}}$ is the internal kinetic energy. For a typical stellar-mass source with $(KE)_{\text{int}} \sim 1M_{\odot}c^2 \approx 1.8 \times 10^{54}$ erg, at a distance of $r \sim 100$ Mpc $\approx 3.1 \times 10^{26}$ cm, and using

$$G/c^4 \approx 8.26 \times 10^{-50}\ \text{cm/erg}, \tag{11.70}$$

we obtain an estimated strain amplitude of $h_{\text{rad}} \sim 10^{-22} \sim 10^{-21}$.

As the preceding discussion indicates, gravitational waves generated by an isolated non-relativistic source are governed by the second time derivative of the energy-density quadrupole moment. In contrast, electromagnetic radiation originates primarily from the time-varying dipole moment of the charge density. Mathematically, this distinction arises because gravitational waves are described by a second-rank tensor field, whereas electromagnetic waves constitute a first-rank tensor field. Physically, the difference is rooted in the universal character of gravity. A time-varying dipole moment corresponds to oscillations of the density center—be it charge density in electromagnetism or energy density in gravity. However, for an isolated system, oscillation of the center of mass is forbidden by momentum conservation. Consequently, dipole radiation is absent in gravity. The quadrupole moment, which characterizes the system's shape and internal mass distribution, is typically much smaller than the dipole moment and is not constrained by these conservation laws.

Gravitational radiation is significantly weaker than its electromagnetic counterpart. Beyond the higher multipolar order of the leading radiative contribution, the fundamental reason lies in the extremely weak coupling strength of gravity compared to electromagnetism. Indeed, gravity is the weakest of all fundamental forces in nature. Although the amplitude of gravitational waves is exceedingly small, making direct detection challenging, the energy they carry can be substantial. The energy loss per unit time due to gravitational radiation is approximately

$$\frac{\mathrm{d}E}{\mathrm{d}t} = \int T^{0i}\mathrm{d}a_i \sim \frac{c^3}{G}r^2\left|\frac{\mathrm{d}h}{\mathrm{d}t}\right|^2. \tag{11.71}$$

For a strain amplitude $h \sim 10^{-22}$ and a frequency of $\sim$ 100 Hz, the radiated power becomes

$$\frac{\mathrm{d}E}{\mathrm{d}t} \sim 10^{53}\ \mathrm{erg/s}\ \sim 10^{20}\ L_\odot, \tag{11.72}$$

where $L_\odot$ denotes the solar luminosity. This represents an enormous energy loss rate, underscoring the dramatic impact gravitational wave emission can have on astrophysical systems, even though the waves themselves are notoriously difficult to detect.

11.3 Energy-Momentum Tensor of Gravitational Radiation

In general relativity, the problem of defining the energy associated with gravitational perturbations has been a long-standing challenge. This difficulty is not merely technical but also conceptual, as there is no broadly accepted framework for defining a local energy-momentum density for the gravitational field itself. The core issue stems from the dynamical nature of the metric: it is inherently difficult to unambiguously decompose the full metric into a fixed background and a dynamical perturbation in a completely general and gauge-invariant manner. Consequently, no unique and locally defined energy-momentum tensor for gravitational perturbations exists in the full nonlinear theory. One approach to address this is through the concept of a quasi-local mass, which aims to define energy and momentum within a finite region of space. However, such definitions often involve subtleties and are not universally adopted. The situation improves significantly in the weak-field approximation. Here, the background spacetime is unambiguously taken to be flat Minkowski space, and the metric perturbation $h_{\mu\nu}$ can be treated as a symmetric tensor field propagating on this fixed background. Within this regime, several mathematical formulations—such as the Landau–Lifshitz pseudotensor or effective energy-momentum tensors derived from the second-order Einstein tensor—provide consistent ways to define an effective energy-momentum tensor $T^{\mu\nu}$ for the gravitational field (often interpreted in terms of gravitons in the quantum context). Fortunately, for well-behaved physical systems like inspiraling binary stars, all these various schemes yield identical results for the total radiated energy and momentum, allowing for robust predictions of gravitational wave emission.

From a technical standpoint, what should the energy-momentum tensor $T^{\mu\nu}$ for the graviton be? By analogy with electromagnetic and scalar fields, whose energy-momentum tensors are quadratic in the field derivatives, we expect the gravitational counterpart to be quadratic in the metric perturbation $h_{\mu\nu}$. This implies the need to expand the Einstein equations to second order in perturbation theory.

Let the perturbed metric be $g_{\mu\nu} = \eta_{\mu\nu} + h_{\mu\nu}$. At first order, the vacuum Einstein equation becomes

$$G^{(1)}_{\mu\nu}[\eta + h] = 0, \tag{11.73}$$

where $G^{(1)}_{\mu\nu}$ is the linearized Einstein tensor in $h_{\mu\nu}$. This determines the first-order perturbation up to a gauge transformation. To incorporate gravitational self-interaction and define an effective energy-momentum tensor, we proceed to second order. We write the metric as

$$g_{\mu\nu} = \eta_{\mu\nu} + h_{\mu\nu} + h^{(2)}_{\mu\nu}, \tag{11.74}$$

with $h_{\mu\nu} = \mathcal{O}(\epsilon)$ and $h^{(2)}_{\mu\nu} = \mathcal{O}(\epsilon^2)$. The Ricci tensor expands order by order $R_{\mu\nu} = R^{(0)}_{\mu\nu} + R^{(1)}_{\mu\nu} + R^{(2)}_{\mu\nu}$, where $R^{(0)}_{\mu\nu} = 0$ (flat background), and $R^{(1)}_{\mu\nu}(h^{(1)}_{\mu\nu}) = 0$ governs the linear dynamics. The second-order vacuum Einstein equation involves both the second-order perturbation $h^{(2)}_{\mu\nu}$ and quadratic terms in $h_{\mu\nu}$:

$$R^{(1)}_{\mu\nu}[h^{(2)}] + R^{(2)}_{\mu\nu}[h] = 0. \tag{11.75}$$

Here the second-order Ricci tensor $R^{(2)}_{\mu\nu}$ is given by

$$\begin{aligned} R^{(2)}_{\mu\nu} &= \frac{1}{2}h^{\rho\sigma}\partial_\mu\partial_\nu h_{\rho\sigma} - h^{\rho\sigma}\partial_\rho\partial_{(\mu}h_{\nu)\sigma} + \frac{1}{4}(\partial_\mu h_{\rho\sigma})\partial_\nu h^{\rho\sigma} + (\partial^\sigma h^\rho_\nu)\partial_{[\sigma}h_{\rho]\mu} \\ &\quad + \frac{1}{2}\partial_\sigma(h^{\rho\sigma}\partial_\rho h_{\mu\nu}) - \frac{1}{4}(\partial_\rho h_{\mu\nu})\partial^\rho h - \left(\partial_\sigma h^{\rho\sigma} - \frac{1}{2}\partial^\rho h\right)\partial_{(\mu}h_{\nu)\rho}. \end{aligned} \tag{11.76}$$

This structure allows us to reinterpret the second-order equation as an effective non-vacuum equation for the background metric:

$$G^{(1)}_{\mu\nu}[\eta + h^{(2)}] = 8\pi G t_{\mu\nu}, \tag{11.77}$$

where $t_{\mu\nu} = -\dfrac{1}{8\pi G}G^{(2)}_{\mu\nu}[\eta + h]$. Note that since $R^{(1)}_{\rho\sigma}[h] = 0$, no linear terms in h appear in $G^{(2)}_{\mu\nu}$. Equation (11.77) indicates that the energy-momentum tensor $t_{\mu\nu}$, built quadratically from the first-order perturbations, acts as a source for the second-order metric correction $h^{(2)}_{\mu\nu}$. This tensor $t_{\mu\nu}$ is interpreted as the effective energy-momentum tensor of the gravitational wave (or graviton) field.

Does $t_{\mu\nu}$ qualify as a genuine energy-momentum tensor? It is symmetric and contains quadratic terms in $h_{\mu\nu}$, and since $\partial_\mu G^{\mu\nu} = 0$, it satisfies the conservation law $\partial_\mu t^{\mu\nu} = 0$. However, it is not true tensor under general coordinate transformations and, more critically, it is not gauge-invariant. A solution is to average $t_{\mu\nu}$ over several gravitational wave wavelengths. Physically, while Riemann normal coordinates allow a local inertial frame where gravitational effects vanish, integrating over a finite region (spanning multiple wavelengths) captures gauge-invariant information about the wave energy. This averaging procedure, rigorously formalized by David Isaacson (1968), defines an effective energy-momentum tensor for gravitational waves. Denoting the average as $\langle \cdot \rangle$, it satisfies

$$\langle \partial_\mu (X) \rangle = 0 \quad \Rightarrow \quad \langle A \partial_\mu B \rangle = -\langle (\partial_\mu A) B \rangle, \tag{11.78}$$

for and function A and B. To simplify calculations, we impose the transverse-traceless gauge: $\partial^\mu h^{\mathrm{TT}}_{\mu\nu} = 0$, $h^{\mathrm{TT}} = 0$. In this gauge, the second-order Ricci tensor simplifies to

$$\begin{aligned} R^{(2)\mathrm{TT}}_{\mu\nu} &= \frac{1}{2} h^{\rho\sigma} \partial_\mu \partial_\nu h_{\rho\sigma} + \frac{1}{4} \partial_\mu h_{\rho\sigma} \partial_\nu h^{\rho\sigma} + \frac{1}{2} \eta^{\rho\lambda} (\partial^\sigma h_{\rho\nu}) \partial_\sigma h_{\lambda\mu} \\ &= \frac{1}{2} \partial^\sigma h_{\rho\nu} \partial^\rho h_{\sigma\mu} - h^{\rho\sigma} \partial_\rho \partial_{(\mu} h_{\nu)\sigma} + \frac{1}{2} h^{\rho\sigma} \partial_s \partial_\rho h_{\mu\nu} \\ \Rightarrow \langle R^{(2)\mathrm{TT}}_{\mu\nu} \rangle &= \left\langle -\frac{1}{4} \partial_\mu h_{\rho\sigma} \partial_\nu h^{\rho\sigma} - \frac{1}{2} \eta^{\rho\lambda} (\Box h_{\rho\nu}) h_{\lambda\mu} \right\rangle = -\frac{1}{4} \langle \partial_\mu h_{\rho\sigma} \partial_\nu h^{\rho\sigma} \rangle. \end{aligned}$$

The effective energy-momentum tensor is defined as

$$t_{\mu\nu} = -\frac{1}{8\pi G} G^{(2)}_{\mu\nu}[\eta + h] = -\frac{1}{8\pi G} \left\{ R^{(2)}_{\mu\nu}[h^{(1)}] - \frac{1}{2} \eta^{\rho\sigma} R^{(2)}_{\rho\sigma}[h^{(1)}] \eta_{\mu\nu} \right\}.$$

Since $0 = \langle \eta^{\mu\nu} R^{(2)\mathrm{TT}}_{\mu\nu} \rangle$ (for $\Box h^{\mathrm{TT}}_{\mu\nu} = 0$), and in the TT gauge, $h^{\mathrm{TT}}_{0\nu} = 0$ and only the (ij)-components contribute, so the averaged tensor becomes

$$\bar{t}_{\mu\nu} = \frac{1}{32\pi G} \langle (\partial_\mu h^{\mathrm{TT}}_{\rho\sigma})(\partial_\nu h^{\rho\sigma}_{\mathrm{TT}}) \rangle.$$

In an arbitrary gauge, the gauge-invariant (averaged) energy-momentum tensor is

$$\begin{aligned} t_{\mu\nu} = \frac{1}{32\pi G} \langle &(\partial_\mu h_{\rho\sigma})(\partial_\nu h^{\rho\sigma}) - \frac{1}{2} (\partial_\mu h)(\partial_\nu h) \\ &- (\partial_\rho h^{\rho\sigma}) \partial_\mu h_{\nu\sigma} - \partial_\rho h^{\rho\sigma} \partial_\nu h_{\mu\sigma} \rangle. \end{aligned} \tag{11.79}$$

This expression is conserved and gauge-invariant when averaged, providing a well-defined measure of the energy and momentum carried by gravitational waves.

For a plane wave in the transverse-traceless gauge, $h_{\mu\nu}^{\mathrm{TT}} = C_{\mu\nu}\sin(k_\lambda x^\lambda)$, the effective energy-momentum tensor becomes

$$t_{\mu\nu} = \frac{1}{32\pi G}k_\mu k_\nu C_{\rho\sigma}C^{\rho\sigma}\langle\cos^2(k_\lambda x^\lambda)\rangle. \tag{11.80}$$

Using the identity

$$\langle\cos^2(k_\lambda x^\lambda)\rangle = \left\langle\frac{1+\cos(2k\cdot\lambda)}{2}\right\rangle = \frac{1}{2}, \tag{11.81}$$

the expression simplifies accordingly. Choose a wave propagating along the z-direction with null wave vector $k^\lambda = (\omega, 0, 0, \omega)$, so $k_\lambda = (-\omega, 0, 0, \omega)$. The squared amplitude is $C_{\rho\sigma}C^{\rho\sigma} = 2(h_+^2 + h_\times^2)$ and $f \equiv \omega/(2\pi)$, we obtain the explicit form of the energy-momentum tensor for a plane gravitational wave:

$$t_{\mu\nu} = \frac{\pi}{8G}f^2(h_+^2 + h_\times^2)\begin{pmatrix} 1 & 0 & 0 & -1 \\ 0 & 0 & 0 & 0 \\ 0 & 0 & 0 & 0 \\ -1 & 0 & 0 & 1 \end{pmatrix}. \tag{11.82}$$

For typical gravitational waves, the frequency ranges from $f \sim 10^{-4} \sim 10^4$ Hz, and the strain amplitude is $h \sim 10^{-22}$. The energy flux in the direction of propagation is given by the component $-T_{0z}$:

$$-T_{0z} \sim 10^{-4}\left(\frac{f}{\mathrm{Hz}}\right)^2\frac{h_+^2 + h_\times^2}{(10^{-21})^2}\ \mathrm{erg}\cdot\mathrm{cm}^{-2}\cdot\mathrm{s}^{-1}. \tag{11.83}$$

This energy flux is not negligible at high frequencies. For example, at $f \sim 10^4$ Hz, it can be significant. In comparison, the electromagnetic radiation from a supernova has a typical energy flux of only $\sim 10^{-9}$ $\mathrm{erg}\cdot\mathrm{cm}^{-2}\cdot\mathrm{s}^{-1}$, suggesting that high-frequency gravitational waves can carry more energy per unit area per unit time. However, it is crucial to note that the duration of emission differs vastly: high-frequency gravitational radiation from a supernova lasts only milliseconds, whereas electromagnetic emission persists for months. Moreover, neutrino radiation during a supernova carries an energy flux of $\sim 10^5$ $\mathrm{erg}\cdot\mathrm{cm}^{-2}\cdot\mathrm{s}^{-1}$ and lasts for about 10 seconds, making it the dominant energy carrier in terms of total integrated energy despite the high instantaneous flux of gravitational waves.

11.4 Energy Loss Due to Gravitational Radiation

Since gravitational radiation carries energy and momentum, its emission leads to energy loss in the source system. This effect provides important indirect evidence for the existence of gravitational waves. Indeed, prior to the first direct detection of gravitational waves in 2015, the most compelling support came from such indirect observational evidence—most notably, the precise measurements of orbital decay in binary star systems. As these systems emit gravitational waves, they lose energy, causing their orbits to gradually shrink and their orbital period to decrease. The observed rate of orbital change in several binary pulsars (most famously the Hulse–Taylor binary pulsar PSR B1913+16) matches the predictions of general relativity with remarkable accuracy. This consistency between theory and observation strongly supports the reality of gravitational radiation and its role in astrophysical dynamics.

The energy loss due to gravitational wave emission can be determined by computing the total energy radiated to infinity. This is expressed as

$$\Delta E = \int P \mathrm{d}t, \tag{11.84}$$

where P is the instantaneous radiation power, given by the flux integral

$$P = \int_{S^2_\infty} t_{0\mu} n^\mu r^2 \mathrm{d}\Omega. \tag{11.85}$$

Here, S^2_∞ denotes the two-dimensional sphere at spatial infinity, and n^μ is the outward unit spacelike vector normal to this sphere. In spherical coordinates (t, r, θ, ϕ), this normal vector is simply

$$n^\mu = (0, 1, 0, 0). \tag{11.86}$$

To isolate the transverse-traceless components of the radiation field, which are the only ones contributing to physical observables at infinity, we introduce the projection operator $P_{ij} = \delta_{ij} - n_i n_j$, which projects tensors onto the subspace orthogonal to the radial direction n_i. For a symmetric spatial tensor X_{kl}, its transverse-traceless projection on the sphere S^2_∞ is given by

$$X^{\mathrm{TT}}_{ij} = \left(P^k_i P^l_j - \frac{1}{2} P_{ij} P^{kl}\right) X_{kl}. \tag{11.87}$$

Given the expression for the gravitational wave perturbation in the transverse-traceless gauge, $\bar{h}^{\mathrm{TT}}_{ij} = h^{\mathrm{TT}}_{ij} = \dfrac{2G}{r} \dfrac{\mathrm{d}^2 q^{\mathrm{TT}}_{ij}}{\mathrm{d}t^2}(t - r)$, the tensor q^{ij} is related to the mass distribution but may involve complicated integrals of the energy density. To simplify the physical description, we introduce the reduced quadrupole moment, defined as

$$J_{ij} = q_{ij} - \frac{1}{3} \delta_{ij} \delta^{kl} q_{kl}, \tag{11.88}$$

which is the traceless part of q^{ij}. This quantity naturally arises in the multipole expansion of the Newtonian gravitational potential:

$$\Phi = -\frac{GM}{r} - \frac{G}{r^3}D_i x^i - \frac{3G}{2r^5}J_{ij}x^i x^j + \cdots , \tag{11.89}$$

where $D_i = \displaystyle\int T^{00}x^i \mathrm{d}x^3$ is the mass dipole moment. In the TT gauge, the quadrupole moment $q_{ij}^{\mathrm{TT}} = J_{ij}^{\mathrm{TT}}$, and the gravitational wave strain simplifies to

$$h_{ij}^{\mathrm{TT}} = \frac{2G}{r}\frac{\mathrm{d}^2 J_{ij}^{\mathrm{TT}}}{\mathrm{d}t^2}(t-r). \tag{11.90}$$

This form explicitly shows that the leading-order gravitational radiation is generated by the second time derivative of the reduced quadrupole moment, projected into the transverse-traceless subspace.

For gravitational waves propagating radially outward, the energy flux is given by $t_{0\mu}n^\mu = t_{0r}$. The derivatives of the gravitational wave strain in terms of the reduced quadrupole moment are

$$\begin{aligned}
\partial_0 h_{ij}^{\mathrm{TT}} &= \frac{2G}{r}\frac{\mathrm{d}^3 J_{ij}^{\mathrm{TT}}}{\mathrm{d}t^3},\\
\partial_r h_{ij}^{\mathrm{TT}} &= -\frac{2G}{r}\frac{\mathrm{d}^3 J_{ij}^{\mathrm{TT}}}{\mathrm{d}t^3} - \frac{2G}{r^2}\frac{\mathrm{d}^2 J_{ij}^{\mathrm{TT}}}{\mathrm{d}t^2} \approx -\frac{2G}{r}\frac{\mathrm{d}^3 J_{ij}^{\mathrm{TT}}}{\mathrm{d}t^3},\\
t_{0r} &= -\frac{G}{8\pi r^2}\left\langle\left(\frac{\mathrm{d}^3 J_{ij}^{\mathrm{TT}}}{\mathrm{d}t^3}\right)\left(\frac{\mathrm{d}^3 J_{ij}^{\mathrm{TT}}}{\mathrm{d}t^3}\right)\right\rangle.
\end{aligned} \tag{11.91}$$

To compute the total radiated power P, we must account for the projection onto the transverse-traceless components. The product of projected tensors satisfies

$$X_{ij}^{\mathrm{TT}}X_{\mathrm{TT}}^{ij} = X^{ij}X_{ij} - 2X_i^j X^{ik}n_j n_k + \frac{1}{2}X^{ij}X^{kl}n_i n_j n_k n_l - \frac{1}{2}X^2 + XX^{ij}n_i n_j,$$

where $X = \delta^{ij}X_{ij}$. Since J_{ij} is traceless, $J = 0$, this simplifies and gives rise to the radiated power

$$\begin{aligned}
J_{ij}^{\mathrm{TT}}J_{\mathrm{TT}}^{ij} &= J^{ij}J_{ij} - 2J_i^j J^{ik}n_j n_k + \frac{1}{2}J^{ij}J^{kl}n_i n_j n_k n_l,\\
P &= -\frac{G}{8\pi}\int_{S_\infty^2}\left\langle\frac{\mathrm{d}^3 J^{ij}}{\mathrm{d}t^3}\frac{\mathrm{d}^3 J_{ij}}{\mathrm{d}t^3} - 2\frac{\mathrm{d}^3 J_i^j}{\mathrm{d}t^3}\frac{\mathrm{d}^3 J^{ik}}{\mathrm{d}t^3}n_j n_k + \frac{1}{2}\frac{\mathrm{d}^3 J^{ij}}{\mathrm{d}t^3}\frac{\mathrm{d}^3 J^{kl}}{\mathrm{d}t^3}n_i n_j n_k n_l\right\rangle.
\end{aligned} \tag{11.92}$$

Using the angular averages

$$\int \mathrm{d}\Omega = 4\pi, \quad \int n_i n_j \mathrm{d}\Omega = \frac{4\pi}{3}\delta_{ij} \tag{11.93}$$

$$\int n_i n_j n_k n_l \mathrm{d}\Omega = \frac{4\pi}{15}(\delta_{ij}\delta_{kl} + \delta_{ik}\delta_{jl} + \delta_{il}\delta_{jk}) \tag{11.94}$$

and noting that J^{ij} is independent of angle, the integration yields we have

$$P = -\frac{G}{5}\left\langle \frac{\mathrm{d}^3 J^{ij}}{\mathrm{d}t^3}\frac{\mathrm{d}^3 J_{ij}}{\mathrm{d}t^3} \right\rangle. \tag{11.95}$$

This is the celebrated quadrupole formula for the total power radiated in gravitational waves.

For a compact binary system with equal masses, the mass quadrupole moment and its third time derivative are given by

$$\begin{aligned} J_{ij} &= \frac{MR^2}{3}\begin{pmatrix} 1+3\cos 2\Omega t & 3\sin 2\Omega t & 0 \\ 3\sin 2\Omega t & 1-3\cos 2\Omega t & 0 \\ 0 & 0 & -2 \end{pmatrix}, \\ \frac{\mathrm{d}^3 J_{ij}}{\mathrm{d}t^3} &= 8MR^2\Omega^3 \begin{pmatrix} \sin 2\Omega t & -\cos 2\Omega t & 0 \\ -\cos 2\Omega t & -\sin 2\Omega t & 0 \\ 0 & 0 & 0 \end{pmatrix}. \end{aligned} \tag{11.96}$$

Substituting into the quadrupole formula for power emission (11.95), we obtain the radiated power

$$P = -\frac{2}{5}\frac{G^4 M^5}{R^5}. \tag{11.97}$$

For a binary system with general masses, $m_1 \neq m_2$ in a circular orbit, the energy loss due to gravitational radiation leads to a decay of the orbital radius and an increase in the orbital frequency. The evolution equations are

$$\begin{aligned} \frac{\mathrm{d}r}{\mathrm{d}t} &= -\frac{64G^4}{5c^5 r^3} m_1 m_2 (m_1 + m_2), \\ \frac{\mathrm{d}\omega}{\mathrm{d}t} &= -\frac{3\omega}{2r}\frac{\mathrm{d}r}{\mathrm{d}t} = \frac{96}{5}\left(\frac{G(m_1+m_2)}{c^2 r}\right)^{\frac{3}{2}} \frac{G^2 m_1 m_2}{c^2 r^4} \end{aligned} \tag{11.98}$$

where ω is the orbital angular frequency, and the expression $\frac{\mathrm{d}\omega}{\mathrm{d}t}$ describes how the orbital frequency changes over time due to gravitational wave emission, which is a key observable in gravitational wave astronomy.

In 1974, Russel A. Hulse and Joseph H. Taylor discovered and began meticulous studies of the binary system PSR1913+16. Both stars in this binary system are relatively light, with masses of about $1.4M_\odot$. One of them is a rapidly rotating pulsar, and the orbital period is about 7.9 hours. Due to gravitational wave radiation, the system loses energy, leading to gradually decrease in the orbital period. The observed rate of change in the orbital frequency is

$$\frac{\Delta\omega}{\Delta t} = 2.4 \times 10^{-12}\ \text{s}, \tag{11.99}$$

while the theoretical prediction from general relativity gives $\dfrac{\Delta\omega}{\Delta t} = 2.38 \times 10^{-12}$ s. The remarkable agreement between theory and observation provided the first strong indirect evidence for the existence of gravitational waves.[3] The gravitational waves emitted by this binary system have frequencies of $10^{-5} \sim 10^{-6}$ Hz, which are too low to be detected by current ground-based gravitational wave detectors (such as LIGO or Virgo). However, the precise measurement of orbital decay serves as a robust indirect validation of gravitational wave radiation.

It is noteworthy that the orbit of PSR B1913+16 is not circular but elliptical, with an eccentricity $e = 0.617$. For a circular orbit, the dominant frequency of gravitational waves is twice the orbital frequency, as the quadrupole moment repeats every half-period. In contrast, for an elliptical orbit, gravitational radiation occurs at harmonics of the orbital frequency ω_0. In the case of PSR B1913+16, the most powerful harmonic (carrying the maximum energy) is at $\omega = 8\omega_0$, while the harmonic with the largest amplitude is at $\omega = 4\omega_0$. This complexity enriches the gravitational wave signal and provides additional tests for general relativity.

Given that the pulsar in the binary system is a rapidly rotating neutron star exhibiting strong relativistic effects, its physical parameters are

$$M \approx 1.44 M_\odot, \quad r \approx 10\ \text{km}, \quad \frac{GM}{rc^2} \approx 0.2. \tag{11.100}$$

One might question whether the Newtonian-level quadrupole formula remains applicable for describing gravitational wave emission from such a system. This issue was addressed in depth by Thibault Damour in the 1980s. Through detailed post-Newtonian calculations and analytic treatments, he demonstrated that even in this moderately strong-field regime, the standard quadrupole formula continues to provide an excellent approximation for the gravitational wave luminosity and the resulting orbital decay [50]. His work showed that the leading-order quadrupole prediction is robust and that post-Newtonian corrections, while present, do not significantly alter the rate of energy loss for the purpose of comparing with observations of systems like PSR B1913+16.

[3] R. A. Hulse and J. H. Taylor were awarded the Nobel Prize in Physics in 1993 for their careful study of binary star systems.

As previously discussed, a time-varying mass quadrupole moment generates gravitational radiation. This implies that any non-spherically symmetric rotating object can emit gravitational waves. A prominent example is a rapidly rotating neutron star (a pulsar) with a rotational period in the range $T = 0.03 \sim 0.3$ s. Such an object, if asymmetric, e.g., due to mountains on its crust or a non-axisymmetric magnetic field structure, will lose energy through gravitational wave emission. In binary systems, gravitational radiation plays a crucial dynamical role. As energy is carried away by gravitational waves, the orbital radius gradually decreases. This causes the two stars to spiral inward, increasing their orbital velocity and angular frequency Ω. Consequently, the frequency of the emitted gravitational waves, which is proportional to Ω also increases. Consequently, the frequency of the emitted gravitational waves also increases. The inspiral continues until the two stars eventually merge, producing a final burst of high-frequency gravitational waves. This merger event emits a powerful, transient gravitational wave signal that is a primary target for ground-based detectors like LIGO, Virgo, and KAGRA.

Another instructive example of gravitational radiation is the case of a small mass m falling from rest at infinity toward a much larger mass M. Consider motion along the z-direction. The power radiated as gravitational waves is given by

$$-\frac{\mathrm{d}E}{\mathrm{d}t} = \frac{2Gm^2}{15c^5}(6\dot{z}\ddot{z} + 2z\dddot{z})^2. \tag{11.101}$$

Here, we may assume that the larger mass remains fixed and consider only the motion of the smaller mass. The smaller mass is in free fall, satisfying the energy conservation equation:

$$\frac{1}{2}m\dot{z}^2 = \frac{GmM}{|z|}. \tag{11.102}$$

From this, we derive

$$\frac{\mathrm{d}z}{\mathrm{d}t} = \frac{1}{|z|^{1/2}}(2GM)^{1/2}, \quad \frac{\mathrm{d}^2z}{\mathrm{d}t^2} = \frac{GM}{z^2}, \quad \frac{\mathrm{d}^3z}{\mathrm{d}t^3} = \frac{(2GM)^{3/2}}{|z|^{7/2}}. \tag{11.103}$$

Substituting into the radiation formula yields the energy loss per unit distance

$$-\mathrm{d}E = \frac{1}{|z|^{9/2}}\frac{2Gm^2}{15c^5}(2GM)^{5/2}\mathrm{d}z. \tag{11.104}$$

Integrating from infinity to a closest approach $z = -R$ gives the total energy radiated

$$-\Delta E = \frac{1}{R^{7/2}}\frac{4Gm^2}{105c^5}(2GM)^{5/2}. \tag{11.105}$$

If the larger mass is a black hole with Schwarzschild radius $R_s = \dfrac{2GM}{c^2}$, and setting $R = R_s$, the non-relativistic estimate gives

$$-\Delta E = 0.019mc^2\frac{m}{M}. \tag{11.106}$$

A more accurate treatment including relativistic corrections yields

$$-\Delta E = 0.0104mc^2\frac{m}{M}. \tag{11.107}$$

This shows that the fractional energy loss scales inversely with the black hole mass $\Delta E \propto \dfrac{1}{M}$. For a typical stellar-mass black hole with $M = 10M_\odot$, $R \approx 30$ km, and an infalling mass $m = M_\odot$, the radiated energy is approximately $-\Delta E \approx 2 \times 10^{51}$ erg. Notably, most of this energy is emitted in the final plunge from about $2R_s$ to R_s, during which the gravitational wave frequency is high. The duration of this burst is roughly

$$\Delta t \approx R_s/\text{particle speed} \approx R_s/c \approx 10^{-4}\ \text{s}, \tag{11.108}$$

characterizing it as an intense, high-frequency gravitational wave signal.

A simple yet illustrative model for gravitational wave emission is the spring-mass oscillator. Consider two masses m connected by a spring along the z-axis, separated by a distance $2b$. Their positions are $z = \pm(b + a\sin\omega t)$, where a is the oscillation amplitude and ω the frequency. Assuming small oscillations, $a \ll b$, we approximate $z^2 \approx b^2 + 2ab\sin\omega t$. The reduced quadrupole moment then varies as

$$J^{kl} \approx \left[1 + \frac{2a}{b}\sin\omega(t-r)\right] J_0^{kl} \tag{11.109}$$

where the constant part is

$$J_0^{kl} = \frac{1}{3}\begin{pmatrix} -2mb^2 & 0 & 0 \\ 0 & -2mb^2 & 0 \\ 0 & 0 & 4mb^2 \end{pmatrix}. \tag{11.110}$$

The gravitational wave strain is

$$h^{kl} = \frac{G}{8\pi r}\frac{2a}{b}\omega^2 \sin\omega(t-r) J_0^{kl}. \tag{11.111}$$

The energy radiated per unit time and solid angle is

$$\begin{aligned} -\frac{\mathrm{d}^2E}{\mathrm{d}t\mathrm{d}\Omega} &= \left(\frac{G}{8\pi}\right)^2 [2mab\omega^3\cos\omega(t-r)]^2\sin^4\theta, \\ -\frac{\mathrm{d}E}{\mathrm{d}t} &= \frac{32G}{15c^5}[mab\omega^3\cos\omega(t-r)]^2, \end{aligned} \tag{11.112}$$

where the full expression includes the factor c^{-5} for correct dimensionality. Doing average gives the mean power radiated:

$$-\left\langle \frac{\mathrm{d}E}{\mathrm{d}t} \right\rangle = \frac{16G\omega^6}{15c^5}(mab)^2, \tag{11.113}$$

which depends strongly on the frequency and the square of the oscillating dipole moment, highlighting the efficiency of high-frequency, large-amplitude motions in generating gravitational radiation.

11.5 Detection of Gravitational Waves

Gravitational waves represent a cornerstone prediction of general relativity. Since the 1960s, sustained efforts have been made to develop techniques for their direct detection. Given that gravity is the weakest fundamental force in nature, measuring gravitational waves directly poses immense experimental challenges. However, this very weakness also implies that gravitational waves interact only minutely with matter, allowing them to propagate across cosmic distances virtually unattenuated and without distortion. Consequently, once detected, gravitational waves provide a pristine and faithful record of their cataclysmic origins. Current astronomical and cosmological observations rely predominantly on electromagnetic radiation. The detection of gravitational waves opens a fundamentally new observational window, offering a unique means to study the universe. This is particularly valuable for probing strong gravitational wave sources that are often "dark", emitting little or no electromagnetic radiation. Many of these sources involve extreme energies and are linked to pivotal astrophysical processes.[4] The study of gravitational wave radiation is deeply connected to a broad spectrum of fundamental physical questions, including the formation and growth of supermassive black holes and their host galaxies, large-scale structure formation in the universe, the density and dynamics of stellar nuclei in galactic centers, the physics and equation of state of compact stars (e.g., neutron stars), the structure and evolution of the Milky Way, tests of general relativity in strong-field and dynamical regimes, primordial gravitational waves and cosmic inflation, Hawking radiation of primordial black holes, and physics beyond the Standard Model, such as extra spatial dimensions. Gravitational wave astronomy thus not only validates general relativity but also serves as a powerful probe for exploring high-energy

[4] See `http://www.livingreviews.org/lrr-2009-2`.

Table 11.1 Some gravitational wave sources and their parameters.

Source	Frequency/Hz	Distance/pc	Amplitude/kA
Binary systems	10^{-4}	10	10^{-20}
Stellar core collapse	$10^{-2} \sim 1$	500	10^{-22}
Rotating neutron stars	60	2k	$< 10^{-24}$
Stellar mergers	$10 \sim 10^3$	100M	10^{-21}
Stars falling into black holes	10^{-4}	10M	10^{-21}
Supernova explosions	10^3	10k	10^{-19}

astrophysics, cosmology, and fundamental physics under conditions inaccessible by other means.

Ideal sources of gravitational waves are characterized by a rapidly changing mass quadrupole moment, which requires both a large amount of mass in rapid, asymmetric motion. Such motion is only feasible in very compact systems, where high orbital or rotational speeds can be achieved. Consequently, the most promising sources of gravitational waves include coalescing black holes and neutron stars, core-collapse supernovae, rapidly rotating neutron stars with non-axisymmetric deformations, and dynamical processes in the very early universe, such as phase transitions or cosmic strings. Table 11.1 summarizes some canonical gravitational wave sources, along with their characteristic frequencies, estimated distances from Earth, and expected strain amplitudes.

Depending on the characteristics of the sources and the detection techniques employed, gravitational waves can be categorized into the following four frequency bands:

(1) **High-frequency band:** 1 Hz $\leq f \leq 10^4$ Hz. This band is primarily targeted by ground-based interferometric detectors, such as LIGO, Virgo, and KAGRA.

(2) **Low-frequency band:** 10^{-5} Hz $\leq f \leq 1$ Hz. This band is accessible to future space-based detectors, like LISA (Laser Interferometer Space Antenna).

(3) **Very low-frequency band:** 10^{-9} Hz $\leq f \leq 10^{-7}$ Hz. Gravitational waves in this range can be detected through precise timing of millisecond pulsars, known as Pulsar Timing Arrays (PTAs).

(4) **Extremely low-frequency band:** $10^{-5}H_0^{-1} \leq \lambda \leq H_0^{-1}$, where H_0 is the Hubble constant. These gravitational waves are expected to leave imprints on the Cosmic Microwave Background (CMB) polarization.

For the high-frequency band, the lower frequency limit is approximately 1 Hz. Signals below this frequency are typically indistinguishable from seismic and other ground-based environmental noise. The upper frequency limit is around 10^4 Hz, which corresponds to the dynamical timescale of the most compact gravitational sources. More precisely, the maximum frequency is determined by the characteristic velocity v and size R of the source:

$$\begin{aligned} f_{\max} &\approx \frac{1}{2\pi}\frac{v}{R} \\ &= \frac{1}{2\pi}\frac{c^3}{GM}\left(v \leq c, R \geq \frac{GM}{c^2}\right) \\ &\sim 10^4\left(\frac{M_\odot}{M}\right)\text{Hz}. \end{aligned} \tag{11.114}$$

Here, M is the typical mass of the source. This expression shows that the maximum frequency scales inversely with the mass, implying that lighter systems (such as stellar-mass black holes or neutron stars) can generate higher-frequency gravitational waves.

The detection of high-frequency gravitational waves is primarily conducted using ground-based observatories. These detectors operate on the principle that a passing gravitational wave causes minute deformations in test masses, which can be resonantly amplified and measured. The earliest such detector was the Weber bar, developed in the 1960s, which used a massive cylindrical bar instrumented with sensors to measure tiny strain signals. However, this technology had limited sensitivity and was unable to reliably confirm gravitational wave signals. Modern gravitational wave detectors employ laser interferometry to achieve vastly improved sensitivity. The most prominent example is LIGO (Laser Interferometer Gravitational-Wave Observatory), which consists of two facilities located in Hanford, Washington and Livingston, Louisiana, USA.[5] Other major interferometric detectors include Virgo (Italy), GEO600 (Germany), and KAGRA (Japan), which together form a global network. High-frequency gravitational wave detectors are designed to observe sources such as:

(1) Inspiraling and merging compact binaries composed of neutron stars and/or black holes;
(2) Quasi-normal modes and vibrations of compact objects like neutron stars;
(3) Stellar core collapse events—the central engines of supernova explosions;

[5] In 2015, Advanced LIGO made the first direct detection of gravitational waves, originating from the merger of two black holes each about 30 times the mass of the Sun.

(4) Stochastic backgrounds or signals from cosmological phase transitions in the early universe.

The low-frequency gravitational wave band spans from approximately 10^{-5} Hz to about 1 Hz. The lower bound is set by practical limitations in controlling spacecraft noise over timescales of several hours, while the upper bound is constrained by photon counting statistics in data transmission to Earth, effectively linking it to the lower sensitivity range of high-frequency detectors. Gravitational waves in this band are primarily targeted by space-based laser interferometers. A leading proposed mission is LISA.[6] These detectors leverage the quiet environment of space to achieve sensitivities unattainable on the ground. Prominent sources of low-frequency gravitational waves include:

(1) Supermassive black hole binaries formed during hierarchical galaxy mergers;
(2) Compact objects undergoing extreme mass-ratio inspirals (EMRIs) into supermassive black holes in galactic nuclei;
(3) Galactic populations of stellar-mass compact binaries.

For the very low-frequency band, it is more practical to use period rather than frequency. The long-period end is approximately 30 years, corresponding to the duration of available supernova timing data. The short-period end is around 3 months, which represents the minimum baseline required to accumulate sufficient data and mitigate confounding noise sources. Detection in this band relies on using exceptionally stable millisecond pulsars as precision celestial clocks, forming what are known as Pulsar Timing Arrays (PTAs). Potential gravitational wave sources in this regime include supermassive binary black hole systems formed during the hierarchical growth of galaxies through mergers.

For the extremely low-frequency band, wavelength provides a more natural description than frequency. The long-wavelength end is on the order of the Hubble radius, H_0^{-1}, representing the cosmological horizon. The short-wavelength end is approximately $10^{-4}H_0^{-1}$, corresponding to the smallest scales at which anisotropies in the cosmic microwave background (CMB) radiation can be resolved. Gravitational waves in this band may originate from primordial spacetime fluctuations generated during cosmic inflation,

[6]Space-based detection poses significant technological challenges. Even for LISA, launch and operation are not anticipated before the 2030s. China has also proposed its own space-based gravitational wave detection mission.

which are predicted to imprint a unique polarization pattern (B-mode polarization) on the CMB. A major scientific objective of the Planck satellite, launched in 2009, was to search for these tensor-mode signatures in the CMB polarization.[7]

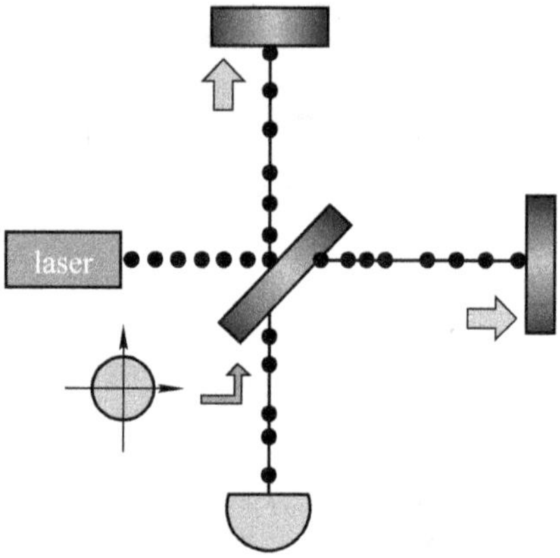

Fig. 11.7 Schematic diagram of laser interferometer.

Finally, we provide a brief introduction to the fundamental operating principle of laser interferometers used in gravitational wave detection. A basic layout is illustrated in Figure 11.7. The core of the detector consists of two test masses, typically highly polished mirrors, oriented along perpendicular arms. A laser beam is directed onto a beam splitter, which divides the light into two beams that travel down the respective arms. Each beam reflects often hundreds of times between the end mirrors and the beam splitter, effectively amplifying the optical path length within the arms. The beams are then recombined at the beam splitter and directed toward a photodetector, where they form an interference pattern. When a gravitational wave passes through the detector perpendicular to the interferometer plane, it induces minute, differential changes in the proper distances between the test masses. This alters the optical path lengths of the two laser beams, which in turn modifies the phase difference between

[7]The Planck mission did not find conclusive evidence for primordial gravitational waves. In recent years, specialized ground-based CMB experiments, such as BICEP and others, have been developed to pursue this goal with higher sensitivity. China has also established a primordial gravitational wave detection platform in Ngari, Tibet, which is the only such observatory in the Northern Hemisphere.

them upon recombination. The resulting shift in the interference pattern is measured at the photodetector, revealing the presence and properties of the gravitational wave. By employing multi-pass (Fabry–Pérot) cavities or delay lines to increase the effective arm length, laser interferometers significantly amplify the tiny strain signals induced by gravitational waves, enabling the detection of distortions on the order of 10^{-22} or smaller.

Appendix 11.1 Resonant Detectors and the Energy of Gravitational Waves

In this appendix, we outline the fundamental principles of resonant detectors for gravitational waves and employ this framework to define the energy carried by such waves. The operating mechanism of a resonant detector can be modeled using a system of two masses connected by a spring. Consider two identical masses m, separated by an equilibrium distance b_0, and linked by a massless spring with spring constant k and damping constant ν. Assuming motion is constrained to the x-direction, the equations of motion are

$$\begin{aligned} m\ddot{x}_1(t) &= -k(x_1 - x_2 + b_0) - \nu(\dot{x}_1 - \dot{x}_2), \\ m\ddot{x}_2(t) &= k(x_1 - x_2 + b_0) + \nu(\dot{x}_1 - \dot{x}_2). \end{aligned} \tag{11.115}$$

Introducing the relative displacement variable

$$\eta = x_2 - x_1 - b_0, \quad \omega_0^2 = \frac{2k}{m}, \quad \gamma = \frac{\nu}{m} \tag{11.116}$$

the system reduces to the equation of a damped harmonic oscillator

$$\ddot{\eta} + 2\gamma\dot{\eta} + \omega_0^2\eta = 0. \tag{11.117}$$

Owing to the damping term, the oscillator's motion gradually diminishes and ultimately ceases.

Next, we consider the response of the above system when a gravitational wave passes through. First, individual masses can be treated as point masses, and the gravitational wave has no effect on them, appearing stationary. Suppose the only non-gravitational force is the elastic force, as discussed above. Consider a gravitational wave propagating along the z-direction. The perturbed metric induces a change in the proper distance between the masses. The instantaneous proper length of the spring becomes:

$$\begin{aligned} b(t) &= \int_{x_1(t)}^{x_2(t)} (1 + h_{xx}^{\mathrm{TT}}(t))^{1/2} \mathrm{d}x \\ &\approx (x_2 - x_1)\left(1 + \frac{1}{2}h_{xx}^{\mathrm{TT}}(t)\right) + O(|h|^2). \end{aligned} \tag{11.118}$$

The equations of motion for the masses, accounting for the spring and damping forces, are

$$\begin{aligned} m\ddot{x}_1 &= -k(b_0 - b) - \nu(\dot{b}_0 - \dot{b}), \\ m\ddot{x}_2 &= k(b_0 - b) + \nu(\dot{b}_0 - \dot{b}). \end{aligned} \tag{11.119}$$

Define the deviation from the equilibrium proper length as

$$\begin{aligned} \eta &= b - b_0 \\ &= x_2 - x_1 - b_0 + \frac{1}{2}(x_2 - x_1)h_{xx}^{\mathrm{TT}}(t) + O(|h|^2). \end{aligned} \tag{11.120}$$

For small deformations, we approximate $x_2 - x_1 = b_0$, obtaining

$$x_2 - x_1 \approx b_0 + \eta - \frac{1}{2}h_{xx}^{\mathrm{TT}}(t)b_0 + O(|h|^2). \tag{11.121}$$

Substituting into the equations of motion leads to the driven oscillator equation

$$\ddot{\eta} + 2\gamma\dot{\eta} + \omega_0^2\eta = \frac{1}{2}b_0\ddot{h}_{xx}^{\mathrm{TT}}(t). \tag{11.122}$$

This describes a damped harmonic oscillator driven by an effective external force proportional to the second time derivative of the gravitational wave strain. This force originates from the tidal acceleration induced by the gravitational wave, consistent with the geodesic deviation equation in general relativity (a connection we leave as an exercise for the reader).

As in a standard forced harmonic oscillator, the tidal acceleration from the gravitational wave drives the system, leading to resonant behavior when the driving frequency matches the natural frequency of the oscillator. Assume a monochromatic gravitational wave of the form

$$h_{xx}^{\mathrm{TT}} = C_+ \cos\Omega_0 t. \tag{11.123}$$

The solution to Eq. (11.122) is then

$$\eta = A\cos(\Omega_0 t + \varphi), \tag{11.124}$$

with amplitude A and phase shift φ given by

$$A = \frac{1}{2}b_0 C_+ \frac{\Omega_0^2}{[(\omega_0^2 - \Omega_0^2)^2 + 4\Omega_0^2\gamma^2]^{1/2}} \tag{11.125}$$

$$\tan\varphi = \frac{2\gamma\Omega_0}{\omega_0^2 - \Omega_0^2}. \tag{11.126}$$

The total energy of the oscillator, comprising kinetic and potential terms, is

$$E = \frac{1}{2}m(\dot{x}_1^2 + \dot{x}_2^2) + \frac{1}{2}k\eta^2. \tag{11.127}$$

Assuming the oscillator starts from rest, this simplifies to

$$\begin{aligned} E &= \frac{1}{2}m(\dot{\eta}^2 + \omega_0^2\eta^2) \\ &= \frac{1}{4}mA^2(\Omega_0^2 \sin^2(\Omega_0 t + \varphi) + \omega_0^2 \cos^2(\Omega_0 t + \varphi)). \end{aligned} \tag{11.128}$$

Taking the average over one period $T_0 = 2\pi/\Omega_0$, we obtain the mean energy

$$\langle E \rangle = \frac{1}{8}mA^2(\Omega_0^2 + \omega_0^2). \tag{11.129}$$

Resonance occurs when $\omega_0 = \Omega_0$, maximizing the response:

$$A_{\text{resonance}} = \frac{1}{4}b_0 C_+ \frac{\Omega_0}{\gamma} \tag{11.130}$$

$$\langle E \rangle_{\text{resonance}} = \frac{1}{64}mb_0^2 C_+^2 \Omega_0^2 \left(\frac{\Omega_0}{\gamma}\right)^2. \tag{11.131}$$

The ratio Ω_0/γ is related to the quality factor Q of the harmonic oscillator, defined for an undriven damped oscillator as

$$Q = \omega_0/2\gamma. \tag{11.132}$$

Thus, the resonant energy becomes

$$\langle E \rangle_{\text{resonance}} = \frac{1}{16}mb_0^2 C_+^2 \Omega_0^2 Q^2. \tag{11.133}$$

In practice, resonant detectors often use massive cylindrical bars, where the elasticity of the material provides the restoring force. In the 1960s, Joseph Weber at the University of Maryland conducted pioneering experiments using an aluminum cylindrical bar with mass $m = 1.4$ tons, length $b_0 = 1.5$ m, resonance frequency $\omega_0 = 10^4$/s, and $Q \sim 10^5$. For a gravitational wave amplitude $C_+ = 10^{-20}$, the expected response is

$$\langle E \rangle_{\text{resonance}} \sim 10^{-20}\ \text{J}, \quad A_{\text{resonance}} \sim 10^{-15}\ \text{m}. \tag{11.134}$$

This amplitude is comparable to the size of an atomic nucleus. However, realistic gravitational wave signals are even weaker and transient, making them insufficient to excite measurable vibrations in such bars.[8]

The foregoing analysis demonstrates that a harmonic oscillator can absorb energy from a passing gravitational wave, confirming that gravitational waves themselves transport energy—energy that ultimately originates from the dynamics of the source. This raises the fundamental question of how

[8] Joseph Weber reported detections, but subsequent analysis indicated that the signals were unlikely to be of gravitational origin. No other resonant bar experiments have confirmed gravitational wave signals.

to define a meaningful energy-momentum tensor for gravitational waves within the framework of general relativity. We approach this problem using the oscillator model within linearized gravity. In the previous treatment, the detector was idealized as a passive probe whose back-reaction on the gravitational field was neglected. However, if the detector gains energy, then by conservation of energy, the gravitational wave must correspondingly lose energy as it passes through the system. This implies that the wave amplitude should decrease after interaction. A more complete physical picture recognizes that the oscillating masses in the detector themselves generate gravitational waves. These emitted waves superpose with the incident wave. Through destructive interference, the total outgoing wave is attenuated compared to the incoming wave (see Figure 11.8). If we attribute this reduction in amplitude to a loss of energy in the gravitational wave, and require that this energy loss equals the energy absorbed by the detector, we can establish a consistent definition for the energy carried by the wave. This line of reasoning provides a physical basis for analyzing both the energy content of gravitational waves and the energy loss rate of their sources.

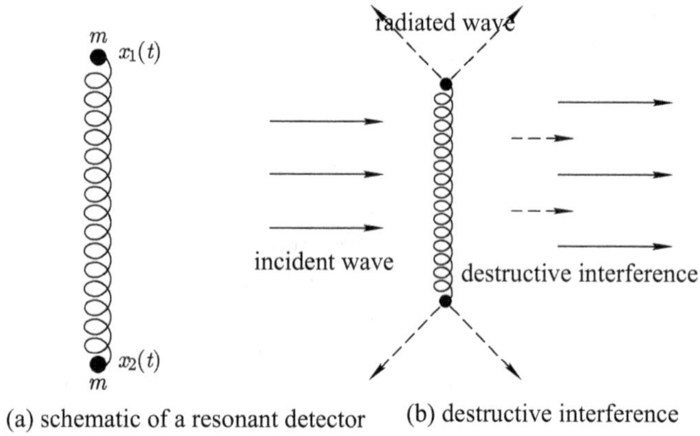

Fig. 11.8 Two mass blocks are connected by a spring to form a resonance detector, which resonates under the excitation of gravitational waves, and the gravitational radiation generated is superimposed with the original gravitational waves, resulting in destructive interference. This means that the gravitational waves passing through the detector are weakened by the loss of energy.

To determine the energy flux of gravitational waves, i.e., the energy carried by gravitational waves per unit time and per unit area, we model a

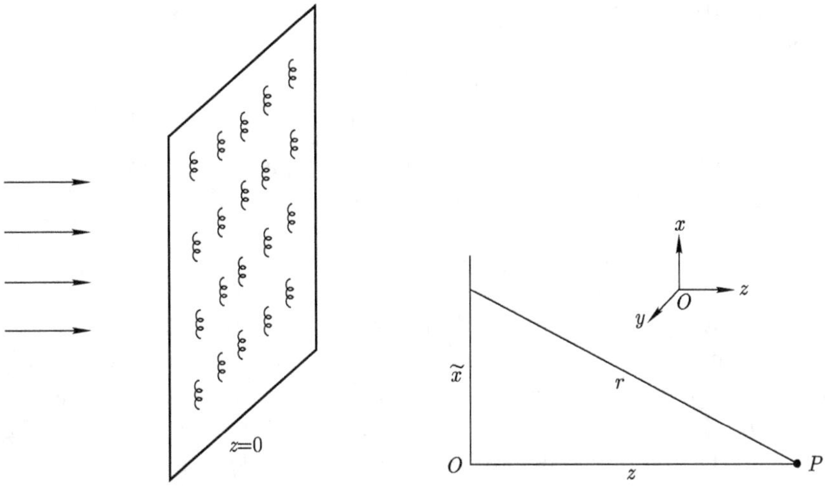

Left: (a) There are s oscillators per unit area. Right: (b) The plane where the oscillators are located is taken in polar coordinates to calculate the gravitational perturbation at point P.

Fig. 11.9 Calculation of gravitational wave energy using resonant resonators.

detection scenario using a distribution of harmonic oscillators. Consider a unit area perpendicular to the direction of propagation containing s identical harmonic oscillators, as depicted in Figure 11.9(a). For a gravitational wave propagating along the z-axis with linear polarization, the metric perturbation in the transverse-traceless gauge is

$$\begin{aligned} \bar{h}_{xx}^{\mathrm{TT}} &= C_+ \cos \Omega_0 (z - t), \\ \bar{h}_{yy}^{\mathrm{TT}} &= -\bar{h}_{xx}^{\mathrm{TT}}. \end{aligned} \tag{11.135}$$

This wave drives each harmonic oscillator into resonant motion described by

$$\eta = A \cos(\Omega_0 t + \varphi), \tag{11.136}$$

where the amplitude A and phase φ are given by the driven oscillator solution previously derived.

The motion of the oscillator reaches a steady state where the energy dissipated through friction is balanced by the work done by the tidal forces of the gravitational wave. The instantaneous power input to each oscillator is given by

$$\frac{\mathrm{d}E}{\mathrm{d}t} = \nu \left(\frac{\mathrm{d}\eta}{\mathrm{d}t} \right)^2 = m\gamma \left(\frac{\mathrm{d}\eta}{\mathrm{d}t} \right)^2. \tag{11.137}$$

Averaging this over one period yields the mean power absorbed per oscillator:

$$\left\langle \frac{\mathrm{d}E}{\mathrm{d}t} \right\rangle_{T_0} = \frac{1}{T_0}\int_0^{T_0} \frac{\mathrm{d}E}{\mathrm{d}t} = \frac{1}{2} m\gamma\Omega_0^2 A^2. \tag{11.138}$$

Consequently, the total energy flux lost by the gravitational wave per unit time per unit area, due to its interaction with s oscillators, is

$$\delta F = -s\left\langle \frac{\mathrm{d}E}{\mathrm{d}t} \right\rangle_{T_0} = -\frac{1}{2} ms\gamma\Omega_0^2 A^2. \tag{11.139}$$

Each harmonic oscillator, with its motion $x = \pm(b_0/2 + a\sin\omega t)$, possesses a time-varying reduced quadrupole moment and thus emits linearly polarized gravitational radiation. Comparing with the resonant oscillation driven by an incident wave, the amplitude of each mass is $A/2$, implying $a = 2A$. The gravitational wave generated by a single oscillator is

$$\delta\bar{h}_{xx} = -2G\Omega_0^2 m b_0 A \frac{\cos(\Omega_0(r-t)-\varphi)}{r}. \tag{11.140}$$

To find the total perturbation due to all oscillators distributed in the plane, we integrate over their contributions

$$\delta\bar{h}_{xx}^{\mathrm{total}} = -2Gm\Omega_0^2 b_0 A 2\pi \int_0^\infty s\cos(\Omega_0(r-t)-\varphi)\frac{\widetilde{x}\mathrm{d}\widetilde{x}}{r}. \tag{11.141}$$

Here, we introduce polar coordinates $(\widetilde{x}, \psi)$ on the plane $z = 0$, where the integral over ψ gives 2π, and the radial coordinate $\widetilde{x}$ is related to r by

$$r = (\widetilde{x}^2 + z^2)^{1/2}. \tag{11.142}$$

The number of oscillators in a ring, $\widetilde{x} \sim \widetilde{x} + \mathrm{d}\widetilde{x}$, is $s2\pi\widetilde{x}\mathrm{d}\widetilde{x}$, thus we obtain Eq. (11.141), as shown in Figure 11.9(b). Since $\widetilde{x}\mathrm{d}\widetilde{x} = r\mathrm{d}r$, the total metric perturbation is

$$\delta\bar{h}_{xx}^{\mathrm{total}} = -4\pi Gm\Omega_0^2 b_0 A \int_z^\infty s\cos(\Omega_0(r-t)-\varphi)\mathrm{d}r. \tag{11.143}$$

As $r \to \infty$, the integrand oscillates rapidly, and physical considerations require that the oscillators do not contribute to the field at infinity.

To handle the divergent integral, we regularize the oscillator density by assuming $s \propto \mathrm{e}^{-\epsilon r}$ and take the limit $\epsilon \to 0$. This yields the total metric perturbation produced by all oscillators

$$\delta\bar{h}_{xx}^{\mathrm{total}} = 4\pi Gms\Omega_0 b_0 A\sin(\Omega_0(z-t)-\varphi). \tag{11.144}$$

Projecting into the transverse-traceless gauge gives

$$\delta\bar{h}_{xx}^{\mathrm{TT}} = -\delta\bar{h}_{yy}^{\mathrm{TT}} = 2\pi Gms\Omega_0 b_0 A\sin(\Omega_0(z-t)-\varphi). \tag{11.145}$$

This is the gravitational perturbation produced by all oscillators. Superposing this with the incident gravitational wave, the net wave becomes

$$\begin{aligned}\bar{h}_{xx}^{\text{net}} &= \bar{h}_{xx}^{\text{TT}} + \delta\bar{h}_{xx}^{\text{TT}} \\ &= (C_+ - 2\pi G m s\Omega_0 b_0 A \sin\varphi)\cos(\Omega_0(z-t)-\theta)\end{aligned} \tag{11.146}$$

where the phase shift θ satisfies,

$$\tan\theta = \frac{2\pi m s\Omega_0 b_0 A}{C_+}\cos\varphi. \tag{11.147}$$

The change in amplitude due to the oscillators is

$$\delta C_+ = -2\pi G m s\Omega_0 b_0 A \sin\varphi. \tag{11.148}$$

By energy conservation, this reduction in wave amplitude corresponds to the energy transferred to the oscillators. Equating the energy loss per unit area δF from the wave to the energy gain by the oscillators leads to

$$\frac{\delta F}{\delta C_+} = \frac{1}{16\pi G}\Omega_0^2 C_+. \tag{11.149}$$

Remarkably, this relation depends only on the frequency and amplitude of the gravitational wave. Integrating gives the energy flux

$$F = \frac{1}{32\pi G}\Omega_0^2 C_+^2. \tag{11.150}$$

Using $\langle(\bar{h}_{xx}^{\text{TT}})^2\rangle = \frac{1}{2}C_+^2$ and generalizing to the full tensor contraction, we obtain the standard expression for the energy flux of a gravitational wave:

$$F = \frac{1}{32\pi G}\Omega_0^2\langle\bar{h}_{\mu\nu}^{\text{TT}}\bar{h}^{\text{TT}\mu\nu}\rangle. \tag{11.151}$$

This result is consistent with the energy density of gravitational waves derived earlier.

Exercises

1. In the TT gauge, prove to the first order in $h_{\mu\nu}$ that

$$\Gamma^\mu_{00} = 0, \quad \Gamma^\mu_{0\nu} = \frac{1}{2}\partial_0(h_{\text{TT}})^\mu_\nu. \tag{11.152}$$

2. Consider a plane gravitational wave of the form $h_{\mu\nu} = C_{\mu\nu}\exp(\mathrm{i}k_\lambda x^\lambda)$.

   (1) Prove that the linearized Riemann tensor is

$$R_{\sigma\mu\nu\rho} = \frac{1}{2}(k_\nu k_\sigma h_{\mu\rho} + k_\rho k_\mu h_{\sigma\nu} - k_\nu k_\mu h_{\sigma\rho} - k_\rho k_\sigma h_{\mu\nu}). \tag{11.153}$$

(2) From this, prove that the linearized Ricci tensor is

$$R_{\mu\nu} = \frac{1}{2}(k_\nu \omega_\mu + k_\mu \omega_\nu - k^2 h_{\mu\nu}) \tag{11.154}$$

where $k^2 = k^\mu k_\mu$, $\omega_\mu = k^\nu \bar{h}_{\mu\nu}$.

(3) Prove that the linearized Einstein equations require

$$k^2 h_{\mu\nu} = k_\nu \omega_\mu + k_\mu \omega_\nu. \tag{11.155}$$

(4) If $k^2 \neq 0$, then we require $R_{\sigma\mu\nu\rho} = 0$. Prove that this case does not correspond to a physical wave but rather a periodic oscillation of the coordinate system.

(5) If $k^2 = 0$, then $k^\nu \bar{h}_{\mu\nu} = 0$. Prove that in this case, the wave vector k^ν is an eigenvector of the Riemann tensor: $R_{\sigma\mu\nu\rho} k^\rho = 0$.

3. Two masses M collide head-on at the event point $(0, 0, 0, 0)$. In the distant past, i.e., as $t \to -\infty$, the two masses start from $x \to \pm\infty$ with zero velocity.

(1) Using Newtonian theory, prove that $x(t) = \pm(9GMt^2/8)^{1/3}$.

(2) Under what separation distance is the Newtonian approximation valid?

(3) Calculate the perturbation $h^{\mathrm{TT}}_{xx}(t)$ at $(x, y, z) = (0, R, 0)$.

4. Consider the line element

$$\mathrm{d}s^2 = -\mathrm{d}t^2 + \mathrm{d}x^2 + f^2(u)\mathrm{d}y^2 + g^2(u)\mathrm{d}z^2 \tag{11.156}$$

where f and g are functions of $u = t - x$. Calculate the connection coefficients and Ricci tensor, and prove that this line element is a solution to the vacuum Einstein equations provided that

$$\frac{f''}{f} + \frac{g''}{g} = 0. \tag{11.157}$$

Prove that this solution can be interpreted as a linearly polarized gravitational wave traveling in the x-direction.

5. Discuss the response of a harmonic oscillator to gravitational waves using the geodesic deviation equation.

6. When a gravitational wave traveling along the z-axis passes through, the spacetime metric can be expressed in the TT gauge as

$$\mathrm{d}s^2 = -\mathrm{d}t^2 + (1 + h_+(z-t))\mathrm{d}x^2 + (1 - h_+(z-t))\mathrm{d}y^2 + \mathrm{d}z^2. \tag{11.158}$$

Assume two objects are located on the x-axis at the origin and $x = L$, respectively. A photon is emitted from the origin along the x-axis, reaches the other object, and is reflected back to the origin. Measure the proper time at the origin and thereby determine the change in distance between the two objects.

(1) Prove that the coordinate velocity of the emitted photon is

$$\left(\frac{\mathrm{d}x}{\mathrm{d}t}\right)^2 = \frac{1}{1+h_+} \tag{11.159}$$

which is not equal to the speed of light. Does this violate relativity?

(2) If h_+ is very small, prove that the coordinate time taken by the photon is

$$t_1 = t_0 + 2L + \frac{1}{2}\int_0^L (h_+(t_0+x)\mathrm{d}x + h_+(t_0+L+x)\mathrm{d}x) \tag{11.160}$$

where t_0 is the time of photon emission and t_1 is the time when the photon returns to the emitter. This fact indicates that if the distance L between the two objects is much smaller than the wavelength of the gravitational wave, or in other words, if h_+ can be treated as constant during the photon's flight, the time taken by the photon to travel back and forth is proportional to the proper distance measured between the two objects in the above metric. This is because if the distance is sufficiently small, we can establish a local inertial frame for measurement.

(3) From Eq. (11.160), prove that

$$\frac{\mathrm{d}t_1}{\mathrm{d}t_0} = 1 + \frac{1}{2}(h_+(t_0+2L) - h_+(t_0)). \tag{11.161}$$

Equation (11.161) shows that the rate of change of the time when the photon returns to the emitter depends only on the spacetime metric at the time of photon emission and when the photon returns to the origin, and not on the spacetime metric at the time of reflection.

(4) Further, consider a continuous electromagnetic wave with frequency ν. Each crest can be treated as a photon, so we have a series of photons traveling back and forth between the two objects. From Eq. (11.161), we get

$$\frac{\mathrm{d}t_1}{\mathrm{d}t_0} = \frac{\nu_1}{\nu_0} \tag{11.162}$$

where ν_0 and ν_1 are the frequencies of the photon at emission and reception, respectively. Thus, we can read out the change in the spacetime metric and the information of the gravitational wave from the change in photon frequency. Suppose the gravitational wave travels at an angle θ to the z-axis within the x-z plane. Derive

$$\begin{aligned}\frac{\mathrm{d}t_1}{\mathrm{d}t_0} = 1 + \frac{1}{2}(&(1-\sin\theta)h_+(t_0+2L) - (1+\sin\theta)h_+(t_0) \\ &+ 2\sin\theta h_+(t_0+(1-\sin\theta)L)).\end{aligned} \tag{11.163}$$

In this case, the derivative of the time when the photon returns to the emitter indeed depends on the amplitude of the gravitational wave at the time of reflection.

(5) Continuing from the previous part, prove that when L is small, t_1 contains an additional proper distance $\delta L = L\cos^2\theta h_+(t)$, explain the origin of the $\cos^2\theta$ factor.

(6) What happens when $\theta = \pi/2$, i.e., when the photon travels in the same direction as the gravitational wave?

7. Consider masses distributed on a circle of radius R in the x-y plane, which can be treated as point particles. If a transmitter at the center of the circle emits photons towards the mass located at $x = R$, and these photons are reflected back to the transmitter at the origin. A gravitational wave travels along the z-direction, and its effect on spacetime is described by the metric in the TT gauge from the previous problem.

   (1) Assuming the radius R is much smaller than the wavelength of the gravitational wave, and the spacetime distortion h_+ can be treated as almost constant during the photon's round trip, prove that the proper time of the transmitter for the photon's round trip is $(2 + h_+)R$.

   (2) The above discussion seems to imply that the proper distance between the transmitter and the mass is $(1 + h_+/2)R$. Is this conclusion correct?

   (3) Prove that if the transmitter performs a similar experiment along the y-axis, the proper time measured by the transmitter for the photon's round trip is $(2 - h_+)R$.

   (4) The difference in the measured times between the two experiments is $2h_+R$, which can be used to measure the amplitude of the gravitational wave. Does this result depend on our use of the TT gauge?

8. Calculate the quadrupole moment and the reduced quadrupole moment for the following systems:

   (1) A spherically symmetric star with energy density $\rho(r,t)$.

   (2) An ellipsoid with uniform density ρ and semi-axes lengths a, b, and c along the x, y, and z directions, respectively (take the center of the ellipsoid as the origin).

   (3) Four masses located at $(a,0,0)$, $(0,a,0)$, $(-a,0,0)$, and $(0,-a,0)$.

   (4) The masses in the previous part begin to rotate counterclockwise around the z-axis with angular velocity ω.

   (5) Two masses m and M ($m \neq M$) connected by a massless spring

with spring constant k and equilibrium length b_0. The system oscillates with natural frequency and amplitude $2A$.

9. For a binary system with different masses m_1 and m_2, calculate the radiation power P and the rate of change of the orbital frequency (assuming circular motion).
10. Consider a mass m falling from infinity with zero initial velocity towards a star of mass $M \gg m$ and radius R. The object can be treated as a point particle, and its motion is strictly radial. Is there gravitational radiation during the fall? If so, how much energy has been carried away by gravitational radiation when the object reaches the surface of the star?
11. Consider two spherically symmetric stars of masses m and M moving in elliptical orbits in the x-y plane, with the orbit characterized by the total energy E and total angular momentum L.

    (1) Using Newtonian gravity, calculate the motion of the two stars relative to their center of mass. Express the orbital period P, minimum distance a, and eccentricity e in terms of E and L.

    (2) Calculate the reduced quadrupole moment of the system.

    (3) Calculate the TT radiation field along the x- and z-axes.

    (4) Prove that the average rate of energy loss over one period is

    $$\langle \mathrm{d}E/\mathrm{d}t\rangle = -\frac{32}{5}\frac{\mu^2(m+M)^3}{a^5(1-e^2)^{7/2}}\left(1+\frac{73}{24}e^2+\frac{37}{96}e^4\right) \tag{11.164}$$

    where $\mu = mM/(m+M)$ is the reduced mass.

    (5) Prove that

    $$\begin{aligned}
    \langle \mathrm{d}a/\mathrm{d}t\rangle &= -\frac{64}{5}\frac{\mu(m+M)^2}{a^3(1-e^2)^{7/2}}\left(1+\frac{73}{24}e^2+\frac{37}{96}e^4\right),\\
    \langle \mathrm{d}e/\mathrm{d}t\rangle &= -\frac{304}{15}\frac{\mu(m+M)^2 e}{a^4(1-e^2)^{5/2}}\left(1+\frac{121}{304}e^2\right),\\
    \langle \mathrm{d}P/\mathrm{d}t\rangle &= -\frac{192\pi}{5}\frac{\mu(m+M)^{3/2}}{a^{5/2}(1-e^2)^{7/2}}\left(1+\frac{73}{24}e^2+\frac{37}{96}e^4\right).
    \end{aligned} \tag{11.165}$$

Bibliography

[1] Carroll S. M. *Spacetime and Geometry: An Introduction to General Relativity.* Addison-Wesley, 2004. (This is a book of moderate difficulty that introduces general relativity using the language of differential geometry.)

[2] Hartle J. B. *Gravity: An Introduction to Einstein's GR.* Addison-Wesley, 2003. (Suitable for undergraduates, with many physical examples and discussions on experiments.)

[3] Schultz B. F. *A First Course in General Relativity.* 2nd ed. Cambridge University Press, 2009. (Suitable for undergraduates, introduces general relativity using the language of differential geometry.)

[4] Wald R. M. *General Relativity.* University of Chicago Press, 1984. (A classic work, very concise and refined, requiring a good mathematical foundation.)

[5] Misner C. W., Thorne K. S., and Wheeler J. A. *Gravitation.* W. H. Freeman and Company, 1973. (A classic work, covering all aspects of gravitation, using the language of differential geometry.)

[6] Weinberg S. *Gravitation and Cosmology: Principles and Applications of the General Theory of Relativity.* Wiley, 1972. (A classic work, understanding gravitation from the perspective of particle physics, with the main mathematical tool being tensor analysis.)

[7] Einstein A. *The Meaning of Relativity.* 5th ed. Princeton University Press, 2004.

[8] Taylor E. F. and Wheeler J. A. *Exploring Black Holes: Introduction to General Relativity.* Addison-Wesley, 2000.

[9] Padmanabhan T. *Gravitation: Foundations and Frontiers.* Cambridge University Press, 2010.

[10] Landau L. D. and Lifshitz E. M. *The Classical Theory of Fields.* 4th ed. Addison-Wesley, 1987.

[11] de Felice F. and Clarke C. J. S. *Relativity on Curved Manifolds.* Cambridge University Press, 1990.

[12] Hawking S. W. and Ellis G. F. R. *The Large Scale Structure of Space-Time.* Cambridge University Press, 1973.

[13] Plebański J. and Krasiński A. *An Introduction to General Relativity and Cosmology.* Cambridge University Press, 2006.

[14] Grøn and Hervik S. *Einstein's General Theory of Relativity: With Modern Applications in Cosmology.* Springer, 2007.

[15] Hobson M. P., Efstathiou G. P., and Lasenby A. N. *General Relativity: An Introduction for Physicists.* Cambridge University Press, 2006.

[16] Rindler W. *Relativity: Special, General, Cosmological.* Oxford University Press, 2006.

[17] Ryder L. *Introduction to General Relativity.* Cambridge University Press, 2009.

[18] Straumannn N. *General Relativity.* Springer, 2013.

[19] Zee A. *Einstein Gravity in a Nutshell.* Princeton University Press, 2013.

[20] DeWitt B. *Bryce DeWitt's Lectures on Gravitation.* Springer, 2011.

[21] Ohanian H. C. and Ruffini R. *Gravitation and Spacetime.* 3rd ed. Cambridge University Press, 2013.

[22] Vannin, Ruffini. *Gravitation and Spacetime.* (In Chinese) Translated by Xiang Shouping, Feng Longlong. Beijing: Science Press, 2006.

[23] Liang Canbin, Zhou Bin. *Differential Geometry and General Relativity: Volume 1.* (In Chinese) 2nd ed. Beijing: Science Press, 2006; Volume 2. 2nd ed. Beijing: Science Press, 2009.

[24] Yu Yunqiang. *Introduction to General Relativity.* (In Chinese) 2nd ed. Beijing: Peking University Press, 1997.

[25] Taylor E. F. and Wheeler J. A. *Spacetime Physics: Introduction to Special Relativity.* 2nd ed. W. H. Freeman and Company, 1992.

[26] Landau L. D. and Lifshitz E. M. *Fluid Mechanics.* 2nd ed. Butterworth-Heinemann, 1987.

[27] Jackson J. D. *Classical Electrodynamics.* Wiley, 1999.

[28] Pais A. *Subtle Is the Lord: The Science and the Life of Albert Einstein.* Oxford University Press, 1982.

[29] Schutz B. *Geometrical Methods of Mathematical Physics.* Cambridge University Press, 1980.

[30] Frankel T. *The Geometry of Physics.* Cambridge University Press, 1997.

[31] Spivak M. *A Comprehensive Introduction to Differential Geometry.* Publish or Perish, 1979.

[32] Boothby W. M. *An Introduction to Differentiable Manifolds and Riemannian Geometry.* Academic Press, 2007.

[33] Chen Shingshen, Chen Weihuan. *Lectures on Differential Geometry.* (In Chinese) 2nd ed. Beijing: Peking University Press, 2001.

[34] Chen Weihuan, Li Xingxiao. *Introduction to Riemannian Geometry: Volume 1.* (In Chinese) Beijing: Peking University Press, 2002; Volume 2. Beijing: Peking University Press, 2004.

[35] Stephani H., Kramer D., MacCallum M., Hoenselaers C., and Herlt E. *Exact Solutions of Einstein's Field Equations.* 2nd ed. Cambridge University Press, 2009.

[36] Griffiths J. B. and Podolský J. *Exact Space-Times in Einstein's General Relativity.* Cambridge University Press, 2012.

[37] Weinberg S. *The Cosmological Constant Problem.* Rev. Mod. Phys., 1989, 61: 1. Padmanabhan T. *Cosmological Constant: The Weight of the Vacuum.*

Phys. Rept., 2003, 380: 235. Peebles P. J. E. and Ratra B. *The Cosmological Constant and Dark Energy.* Rev. Mod. Phys., 2003, 75: 559.

[38] de Rham C., Deskins J. T., Tolley A. J., and Zhou S. Y. *Graviton Mass Bounds.* Rev. Mod. Phys., 2017, 89: 025004.

[39] Will C. M. *Theory and Experiment in Gravitational Physics.* Cambridge University Press, 1993.

[40] Will C. M. *The Confrontation between General Relativity and Experiment.* Living Rev. Relativ., 2006, 9: 3.

[41] Chandrasekhar S. *The Mathematical Theory of Black Holes.* Oxford University Press, 1983.

[42] Arnowitt R., Deser S., and Misner C. W. *The Dynamics of General Relativity.* Witten L. *Gravitation: An Introduction to Current Research.* Wiley, 1962.

[43] Poisson E. *A Relativist's Toolkit: The Mathematics of Black Hole Mechanics.* Cambridge University Press, 2004.

[44] Townsend P. K. *Black Holes.* arXiv: gr-qc/9707012.

[45] Frolov V. P. and Zelnikov A. *Introduction to Black Hole Physics.* Oxford University Press, 2011.

[46] Frolov V. P. and Novikov I. D. *Black Hole Physics: Basic Concepts and New Developments.* Kluwer, 1998.

[47] Birrell N. D. and Davies P. C. W. *Quantum Fields in Curved Space.* Cambridge University Press, 1982.

[48] Wald R. M. *Quantum Field Theory in Curved Spacetime and Black Hole Thermodynamics.* University of Chicago Press, 1994.

[49] Maggiore M. *Gravitational Waves: Vol. 1.* Oxford University Press, 2008.

[50] Darmour T. *The Problem of Motion in Newtonian and Einstein Gravity.* Hawking S. and Israel W. *300 Years of Gravitation.* Cambridge University Press, 1987.

[51] Poisson E. and Will C. M. *Gravity: Newtonian, Post-Newtonian, Relativistic.* Cambridge University Press, 2014.

[52] Padmanabhan T. *Structure Formation in the Universe.* Cambridge University Press, 1993.

[53] Alcubierre M. *Introduction to 3+1 Numerical Relativity.* Oxford University Press, 2008.

[54] Baumgarte T. W. and Shapiro T. L. *Numerical Relativity.* Cambridge University Press, 2010.

[55] Bondi H., van der Burg M. G. J., and Metzner A. W. K. *Gravitational Waves in General Relativity. 7. Waves from Axisymmetric Isolated Systems.* Proc. Roy. Soc. Lond. A, 1962, 269: 21. Sachs R. K. *Gravitational Waves in General Relativity. 8. Waves in Asymptotically Flat Space-Times.* Proc. Roy. Soc. Lond. A, 1962, 270: 103.

[56] Poisson E. and Israel W. *The Internal Structure of Black Holes.* Phys. Rev. D, 1990, 41: 1796. Burko L. M. and Ori A. *Internal Structure of Black Holes and Spacetime Singularities.* An International Research Workshop, Haifa, June 29–July 3, 1997.

[57] Witten E. *Quantum Gravity in de Sitter Space.* arXiv: hep-th/0106109.

[58] Bousso R. *The Holographic Principle.* Rev. Mod. Phys., 2002, 74: 825.
[59] Maldacena J. M. *The Large N Limit of Superconformal Field Theories and Supergravity.* Int. J. Theor. Phys., 1999, 38: 1113. (Adv. Theor. Math. Phys., 1998, 2: 231; arXiv: hep-th/9711200)
[60] Aharony O., Gubser S. S., Maldacena J. M., Ooguri H., and Oz Y. *Large N Field Theories, String Theory and Gravity.* Phys. Rept., 2000, 323: 183. (arXiv: hep-th/9905111)
[61] Susskind L. and Lindesay J. *An Introduction to Black Holes, Information and the String Theory Revolution: The Holographic Universe.* World Scientific, 2004.
[62] Mathur S. D. *The Information Paradox: A Pedagogical Introduction.* Class. Quant. Grav., 2009, 26: 224001. (arXiv: 0909.1038)
[63] Harlow D. *Jerusalem Lectures on Black Holes and Quantum Information.* Rev. Mod. Phys., 2016, 88: 15002. (arXiv: 1409.1231)
[64] Chen Bin. *General Relativity.* (In Chinese), Beijing: Peking University Press, 2018.

Index

O

P

Q

R

S

www.ingramcontent.com/pod-product-compliance
Lightning Source LLC
LaVergne TN
LVHW010340220926
848484LV00004BB/941